ADVANCED
M DERN
PHYSICS
Theoretical Foundations

ADVANCED
M⊙DERN
PHYSICS
Theoretical Foundations

John Dirk Walecka
College of William and Mary, USA

World Scientific

NEW JERSEY · LONDON · SINGAPORE · BEIJING · SHANGHAI · HONG KONG · TAIPEI · CHENNAI

Published by

World Scientific Publishing Co. Pte. Ltd.

5 Toh Tuck Link, Singapore 596224

USA office: 27 Warren Street, Suite 401-402, Hackensack, NJ 07601

UK office: 57 Shelton Street, Covent Garden, London WC2H 9HE

British Library Cataloguing-in-Publication Data
A catalogue record for this book is available from the British Library.

ADVANCED MODERN PHYSICS
Theoretical Foundations

ISBN-13 978-981-4291-51-4
ISBN-10 981-4291-51-X
ISBN-13 978-981-4291-52-1 (pbk)
ISBN-10 981-4291-52-8 (pbk)

Printed in Singapore.

*Dedicated to Kenneth Bainbridge, Sidney Drell, Sergio Fubini,
Francis Low, Norman Ramsey, Felix Villars, Viki Weisskopf,
and all those who taught me physics*

Preface

World Scientific Publishing Company recently published my book entitled *Introduction to Modern Physics: Theoretical Foundations*. The book is aimed at the very best students, with a goal of exposing them to the foundations and frontiers of today's physics. Typically, students have to wade through several courses to see many of these topics, and I wanted them to have some idea of where they were going, and how things fit together, as they went along. Hopefully, they will then see more inter-relationships, and get more original insights, as they progress. The book assumes the reader has had a good one-year, calculus-based freshman physics course, along with a good one-year course in calculus. While it is assumed that mathematical skills will continue to develop, several appendices are included to bring the reader up to speed on any additional mathematics required at the outset. With very few exceptions, the reader should then find the material to be self-contained. Many problems are included, some for each chapter. Although the book is designed so that one can, in principle, read and follow the text without doing any of the problems, the reader is strongly urged to attempt as many of them as possible in order to obtain some confidence in his or her understanding of the basics of modern physics and to hone working skills.

After completing that book, it occurred to me that a second volume could be prepared that would significantly extend the coverage, while furthering the stated goals. The ground rules would be that anything covered in the text and appendices of the first volume would be fair game, while anything covered in the problems would first be re-summarized. Those few results quoted without proof in Vol. I would now be derived. The topics chosen would be those of wide applicability in all areas of physics. Again, an important goal would be to keep the entire coverage self-contained. The present book is the outcome of those musings. All of the material in this book is taken from course lectures given over the years by the author at either Stanford University or the College of William and Mary.

Quantum mechanics is first reformulated in abstract Hilbert space, which allows one to focus on the general structure of the theory. The book then covers the following topics: angular momentum, scattering theory, lagrangian field theory, symmetries, Feynman rules, quantum electrodynamics, path integrals, and canonical transformations for quantum systems. Several appendices are included with important details. When finished, the reader should have an elementary working knowledge in the principal areas of theoretical physics of the twentieth century. With this overview in hand, development in depth and reach in these areas can then be obtained from more advanced physics courses.

I was again delighted when World Scientific Publishing Company, which had done an exceptional job with four of my previous books, showed enthusiasm for publishing this new one. I would like to thank Dr. K. K. Phua, Executive Chairman of World Scientific Publishing Company, and my editor Ms. Lakshmi Narayanan, for their help and support on this project. I am greatly indebted to my colleagues Paolo Amore and Alfredo Aranda for their reading of the manuscript.

Williamsburg, Virginia   *John Dirk Walecka*
December 1, 2009   *Governor's Distinguished CEBAF*
                                    Professor of Physics, emeritus
                                    College of William and Mary

Contents

Chapter 1

Introduction

The goal of this book is to provide an extension of the previous book *Introduction to Modern Physics: Theoretical Foundations*, refered to as Vol. I. That volume develops the underlying concepts in twentieth-century physics: quantum mechanics, special relativity, and general relativity. Included in it are applications in atomic, nuclear, particle, and condensed matter physics. It is assumed in Vol. I that readers have had a good calculus-based introductory physics course together with a good course in calculus. Several appendices then provide sufficient background so that, with very few exceptions, the presentation is self-contained. Many of the topics covered in that work are more advanced than in the usual introductory modern physics books. It was the author's intention to provide the best students with an overview of the subject, so that they are aware of the overall picture and can see how things fit together as they progress.

As projected in Vol. I, it is now assumed that mathematical skills have continued to develop. In this volume, readers are expected to be familiar with multi-variable calculus, in particular, with multiple integrals. It is also assumed that readers have some familiarity with the essentials of linear algebra. An appendix is included here on functions of a complex variable, since complex integration plays a key role in the analysis. The ground rules now are that anything covered in the text and appendices in Vol. I is assumed to be mastered, while anything covered in the problems in Vol. I will be re-summarized. Within this framework, readers should again find Vol. II to be self-contained.

There are over 175 problems in this book, some after each chapter and appendix. The problems are not meant to baffle the reader, but rather to enhance the coverage and to provide exercises on working skills. The problems for the most part are not difficult, and in most cases the steps are clearly laid out. Those problems that may involve somewhat more algebra are so noted. The reader is urged to attempt as many problems as possible in order to obtain some confidence in his or her understanding of the framework of modern theoretical physics.

In chapter 2 we revisit quantum mechanics and reformulate the theory in terms of linear hermitian operators acting in an abstract Hilbert space. Once we know how to compute inner products, and have the completeness relation, we understand the essentials of operating in this space. The basic elements of measurement theory are also covered. We are then able to present quantum mechanics in terms of a set of postulates within this framework. The quantum fields of Vol. I are operators acting in the abstract many-particle Hilbert space.

Chapter 3 is devoted to the quantum theory of angular momentum, and this subject is covered in some depth. There are a variety of motivations here: this theory governs the behavior of any isolated, finite quantum mechanical system and lies at the heart of most of the applications in Vol. I;[1] it provides a detailed illustration of the consequences of a continuous symmetry in quantum mechanics, in this case the very deep symmetry of the isotropy of space; furthermore, it provides an extensive introduction to the theory of Lie groups, here the special unitary group in two dimensions SU(2), which finds wide applicability in internal symmetries. An appendix explores the use of angular momentum theory in the multipole analysis of the radiation field, which is applicable to transitions in any finite quantum mechanical system.

Chapter 4 is devoted to scattering theory. The Schrödinger equation is solved in terms of a time-development operator in the abstract Hilbert space, and the scattering operator is identified. The interaction is turned on and off "adiabatically", which allows a simple construction of initial and final states, and the S-matrix elements then follow immediately. Although inappropriate for developing a covariant scattering analysis, the time integrations in the scattering operator can be explicitly performed and contact made with time-independent scattering theory. It is shown how adiabatic damping puts the correct boundary conditions into the propagators. A general expression is derived for the quantum mechanical transition rate. Non-relativistic scattering from a static potential provides a nice example of the time-independent analysis. If the time is left in the scattering operator, one has a basis for the subsequent analysis in terms of Feynman

---

[1] For example, here we validate the "vector model" used there.

diagrams and Feynman rules. The tools developed in this chapter allow one to analyze any scattering or reaction process in quantum mechanics.

Lagrangian field theory provides the dynamical framework for a consistent, covariant, quantum mechanical description of many interacting particles, and this is the topic in chapter 4. We first review classical lagrangian particle mechanics, and then classical lagrangian continuum mechanics, using our paradigm of the transverse planar oscillations of a string. The string mechanics can be expressed in terms of "two-vectors" (x, ict) where c is the sound velocity in the string. We then discuss the quantization of these classical mechanical systems obtained by imposing canonical quantization relations on the operators in the abstract Hilbert space.

The appending of two additional spatial dimensions to obtain four-vectors $(\mathbf{x}, ict)$, where c is now the speed of light, leads immediately to a covariant, continuum lagrangian mechanics for a scalar field in Minkowski space, which is then quantized with the same procedure used for the string. We develop a covariant, continuum lagrangian mechanics for the Dirac field, and discuss how anticommutation relations must be imposed when quantizing in this case. A general expression is derived for the energy-momentum tensor, and Noether's theorem is proven, which states that for every continuous symmetry of the lagrangian density there is an associated conserved current. A full appendix is dedicated to the lagrangian field theory of the electromagnetic field.

Symmetries play a central role in developing covariant lagrangian densities for various interacting systems, and chapter 6 is devoted to symmetries. The discussion starts with spatial rotations and the internal symmetry of isospin, and it builds on the analysis of SU(2) in chapter 3. Here isospin is developed in terms of global SU(2) transformations of the nucleon field $\psi = (\psi_p, \psi_n)$. The internal symmetry is generalized to SU(3) within the framework of the Sakata model with a baryon field $\psi = (\psi_p, \psi_n, \psi_\Lambda)$.[2]

It is also shown in chapter 6 how the imposition of invariance under local phase transformations of the charged Dirac field, where the transformation parameter depends on the space-time point x, necessitates the introduction of a photon (gauge) field $A_\mu(x)$ and leads to quantum electrodynamics (QED), the most accurate theory known. Yang-Mills theory, which extends this idea to invariance under local internal symmetry transformations of the Dirac field, and necessitates the introduction of corresponding gauge bosons, is developed in detail. These gauge bosons must be massless, and to

---

[2]Wigner's supermultiplet theory based on internal SU(4) transformations of the nucleon field $\psi = (\psi_{p\uparrow}, \psi_{p\downarrow}, \psi_{n\uparrow}, \psi_{n\downarrow})$ is also touched on.

understand the very successful physical application of Yang-Mills theories, it is necessary to understand how mass is generated in relativistic quantum field theories.[3]

We do this within the framework of the σ-model, a very simple model which has had a profound effect on the development of modern physics. A massless Dirac field has an additional chiral invariance under a global transformation that also mixes the components of the Dirac field. The corresponding conserved axial-vector current, which augments the conserved vector current arising from global isospin invariance, corresponds closely to what is observed experimentally in the weak interactions. The σ-model extends the massless Dirac lagrangian through a chiral-invariant interaction with a pion and scalar field $(\boldsymbol{\pi}, \sigma)$. A choice of shape of the chiral-invariant meson potential $\mathcal{V}(\boldsymbol{\pi}^2 + \sigma^2)$ then leads to a vacuum expectation value for the scalar field that gives rise to a mass for the Dirac particle while maintaining chiral invariance of the lagrangian. This *spontaneous symmetry breaking* illustrates how observed states do not necessarily reflect the symmetry of the underlying lagrangian. Generating mass through the expectation value of a scalar field, in one way or another, now underlies most modern theories of particle interactions.[4]

The most fundamental symmetry in nature is Lorentz invariance. One must obtain the same physics in any Lorentz frame. The Lorentz transformation properties of the scalar and Dirac fields are detailed in an appendix. Some very useful tools are provided in another appendix devoted to the irreducible representations of SU(n).

Chapter 7 is concerned with the derivation of the Feynman rules, and to focus on the method, they are developed for the simplest theory of a Dirac particle interacting with a neutral, massive, scalar field. Wick's theorem is proven. This allows one to convert a time-ordered product of fields in the interaction picture, where the time dependence is that of free fields, into a normal-ordered product where the destruction operators sit to the right of the creation operators for all times. It is the time-ordered product that occurs naturally in the scattering operator, and it is the normal-ordered product from which it is straightforward to compute any required matrix elements. Wick's theorem introduces the vacuum expectation value

---

[3]Both quantum chromodynamics (QCD) and the Standard Model of electroweak interactions are Yang-Mills theories built on internal symmetry groups, the former on an internal color $SU(3)_C$ symmetry and the latter on an internal weak $SU(2)_W \otimes U(1)_W$.

[4]In the Standard Model, it provides the basis for the "Higgs mechanism" (see, for example, [Walecka (2004)]).

of the time-ordered product of pairs of interaction-picture fields— these are the *Feynman propagators*. An appendix provides a thorough discussion of these Green's functions, as well as other singular functions, for the scalar, Dirac, and electromagnetic fields.[5] The lowest-order scattering amplitudes, self-energies, and vacuum amplitude are all calculated for the Dirac-scalar theory, and then interpreted in terms of Feynman diagrams and Feynman rules. The cancellation of the disconnected diagrams is demonstrated in this chapter, as is the requisite procedure for mass renormalization.

In chapter 8 these techniques are applied to a theory with immediate experimental implications. That theory is quantum electrodynamics (QED), where the fine-structure constant $\alpha = e^2/4\pi\hbar c\varepsilon_0 = 1/137.04$ provides a meaningful dimensionless expansion parameter. The point of departure here is the derived QED hamiltonian in the Coulomb gauge, where $\nabla \cdot \mathbf{A}(x) = 0$ and there is a one-to-one correspondence between the degrees of freedom in the vector potential and transverse photons. The interaction of the electron current and vector potential is combined with the instantaneous Coulomb interaction to produce a photon propagator, and then conservation of the interaction-picture current is invoked to reduce this to an effective photon propagator with a Fourier transform in Minkowski space of $\tilde{D}_{\mu\nu}(q) = \delta_{\mu\nu}/q^2$. One thereby recovers covariance and gauge invariance in the electromagnetic interaction.

The steps leading from an S-matrix element to a cross section are covered in detail in two examples, $\mu^- + e^- \to \mu^- + e^-$ and $e^+ + e^- \to \mu^+ + \mu^-$. Expressions are obtained in the center-of-momentum (C-M) frame that are exact to $O(\alpha^2)$. The scattering operator is extended to include an interaction with a specified external field, and the lowest-order amplitudes for bremsstrahlung and pair production are obtained. The Feynman diagrams and Feynman rules in these examples serve to provide us with the Feynman diagrams and Feynman rules for QED.

Chapter 9 presents an introduction to the calculation of various virtual processes in relativistic quantum field theory, and again, to keep close contact with experiment, we focus on QED. Calculations of the $O(\alpha)$ corrections to the scattering amplitude for an electron in an external field provide an introduction to the relevant lowest-order "loop" contributions, where there is an integral over one virtual four-momentum. The insertions here are characterized through the electron self-energy, vertex modification, and vacuum polarization (photon self-energy) diagrams.

---

[5]The neutral, massive vector meson field is covered in the problems.

Dimensional regularization, detailed in an appendix, serves as a technique that gives mathematical meaning to originally ill-defined integrals. Here one works in the complex n-plane, where n is the dimension, and any potential singularity is then isolated at the point $n \to 4$. The contribution of each of the above diagrams is cast into a general form that isolates such singular pieces and leaves additional well-defined convergent expressions.

Some care must be taken with the contribution of the self-energy insertions on the external legs ("wavefunction renormalization"), and we do so. It is then shown how Ward's identity, which relates the electron self-energy and vertex insertion, leads to a *cancellation* in the scattering amplitude of the singular parts of these insertions. Vacuum polarization then leads to a shielding of the charge in QED and to charge renormalization. The two remaining singular terms in the theory are removed by mass and charge renormalization, and if the scattering amplitude is consistently expressed in terms of the renormalized mass and charge (m, e) one is left with finite, calculable, $O(\alpha)$ corrections to the scattering amplitude. The Schwinger term in the anomalous magnetic moment of the electron is calculated here. Higher-order corrections are summarized in terms of Dyson's and Ward's equations, and it is demonstrated through Ward's identities how the multiplicative renormalizability of QED holds to all orders.

With the techniques developed in chapter 9, one has the tools with which to examine loop contributions in any relativistic quantum field theory.

Chapter 10 is on path integrals. There are many reasons for becoming familiar with the techniques here, which underly much of what now goes on in theoretical physics, for example: this approach provides an alternative to canonical quantization, which, with derivative couplings, can become prohibitively difficult; here one deals entirely with classical quantities, in particular the classical lagrangian and classical action; and the classical limit $\hbar \to 0$ leads immediately to Hamilton's principle of stationary action.

We start from the analysis of a non-relativistic particle moving in a potential in one dimension and show how the quantum mechanical transition amplitude can be exactly expressed as an integral over all possible paths between the initial and final space-time points.[6] We then make the transition to a system with many degrees of freedom, and then to field theory.

The addition of an arbitrary source term, together with the crucial theorem of Abers and Lee, allows one to construct the generating functional as a ratio of two path integrals, one a transition amplitude containing the

---

[6]It is shown in a problem how the partition function of statistical mechanics in the microcanonical ensemble can also be expressed as a path integral.

source and the second a vacuum-vacuum amplitude without it. The connected Green's functions can then be determined from the generating functional by functional differentiation with respect to the source, as detailed here. The generating functional is calculated for the free scalar field using gaussian integration, and it is shown how the Feynman propagator and Wick's theorem are reproduced in this case. The treatment of the Dirac field necessitates the introduction of Grassmann variables, which are anti-commuting c-numbers. The generating functional is computed for the free Dirac field, and the Feynman propagator and Wick's theorem again recovered. It is shown how to include interactions and express the full generating functional in terms of those already computed.

An appendix describes how one uses the Faddeev-Popov method in a gauge theory, at least for QED, to factor the measure in the path integral into one part that is an integral over all gauge functions and a second part that is gauge invariant. With a gauge-invariant action, the path integral over the gauge functions then factors and cancels in the generating functional ratio. It is shown how the accompanying Faddeev-Popov determinant can be expressed in terms of ghost fields, which also factor and disappear from the generating functional in the case of QED. The generating functional for the free electromagnetic field is calculated here.

Although abbreviated, the discussion in chapter 10 should allow one to use path integrals with some facility, and to read with some understanding material that starts from path integrals.

The final chapter 11 deals with canonical transformations for quantum systems. Chapter 11 of Vol. I provides an introduction to the properties of superfluid Bose systems and superconducting Fermi systems. In both cases, in order to obtain a theoretical description of the properties of the quantum fluids, it is necessary to include interactions. A technique that has proven invaluable for the treatment of such systems is that of canonical transformations. Here one makes use of the fact that the properties of the creation and destruction operators follow entirely from the canonical (anti)commutation relations in the abstract Hilbert space. By introducing new "quasiparticle" operators that are linear combinations of the original operators, and that preserve these (anti)commutation relations, one is able to obtain exact descriptions of some interacting systems, both in model problems and in a starting hamiltonian.

The problem of a weakly interacting Bose gas with a repulsive interaction between the particles is solved with the Bogoliubov transformation. A phonon spectrum is obtained for the many-body system, which, as shown in Vol. I, allows one to understand superfluidity. Motivated by the Cooper pairs obtained in Vol. I, a Fermi system with an attractive interaction between those particles at the Fermi surface is analyzed with the Bogoliubov-Valatin transformation. The very successful BCS theory of superconductivity is obtained in the case that the residual quasiparticle interactions can be neglected.

A problem takes the reader through the Bloch-Nordsieck transformation, which examines the quantized electromagnetic field interacting with a specified, time-independent current source. A key insight into the infrared problem in QED is thereby obtained. A second problem guides the reader through the analysis of a quantized, massive, neutral scalar field interacting with a classical, specified, time-independent source. The result is an exact derivation of the Yukawa interaction of nuclear physics.

This book is designed to further the goals of Vol. I and to build on the foundation laid there. Volume II covers in more depth those topics that form the essential framework of modern theoretical physics.[7] Readers should now be in a position to go on to more advanced texts, such as [Bjorken and Drell (1964); Bjorken and Drell (1965); Schiff (1968); Itzykson and Zuber(1980); Cheng and Li (1984); Donoghue, Golowich, and Holstein (1993); Merzbacher (1998); Fetter and Walecka (2003a); Walecka (2004); Banks (2008)], with a deeper sense of appreciation and understanding.

Modern theoretical physics provides a basic understanding of the physical world and serves as a platform for future developments. When finished with this book, readers should have an elementary working knowledge in the principal areas of theoretical physics of the twentieth-century.

---

[7]The author considered also including in Vol. II a chapter on solutions to the Einstein field equations in general relativity; however, given the existence of [Walecka (2007)], it was deemed sufficient to simply refer readers to that book.

Chapter 2

Quantum Mechanics (Revisited)

In this chapter we formalize some of the analysis of quantum mechanics in Vol. I, which will allow us to focus on the general structure of the theory. We start the discussion with a review of linear vector spaces.

2.1 Linear Vector Spaces

Consider the ordinary three-dimensional linear vector space in which we live.

2.1.1 *Three-Dimensional Vectors*

Introduce an orthonormal set of basis vectors $\mathbf{e}_i$ with $i = 1, 2, 3$ satisfying

$$\mathbf{e}_i \cdot \mathbf{e}_j = \delta_{ij} \qquad ; (i, j) = 1, 2, 3 \qquad (2.1)$$

An arbitrary vector $\mathbf{v}$ is a physical quantity that has a direction and length in this space. It can be expanded in the basis $\mathbf{e}_i$ according to

$$\mathbf{v} = \sum_{i=1}^{3} v_i \mathbf{e}_i \qquad ; v_i = \mathbf{e}_i \cdot \mathbf{v} \qquad (2.2)$$

$\mathbf{v}$ can now be characterized by its components (v_1, v_2, v_3) in this basis.[1] Vectors have the following properties:

(1) Addition of vectors, and multiplication of a vector by a constant, are

---

[1] This characterization will be denoted by $\mathbf{v} : (v_1, v_2, v_3)$.

9

expressed in terms of the components by

$$\mathbf{a} + \mathbf{b} : \ (a_1 + b_1, a_2 + b_2, a_3 + b_3)$$
$$\gamma \mathbf{a} : \ (\gamma a_1, \gamma a_2, \gamma a_3) \qquad ; \text{ linear space} \qquad (2.3)$$

These properties characterize a *linear space*;

(2) The dot product, or inner product, of two vectors is defined by

$$\mathbf{a} \cdot \mathbf{b} \equiv a_1 b_1 + a_2 b_2 + a_3 b_3 \qquad ; \text{ dot product} \qquad (2.4)$$

The *length* of the vector is then determined by

$$|\mathbf{v}| = \sqrt{\mathbf{v}^2} = \sqrt{\mathbf{v} \cdot \mathbf{v}} = (v_1^2 + v_2^2 + v_3^2)^{1/2} \qquad ; \text{ length} \qquad (2.5)$$

One says that there is an *inner-product norm* in the space.

(3) Suppose one goes to a new orthonormal basis $\boldsymbol{\alpha}_i$ where the vector $\mathbf{v}$ has the components $\mathbf{v} : (\bar{v}_1, \bar{v}_2, \bar{v}_3)$. Then the components are evidently related by

$$\mathbf{v} = \sum_{i=1}^{3} v_i \mathbf{e}_i = \sum_{i=1}^{3} \bar{v}_i \boldsymbol{\alpha}_i$$
$$\Rightarrow \qquad v_i = \sum_{j=1}^{3} \bar{v}_j (\mathbf{e}_i \cdot \boldsymbol{\alpha}_j) = \sum_{j=1}^{3} \bar{v}_j [\boldsymbol{\alpha}_j]_i \qquad (2.6)$$

2.1.2   *n-Dimensions*

These arguments are readily extended to n-dimensions by simply increasing the number of components

$$\mathbf{v} : \ (v_1, v_2, v_3, \cdots, v_n) \qquad ; \ n\text{-dimensions} \qquad (2.7)$$

The extension to *complex vectors* is accomplished through the use of the linear multiplication property with a complex γ. The positive-definite norm is then correspondingly defined through $|\mathbf{v}|^2 \equiv \mathbf{v}^\star \cdot \mathbf{v}$,

$$\gamma \mathbf{v} : \ (\gamma v_1, \gamma v_2, \gamma v_3, \cdots, \gamma v_n) \qquad ; \text{ complex vectors}$$
$$|\mathbf{v}|^2 \equiv \mathbf{v}^\star \cdot \mathbf{v} = |v_1|^2 + |v_2|^2 + \cdots + |v_n|^2 \qquad (2.8)$$

2.2 Hilbert Space

The notion of a *Hilbert space* involves the generalization of these concepts to a space with an infinite number of dimensions. Let us start with an example.

2.2.1 *Example*

Recall the set of plane waves in one spatial dimension in an interval of length L satisfying periodic boundary conditions

$$\phi_n(x) = \frac{1}{\sqrt{L}} e^{ik_n x} \qquad ; \; k_n = \frac{2\pi n}{L} \qquad ; \; n = 0, \pm 1, \pm 2, \cdots \qquad ;$$

$$\text{basis vectors} \qquad (2.9)$$

These will be referred to as the *basis vectors*. They are orthonormal and satisfy

$$\int_0^L dx \, \phi_m^\star(x)\phi_n(x) = \delta_{mn} \qquad ; \; \text{orthonormal}$$

$$\equiv \langle \phi_m | \phi_n \rangle \qquad ; \; \text{inner product} \qquad (2.10)$$

This relation allows us to define the *inner product* of two basis vectors, denoted in the second line by $\langle \phi_m | \phi_n \rangle$,[2] and the positive-definite *inner-product norm* of the basis vectors is then given by

$$|\phi_n|^2 = \langle \phi_n | \phi_n \rangle = \int_0^L dx \, |\phi_n(x)|^2 \qquad ; \; (\text{``length''})^2 \qquad (2.11)$$

An arbitrary function $\psi(x)$ can be expanded in this basis according to

$$\psi(x) = \sum_{n=-\infty}^{\infty} c_n \phi_n(x) \qquad ; \; \text{expansion in complete set} \qquad (2.12)$$

This is, after all, just a complex Fourier series. The orthonormality of the basis vectors allows one to solve for the coefficients c_n

$$c_n = \langle \phi_n | \psi \rangle = \int_0^L dx \, \phi_n^\star(x)\psi(x) \qquad (2.13)$$

---

[2] The notation, and most of the analysis in this chapter, is due to Dirac [Dirac (1947)].

Any piecewise continuous function can actually be expanded in this set, and the basis functions are *complete* in the sense that[3]

$$\text{Lim}_{N \to \infty} \int_0^L dx \left| \psi(x) - \sum_{n=-N}^N c_n \phi_n(x) \right|^2 = 0 \qquad ; \text{completeness} \quad (2.14)$$

Just as with an ordinary vector, the function $\psi(x)$ can now be characterized by the expansion coefficients c_n

$$\psi(x) \ : (c_{-\infty}, \cdots, c_{-1}, c_0, c_1, \cdots, c_\infty)$$
$$\text{or;} \qquad \psi \ : \{c_n\} \qquad\qquad\qquad\qquad (2.15)$$

Addition of functions and multiplication by constants are defined in terms of the coefficients by

$$\psi^{(1)} + \psi^{(2)} \ : \{c_n^{(1)} + c_n^{(2)}\}$$
$$\gamma\psi \ : \{\gamma c_n\} \qquad\qquad ; \text{linear space} \quad (2.16)$$

This function space is again a *linear space*. The norm of ψ is given by

$$|\psi|^2 = \langle \psi | \psi \rangle = \int_0^L dx \, |\psi(x)|^2 = \sum_{n=-\infty}^{\infty} |c_n|^2 \qquad ; (\text{norm})^2 \quad (2.17)$$

which, in the case of Fourier series, is just Parseval's theorem.

2.2.2   *Definition*

The function $\psi(x)$ in Eq. (2.17) is said to be square-integrable. The set of all square-integrable functions ($\mathcal{L}^2$) forms a *Hilbert space*. Mathematicians define a Hilbert space as follows:

(1) It is a linear space;
(2) There is an inner-product norm;
(3) The space is complete in the sense that every Cauchy sequence converges to an element in the space.

The above analysis demonstrates, through the expansion coefficients c_n, the isomorphism between the space of all square-integrable functions ($\mathcal{L}^2$) and the ordinary infinite-dimensional complex linear vector space (l^2) discussed at the beginning of this section.

---

[3]This is all the completeness we will need for the physics in this volume.

2.2.3   *Relation to Linear Vector Space*

A more direct analogy to the infinite-dimensional complex linear vector space (l^2) is obtained through the following identification

$$v_i \rightarrow \psi_x \qquad\qquad ;\ \text{coordinate space} \qquad\qquad (2.18)$$

We now use the coordinate x as a subscript, and we note that it is here a *continuous* index. The square of the norm then becomes

$$\sum_i v_i^{\star} v_i \rightarrow \sum_x \psi_x^{\star} \psi_x \equiv \int dx\, \psi^{\star}(x)\psi(x) \qquad\qquad (2.19)$$

The sum over the continuous index has here been appropriately defined through a familiar integral. With this notation, the starting expansion in Eq. (2.12) takes the form

$$\psi_x = \sum_{n=-\infty}^{\infty} c_n [\phi_n]_x \qquad\qquad (2.20)$$

2.2.4   *Abstract State Vector*

Equation (2.20) can be interpreted in the following manner:

This is just one component of the abstract vector relation

$$|\psi\rangle = \sum_n c_n |\phi_n\rangle \qquad\qquad ;\ abstract\ vector\ relation \qquad\qquad (2.21)$$

The quantity $|\psi\rangle$ is now interpreted as a vector in an infinite-dimensional, abstract Hilbert space. It can be given a concrete representation through the component form in Eqs. (2.20) and (2.12), using the particular set of basis vectors in Eqs. (2.9).

As before, one solves for the expansion coefficients c_n by simply using the orthonormality of the basis vectors in Eq. (2.10)

$$c_n = \langle \phi_n | \psi \rangle = \sum_x [\phi_n]_x^{\star} \psi_x = \int dx\, \phi_n^{\star}(x)\psi(x) \qquad\qquad (2.22)$$

2.3   Linear Hermitian Operators

Consider an operator L in Hilbert space. Given $\psi(x)$, then $L\psi(x)$ is some new state in the space. L is a *linear operator* if it satisfies the condition

$$L(\alpha\phi_1 + \beta\phi_2) = \alpha(L\phi_1) + \beta(L\phi_2) \qquad \text{; linear operator} \qquad (2.23)$$

for any (ϕ_1, ϕ_2) in the space. L is *hermitian* if it satisfies the relation[4]

$$\int dx\, \phi_1^\star(x) L\phi_2(x) = \int dx\, [L\phi_1(x)]^\star \phi_2(x) = \left[\int dx\, \phi_2^\star(x) L\phi_1(x) \right]^\star \quad ;$$

$$\text{hermitian} \qquad (2.24)$$

A shorthand for these relations is as follows

$$\langle \phi_1 | L | \phi_2 \rangle = \langle L\phi_1 | \phi_2 \rangle = \langle \phi_2 | L | \phi_1 \rangle^\star \qquad \text{; shorthand} \qquad (2.25)$$

We now make the important observation that *if one knows the matrix elements of* L

$$L_{mn} \equiv \int dx\, \phi_m^\star(x) L\phi_n(x) \equiv \langle m | L | n \rangle \qquad \text{; matrix elements} \qquad (2.26)$$

in any complete basis, then one knows the operator L. Let us prove this assertion. Let $\psi(x)$ be an arbitrary state in the space. If one knows the corresponding $L\psi(x)$, then L is determined. Expand $\psi(x)$ in the complete basis

$$\psi(x) = \sum_n c_n \phi_n(x) \qquad \text{; complete basis} \qquad (2.27)$$

As above, the coefficients c_n follow from the orthonormality of the eigenfunctions ϕ_n

$$c_n = \int dx\, \phi_n^\star(x) \psi(x) \qquad \text{; known} \qquad (2.28)$$

These coefficients are thus determined for any given ψ. Now compute[5]

$$L\psi(x) = \sum_n c_n [L\phi_n(x)] \qquad (2.29)$$

---

[4]See Probl. 4.5—the notation "Probl" refers to the problems in Vol. I.

[5]It is assumed here that there is enough convergence that one can operate on this series term by term.

The expansion in a complete basis can again be invoked to write the state $L\phi_n(x)$ as

$$L\phi_n(x) = \sum_m \beta_{mn}\phi_m(x) \qquad ; \text{ complete basis} \qquad (2.30)$$

and the orthonormality of the eigenfunctions allows one to identify

$$\beta_{mn} = \int dx\, \phi_m^\star(x)L\phi_n(x) = L_{mn} \qquad (2.31)$$

Hence

$$L\phi_n(x) = \sum_m \phi_m(x)L_{mn}$$

$$\Rightarrow \quad L\psi(x) = \sum_n \sum_m \phi_m(x)L_{mn}c_n \qquad ; \text{ known} \qquad (2.32)$$

This is now a *known quantity*, and thus we have established the equivalence[6]

$$L \longleftrightarrow L_{mn} \qquad ; \text{ equivalent} \qquad (2.33)$$

2.3.1 *Eigenfunctions*

The *eigenfunctions* of a linear operator are defined by the relation

$$L\phi_\lambda(x) = \lambda\phi_\lambda(x) \qquad ; \text{ eigenfunctions}$$
$$\lambda \text{ is eigenvalue} \qquad (2.34)$$

Here the operator simply reproduces the function and multiplies it by a constant, the *eigenvalue*. If L is an *hermitian* operator, then the following results hold:

- The eigenvalues λ are real (ProbI. 4.6);
- The eigenfunctions corresponding to different eigenvalues are orthogonal.[7]

We give two examples from Vol. I:

(1) *Momentum.* The momentum operator in one dimension in coordinate space is

$$p = \frac{\hbar}{i}\frac{\partial}{\partial x} \qquad ; \text{ momentum} \qquad (2.35)$$

---

[6]This equivalence is the basis of *matrix mechanics* (compare Prob. 2.8).

[7]The proof here is essentially that of ProbI. H.4; dedicated readers can supply it.

With periodic boundary conditions, the eigenfunctions are just those of Eq. (2.9), and

$$p\phi_k(x) = \hbar k\, \phi_k(x) \qquad ; k = \frac{2\pi n}{L} \qquad ; n = 0, \pm 1, \cdots \quad (2.36)$$

p is hermitian with these boundary conditions (Probl. 4.5), and as we have seen, these eigenfunctions are both orthonormal and complete.

(2) *Hamiltonian.* In one dimension in coordinate space the hamiltonian is given by

$$H = \frac{-\hbar^2}{2m} \frac{\partial^2}{\partial x^2} + V(x) \qquad ; \text{hamiltonian} \qquad (2.37)$$

We assume that $V(x)$ is real. The eigenstates are

$$H u_{E_n}(x) = E_n\, u_{E_n}(x) \qquad ; \text{eigenstates} \qquad (2.38)$$

In general, there will be both bound-state and continuum solutions to this equation. With the choice of periodic boundary conditions in the continuum, the hamiltonian is hermitian (Probl. 4.5), and the energy eigenvalues E_n are real (Probl. 4.6). The eigenstates of this hermitian operator also form a *complete set*, so that one can similarly expand an arbitrary $\psi(x)$ as

$$\psi(x) = \sum_n a_n\, u_{E_n}(x) \qquad ; \text{complete set} \qquad (2.39)$$

For the present purposes, one can simply take two of the postulates of quantum mechanics to be:

(1) Observables are represented with linear hermitian operators;
(2) The eigenfunctions of any linear hermitian operator form a complete set.[8]

2.3.2 *Eigenstates of Position*

The position operator x in one dimension is an hermitian operator. Consider the eigenstates of x with eigenvalues ξ so that

$$x\psi_\xi(x) = \xi\, \psi_\xi(x) \qquad ; \text{position operator} \qquad (2.40)$$

---

[8]A proof of completeness for any operator of the Sturm-Liouville type is contained in [Fetter and Walecka (2003)]. The use of ordinary riemannian integration in the definition of the inner product in Eq. (2.19), and the notion of completeness expressed in Eq. (2.14), represent the extent of the mathematical rigor in the present discussion.

The solution to this equation, in coordinate space, is just a Dirac delta function

$$\psi_\xi(x) = \delta(x - \xi) \qquad \text{; eigenstates of position} \qquad (2.41)$$

It is readily verified that

$$x\psi_\xi(x) = x\delta(x - \xi) = \xi\delta(x - \xi) = \xi\,\psi_\xi(x) \qquad (2.42)$$

On the interval $[0, L]$, with periodic boundary conditions, the eigenvalues ξ run continuously over this interval. As to the orthonormality of these eigenfunctions, one can just compute

$$\int dx\,\psi_{\xi'}^\star(x)\psi_\xi(x) = \int dx\,\delta(x - \xi')\delta(x - \xi) = \delta(\xi - \xi') \qquad (2.43)$$

Hence

$$\int dx\,\psi_{\xi'}^\star(x)\psi_\xi(x) = \delta(\xi - \xi') \qquad \text{; orthonormality} \qquad (2.44)$$

We make some comments on this result:

- One cannot avoid a continuum normalization here, since the position eigenvalue ξ is truly continuous;
- In contrast, in one dimension with periodic boundary conditions on this interval, the eigenfunctions of momentum in Eq. (2.36) have a *denumerably infinite set of discrete eigenvalues*. This proved to be an essential calculational tool in Vol. I;
- To make the analogy between coordinate space and momentum space closer, one can take L to infinity.[9] Define

$$\psi_k(x) = \left(\frac{L}{2\pi}\right)^{1/2}\phi_k(x) = \frac{1}{\sqrt{2\pi}}e^{ikx} \qquad (2.45)$$

Then

$$\int dx\,\psi_{k'}^\star(x)\psi_k(x) = \frac{1}{2\pi}\int dx\,e^{i(k-k')x}$$
$$\to \delta(k - k') \qquad \text{; } L \to \infty \qquad (2.46)$$

In this limit *both* the momentum and position eigenfunctions have a continuum norm.

---

[9] As shown in Vol. I, Fourier series are converted to Fourier integrals in this limit; one first uses the p.b.c. to convert the interval to $[-L/2, L/2]$.

2.4   Abstract Hilbert Space

Recall Eqs. (2.12) and (2.20) from above, which represent an expansion in a complete set,

$$\psi(x) = \sum_n c_n \phi_n(x)$$

$$\text{or} \ ; \qquad \psi_x = \sum_n c_n [\phi_n]_x \tag{2.47}$$

This can be viewed as the component form of the abstract vector relation

$$|\psi\rangle = \sum_n c_n |\phi_n\rangle \qquad ; \ \text{abstract vector relation} \tag{2.48}$$

Just as an ordinary three-dimensional vector $\mathbf{v}$ has meaning independent of the basis vectors in which it is being decomposed, one can think of this as a vector pointing in some direction in the abstract, infinite-dimensional Hilbert space. Equations (2.47) then provide a component form of this abstract vector relation.

2.4.1   *Inner Product*

The inner product in this space is provided by Eq. (2.19)

$$\langle \psi_a | \psi_b \rangle = \sum_x [\psi_a]_x^\star [\psi_b]_x \equiv \int dx \, \psi_a^\star(x) \psi_b(x) \qquad ; \ \text{inner product} \tag{2.49}$$

Thus, from Eqs. (2.47)

$$c_n = \langle \phi_n | \psi \rangle = \sum_x [\phi_n]_x^\star \psi_x \equiv \int dx \, \phi_n^\star(x) \psi(x) \tag{2.50}$$

We note the following important inner products:

$$\langle \xi' | \xi \rangle = \int dx \, \psi_{\xi'}^\star(x) \psi_\xi(x) = \delta(\xi' - \xi)$$

$$\langle k' | k \rangle = \int dx \, \phi_{k'}^\star(x) \phi_k(x) = \delta_{kk'} \qquad ; \ \text{with p.b.c.}$$

$$\langle \xi | k \rangle = \int dx \, \psi_\xi^\star(x) \phi_k(x) = \frac{1}{\sqrt{L}} e^{ik\xi} \tag{2.51}$$

The last relation follows directly from the wave functions in Eqs. (2.36) and (2.41).[10]

---

[10]See also Eq. (2.9); note that the subscript n on $k_n = 2\pi n/L$ is suppressed.

2.4.2 Completeness

As established in Vol. I, the statement of completeness with the set of coordinate space eigenfunctions $\phi_p(x)$, where p denotes the eigenvalues of a linear hermitian operator, is

$$\sum_p \phi_p(x)\phi_p^\star(y) = \delta(x-y) \qquad ; \text{ completeness} \qquad (2.52)$$

Insert this relation in the definition of the inner product in Eq. (2.49)

$$\begin{aligned}
\langle\psi_a|\psi_b\rangle &= \int dx\,\psi_a^\star(x)\psi_b(x) \equiv \int dxdy\,\psi_a^\star(x)\delta(x-y)\psi_b(y) \\
&= \sum_p \int dx\,\psi_a^\star(x)\phi_p(x) \int dy\,\phi_p^\star(y)\psi_b(y) \\
&= \sum_p \langle\psi_a|\phi_p\rangle\langle\phi_p|\psi_b\rangle
\end{aligned} \qquad (2.53)$$

Here Eq. (2.52) has been used in the second line, and the definition of the inner product used in the third. This relation can be summarized by writing the abstract vector relation

$$\sum_p |\phi_p\rangle\langle\phi_p| = 1_{\text{op}} \qquad ; \text{ completeness} \qquad (2.54)$$

This unit operator 1_{op} can be inserted into any inner product, leaving that inner product unchanged. This relation follows from the completeness of the wave functions $\phi_p(x)$ providing the coordinate space components of the abstract state vectors $|\phi_p\rangle$.

2.4.3 Linear Hermitian Operators

In Vol. I, quantum mechanics was introduced in coordinate space, where the momentum p is given by $p = (\hbar/i)\partial/\partial x$. It was observed in Probl. 4.8 that one could equally well work in momentum space, where the position x is given by $x = i\hbar\partial/\partial p$. It was also observed there that the commutation relation $[p, x] = \hbar/i$ is independent of the particular representation. Our goal in this section is to similarly *abstract the Schrödinger equation* and free it from any particular component representation.

2.4.3.1 *Eigenstates*

A linear hermitian operator L_{op} takes one abstract vector $|\psi\rangle$ into another $L_{op}|\psi\rangle$. The eigenstates of L_{op}, as before, are defined by

$$L_{op}|\phi_\lambda\rangle = \lambda|\phi_\lambda\rangle \qquad ; \text{ eigenstates} \qquad (2.55)$$

For example:

$$
\begin{aligned}
p_{op}|k\rangle &= \hbar k|k\rangle & &; \text{ momentum} \\
x_{op}|\xi\rangle &= \xi|\xi\rangle & &; \text{ position} \\
(L_z)_{op}|m\rangle &= m|m\rangle & &; \text{ z-component of angular momentum} \\
H_{op}|\psi\rangle &= E|\psi\rangle & &; \text{ hamiltonian} \qquad (2.56)
\end{aligned}
$$

2.4.3.2 *Adjoint Operators*

In coordinate space, the adjoint operator $L^\dagger$ is defined by

$$\int d\xi\, \psi_a^\star(\xi)L^\dagger\psi_b(\xi) \equiv \int d\xi\, [L\psi_a(\xi)]^\star\psi_b(\xi) = \left[\int d\xi\, \psi_b^\star(\xi)L\psi_a(\xi)\right]^\star \quad (2.57)$$

The adjoint operator in the abstract Hilbert space is defined in exactly the same manner

$$\langle\psi_a|L_{op}^\dagger|\psi_b\rangle \equiv \langle L_{op}\psi_a|\psi_b\rangle = \langle\psi_b|L_{op}|\psi_a\rangle^\star \qquad ; \text{ adjoint} \qquad (2.58)$$

Note that it follows from this definition that if γ is some complex number, then

$$[\gamma L_{op}]^\dagger = \gamma^* L_{op}^\dagger \qquad (2.59)$$

An operator is *hermitian* if it is equal to its adjoint

$$
\begin{aligned}
& L_{op}^\dagger = L_{op} & &; \text{ hermitian} \\
\Rightarrow \quad & \langle\psi_a|L_{op}|\psi_b\rangle = \langle L_{op}\psi_a|\psi_b\rangle = \langle\psi_b|L_{op}|\psi_a\rangle^\star & & \qquad (2.60)
\end{aligned}
$$

With an hermitian operator, one can just let it act on the state on the left when calculating matrix elements.

2.4.4 *Schrödinger Equation*

To get the time-independent Schrödinger equation in the coordinate representation, one projects the abstract operator relation $H_{op}|\psi\rangle = E|\psi\rangle$ onto the basis of eigenstates of position $|\xi\rangle$.

We show this through the following set of steps:

(1) First project $|\psi\rangle$ onto an eigenstate of position $|\xi\rangle$

$$\langle\xi|\psi\rangle = \sum_x [\psi_\xi]^\star_x [\psi]_x = \int dx\, \psi^\star_\xi(x)\psi(x) = \int dx\, \delta(\xi - x)\psi(x)$$

$$\langle\xi|\psi\rangle = \psi(\xi) \qquad \text{; wave function} \qquad (2.61)$$

This is simply the familiar coordinate space wave function $\psi(\xi)$;

(2) Compute the matrix element of the potential $V_{\rm op} = V(x_{\rm op})$ between eigenstates of position

$$\langle\xi|V_{\rm op}|\xi'\rangle = \langle\xi|V(x_{\rm op})|\xi'\rangle = V(\xi')\langle\xi|\xi'\rangle = V(\xi)\delta(\xi - \xi') \quad (2.62)$$

(3) Similarly, compute the matrix elements of the kinetic energy $T_{\rm op}$. This is readily accomplished by invoking the completeness relation for the eigenstates of momentum [see Eq. (2.54)]

$$\sum_k |k\rangle\langle k| = 1_{\rm op} \qquad \text{; completeness} \qquad (2.63)$$

With the insertion of this relation (twice), one finds

$$\langle\xi|T_{\rm op}|\xi'\rangle = \frac{1}{2m}\langle\xi|p^2_{\rm op}|\xi'\rangle = \frac{1}{2m}\sum_k\sum_{k'}\langle\xi|k\rangle\langle k|p^2_{\rm op}|k'\rangle\langle k'|\xi'\rangle$$

$$= \frac{\hbar^2}{2m}\sum_k\sum_{k'}\langle\xi|k\rangle k^2\delta_{kk'}\langle k'|\xi'\rangle = \frac{\hbar^2}{2m}\sum_k\frac{k^2}{L}e^{ik(\xi-\xi')}$$

$$= -\frac{\hbar^2}{2m}\frac{\partial^2}{\partial\xi^2}\sum_k\frac{1}{L}e^{ik(\xi-\xi')} = -\frac{\hbar^2}{2m}\frac{\partial^2}{\partial\xi^2}\delta(\xi - \xi') \quad (2.64)$$

The final relation follows from the completeness of the momentum wave functions.

(4) Make use of the statement of completeness of the abstract eigenstates of position, which is

$$\int d\xi\, |\xi\rangle\langle\xi| = 1_{\rm op} \qquad \text{; completeness} \qquad (2.65)$$

Note that the sum here is actually an integral because the position eigenvalues are continuous.[11]

---

[11]See Prob. 2.2.

(5) The operator form of the time-independent Schrödinger equation is

$$H_{op}|\psi\rangle = (T_{op} + V_{op})|\psi\rangle = E|\psi\rangle \qquad ; \text{S-equation} \qquad (2.66)$$

A projection of this equation on the eigenstates of position gives

$$\langle\xi|H_{op}|\psi\rangle = E\langle\xi|\psi\rangle = E\psi(\xi) \qquad (2.67)$$

Now insert Eq. (2.65) in the expression on the l.h.s., and use the results from Eqs. (2.62) and (2.64)

$$\begin{aligned}
\langle\xi|H_{op}|\psi\rangle &= \int d\xi' \, \langle\xi|H_{op}|\xi'\rangle\langle\xi'|\psi\rangle \\
&= \int d\xi' \left[-\frac{\hbar^2}{2m}\frac{\partial^2}{\partial\xi^2} + V(\xi) \right] \delta(\xi - \xi')\psi(\xi') \\
&= \left[-\frac{\hbar^2}{2m}\frac{\partial^2}{\partial\xi^2} + V(\xi) \right] \int d\xi' \, \delta(\xi - \xi')\psi(\xi') \\
&= \left[-\frac{\hbar^2}{2m}\frac{\partial^2}{\partial\xi^2} + V(\xi) \right] \psi(\xi) \qquad (2.68)
\end{aligned}$$

Thus, in summary,

$$\left[-\frac{\hbar^2}{2m}\frac{\partial^2}{\partial\xi^2} + V(\xi) \right] \psi(\xi) = E\,\psi(\xi) \qquad ; \text{S-equation} \qquad (2.69)$$

This is just the time-independent Schrödinger equation in the coordinate representation. It is the component form of the operator relation of Eq. (2.66) in a basis of eigenstates of position.[12]

For the *time-dependent Schrödinger equation*, the state vector $|\Psi(t)\rangle$ simply moves in the abstract Hilbert space with a time dependence generated by the hamiltonian. Quantum dynamics is thus summarized in the following relations

$$i\hbar\frac{\partial}{\partial t}|\Psi(t)\rangle = \hat{H}|\Psi(t)\rangle \qquad ; \text{S-equation}$$

$$[\hat{p}, \hat{x}] = \frac{\hbar}{i} \qquad ; \text{C.C.R.} \qquad (2.70)$$

We make several comments:

---

[12]The time-independent Schrödinger equation in the momentum representation is obtained by projecting Eq. (2.66) onto the states $|k\rangle$. This gives the components of the operator relation in a basis of eigenstates of momentum (see Prob. 2.9).

- Here, and henceforth, we shall use a caret over a symbol to denote an operator in the abstract Hilbert space;[13]
- The first equation is the abstract form of the time-dependent Schrödinger equation;
- The second equation is the canonical commutation relation for the momentum and position operators;
- As shown above, the usual Schrödinger equation in coordinate space is obtained by projecting the first relation on eigenstates of position; however, these relations are now *independent of the particular basis in which we choose to express their components.*

2.4.4.1 *Stationary States*

With a time-independent potential, one can again look for normal-mode solutions to the time-dependent Schrödinger Eq. (2.70) of the form

$$|\Psi(t)\rangle = e^{-iEt/\hbar}|\psi\rangle \qquad ; \text{ normal modes} \qquad (2.71)$$

Substitution into the first of Eqs. (2.70), and cancellation of a factor $\exp\left(-iEt/\hbar\right)$, leads to the stationary-state Schrödinger equation

$$\hat{H}|\psi\rangle = E|\psi\rangle \qquad ; \text{ stationary-state S-eqn} \qquad (2.72)$$

2.5 Measurements

We must establish the relation between these formal developments and physical measurements. Measurement theory is a deep and extensive topic, and we certainly shall not do justice to it here. No attempt is made to consider implications for very complex objects with a myriad of degrees of freedom.[14] Rather, the discussion here focuses on simple systems where measurement theory is really quite intuitive.

2.5.1 *Coordinate Space*

We start in coordinate space and abstract later. An observable F is represented by a linear hermitian operator $(H, p, x, L_z, etc.)$ with an (assumed)

---

[13]Except for the creation and destruction operators, where their operator nature is evident (see later).

[14]Schrödinger's cat, for example (see [Wikipedia (2009)]).

complete set of eigenstates[15]

$$Fu_{f_n}(x) = f_n u_{f_n}(x) \qquad ; \text{ eigenstates}$$
$$\text{eigenvalues } f_1, f_2, \cdots, f_\infty \qquad (2.73)$$

Let $\Psi(x,t)$ be an arbitrary wave function. At a given time t, its spatial dependence can be expanded in the complete set of wave functions $u_{f_n}(x)$

$$\Psi(x,t) = \sum_n a_{f_n}(t)\, u_{f_n}(x) \qquad (2.74)$$

The wave function is assumed to be normalized so that

$$\int dx\, |\Psi(x,t)|^2 = \sum_n |a_{f_n}(t)|^2 = 1 \qquad (2.75)$$

We can *measure* the expectation value of F given by[16]

$$\langle F \rangle = \int dx\, \Psi^*(x,t)F\Psi(x,t) = \sum_n \sum_{n'} a_{f_n}^*(t) a_{f_{n'}}(t) \int dx\, u_{f_n}^*(x) F u_{f_{n'}}(x)$$
$$= \sum_n \sum_{n'} a_{f_n}^*(t) a_{f_{n'}}(t) f_n \delta_{nn'} \qquad (2.76)$$

Hence

$$\langle F \rangle = \sum_n |a_{f_n}(t)|^2 f_n \qquad ; \text{ expectation value} \qquad (2.77)$$

If one is in a stationary state so that

$$\Psi(x,t) = \psi(x)e^{-iEt/\hbar} \qquad ; \text{ stationary state} \qquad (2.78)$$

then the wave function $\psi(x)$ can be expanded in the $u_{f_n}(x)$ with time-independent coefficients a_{f_n}

$$\psi(x) = \sum_n a_{f_n} u_{f_n}(x) \qquad ; \text{ completeness} \qquad (2.79)$$

It follows as above that in this case

$$1 = \sum_n |a_{f_n}|^2$$
$$\langle F \rangle = \sum_n |a_{f_n}|^2 f_n \qquad ; \text{ stationary state} \qquad (2.80)$$

---

[15] For clarity, we present the following arguments in one dimension.
[16] See Probl. 4.5.

If one is also in an eigenstate of F, then

$$\langle F \rangle = f_n \qquad\qquad ; \text{ in eigenstate} \qquad (2.81)$$

Equations (2.77) and (2.75) suggest that one should interpret the quantity $|a_{f_n}(t)|^2$ as the *probability of measuring the value f_n at the time t if a system is in the state* $\Psi(x,t)$. Based on this argument, we make the following *measurement postulates*:

(1) If one makes a precise measurement of F, then one *must observe one of the eigenvalues f_n*;
(2) If one is in an arbitrary state $\Psi(x,t)$, then $|a_{f_n}(t)|^2$ is the probability that one will observe the value f_n for F at the time t, where[17]

$$a_{f_n}(t) = \int dx\, u_{f_n}^\star(x)\Psi(x,t) \qquad (2.82)$$

As an example, consider the free-particle wave packet of Vol. I

$$\Psi(x,t) = \frac{1}{\sqrt{2\pi}} \int dk\, A(k) e^{i(kx - \omega_k t)} \qquad ; \text{ free particle} \qquad (2.83)$$

The probability density in coordinate space is $|\Psi(x,t)|^2$. The Fourier transform of this relation gives

$$\frac{1}{\sqrt{2\pi}} \int dx\, e^{-ikx}\Psi(x,t) = A(k) e^{-i\omega_k t} \qquad (2.84)$$

For localized wave packets, one can take $u_p(x) = e^{ikx}/\sqrt{2\pi}$ as the eigenstates of momentum. Then, consistent with our interpretation in Vol. I,

$$|A(k)|^2 = \left| \int dx\, u_p^\star(x)\Psi(x,t) \right|^2 \qquad ; \, p = \hbar k \qquad (2.85)$$

is the *probability density in momentum space* (see Probl. 4.8).

2.5.2 *Abstract Form*

One can now proceed to *abstract* these results:

(1) The quantity F is represented with a linear hermitian operator $\hat{F}$ with eigenstates

$$\hat{F}|f_n\rangle = f_n|f_n\rangle \qquad (2.86)$$

---

[17]Alternatively, if one has a large number of identical systems with wave function $\Psi(x,t)$, then the fraction of measurements yielding f_n will be $|a_{f_n}(t)|^2$.

If one makes a precise measurement of F, then one will observe one of the eigenvalues f_n.

(2) An arbitrary state $|\Psi(t)\rangle$ can be expanded in the (assumed) complete set of eigenstates of $\hat{F}$ according to

$$|\Psi(t)\rangle = \sum_n a_{f_n}(t)|f_n\rangle \qquad (2.87)$$

Then the probability that a measurement will yield the value f_n is

$$|a_{f_n}(t)|^2 = |\langle f_n|\Psi(t)\rangle|^2 \qquad (2.88)$$

In particular, $|\langle \xi|\Psi(t)\rangle|^2 = |\Psi(\xi, t)|^2$ is the *probability density that one will observe the value ξ if one makes a measurement of the position x*. This is how we have used the wave function $\Psi(\xi, t)$. Now everything stands on the same footing, and the above contains all our previous assumptions concerning the physical interpretation of the theory.

2.5.3 *Reduction of the Wave Packet*

If a particle moves in a classical orbit, its position can be measured and one finds a value q. If the measurement is repeated a short time Δt later, such that $|\Delta q| \ll |q|$, one must again find the value q. *Measurements must be reproducible.* How does this show up in quantum mechanics?

If one measures the quantity F at the time t and finds a value f_n, then if F is measured again right away, must again find the value f_n. *This is an assumption of the reproducibility of measurements.*

Suppose one is in the state

$$\Psi(x, t) = \sum_n a_{f_n}(t)u_{f_n}(x) \qquad (2.89)$$

If one measures F at the time t_0 and finds a value f_n, then right after this measurement, the wave function must be such as *to again give the value f_n*, and it must be normalized. Thus, with no degeneracy, the effect of this measurement is to *reduce* the wave function to the form[18]

$$\Psi(x, t_0)' = \frac{a_{f_n}(t_0)}{|a_{f_n}(t_0)|}u_{f_n}(x) \qquad (2.90)$$

This result can be abstracted and extended to lead to an additional measurement postulate:

---

[18]We speak here of "pure pass measurements" that do not modify the coefficients $a_{f_n}(t)$.

(3) If, at the time t_0, one observes a value f for the quantity F which lies in the inteval $f' \leq f \leq f''$, then the state vector is reduced to

$$|\Psi(t_0)\rangle' = \frac{\sum'_n a_{f_n}(t_0)|f_n\rangle}{\left(\sum'_n |a_{f_n}(t_0)|^2\right)^{1/2}} \qquad ; \text{ where } f' \leq f_n \leq f'' \qquad (2.91)$$

Here $\sum'_n$ implies $f' \leq f_n \leq f''$.

Although this postulate may at first seem very mysterious, a little reflection will convince the reader that a measurement does indeed provide a great deal of information about a system, in particular, this type of information. We briefly discuss, as an example, the classic Stern-Gerlach experiment.

2.5.4 *Stern-Gerlach Experiment*

The first moral here is that in applying measurement theory, one must always discuss the specific measurement in detail.[19] Consider, for illustration, a spinless, positively-charged particle in a metastable p-state in a neutral atom, where there is no Lorentz force on the atom. There are three possible values of L_z, the angular momentum in the z-direction, $m = 0, \pm 1$. This atom has a magnetic moment, and if placed in a magnetic field which determines the z-direction, and which also *varies* in the z-direction, it will feel a force in the z-direction of

$$F_z = \mu_z \frac{dB_z}{dz} \qquad (2.92)$$

This force acts differently on the different m components, and can be used to separate them. Suppose a beam of these atoms, produced, say, in an oven, is passed through an appropriate inhomogeneous magnet as sketched in Fig. 2.1. We then note the following:

- The beam will subsequently *split into three separate components with* $m = 0, \pm 1$. Each beam can be caused to pass through a separate slit as shown in Fig. 2.1. *This illustrates that one observes the eigenvalues of L_z.*
- Initially, the internal wave function of an atom can be written

$$\psi_{\text{int}}(\mathbf{x}, t) = R_{np}(r) \sum_{m=0,\pm 1} c_m(t) Y_{1m}(\theta, \phi) \qquad (2.93)$$

---

[19]See chapter IV of [Gottfried (1966)] for a thorough discussion of the measurement process.

If the center-of-mass of the atom goes through the top slit (this will happen with probability $|c_{+1}(t_0)|^2$ where t_0 is the time it goes through the magnet), then the *internal* wave function of the atom must be[20]

$$\psi_{\text{int}}(\mathbf{x}, t) = \frac{c_{+1}(t_0)}{|c_{+1}(t_0)|} R_{np}(r) Y_{11}(\theta, \phi) e^{-iE_{np}(t-t_0)/\hbar} \qquad (2.94)$$

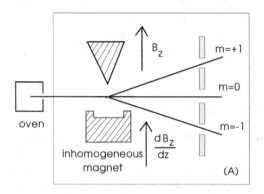

Fig. 2.1   Sketch of the Stern-Gerlach experiment. We will refer to the entire boxed unit as detector (A).

If a second detector identical to (A) in Fig. 2.1 is placed after the top slit, the beam will be observed to pass through and emerge from *its* top slit with unit probability (see Fig. 2.2).   *This illustrates the reproduciblity of the measurement.*

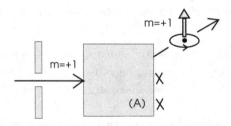

Fig. 2.2   Detector (A) placed after the upper beam with $m = +1$ in Fig. 2.1.

---

[20]Again, we assume a "pure pass measurement" here.

- If one looks for a beam emerging from the middle and bottom slits of the second detector, there will be none. *This illustrates the reduction of the wave packet by the first measurement.*

Whenever you run into apparent paradoxes in discussing the measurement process, you should always return to this simple and fundamental example of the analysis.

2.6 Quantum Mechanics Postulates

Here we *summarize* the quantum mechanics postulates arrived at in the previous discussion. They are formulated in the abstract Hilbert space.

(1) There is a state vector $|\Psi(t)\rangle$ that provides a complete dynamical description of a system;

(2) An observable F is represented by a linear hermitian operator $\hat{F}$;

(3) The operators obey canonical commutation relations, in particular

$$[\hat{p}, \hat{x}] = \frac{\hbar}{i} \tag{2.95}$$

(4) The dynamics is given by the Schrödinger equation

$$i\hbar \frac{\partial}{\partial t} |\Psi(t)\rangle = \hat{H} |\Psi(t)\rangle \tag{2.96}$$

(5) The eigenstates of a linear hermitian operator form a complete set

$$\hat{F} |f_n\rangle = f_n |f_n\rangle \qquad ; \quad \sum_n |f_n\rangle\langle f_n| = \hat{1} \tag{2.97}$$

(6) Measurement postulate:

  (a) A precise measurement of F must yield one of the eigenvalues f_n;

  (b) The probability of observing an eigenvalue f_n at the time t is $|\langle f_n|\Psi(t)\rangle|^2$;

  (c) A measurement $f' \leq f \leq f''$ at time t_0 reduces the state vector to

$$|\Psi(t_0)\rangle' = \frac{\sum_n' a_{f_n}(t_0) |f_n\rangle}{\left(\sum_n' |a_{f_n}(t_0)|^2\right)^{1/2}} \qquad ; \text{ where } f' \leq f_n \leq f'' \tag{2.98}$$

Through his many years in physics, the author has found this to be a complete and essential set of postulates for the implementation of quantum mechanics.

2.7   Many-Particle Hilbert Space

The previous discussion has effectively focused on the quantum mechanics of a single particle. Most of the applications discussed in Vol. I involve many-body systems: atoms, nuclei, hadrons, and quantum fluids. The goal of this section is to extend the previous analysis to the *abstract many-particle Hilbert space*, and to make a connection with quantum field theory as presented in chapter 12 of Vol. I. We start with a summary of the one-dimensional simple harmonic oscillator in abstract Hilbert space.

2.7.1   *Simple Harmonic Oscillator*

The operator analysis of the one-dimensional simple harmonic oscillator is, in fact, carried out in ProbsI. 4.17–4.18.[21] The creation and destruction operators $(a^\dagger, a)$ are first defined as linear combinations of the momentum and coordinate $(\hat{p}, \hat{q})$. The canonical commutation relations for $(\hat{p}, \hat{q})$ imply that[22]

$$[a, a^\dagger] = 1 \tag{2.99}$$

The hermitian number operator is defined as

$$
\begin{aligned}
\hat{N} &\equiv a^\dagger a && \text{; number operator} \\
\hat{H} &= \hbar\omega(\hat{N} + 1/2) && \text{; hamiltonian}
\end{aligned}
\tag{2.100}
$$

The second line expresses the hamiltonian in terms of the number operator. As demonstrated in ProbsI. 4.17–4.18, *it follows entirely from the general properties of the linear hermitian operators involved that the spectrum of the number operator consists of the positive integers and zero*

$$
\begin{aligned}
\hat{N}|n\rangle &= n|n\rangle && \text{; } n = 0, 1, 2, \cdots, \infty \\
\hat{N}|0\rangle &= 0 && \text{; ground state}
\end{aligned}
\tag{2.101}
$$

The last relation defines the ground state. It further follows that

$$
\begin{aligned}
a|n\rangle &= \sqrt{n}\,|n-1\rangle && \text{; destruction operator} \\
a^\dagger|n\rangle &= \sqrt{n+1}\,|n+1\rangle && \text{; creation operator}
\end{aligned}
\tag{2.102}
$$

---

[21]The reader is again strongly urged to work through those problems (see Prob. 2.1).

[22]We suppress the carets on the creation and destruction operators, since it will henceforth be obvious that they act in the abstract occupation-number Hilbert space.

As shown in ProbI. 4.18, the eigenstates $|n\rangle$ can be explicitly constructed as

$$|n\rangle = \frac{1}{\sqrt{n!}}(a^\dagger)^n|0\rangle \tag{2.103}$$

We note that this construction involves a relative phase convention.

The eigenstates of the simple harmonic oscillator in the abstract Hilbert space are both orthonormal and complete

$$\langle n|n'\rangle = \delta_{nn'} \qquad ; \text{ orthonormal}$$

$$\sum_n |n\rangle\langle n| = \hat{1} \qquad ; \text{ complete} \tag{2.104}$$

2.7.2 *Bosons*

With many identical bosons, one introduces a set of creation and destruction operators satisfying

$$[a_k, a_{k'}^\dagger] = \delta_{kk'} \tag{2.105}$$

Here k denotes a complete set of single-particle quantum numbers appropriate to the problem at hand. The basis vectors in the abstract many-particle Hilbert space are then constructed as the direct product of the basis vectors for each of the single-particle states

$$|n_1 n_2 \cdots n_\infty\rangle \equiv |n_1\rangle |n_2\rangle \cdots |n_\infty\rangle \qquad ; \text{ many-body basis states} \tag{2.106}$$

Here the subscripts $\{1, 2, \cdots, \infty\}$ simply represent an ordering of all possible values of k. The effects of the creation and destruction operators for any given mode now follow from the above discussion of the simple harmonic oscillator, and as the operators for the different modes commute, it does not matter where one sits relative to the others.

Quantum fields are then operators in this abstract many-particle Hilbert space. We give three examples from Vol. I:

(1) The normal modes for the transverse oscillations of a continuous string of length L with periodic boundary conditions are given by[23]

$$\phi_k(x) = \frac{1}{\sqrt{L}}e^{ikx} \qquad ; k = \frac{2\pi m}{L} \qquad ; m = 0, \pm 1, \pm 2, \cdots \tag{2.107}$$

---

[23] We again suppress the subscript m on $k_m = 2\pi m/L$.

The string energy, which plays the role of free-field hamiltonian, is then found in terms of the quantum field of the string $\hat{q}(x,t)$, obtained from its classical transverse displacement, and the corresponding quantum momentum density $\hat{\pi}(x,t) = \sigma \partial \hat{q}(x,t)/\partial t$, obtained from its classical transverse motion

$$\hat{q}(x,t) = \sum_k \left(\frac{\hbar}{2\omega_k \sigma L}\right)^{1/2} \left[a_k e^{i(kx-\omega_k t)} + a_k^\dagger e^{-i(kx-\omega_k t)}\right] \quad ; \quad \omega_k = |k|c$$

$$\hat{\pi}(x,t) = \frac{1}{i}\sum_k \left(\frac{\hbar \omega_k \sigma}{2L}\right)^{1/2} \left[a_k e^{i(kx-\omega_k t)} - a_k^\dagger e^{-i(kx-\omega_k t)}\right] \tag{2.108}$$

Here σ is the mass density, and c is the sound velocity. These operators, which here carry the free-field time dependence, satisfy the canonical equal-time commutation relations

$$[\hat{q}(x,t), \hat{\pi}(x',t')]_{t=t'} = i\hbar\delta(x-x') \tag{2.109}$$

The free-field hamiltonian is[24]

$$\hat{H} = \frac{\sigma}{2}\int_0^L dx \left\{\left[\frac{\partial \hat{q}(x,t)}{\partial t}\right]^2 + c^2 \left[\frac{\partial \hat{q}(x,t)}{\partial x}\right]^2\right\} \tag{2.110}$$

Substitution of the expressions in Eqs. (2.108) gives

$$\hat{H} = \sum_k \hbar\omega_k(a_k^\dagger a_k + a_k a_k^\dagger) = \sum_k \hbar\omega_k(\hat{N}_k + 1/2) \tag{2.111}$$

This represents an infinite collection of uncoupled simple harmonic oscillators, as discussed above.

The energy eigenvalues for the whole system are given by[25]

$$\hat{H}|n_1 n_2 \cdots n_\infty\rangle = E_{n_1 n_2 \cdots n_\infty}|n_1 n_2 \cdots n_\infty\rangle$$

$$E_{n_1 n_2 \cdots n_\infty} = \sum_k \hbar\omega_k(n_k + 1/2) \tag{2.112}$$

The quantity n_k is the number of quanta in the kth mode, and in analogy to the quantization of light, we refer to these quanta of the sound waves in a string as *phonons*.

---

[24]Note $c^2 = \tau/\sigma$ where τ is the tension.

[25]Since the subscripts $\{1, 2, \cdots, \infty\}$ on $(n_1, n_2, \cdots, n_\infty)$ simply label the ordered members of the set $k = (0, \pm 2\pi/L, \pm 4\pi/L, \cdots)$, the second of Eqs. (2.112) can equally well be written as $E_{n_1 n_2 \cdots n_\infty} = \sum_{i=1}^\infty \hbar\omega_i(n_i + 1/2)$.

Various interaction terms (non-linearity in the string, a spring attached to the string, *etc.*) can now be written in terms of the fields. Since these interactions do not conserve the number of phonons, they will connect one state to any other in the many-particle Hilbert space.

(2) The quantization of the electromagnetic field in Vol. I follows in an analogous fashion.

(3) The non-relativistic many-body hamiltonian for a collection of identical, massive, spin-zero bosons, each with kinetic energy $T = \mathbf{p}^2/2m = -\hbar^2 \mathbf{\nabla}^2/2m$, and interacting through an instantaneous two-body potential of the form $V(\mathbf{x}, \mathbf{y})$, can be written as

$$\hat{H} = \int d^3x\, \hat{\psi}^\dagger(\mathbf{x}) T \hat{\psi}(\mathbf{x}) + \frac{1}{2} \int d^3x \int d^3y\, \hat{\psi}^\dagger(\mathbf{x}) \hat{\psi}^\dagger(\mathbf{y}) V(\mathbf{x}, \mathbf{y}) \hat{\psi}(\mathbf{y}) \hat{\psi}(\mathbf{x})$$

$$(2.113)$$

Here the quantum field is defined by

$$\hat{\psi}(\mathbf{x}) \equiv \sum_k a_k\, \phi_k(\mathbf{x}) \qquad (2.114)$$

where the $\phi_k(\mathbf{x})$ form a complete set of solutions to a one-body Schrödinger equation appropriate, as a starting basis, for the problem at hand. The fields satisfy the canonical commutation relation

$$[\hat{\psi}(\mathbf{x}), \hat{\psi}^\dagger(\mathbf{x}')] = \delta^{(3)}(\mathbf{x} - \mathbf{x}') \qquad (2.115)$$

The time evolution of the many-particle system is now governed by the many-body Schrödinger equation.[26] Here, as in our original formulation of quantum mechanics, the operators in this Schrödinger picture are taken to be time-independent, and all the time dependence derives from the Schrödinger equation. When the number of bosons is a constant of the motion, as in liquid ^{4}He, then this hamiltonian never takes one out of the subspace with given N (See Prob. 2.4).

2.7.3 *Fermions*

In the case of fermions, in order to satisfy the Pauli exclusion principle, one quantizes with *anticommutation* relations instead of commutation relations.

---

[26]This is called "second quantization", since what were previously single-particle wave functions now become field operators in the abstract many-particle Hilbert space. The formulation of the many-body problem in second quantization is carried out in detail in chapter 1 of [Fetter and Walecka (2003a)].

For a single mode, one then has

$$\{a, a^\dagger\} \equiv aa^\dagger + a^\dagger a = 1$$
$$\{a, a\} = \{a^\dagger, a^\dagger\} = 0 \qquad (2.116)$$

The number operator is again defined as

$$\hat{N} \equiv a^\dagger a \qquad ; \text{ number operator} \qquad (2.117)$$

It follows that this number operator has eigenvalues 0 and 1 (see Vol. I)

$$\hat{N}|n\rangle = n|n\rangle \qquad ; n = 0, 1 \qquad (2.118)$$

Furthermore (see Probl 12.8)

$$a|1\rangle = |0\rangle \qquad ; a|0\rangle = 0$$
$$a^\dagger|0\rangle = |1\rangle \qquad ; a^\dagger|1\rangle = 0 \qquad (2.119)$$

The basis states in the abstract Hilbert space are again formed from the direct product of the single-particle states as in Eq. (2.106); however, since the operators for the different single-particle modes now *anticommute*, one has to keep careful track of the ordering of various terms.

As an example, the non-relativistic many-body hamiltonian for a collection of identical spin-1/2 fermions, each with kinetic energy $T = \mathbf{p}^2/2m = -\hbar^2 \boldsymbol{\nabla}^2/2m$, and interacting through an instantaneous two-body spin-independent potential of the form $V(\mathbf{x}, \mathbf{y})$, can again be written as

$$\hat{H} = \int d^3x \, \underline{\hat{\psi}}^\dagger(\mathbf{x}) T \, \underline{\hat{\psi}}(\mathbf{x}) + \frac{1}{2} \int d^3x \int d^3y \, \underline{\hat{\psi}}^\dagger(\mathbf{x}) \underline{\hat{\psi}}^\dagger(\mathbf{y}) V(\mathbf{x}, \mathbf{y}) \, \underline{\hat{\psi}}(\mathbf{y}) \underline{\hat{\psi}}(\mathbf{x})$$

$$(2.120)$$

Here the quantum field is defined by

$$\underline{\hat{\psi}}(\mathbf{x}) \equiv \sum_{k\lambda} a_{k\lambda} \, \underline{\phi}_{k\lambda}(\mathbf{x}) \qquad (2.121)$$

where the two-component spinors $\underline{\phi}_{k\lambda}(\mathbf{x})$ form a complete set of solutions to a one-body Schrödinger equation again appropriate, as a starting basis, for the problem at hand. The index $\lambda = (\uparrow, \downarrow)$ denotes the two spin projections.[27] The components of the field, in this case, now satisfy the canonical *anticommutation* relation

$$\{\hat{\psi}_\alpha(\mathbf{x}), \hat{\psi}_\beta^\dagger(\mathbf{x}')\} = \delta_{\alpha\beta} \, \delta^{(3)}(\mathbf{x} - \mathbf{x}') \qquad (2.122)$$

---

[27]The spinors with the same coordinate label are to be paired in Eq. (2.120).

Chapter 3

Angular Momentum

The theory of angular momentum in quantum mechanics is of great utility, as it provides the basis for the calculation of matrix elements in any finite system.[1] Furthermore, it provides an example of the role of a continuous symmetry in quantum mechanics, in this case the very deep symmetry of the isotropy of space and invariance under rotations.

3.1 Translations

The translation operator in quantum mechanics is analyzed in Probs. 2.6-2.7. Let us arrive at this operator in a more systematic fashion. Consider a classical coordinate translation to a new primed coordinate system translated by a distance a along the x-axis from the unprimed one, so that

$$x' = x - a \qquad ; \text{ translation}$$
$$p' = p \qquad\qquad\qquad\qquad (3.1)$$

The second relation follows since $dx'/dt = dx/dt$. In quantum mechanics $(\hat{p}, \hat{x})$ are *operators*. What do we mean by a translation in quantum mechanics? Let us ask the following question:

> *Is there an operator which induces a transformation on our operators and state vectors such that the new quantities can be put into a one-to-one correspondence with the translated coordinate system?*

We start with an infinitesimal transformation where $a \equiv \varepsilon \to 0$; the finite transformation will then be built up by repeated application of in-

---

[1] See [Edmonds (1974)]; see also appendix B in [Fetter and Walecka (2003a)].

finitesimals. We look for an operator of the form

$$\hat{U} = 1 - \frac{i}{\hbar}\varepsilon\hat{K} \qquad \text{; infinitesimal } a \equiv \varepsilon \to 0$$

$$\hat{U}^{-1} = 1 + \frac{i}{\hbar}\varepsilon\hat{K}$$

$$\hat{U}\hat{U}^{-1} = 1 + O(\varepsilon^2) \tag{3.2}$$

$\hat{K}$ is referred to as the *generator* of the transformation. We require that, to $O(\varepsilon)$, the transformation should satisfy

$$\hat{x}' = \hat{U}\hat{x}\hat{U}^{-1} = \left(1 - \frac{i}{\hbar}\varepsilon\hat{K}\right)\hat{x}\left(1 + \frac{i}{\hbar}\varepsilon\hat{K}\right) = \hat{x} - \frac{i}{\hbar}\varepsilon[\hat{K},\hat{x}] = \hat{x} - \varepsilon$$

$$\hat{p}' = \hat{U}\hat{p}\hat{U}^{-1} = \left(1 - \frac{i}{\hbar}\varepsilon\hat{K}\right)\hat{p}\left(1 + \frac{i}{\hbar}\varepsilon\hat{K}\right) = \hat{p} - \frac{i}{\hbar}\varepsilon[\hat{K},\hat{p}] = \hat{p} \tag{3.3}$$

Here the last equalities in each line are the infinitesimal form of the required result in Eq. (3.1). These equations imply that the generator should satisfy

$$\frac{i}{\hbar}[\hat{K},\hat{x}] = 1 \qquad\qquad ; \frac{i}{\hbar}[\hat{K},\hat{p}] = 0 \tag{3.4}$$

Can one find a solution to these equations? The answer is *yes*; just take

$$\hat{K} = \hat{p} \qquad \text{; generator of translations} \tag{3.5}$$

The momentum is thus the generator of translations.

The transformation through a finite distance a is obtained by repeating the infinitesimal transformation N times, so that $a = N\varepsilon$, and then using the binomial theorem to write

$$\left(1 - \frac{i}{\hbar}\varepsilon\hat{p}\right)^N = 1 + N\left(-\frac{i}{\hbar}\varepsilon\hat{p}\right) + \frac{N(N-1)}{2!}\left(-\frac{i}{\hbar}\varepsilon\hat{p}\right)^2 + \cdots$$

$$= 1 + N\left(-\frac{i}{\hbar}\varepsilon\hat{p}\right) + \frac{N^2}{2!}\left(-\frac{i}{\hbar}\varepsilon\hat{p}\right)^2 + \cdots + O(\varepsilon a) \tag{3.6}$$

Up to $O(\varepsilon a)$ this is identical with the exponential series, and hence in the limit as $N \to \infty$ at fixed $a = N\varepsilon$ (which implies $\varepsilon a \to 0$), we have the important relation

$$\text{Lim}_{N\to\infty}\left(1 - \frac{i}{\hbar}\varepsilon\hat{p}\right)^N = \exp\left\{-\frac{i}{\hbar}\hat{p}a\right\} \qquad ; a = N\varepsilon \text{ fixed} \tag{3.7}$$

Thus

$$\hat{U}(a) = \exp\left\{-\frac{i}{\hbar}\hat{p}a\right\} \qquad ; \text{ finite translation} \qquad (3.8)$$

We comment on this result:

- One can obtain the effects of the finite transformation through the use of the crucial operator relation developed in Prob. 2.5

$$e^{i\hat{A}}\hat{B}e^{-i\hat{A}} = \hat{B} + i[\hat{A}, \hat{B}] + \frac{i^2}{2!}[\hat{A}, [\hat{A}, \hat{B}]] + \frac{i^3}{3!}[\hat{A}, [\hat{A}, [\hat{A}, \hat{B}]]] + \cdots \quad (3.9)$$

  Indeed, Eqs. (3.1) are now reproduced (Prob. 2.6).
- The operator $\hat{U}(a)$ is unitary since $\hat{p}$ is hermitian;
- All that has been used in this derivation is the set of *commutation relations* of the operators in the abstract Hilbert space; one never needs an explicit representation of these operators;
- In three dimensions the translation operator is evidently[2]

$$\hat{U}(\mathbf{a}) = \exp\left\{-\frac{i}{\hbar}\hat{\mathbf{p}}\cdot\mathbf{a}\right\} \qquad ; \text{ three dimensions} \qquad (3.10)$$

3.2 Rotations

We start with a rotation about the 3-axis, and use the same arguments as in the previous section.

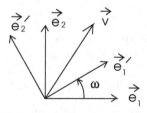

Fig. 3.1 Rotation of the coordinate system by an angle ω about the 3-axis, which comes out of the page.

The classical situation is shown in Fig. 3.1. The basis vectors are evi-

---

[2]See Prob. 3.2.

dently related to each other by

$$\mathbf{e}'_1 = \mathbf{e}_1 \cos\omega + \mathbf{e}_2 \sin\omega$$
$$\mathbf{e}'_2 = -\mathbf{e}_1 \sin\omega + \mathbf{e}_2 \cos\omega \tag{3.11}$$

The components of an arbitrary vector $\mathbf{v}$ in the new primed system are related to those in the original unprimed system by

$$\mathbf{v} \cdot \mathbf{e}'_1 = v'_1 = v_1 \cos\omega + v_2 \sin\omega$$
$$\mathbf{v} \cdot \mathbf{e}'_2 = v'_2 = -v_1 \sin\omega + v_2 \cos\omega \tag{3.12}$$

In the quantum mechanics of a single non-relativistic spinless particle, the *vectors* are $\hat{\mathbf{v}} = (\hat{\mathbf{p}}, \hat{\mathbf{x}})$. Both are operators; both change under rotations. Can one find an operator $\hat{\mathcal{R}}$ which induces the above transformation? We again start with infinitesimals where $\omega \equiv \varepsilon \to 0$, and look for

$$\hat{\mathcal{R}} = 1 - \frac{i}{\hbar}\varepsilon\hat{K} \qquad ; \ \hat{\mathcal{R}}^{-1} = 1 + \frac{i}{\hbar}\varepsilon\hat{K} \tag{3.13}$$

A repetition of the calculation in Eqs. (3.3) gives, to $O(\varepsilon)$,

$$v'_1 = \hat{\mathcal{R}}\hat{v}_1\hat{\mathcal{R}}^{-1} = \hat{v}_1 - \frac{i}{\hbar}\varepsilon[\hat{K}, \hat{v}_1] = \hat{v}_1 + \varepsilon\hat{v}_2$$
$$v'_2 = \hat{\mathcal{R}}\hat{v}_2\hat{\mathcal{R}}^{-1} = \hat{v}_2 - \frac{i}{\hbar}\varepsilon[\hat{K}, \hat{v}_2] = \hat{v}_2 - \varepsilon\hat{v}_1 \tag{3.14}$$

where the final equalities in each line are the infinitesimal form of the required transformation in Eqs. (3.12). The generator of rotations about the 3-axis must therefore satisfy

$$\frac{i}{\hbar}[\hat{K}, \hat{v}_1] = -\hat{v}_2 \qquad ; \ \frac{i}{\hbar}[\hat{K}, \hat{v}_2] = \hat{v}_1 \tag{3.15}$$

Can one find a $\hat{K}$ with these commutation rules with $\hat{\mathbf{v}} = (\hat{\mathbf{p}}, \hat{\mathbf{x}})$? The answer is again *yes*. Take $\hat{K} = \hat{J}_3$ where

$$\hat{J}_3 = \hat{x}_1\hat{p}_2 - \hat{x}_2\hat{p}_1 = (\hat{\mathbf{x}} \times \hat{\mathbf{p}})_3 \tag{3.16}$$

This is just the third component of the angular momentum! It is readily verified with the aid of the canonical commutation relations

$$[\hat{p}_i, \hat{x}_j] = \frac{\hbar}{i}\delta_{ij} \qquad ; \ \text{C.C.R.} \tag{3.17}$$

that Eqs. (3.15) are indeed satisfied (Prob. 3.1). We note that the angular momentum defined here is an *hermitian operator*.

It is convenient at this point to again measure angular momenta in units of $\hbar$ and define[3]

$$\hat{J}_3 \equiv \hbar \hat{J}_3 \qquad ; \text{ units of } \hbar \qquad (3.18)$$

Equations (3.15) then read

$$i[\hat{J}_3, \hat{v}_1] = -\hat{v}_2 \qquad ; i[\hat{J}_3, \hat{v}_2] = \hat{v}_1 \qquad (3.19)$$

A repetition of the above arguments leads to a finite transformation of the form (see Prob. 3.2)

$$\hat{\mathcal{R}}(\omega) = \exp\left\{-i\omega \hat{J}_3\right\} \qquad ; \text{ finite rotation} \qquad (3.20)$$

Define the rotation matrix for a rotation about the 3-axis by

$$\underline{a}(\omega) = \begin{pmatrix} \cos\omega & \sin\omega & 0 \\ -\sin\omega & \cos\omega & 0 \\ 0 & 0 & 1 \end{pmatrix} \qquad ; \text{ rotation matrix} \qquad (3.21)$$

Then, with the aid of Eq. (3.9), one readily establishes that Eq. (3.20) produces the correct finite transformation (Prob. 3.2)

$$\hat{\mathcal{R}}(\omega)\hat{v}_i\hat{\mathcal{R}}(\omega)^{-1} = \sum_{j=1}^{3} a_{ij}(\omega)\hat{v}_j \qquad ; i = 1, 2, 3 \qquad (3.22)$$

This result again follows *entirely from the commutation relations*.

3.3 Angular Momentum Operator

Given the definition of the angular momentum operator $\hbar \hat{\mathbf{J}} \equiv \hat{\mathbf{x}} \times \hat{\mathbf{p}}$ in Eq. (3.16), and the canonical commutation relations in Eq. (3.17), one can proceed to compute the commutation relations for the components of the angular momentum itself, and for its total square

$$\hat{\mathbf{J}}^2 = \hat{J}_1^2 + \hat{J}_2^2 + \hat{J}_3^2 \qquad ; \text{ square of total} \qquad (3.23)$$

Exactly as in Vol. I, one finds[4]

---

[3] Angular momenta measured in units of $\hbar$ will again be denoted with Latin letters.
[4] See Probl. 4.25.

$$[\hat{J}_1, \hat{J}_2] = i\hat{J}_3 \qquad ;\text{ and cylic permutations of } (1,2,3)$$
$$[\hat{\mathbf{J}}^2, \hat{J}_i] = 0 \qquad ; i = (1,2,3) \qquad (3.24)$$

The commutator of two components of the angular momentum is again an angular momentum, and the square of the total angular momentum commutes with all the components.

If ϵ_{ijk} denotes the completely antisymmetric tensor in three-dimensions, and if we introduce the convention that repeated Latin indices are summed from 1 to 3, then Eqs. (3.24) can be re-written as (see Prob. 3.1)

$$[\hat{J}_i, \hat{J}_j] = i\epsilon_{ijk}\hat{J}_k \qquad ; (i,j,k) = (1,2,3)$$
$$[\hat{\mathbf{J}}^2, \hat{J}_i] = 0 \qquad (3.25)$$

The generalization of Eqs. (3.19) for an arbitrary rotation is evidently

$$[\hat{J}_i, \hat{v}_j] = i\epsilon_{ijk}\hat{v}_k \qquad ; \text{ vector operator} \qquad (3.26)$$

Given the angular momentum $\hat{\mathbf{J}}$, this relation *defines a vector operator* in quantum mechanics.[5]

3.4   Eigenvalue Spectrum

Let us see how far we can get assuming only the above commutation relations for the hermitian angular momentum operator acting in the abstract Hilbert space.

As with the central force problem in Vol. I, one can label the eigenstates of angular momentum with the eigenvalues of the square of the total angular momentum and one, and only one, of the generators, which fail to commute among themselves. We choose $\hat{J}_3$, and write these eigenstates as[6]

$$\hat{\mathbf{J}}^2|\lambda m\rangle = \lambda|\lambda m\rangle \qquad ; \text{ eigenstates}$$
$$\hat{J}_3|\lambda m\rangle = m|\lambda m\rangle \qquad (3.27)$$

---

[5] An algebra of generators closed under commutation is said to form a *Lie algebra*, and the corresponding finite transformations form a *Lie group*. The ϵ_{ijk} are the *structure constants* of that Lie group, here $SU(2)$. A quantity such as $\hat{\mathbf{J}}^2$, which commutes with all the generators, is known as a *Casimir operator*.

[6] There is thus an uncertainty relation between the various components of the angular momentum, which fail to commute (Probl. 4.25). Here $(\hat{\mathbf{J}}^2, \hat{J}_3)$ form a complete set of *mutually commuting* hermitian operators for this problem.

By our basic assumptions, the eigenvalues (λ, m) are real numbers, and these states are both orthonormal and complete

$$\langle \lambda m | \lambda' m' \rangle = \delta_{\lambda \lambda'} \delta_{mm'} \quad ; \text{ orthonormal}$$

$$\sum_{\lambda m} | \lambda m \rangle \langle \lambda m | = \hat{1} \quad\quad ; \text{ complete} \quad\quad (3.28)$$

Now define *raising and lowering* operators by

$$\hat{J}_{\pm} \equiv \hat{J}_1 \pm i\hat{J}_2 \quad\quad ; \text{ raising and lowering operators} \quad (3.29)$$

The following commutation relations follow immediately from Eq. (3.24)[7]

$$[\hat{J}_{\pm}, \hat{J}_3] = \mp \hat{J}_{\pm}$$

$$[\hat{J}_{+}, \hat{J}_{-}] = 2\hat{J}_3 \quad\quad (3.30)$$

Various results can now be established from these relations.

1) *The operators $\hat{J}_i$ only mix different m components.*

This result follows by taking matrix elements of the second of Eqs. (3.24)

$$\langle \lambda m | (\hat{\mathbf{J}}^2 \hat{J}_i - \hat{J}_i \hat{\mathbf{J}}^2) | \lambda' m' \rangle = 0 \quad\quad (3.31)$$

Let $\hat{\mathbf{J}}^2$ act on the neighboring eigenstates. This gives

$$(\lambda - \lambda') \langle \lambda m | \hat{J}_i | \lambda' m' \rangle = 0 \quad\quad (3.32)$$

The matrix element therefore vanishes unless $\lambda = \lambda'$. Thus for the action of one of the components of the angular momentum on the eigenstates in Eqs. (3.27) one can write

$$\hat{J}_i | \lambda m \rangle = \sum_{\lambda', m'} | \lambda' m' \rangle \langle \lambda' m' | \hat{J}_i | \lambda m \rangle = \sum_{m'} | \lambda m' \rangle \langle \lambda m' | \hat{J}_i | \lambda m \rangle \quad (3.33)$$

The first equality follows from the completeness relation in Eqs. (3.28) and the second from Eq. (3.32). This is an important relation because it tells us that no matter how many times the generators act on these eigenstates, one *remains in the subspace of given λ.*

2) *The operators $\hat{J}_{\pm}$ raise and lower m by 1.*

The proof of this statement follows by taking matrix elements of the first of Eqs. (3.30), for example

$$\langle \lambda m | (\hat{J}_{+} \hat{J}_3 - \hat{J}_3 \hat{J}_{+}) | \lambda m' \rangle = -\langle \lambda m | \hat{J}_{+} | \lambda m' \rangle \quad\quad (3.34)$$

---

[7]See Prob. 3.1.

Now use completeness again on the l.h.s., replace $\hat{J}_3$ by its eigenvalue where appropriate, and use the orthonormality of the eigenstates

$$
\begin{aligned}
\text{l.h.s.} &= \sum_{m''}\left[\langle\lambda m|\hat{J}_+|\lambda m''\rangle\langle\lambda m''|\hat{J}_3|\lambda m'\rangle - \langle\lambda m|\hat{J}_3|\lambda m''\rangle\langle\lambda m''|\hat{J}_+|\lambda m'\rangle\right] \\
&= \sum_{m''}\left[\langle\lambda m|\hat{J}_+|\lambda m''\rangle m'\delta_{m''m'} - m''\delta_{mm''}\langle\lambda m''|\hat{J}_+|\lambda m'\rangle\right] \\
&= (m' - m)\langle\lambda m|\hat{J}_+|\lambda m'\rangle
\end{aligned}
\tag{3.35}
$$

Thus Eq. (3.34) becomes

$$
(m' - m + 1)\langle\lambda m|\hat{J}_+|\lambda m'\rangle = 0
\tag{3.36}
$$

Hence the matrix element of $\hat{J}_+$ vanishes unless $m = m' + 1$, which is the stated result. An exactly analogous proof states that the matrix element of $\hat{J}_-$ vanishes unless $m = m' - 1$. These results can be summarized as[8]

$$
\begin{aligned}
\hat{J}_+|\lambda m\rangle &= \mathcal{A}(\lambda, m)|\lambda, m + 1\rangle \\
\hat{J}_-|\lambda m\rangle &= \mathcal{B}(\lambda, m)|\lambda, m - 1\rangle
\end{aligned}
\tag{3.37}
$$

We speak of $\hat{J}_\pm$ as the *raising and lowering operators*.

Since $\hat{J}_+^\dagger = \hat{J}_-$, the coefficients $\mathcal{A}$ and $\mathcal{B}$ in Eqs. (3.37) are related by

$$
\begin{aligned}
\mathcal{A}(\lambda, m)^\star &= \langle\lambda, m + 1|\hat{J}_+|\lambda m\rangle^\star = \langle\lambda m|\hat{J}_+^\dagger|\lambda, m + 1\rangle \\
&= \langle\lambda m|\hat{J}_-|\lambda, m + 1\rangle = \mathcal{B}(\lambda, m + 1)
\end{aligned}
\tag{3.38}
$$

Hence Eqs. (3.37) can be re-written as

$$
\begin{aligned}
\hat{J}_+|\lambda m\rangle &= \mathcal{A}(\lambda, m)|\lambda, m + 1\rangle \\
\hat{J}_-|\lambda m\rangle &= \mathcal{A}(\lambda, m - 1)^\star|\lambda, m - 1\rangle
\end{aligned}
\tag{3.39}
$$

3) *The coefficients $\mathcal{A}(\lambda, m)$ must satisfy the relation*

$$
|\mathcal{A}(\lambda, m)|^2 = \mathcal{C}_\lambda - m(m + 1)
\tag{3.40}
$$

where $\mathcal{C}_\lambda$ is a constant independent of m.

This is shown by taking the diagonal matrix element of the second of Eqs. (3.30)

$$
\langle\lambda m|(\hat{J}_+\hat{J}_- - \hat{J}_-\hat{J}_+)|\lambda m\rangle = \langle\lambda m|2\hat{J}_3|\lambda m\rangle = 2m
\tag{3.41}
$$

---

[8]We use commas to separate quantum numbers, indices, and arguments only when necessary to avoid ambiguity.

The use of completeness then gives for the l.h.s.[9]

$$\text{l.h.s.} = \sum_{m'} \left[\langle \lambda m | \hat{J}_+ | \lambda m' \rangle \langle \lambda m' | \hat{J}_- | \lambda m \rangle - \langle \lambda m | \hat{J}_- | \lambda m' \rangle \langle \lambda m' | \hat{J}_+ | \lambda m \rangle \right] \tag{3.42}$$

Only $m' = m - 1$ contributes in the first term, and $m' = m + 1$ in the second. The use of Eqs. (3.39) then gives

$$\text{l.h.s.} = |\mathcal{A}(\lambda, m - 1)|^2 - |\mathcal{A}(\lambda, m)|^2 \tag{3.43}$$

Hence Eq. (3.41) becomes

$$|\mathcal{A}(\lambda, m - 1)|^2 - |\mathcal{A}(\lambda, m)|^2 = 2m \tag{3.44}$$

This is a one-term recursion relation for the quantity $|\mathcal{A}(\lambda, m)|^2$. A simple substitution shows that Eq. (3.40), with the single constant $\mathcal{C}_\lambda$, provides the general solution to this recursion relation

$$[\mathcal{C}_\lambda - (m - 1)(m - 1 + 1)] - [\mathcal{C}_\lambda - m(m + 1)] = 2m \tag{3.45}$$

4) *The eigenvalue λ is non-negative.*

Use the following identity to rewrite the square of the total angular momentum $\hat{\mathbf{J}}^2$

$$\begin{aligned}
\hat{\mathbf{J}}^2 &= \hat{J}_1^2 + \hat{J}_2^2 + \hat{J}_3^2 \\
&\equiv \frac{1}{2} \left[(\hat{J}_1 + i\hat{J}_2)(\hat{J}_1 - i\hat{J}_2) + (\hat{J}_1 - i\hat{J}_2)(\hat{J}_1 + i\hat{J}_2) \right] + \hat{J}_3^2 \\
&= \frac{1}{2}(\hat{J}_+ \hat{J}_- + \hat{J}_- \hat{J}_+) + \hat{J}_3^2
\end{aligned} \tag{3.46}$$

The diagonal matrix element of this relation, and a repetition of the arguments in Eqs. (3.42)–(3.44), give

$$\begin{aligned}
\langle \lambda m | \hat{\mathbf{J}}^2 | \lambda m \rangle &= \lambda \\
&= \frac{1}{2} \left[|\mathcal{A}(\lambda, m - 1)|^2 + |\mathcal{A}(\lambda, m)|^2 \right] + m^2 \geq 0
\end{aligned} \tag{3.47}$$

This proves the assertion that λ is non-negative.

5) *The constant $\mathcal{C}_\lambda = \lambda$.*

The substitution of Eq. (3.40) into Eq. (3.47) allows us to identify

$$\lambda = \frac{1}{2} \left\{ [\mathcal{C}_\lambda - (m - 1)m] + [\mathcal{C}_\lambda - m(m + 1)] \right\} + m^2 = \mathcal{C}_\lambda \tag{3.48}$$

---

[9] As before, only intermediate states with $\lambda' = \lambda$ contribute here.

Hence, in summary,

$$\lambda \geq 0$$
$$|\mathcal{A}(\lambda, m)|^2 = \lambda - m(m+1) \geq 0 \qquad (3.49)$$

We now make the following observations:

(1) The operators $\hat{J}_\pm$ can raise and lower m by one unit [Eqs. (3.39)];
(2) The positivity condition in the second of Eqs. (3.49) is violated as $m \to \pm\infty$; hence the process in Eqs. (3.39) must *terminate*, or one will produce states that violate this positivity condition;
(3) We choose to parameterize the positive quantity λ in Eqs. (3.49) as

$$\lambda \equiv j(j+1) \qquad ; \text{parameterize} \qquad (3.50)$$

Here j is non-negative.

The condition that the process in Eqs. (3.39) should terminate is then that

$$|\mathcal{A}(\lambda, m)|^2 = j(j+1) - m(m+1) = 0 \qquad (3.51)$$

This equation has two solutions

$$m_{\max} = j$$
$$m_{\min} = -j - 1 \qquad (3.52)$$

It follows that

$$m_{\max} - m_{\min} = 2j + 1 \qquad (3.53)$$

This is just the number of intervals of integer spacing between $m_{\max}$ and $m_{\min}$ (see Fig. 3.2), and hence this quantity is again an integer[10]

$$2j + 1 = \text{number of intervals} = \text{integer} \qquad (3.54)$$

It follows that the allowed values of the non-negative j are

$$j = 0, 1, 2, 3, \cdots$$
$$1/2, 3/2, 5/2, \cdots \qquad ; \text{allowed values} \qquad (3.55)$$

This is a *remarkable result* that follows entirely from our general considerations.

---

[10]Note that the condition that the lowering process in Eqs. (3.39) terminate is that $|\mathcal{A}(\lambda, m-1)|^2 = 0$, so in this case $m - 1 = m_{\min} = -j - 1$, or $m = -j$.

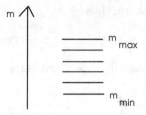

Fig. 3.2   Set of values of m, with integer spacing, for which $|\mathcal{A}(\lambda, m)|^2$ vanishes at $m_{\max}$ and $m_{\min}$ and is non-zero in between.

It is convenient at this point to re-label the eigenstates by $|jm\rangle$, in which case

$$\hat{\mathbf{J}}^2|jm\rangle = j(j+1)|jm\rangle \qquad ; \text{ eigenstates}$$
$$\hat{J}_3|jm\rangle = m|jm\rangle \qquad\qquad (3.56)$$

One also defines [Edmonds (1974)]

$$A(j,m) \equiv \sqrt{(j+m)(j-m+1)}$$
$$A(j,m+1) = \sqrt{(j+m+1)(j-m)} = \sqrt{j(j+1) - m(m+1)}$$
$$= A(j,-m) \qquad\qquad (3.57)$$

At this point we introduce a *phase convention* by taking the real, positive square root in the second of Eqs. (3.49), which gives[11]

$$\mathcal{A}(\lambda, m) = \sqrt{j(j+1) - m(m+1)} = A(j,-m) \qquad (3.58)$$

Equations (3.39) then take the nice symmetric form

$$\hat{J}_+|jm\rangle = A(j,-m)|j,m+1\rangle$$
$$\hat{J}_-|jm\rangle = A(j,m)|j,m-1\rangle \qquad\qquad (3.59)$$

The condition that this raising and lowering process terminates is then

$$\hat{J}_+|j,j\rangle = A(j,-j)|j,j+1\rangle = 0$$
$$\hat{J}_-|j,-j\rangle = A(j,-j)|j,-j-1\rangle = 0 \qquad (3.60)$$

---

[11]It is important to realize that, given $\hat{\mathbf{J}}$, this convention fixes the relative phases of the states. In relating to the work of others, it is always *essential* to first verify that the same phase conventions are employed. Here we follow the phase conventions of [Edmonds (1974)].

The allowed values of m are thus

$$-j \leq m \leq j \qquad ; \text{ integer steps} \qquad (3.61)$$

These results for the eigenvalue spectrum of the angular momentum played a central role in Vol. I.

3.5   Coupling of Angular Momenta

Consider a problem with *two* angular momenta $\hat{\mathbf{L}}$ and $\hat{\mathbf{S}}$, for example the orbital angular momentum and spin, which entered into most of the applications in Vol. I. These operators obey the commutation relations

$$[\hat{L}_i, \hat{L}_j] = i\epsilon_{ijk}\hat{L}_k \qquad ; \; [\hat{S}_i, \hat{S}_j] = i\epsilon_{ijk}\hat{S}_k \qquad (3.62)$$

They each have a set of eigenstates

$$\hat{\mathbf{L}}^2|lm_l\rangle = l(l+1)|lm_l\rangle \qquad ; \; \hat{\mathbf{S}}^2|sm_s\rangle = s(s+1)|sm_s\rangle$$
$$\hat{L}_3|lm_l\rangle = m_l|lm_l\rangle \qquad ; \; \hat{S}_3|sm_s\rangle = m_s|sm_s\rangle \qquad (3.63)$$

The two angular momenta are then described in the *direct-product space*

$$|lm_l sm_s\rangle \equiv |lm_l\rangle|sm_s\rangle \qquad ; \text{ direct-product space} \qquad (3.64)$$

Since they act in different spaces, the operators $\hat{\mathbf{L}}$ and $\hat{\mathbf{S}}$ commute with each other

$$[\hat{L}_i, \hat{S}_j] = 0 \qquad ; \text{ different spaces} \qquad (3.65)$$

Thus the operators $\{\hat{\mathbf{L}}^2, \hat{L}_3, \hat{\mathbf{S}}^2, \hat{S}_3\}$ all commute with each other

$$\{\hat{\mathbf{L}}^2, \hat{L}_3, \hat{\mathbf{S}}^2, \hat{S}_3\} \qquad ; \text{ mutually commuting} \qquad (3.66)$$

The direct-product states are evidently eigenstates of these mutually commuting hermitian operators

$$\hat{\mathbf{L}}^2|lm_l sm_s\rangle = l(l+1)|lm_l sm_s\rangle \quad ; \; \hat{L}_3|lm_l sm_s\rangle = m_l|lm_l sm_s\rangle$$
$$\hat{\mathbf{S}}^2|lm_l sm_s\rangle = s(s+1)|lm_l sm_s\rangle \quad ; \; \hat{S}_3|lm_l sm_s\rangle = m_s|lm_l sm_s\rangle \quad (3.67)$$

There are $(2l+1) \times (2s+1)$ states in this direct-product basis.

Now introduce the *total angular momentum*

$$\hat{\mathbf{J}} \equiv \hat{\mathbf{L}} + \hat{\mathbf{S}} \qquad (3.68)$$

As always

$$\hat{\mathbf{J}}^2 = \hat{J}_1^2 + \hat{J}_2^2 + \hat{J}_3^2 \tag{3.69}$$

The commutation relations for $\hat{\mathbf{J}}$ follow immediately from the above

$$[\hat{J}_i, \hat{J}_j] = i\epsilon_{ijk}\hat{J}_k$$
$$[\hat{\mathbf{J}}^2, \hat{J}_i] = 0 \tag{3.70}$$

The square of the total angular momentum is now also given by

$$\hat{\mathbf{J}}^2 = \hat{\mathbf{L}}^2 + \hat{\mathbf{S}}^2 + 2\hat{\mathbf{L}} \cdot \hat{\mathbf{S}} \tag{3.71}$$

Since the square of an angular momentum commutes with all of its components, the operators $\{\hat{\mathbf{L}}^2, \hat{\mathbf{S}}^2, \hat{\mathbf{J}}^2, \hat{J}_3\}$ all commute with each other

$$\{\hat{\mathbf{L}}^2, \hat{\mathbf{S}}^2, \hat{\mathbf{J}}^2, \hat{J}_3\} \qquad ; \text{ mutually commuting} \tag{3.72}$$

Thus one can equivalently introduce a set of basis states $|lsjm_j\rangle$, which are linear combinations of the direct-product states in Eq. (3.64), and which are eigenstates of this set of mutually commuting hermitian operators

$$\hat{\mathbf{J}}^2|lsjm_j\rangle = j(j+1)|lsjm_j\rangle \quad ; \quad \hat{J}_3|lsjm_j\rangle = m_j|lsjm_j\rangle$$
$$\hat{\mathbf{L}}^2|lsjm_j\rangle = l(l+1)|lsjm_j\rangle \quad ; \quad \hat{\mathbf{S}}^2|lsjm_j\rangle = s(s+1)|lsjm_j\rangle \tag{3.73}$$

We make some comments on this result:

- The spectrum of eigenvalues of the total angular momentum follows from our general analysis in the previous section;
- The eigenvalues of $(\hat{\mathbf{S}}^2, \hat{\mathbf{L}}^2)$ must remain the same due to the orthogonality of the eigenstates of hermitian operators;
- The goal now is to express the eigenstates of total angular momentum in terms of the direct product basis in Eqs. (3.64)

$$|lsjm_j\rangle = \sum_{m_l, m_s} \langle lm_l sm_s|lsjm_j\rangle \, |lm_l sm_s\rangle \quad ; \text{C-G coefficients} \tag{3.74}$$

The numerical transformation coefficients $\langle lm_l sm_s|lsjm_j\rangle$ in this expression are known as the *Clebsch-Gordan (C-G) coefficients*.

- The eigenstates appearing in Eq. (3.74) are both orthonormal and complete. This implies that the C-G coefficients must satisfy the relations[12]

$$\sum_{m_l,m_s} \langle lm_l sm_s|lsjm_j\rangle\langle lm_l sm_s|lsj'm_j'\rangle^\star = \delta_{jj'}\delta_{m_j m_j'}$$

$$\sum_{j,m_j} \langle lm_l sm_s|lsjm_j\rangle\langle lm_l'sm_s'|lsjm_j\rangle^\star = \delta_{m_l m_l'}\delta_{m_s m_s'} \quad (3.75)$$

- We proceed to explicitly construct the C-G coefficients, again using only general properties of the theory; in so doing, we shall determine the allowed values of j.

The C-G coefficients can be obtained by diagonalizing the operators

$$\hat{J}_3 = \hat{L}_3 + \hat{S}_3$$
$$\hat{\mathbf{J}}^2 = \hat{\mathbf{L}}^2 + \hat{\mathbf{S}}^2 + 2\hat{\mathbf{L}}\cdot\hat{\mathbf{S}} \quad (3.76)$$

in the new basis in Eq. (3.74).

(1) The first operator is readily diagonalized by simply letting it act on both sides of Eq. (3.74), which leads to

$$\hat{J}_3|lsjm_j\rangle = m_j \sum_{m_l,m_s} \langle lm_l sm_s|lsjm_j\rangle\,|lm_l sm_s\rangle$$

$$= \sum_{m_l,m_s} \langle lm_l sm_s|lsjm_j\rangle(m_l + m_s)\,|lm_l sm_s\rangle \quad (3.77)$$

The orthonormality of the states then gives

$$[m_j - (m_l + m_s)]\langle lm_l sm_s|lsjm_j\rangle = 0 \quad (3.78)$$

This expression implies that the C-G coefficients must vanish unless $m_j = m_l + m_s$

$$m_j = m_l + m_s \qquad ; \text{ C-G coefficients} \quad (3.79)$$

(2) To diagonalize $\hat{\mathbf{J}}^2$, we first note that both sides of Eq. (3.74) are already eigenstates of $\hat{\mathbf{L}}^2$ and $\hat{\mathbf{S}}^2$. Thus it just remains to diagonalize

$$2\hat{\mathbf{L}}\cdot\hat{\mathbf{S}} = \hat{\mathbf{J}}^2 - \hat{\mathbf{L}}^2 - \hat{\mathbf{S}}^2$$

$$\doteq j(j+1) - l(l+1) - s(s+1) \quad (3.80)$$

---

[12]See Prob. 3.5.

The second line is the form that operator must take when acting on the l.h.s. of Eq. (3.74). If the eigenvalues of this operator are denoted by α_j, then one must ensure that[13]

$$2\hat{\mathbf{L}} \cdot \hat{\mathbf{S}}|lsjm_j\rangle = \alpha_j|lsjm_j\rangle$$
$$= \alpha_j \sum_{m_l,m_s} \langle lm_lsm_s|lsjm_j\rangle |lm_lsm_s\rangle \qquad (3.81)$$

To accomplish this, we use the following identity

$$2\hat{\mathbf{L}} \cdot \hat{\mathbf{S}} \equiv (\hat{L}_1 + i\hat{L}_2)(\hat{S}_1 - i\hat{S}_2) + (\hat{L}_1 - i\hat{L}_2)(\hat{S}_1 + i\hat{S}_2) + 2\hat{L}_3\hat{S}_3$$
$$= \hat{L}_+\hat{S}_- + \hat{L}_-\hat{S}_+ + 2\hat{L}_3\hat{S}_3 \qquad (3.82)$$

Through the use of the previous results in Eqs. (3.59), we know what each of these operators does when applied to the r.h.s. of Eq. (3.74). Thus

$$2\hat{\mathbf{L}} \cdot \hat{\mathbf{S}} \sum_{m_l,m_s} \langle lm_lsm_s|lsjm_j\rangle |lm_lsm_s\rangle = \sum_{m_l,m_s} \langle lm_lsm_s|lsjm_j\rangle \times$$
$$\left\{ 2m_lm_s |lm_lsm_s\rangle + A(l,-m_l)A(s,m_s) |l,m_l+1,s,m_s-1\rangle \right.$$
$$\left. +A(l,m_l)A(s,-m_s) |l,m_l-1,s,m_s+1\rangle \right\} \quad (3.83)$$

Now carry out the following operations on the r.h.s. of this expression:

(1) Change the dummy summation variables in the last two terms so that the same state vector appears as on the l.h.s.;
(2) Use the fact that the coefficients $A(j, m_j)$ satisfy $A(j,-j) = A(j,j+1) = 0$ to restore the limits on the sums so that they are the same as occur on the l.h.s.

As an example of this process, consider what happens to the last term in

---

[13]When combined with the second of Eqs. (3.80), one has $\alpha_j = j(j+1) - l(l+1) - s(s+1)$; thus determining the eigenvalue α_j is equivalent to determining j.

Eq. (3.83). Define $m_l - 1 \equiv m_l'$ and $m_s + 1 \equiv m_s'$. Then

$$\sum_{m_l, m_s} \langle lm_l sm_s | lsjm_j \rangle A(l, m_l) A(s, -m_s) | l, m_l - 1, s, m_s + 1 \rangle$$

$$= \sum_{m_l' = -l-1}^{l-1} \sum_{m_s' = -s+1}^{s+1} | lm_l' sm_s' \rangle \big\{ A(l, m_l' + 1) A(s, -m_s' + 1) \times$$

$$\langle l, m_l' + 1, s, m_s' - 1 | lsjm_j \rangle \big\}$$

$$= \sum_{m_l' = -l}^{l} \sum_{m_s' = -s}^{s} | lm_l' sm_s' \rangle \big\{ A(l, m_l' + 1) A(s, -m_s' + 1) \times$$

$$\langle l, m_l' + 1, s, m_s' - 1 | lsjm_j \rangle \big\} \tag{3.84}$$

In this way, Eqs. (3.81) and (3.83) are reduced to the form

$$\sum_{m_l m_s} \big\{ (2m_l m_s - \alpha_j) \langle m_l m_s | jm_j \rangle$$

$$+ A(l, m_l + 1) A(s, -m_s + 1) \langle m_l + 1, m_s - 1 | jm_j \rangle$$

$$+ A(l, -m_l + 1) A(s, m_s + 1) \langle m_l - 1, m_s + 1 | jm_j \rangle \big\} | lm_l sm_s \rangle = 0 \tag{3.85}$$

where we now employ the shorthand

$$\langle lm_l sm_s | lsjm_j \rangle \equiv \langle m_l m_s | jm_j \rangle \qquad ; \text{ shorthand} \tag{3.86}$$

The orthonormality of the eigenstates then leads to the following set of equations

$$(2m_l m_s - \alpha_j) \langle m_l m_s | jm_j \rangle +$$

$$A(l, m_l + 1) A(s, -m_s + 1) \langle m_l + 1, m_s - 1 | jm_j \rangle +$$

$$A(l, -m_l + 1) A(s, m_s + 1) \langle m_l - 1, m_s + 1 | jm_j \rangle = 0 \tag{3.87}$$

Several comments:

- The coefficients $A(j, m_j)$ are determined from Eqs. (3.57)

$$A(j, m) = [(j + m)(j - m + 1)]^{1/2} \tag{3.88}$$

- There is one equation of the form in Eq. (3.87) for each pair of values (m_l, m_s) satisfying Eq. (3.79);
- Equations (3.87) then provide *coupled, linear, homogeneous algebraic equations* for the C-G coefficients $\langle m_l m_s | jm_j \rangle$;

- These equations will only have a non-trivial solution for the C-G coefficients if the determinant of their coefficients vanishes, and this will only happen for certain eigenvalues α_j;
- The corresponding allowed values of j are determined from α_j through the relation

$$\alpha_j = j(j+1) - l(l+1) - s(s+1) \tag{3.89}$$

- To see how this works, we study the example of $s = 1/2$.

If $s = 1/2$, then $m_s = \pm 1/2$, and $m_l = m_j - m_s$. The required coefficients in this case follow from Eq. (3.88) as

$$A\left(\frac{1}{2}, \frac{3}{2}\right) = 0 \qquad ; A\left(\frac{1}{2}, \frac{1}{2}\right) = 1 \qquad ; A\left(\frac{1}{2}, -\frac{1}{2}\right) = 0$$

$$A\left(l, m+\frac{1}{2}\right) = A\left(l, -m+\frac{1}{2}\right) = \left[\left(l+\frac{1}{2}\right)^2 - m^2\right]^{1/2} \tag{3.90}$$

where we now simply define $m_j \equiv m$. There are then two linear, homogeneous, algebraic equations for the C-G coefficients

$$\left(m - \frac{1}{2} - \alpha_j\right)\left\langle m - \frac{1}{2}, \frac{1}{2}\middle| jm\right\rangle + \left[\left(l+\frac{1}{2}\right)^2 - m^2\right]^{1/2}\left\langle m + \frac{1}{2}, -\frac{1}{2}\middle| jm\right\rangle$$

$$= 0 \qquad ; m_s = \frac{1}{2}$$

$$\left[\left(l+\frac{1}{2}\right)^2 - m^2\right]^{1/2}\left\langle m - \frac{1}{2}, \frac{1}{2}\middle| jm\right\rangle - \left(m + \frac{1}{2} + \alpha_j\right)\left\langle m + \frac{1}{2}, -\frac{1}{2}\middle| jm\right\rangle$$

$$= 0 \qquad ; m_s = -\frac{1}{2} \tag{3.91}$$

The determinant of the coefficients must vanish if these equation are to have a non-trivial solution, and hence

$$\left(\alpha_j + \frac{1}{2}\right)^2 - m^2 = \left(l + \frac{1}{2}\right)^2 - m^2 \tag{3.92}$$

There are two solutions to this equation

$$\alpha_j = l \qquad\qquad \Rightarrow j = l + \frac{1}{2}$$

$$\alpha_j = -l - 1 \qquad\qquad \Rightarrow j = l - \frac{1}{2} \tag{3.93}$$

We thus derive the rule for the addition of angular momenta $j = |l - s|, |l - s| + 1, \cdots, l + s$ used in Vol. I, at least in this case.

If the determinant of the coefficients vanishes in Eqs. (3.91), then the equations are not linearly independent, and either one of them can be used to determine the *ratio* of the C-G coefficients. Since the linear Eqs. (3.91) are real, one can choose the C-G coefficients to also be real. The normalization condition in Eqs. (3.75) determines their overall magnitude, and the C-G coefficients in this case are shown in Table 3.1. We note the important point that the *overall sign* of the C-G coeffients is not determined by these arguments, but must again be determined by a set of *phase conventions*. We choose to follow the conventions of [Edmonds (1974)].[14]

Table 3.1 Clebsch-Gordan coefficients for $s = 1/2$ in the form $\langle l, m - m_s, 1/2, m_s | l, 1/2, j, m \rangle$.

	$m_s = 1/2$	$m_s = -1/2$
$j = l + 1/2$	$\left[\frac{l+1/2+m}{2l+1}\right]^{1/2}$	$\left[\frac{l+1/2-m}{2l+1}\right]^{1/2}$
$j = l - 1/2$	$-\left[\frac{l+1/2-m}{2l+1}\right]^{1/2}$	$\left[\frac{l+1/2+m}{2l+1}\right]^{1/2}$

There are various *symmetry properties* of the C-G coefficients that follow from the defining Eqs. (3.87) and the phase conventions. These symmetry properties are most usefully summarized in terms of Wigner's 3-j symbol defined by

$$\begin{pmatrix} j_1 & j_2 & j_3 \\ m_1 & m_2 & m_3 \end{pmatrix} \equiv (-1)^{j_1 - j_2 - m_3} \frac{1}{\sqrt{2j_3 + 1}} \langle j_1 m_1 j_2 m_2 | j_1 j_2 j_3, -m_3 \rangle \quad (3.94)$$

The properties of the 3-j symbols are [Edmonds (1974)]

- They are invariant under any even permutation of the columns;
- Any odd permutation of the columns is equivalent to multiplication of the 3-j symbol by $(-1)^{j_1 + j_2 + j_3}$;
- They vanish unless $m_1 + m_2 + m_3 = 0$;
- They vanish unless $j_1 + j_2 \geq j_3 \geq |j_1 - j_2|$, which is the *addition law* for angular momenta.[15]

---

[14]Other authors may follow different phase conventions, and, again, the reader is strongly urged to first check the set of phase conventions employed when using the work of others.

[15]Problem 3.15 deals with this addition law.

A numerical tabulation of the 3-j symbols can be found in [Rotenberg *et al.* (1959)], which forms an indispensable resource.[16]

3.6 Recoupling

Suppose there is a *third* angular momentum in the problem, for example, the nuclear spin **I**, so that the total angular momentum operator for an atom becomes

$$\hat{\mathbf{F}} = \hat{\mathbf{J}} + \hat{\mathbf{I}} = \hat{\mathbf{L}} + \hat{\mathbf{S}} + \hat{\mathbf{I}} \qquad ; \text{total angular momentum} \qquad (3.95)$$

How do we proceed in this case? One way would be to couple the first two up to some intermediate angular momentum (we now know how to do this), and then couple that to the third to get the total. Thus, in a general notation,

$$|(j_1 j_2) j_{12} j_3 j m\rangle = \sum_{m_1, m_2, m_3, m_{12}} \langle j_1 m_1 j_2 m_2 | j_1 j_2 j_{12} m_{12}\rangle \times$$
$$\langle j_{12} m_{12} j_3 m_3 | j_{12} j_3 j m\rangle \, |j_1 m_1 j_2 m_2 j_3 m_3\rangle \qquad (3.96)$$

Now simply using the properties of the direct-product basis states $|j_1 m_1 j_2 m_2 j_3 m_3\rangle$, and the properties of the (real) C-G coefficients in Eqs. (3.75), one observes that these new states are both orthonormal and complete[17]

$$\langle (j_1 j_2) j_{12} j_3 j m | (j_1 j_2) j'_{12} j_3 j' m'\rangle = \delta_{j_{12} \, j'_{12}} \delta_{j \, j'} \delta_{m \, m'}$$

$$\sum_{j, m, j_{12}} |(j_1 j_2) j_{12} j_3 j m\rangle \langle (j_1 j_2) j_{12} j_3 j m| =$$

$$\sum_{m_1, m_2, m_3} |j_1 m_1 j_2 m_2 j_3 m_3\rangle \langle j_1 m_1 j_2 m_2 j_3 m_3| \qquad (3.97)$$

Another way to achieve the same goal is to first couple the second and third angular momenta to some intermediate value, and then couple the first to that to achieve a given total angular momentum. Thus

$$|j_1 (j_2 j_3) j_{23} j m\rangle = \sum_{m_1, m_2, m_3, m_{23}} \langle j_1 m_1 j_{23} m_{23} | j_1 j_{23} j m\rangle \times$$
$$\langle j_2 m_2 j_3 m_3 | j_2 j_3 j_{23} m_{23}\rangle \, |j_1 m_1 j_2 m_2 j_3 m_3\rangle \qquad (3.98)$$

---

[16]One must pay attention when calculating with half-integral angular momenta, since if j is half-integral, then $(-1)^j \neq (-1)^{-j}$; however, $(-1)^{2j} = (-1)^{-2j}$ always holds, for example, and if (j_1, j_2, j_3) add as angular momenta, then $(-1)^{j_1+j_2+j_3} = (-1)^{-j_1-j_2-j_3}$.

[17]See Prob. 3.6.

These states are *again* orthonormal and complete, by the same argument as given above. Since *both* sets of states are orthonormal and complete, there must be a unitary transformation between them. The transformation coefficients are defined in terms of Wigner's 6-j symbol according to

$$\langle (j_1 j_2) j_{12} j_3 jm | j_1 (j_2 j_3) j_{23} jm \rangle \equiv$$

$$(-1)^{j_1+j_2+j_3+j} \sqrt{(2j_{12}+1)(2j_{23}+1)} \begin{Bmatrix} j_1 & j_2 & j_{12} \\ j_3 & j & j_{23} \end{Bmatrix} \quad ; \text{ 6-}j \text{ symbol} \quad (3.99)$$

We make several comments:

- The inner product of the two states in Eqs. (3.96) and (3.98) expresses the 6-j symbol as a sum over four C-G coefficients[18]

$$\langle (j_1 j_2) j_{12} j_3 jm | j_1 (j_2 j_3) j_{23} jm \rangle =$$

$$(-1)^{j_1+j_2+j_3+j} \sqrt{(2j_{12}+1)(2j_{23}+1)} \begin{Bmatrix} j_1 & j_2 & j_{12} \\ j_3 & j & j_{23} \end{Bmatrix} =$$

$$\sum_{m_1,m_2,m_3,m_{12},m_{23}} \langle j_1 m_1 j_2 m_2 | j_1 j_2 j_{12} m_{12} \rangle \langle j_{12} m_{12} j_3 m_3 | j_{12} j_3 jm \rangle \times$$

$$\langle j_1 m_1 j_{23} m_{23} | j_1 j_{23} jm \rangle \langle j_2 m_2 j_3 m_3 | j_2 j_3 j_{23} m_{23} \rangle \quad (3.100)$$

- The transformation coefficients in Eq. (3.99) are *independent of m.*[19] The proof is very simple, but also very important. Consider the general transformation from one set of states to the other

$$|(j_1 j_2) j_{12} j_3 jm \rangle = \sum_{j_{23},j',m'} \langle j_1 (j_2 j_3) j_{23} j' m' | (j_1 j_2) j_{12} j_3 jm \rangle \times$$

$$|j_1 (j_2 j_3) j_{23} j' m' \rangle \quad (3.101)$$

This is simply the expansion in a complete basis. The orthogonality of eigenstates of different (j, m) reduces this to

$$|(j_1 j_2) j_{12} j_3 jm \rangle = \sum_{j_{23}} \langle j_1 (j_2 j_3) j_{23} jm | (j_1 j_2) j_{12} j_3 jm \rangle \times$$

$$|j_1 (j_2 j_3) j_{23} jm \rangle \quad (3.102)$$

---

[18]Note that $\langle (j_1 j_2) j_{12} j_3 jm | j_1 (j_2 j_3) j_{23} jm \rangle = \langle j_1 (j_2 j_3) j_{23} jm | (j_1 j_2) j_{12} j_3 jm \rangle$, which follows from the reality of the C-G coefficients.

[19]They are *scalars*, independent of the orientation of the system. This often serves as a nice check on a calculation which involves several C-G coefficients where one knows that the final result must be a scalar; it must be possible to re-express any such sum over four of them, for example, in this form. Problems 3.7–3.10 are very useful in this regard.

Both sides are now eigenstates of $(\hat{\mathbf{J}}^2, \hat{J}_3)$. Operate on this equation with $\hat{J}_+/A(j, -m)$. This gives (in a shorthand notation)

$$|(j_1 j_2) j_{12} j_3 j, m+1\rangle = \sum_{j_{23}} \langle \cdots m | \cdots m \rangle \times$$

$$|j_1 (j_2 j_3) j_{23} j, m+1\rangle \qquad (3.103)$$

But by definition

$$|(j_1 j_2) j_{12} j_3 j, m+1\rangle = \sum_{j_{23}} \langle \cdots, m+1 | \cdots, m+1 \rangle \times$$

$$|j_1 (j_2 j_3) j_{23} j, m+1\rangle \qquad (3.104)$$

Therefore, the coefficients are indeed independent of m.

- The 6-j symbols have nice symmetry properties [Edmonds (1974)]:

  - They are invariant under the interchange of columns;
  - They are invariant under the interchange of the upper and lower arguments of any two columns;
  - They vanish unless the angular momentum addition rules are satisfied in each of the following four directions

$$\left\{ \begin{matrix} j & j & j \\ & & \end{matrix} \right\} \quad ; \quad \left\{ \begin{matrix} j & & \\ & j & j \end{matrix} \right\} \quad ; \quad \left\{ \begin{matrix} & j & \\ j & & j \end{matrix} \right\} \quad ; \quad \left\{ \begin{matrix} & & j \\ j & j & \end{matrix} \right\} \qquad (3.105)$$

- An invaluable numerical table of 6-j symbols is contained in [Rotenberg *et al.* (1959)].

3.7 Irreducible Tensor Operators

We have already defined a vector operator in quantum mechanics by its commutation relations with the angular momentum [Eq. (3.26)]. Let us generalize this idea

The set of $2\kappa + 1$ operators $\hat{T}(\kappa, q)$ with $-\kappa \leq q \leq \kappa$ forms an irreducible tensor operator (ITO) of rank κ if

$$[\hat{J}_\pm, \hat{T}(\kappa, q)] = A(\kappa, \mp q) \hat{T}(\kappa, q \pm 1) \qquad ; \text{ ITO}$$

$$[\hat{J}_3, \hat{T}(\kappa, q)] = q \hat{T}(\kappa, q) \qquad (3.106)$$

We refer to these as the *spherical components* of the ITO. An ITO is thus defined entirely through its commutation relations with the angular momentum.[20] We give some examples:

- Work in the coordinate representation, where $\mathbf{J} \to \mathbf{L} = -i\mathbf{x} \times \mathbf{\nabla}$ becomes a differential operator. Then

(1) It is readily established by direct differentiation in (x, y, z) that the spherical components of the vector $\mathbf{x}$ are given by (Prob. 3.6)

$$x_{1,\pm 1} = \mp\frac{1}{\sqrt{2}}(x \pm iy) \qquad ; \text{ ITO of rank one}$$

$$x_{1,0} = z \tag{3.107}$$

These expressions can be re-written in spherical coordinates, where they become proportional to the components of the spherical harmonic $Y_{1,m}(\theta, \phi)$

$$x_{1,m} = \left(\frac{4\pi}{3}\right)^{1/2} rY_{1,m}(\theta, \phi) \tag{3.108}$$

(2) More generally, $\mathbf{L}$ is a differential operator in (θ, ϕ) and the spherical harmonics satisfy [Edmonds (1974)]

$$[L_\pm, Y_{l,m}(\theta, \phi)] = A(l, \mp m)Y_{l,m\pm 1}(\theta, \phi)$$

$$[L_3, Y_{l,m}(\theta, \phi)] = mY_{l,m}(\theta, \phi) \tag{3.109}$$

The quantity $f(r)Y_{l,m}(\theta, \phi)$ is the prototype for an ITO of rank l in quantum mechanics.

(3) Work with a matrix representation in spin space for a spin-1/2 particle where $\mathbf{J} \to \mathbf{S} = \boldsymbol{\sigma}/2$. Now $\mathbf{S}$ is a set of 2×2 matrices. The spherical components of $\boldsymbol{\sigma}$ are then given by

$$\sigma_{1,\pm 1} = \mp\frac{1}{\sqrt{2}}(\sigma_x \pm i\sigma_y) \qquad ; \text{ ITO of rank one}$$

$$\sigma_{1,0} = \sigma_z \tag{3.110}$$

It is readily established from the commutation relations for the Pauli matrices that (Prob. 3.6)

$$[S_\pm, \sigma_{1,q}] = A(1, \mp q)\sigma_{1,q\pm 1}$$

$$[S_3, \sigma_{1,q}] = q\sigma_{1,q} \tag{3.111}$$

---

[20]Note that $A(\kappa, \mp q) = \sqrt{(\kappa \mp q)(\kappa \pm q + 1)}$.

Thus $\sigma_{1,q}$ forms an ITO of rank 1.

It is readily established that the definition of an ITO in Eqs. (3.106) is fully equivalent to the following definition

$$[\hat{J}_i, \hat{T}(\kappa, q)] = \sum_{q'} \langle \kappa q' | \hat{J}_i | \kappa q \rangle \, \hat{T}(\kappa, q') \qquad ; \text{ defines ITO} \quad (3.112)$$

where $i = (1, 2, 3)$.[21]

3.8 The Wigner-Eckart Theorem

The Wigner-Eckart theorem concerns the matrix elements of an ITO taken between eigenstates of angular momentum. Based entirely on general considerations, it allows one to:

(1) Extract the *dependence on orientation*, that is on (m', q, m), of any matrix element;
(2) Extract the *angular momentum selection rules* relating (j', κ, j) for any matrix element.

It is a very powerful and useful result.

Formally, the Wigner-Eckart theorem (W-E) states that

$$\langle \gamma' j' m' | \hat{T}(\kappa, q) | \gamma j m \rangle = \frac{(-1)^{\kappa - j + j'}}{\sqrt{2j' + 1}} \langle \kappa q j m | \kappa j j' m' \rangle \, \langle \gamma' j' || T(\kappa) || \gamma j \rangle$$

$$= (-1)^{j' - m'} \begin{pmatrix} j' & \kappa & j \\ -m' & q & m \end{pmatrix} \langle \gamma' j' || T(\kappa) || \gamma j \rangle \qquad ;$$

Wigner-Eckart theorem (3.113)

The m-dependence and angular momentum selection rules are now contained in the C-G coefficient, or equivalently, in the 3-j symbol. The *reduced matrix element* $\langle \gamma' j' || T(\kappa) || \gamma j \rangle$ is simply what is left over after this dependence has been extracted; it is generally determined by explicitly computing one specific matrix element $\langle \gamma' j' m' | \hat{T}(\kappa, q) | \gamma j m \rangle$ and then writing it in this form. Here γ denotes the remaining set of quantum numbers characterizing the state vector $|\gamma j m \rangle$.[22]

---

[21]See Prob. 3.16; in these arguments $i = (1, 2, 3)$ is the same as $i = (x, y, z)$.

[22]The quantum numbers (γ, j, m) are assumed to be eigenvalues of a set of mutually commuting hermitian operators $(\hat{\mathbf{\Gamma}}, \hat{\mathbf{J}}^2, \hat{J}_3)$, and the eigenstates $|\gamma j m\rangle$ are correspondingly assumed to be both complete and orthonormal.

We proceed to a proof of the Wigner-Eckart theorem. Consider the following state vector, where the ITO and the initial state have been coupled with a C-G coefficient,

$$|\Psi_{j'm'}\rangle \equiv \sum_{qm} \langle \kappa q j m | \kappa j j' m' \rangle \, \hat{T}(\kappa, q) |\gamma j m\rangle \tag{3.114}$$

One now establishes that this state is indeed an eigenstate of angular momentum satisfying

$$\hat{J}_3 |\Psi_{j'm'}\rangle = m' |\Psi_{j'm'}\rangle$$
$$\hat{J}_\pm |\Psi_{j'm'}\rangle = A(j', \mp m') |\Psi_{j'm'\pm 1}\rangle$$
$$\hat{\mathbf{J}}^2 |\Psi_{j'm'}\rangle = j'(j'+1) |\Psi_{j'm'}\rangle \tag{3.115}$$

The proof of these relations involves straightforward algebra using the assumed commutation relations and the defining equations for the C-G coefficients.[23] We give a simpler proof of these relations based on finite rotations in the next section.

Expand the state vector in Eq. (3.114) in the complete basis $|\gamma j m\rangle$

$$|\Psi_{j'm'}\rangle = \sum_{\gamma', j'', m''} \langle \gamma' j'' m'' | \Psi_{j'm'}\rangle |\gamma' j'' m''\rangle$$
$$= \sum_{\gamma'} \langle \gamma' j' m' | \Psi_{j'm'}\rangle |\gamma' j' m'\rangle \tag{3.116}$$

The second line follows from the orthogonality of eigenstates of angular momentum. Now the coefficients in this expression must be independent of m' by exactly the same proof as in Eqs. (3.102)–(3.104)! Thus

$$|\Psi_{j'm'}\rangle = \sum_{\gamma'} \langle \gamma' j' | \Psi_{j'}\rangle |\gamma' j' m'\rangle \tag{3.117}$$

Use the orthogonality of the C-G coefficients [Eqs. (3.75)] to invert Eq. (3.114), and then insert Eq. (3.117)

$$\hat{T}(\kappa, q)|\gamma j m\rangle = \sum_{j', m'} \langle \kappa q j m | \kappa j j' m' \rangle \, |\Psi_{j'm'}\rangle$$
$$= \sum_{j', m', \gamma'} \langle \kappa q j m | \kappa j j' m' \rangle \langle \gamma' j' | \Psi_{j'}\rangle |\gamma' j' m'\rangle \tag{3.118}$$

---

[23] See, for example, [Schiff (1968)].

Finally, take the matrix element with the state $|\gamma'j'm'\rangle$

$$\langle\gamma'j'm'|\hat{T}(\kappa,q)|\gamma jm\rangle = \langle\kappa qjm|\kappa jj'm'\rangle\,\langle\gamma'j'|\Psi_{j'}\rangle \qquad (3.119)$$

The C-G coefficient now appears explicitly, and last factor is independent of m'. Since any m-independent factor can be extracted in defining the reduced matrix element, this is the W-E theorem.

3.9   Finite Rotations

Let us now talk about *active* rotations where the operator rotates the physical state vector through a set of angles. The classical Euler angles are defined in Fig. 3.4.

We claim that the following operator

$$\hat{\mathcal{R}}_{-\alpha,-\beta,-\gamma} \equiv e^{-i\alpha\hat{J}_z}e^{-i\beta\hat{J}_y}e^{-i\gamma\hat{J}_z} \qquad (3.120)$$

rotates the physical state vector through the Euler angles (α,β,γ).

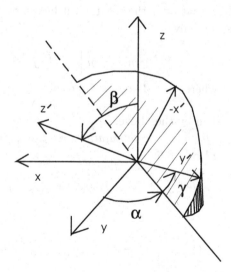

Fig. 3.3   The classical Euler angles (α,β,γ).

To prove this assertion, we first demonstrate that the operator $\exp\{-i\beta\hat{J}_y\}$ rotates the physical state vector through the angle β about the y-axis, as illustrated in Fig. 3.4.

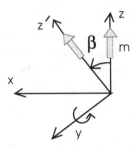

Fig. 3.4   The operator $\exp\{-i\beta\hat{J}_y\}$ rotates the physical system by an angle β about the y-axis.

The basis vectors in this figure are evidently related by

$$\mathbf{e}_{z'} = \mathbf{e}_z \cos\beta + \mathbf{e}_x \sin\beta \tag{3.121}$$

Suppose the system originally has an angular momentum m along the z-axis. We shall show that after the rotation, it has this angular momentum along the z'-axis, or equivalently

$$\hat{\mathbf{J}} \cdot \mathbf{e}_{z'} \left[e^{-i\beta\hat{J}_y} |jm\rangle \right] = m \left[e^{-i\beta\hat{J}_y} |jm\rangle \right]$$
$$\text{where;} \qquad \hat{\mathbf{J}} \cdot \mathbf{e}_{z'} = \hat{J}_z \cos\beta + \hat{J}_x \sin\beta \tag{3.122}$$

This is what we mean by the above.

Now, *entirely from the commutation relations*, one has [compare Eq. (3.22), and see Prob. 3.6]

$$e^{-i\beta\hat{J}_y} \hat{J}_z \, e^{i\beta\hat{J}_y} = \hat{J}_z \cos\beta + \hat{J}_x \sin\beta \tag{3.123}$$

Therefore

$$\hat{\mathbf{J}} \cdot \mathbf{e}_{z'} \left[e^{-i\beta\hat{J}_y} |jm\rangle \right] = e^{-i\beta\hat{J}_y} \hat{J}_z \, |jm\rangle = m \left[e^{-i\beta\hat{J}_y} |jm\rangle \right] \tag{3.124}$$

which was the result to be established, and since the state vectors $|\rho jm\rangle$ provide a complete basis, we have also established the generality of our claim.[24]

---

[24]We now use ρ for the other quantum numbers, and reserve γ for the Euler angle.

Finally, by carrying out the rotations in the indicated order in Eq. (3.120), that is, the one on the right first, it is evident that the operator in Eq. (3.120) takes the physical system from the unprimed to the primed configuration in Fig. 3.3.[25]

Now that the proper physical interpretation of the rotation operator has been established, let us consider quite generally

$$\hat{\mathcal{R}}_{\alpha\beta\gamma} \equiv e^{i\alpha\hat{J}_z} e^{i\beta\hat{J}_y} e^{i\gamma\hat{J}_z} \qquad\qquad ; \text{ rotation operator}$$

$$\hat{\mathcal{R}}_{\alpha\beta\gamma}|\rho j m\rangle = \sum_{\rho',j',m'} \langle \rho'j'm'|\hat{\mathcal{R}}_{\alpha\beta\gamma}|\rho j m\rangle \, |\rho'j'm'\rangle \qquad (3.125)$$

Here the completeness of the states $|\rho j m\rangle$ has been used in the second line. We observe that:

- The operators $\hat{J}_i$ cannot change j, and thus the matrix element is diagonal in j;
- $\hat{R}$ has terms that raise, lower, and reproduce $|jm\rangle$ with known coefficients which depend on j and m;
- Thus the matrix element is also diagonal in ρ, and independent of it.

Hence one can write

$$\hat{\mathcal{R}}_{\alpha\beta\gamma}|\rho j m\rangle = \sum_{m'} \mathcal{D}^j_{m'm}(\alpha\beta\gamma) \, |\rho j m'\rangle$$

$$\mathcal{D}^j_{m'm}(\alpha\beta\gamma) \equiv \langle jm'|\hat{\mathcal{R}}_{\alpha\beta\gamma}|jm\rangle$$

$$= \langle jm'|e^{i\alpha\hat{J}_z} e^{i\beta\hat{J}_y} e^{i\gamma\hat{J}_z}|jm\rangle \quad ; \text{ rotation matrices} \quad (3.126)$$

The last line defines the *rotation matrices*.

3.9.1 *Properties*

We summarize some of the properties of the above quantities:

(1) The rotation operator is unitary[26]

$$\hat{\mathcal{R}}^\dagger_{\alpha\beta\gamma} = e^{-i\gamma\hat{J}_z} e^{-i\beta\hat{J}_y} e^{-i\alpha\hat{J}_z} = \hat{\mathcal{R}}^{-1}_{\alpha\beta\gamma} \qquad (3.127)$$

---

[25]The reader should stare at Fig. 3.3 for awhile and be convinced that by carrying out the rotations in this order, always with respect to the unprimed, fixed, laboratory coordinate system, one does indeed arrive at the configuration shown in that figure.

[26]Recall that the adjoint of a product is the product of adjoints in reverse order.

Matrix elements derived from this relation then give

$$\langle jm'|\hat{\mathcal{R}}_\omega \hat{\mathcal{R}}_\omega^{-1}|jm\rangle = \sum_{m''}\langle jm'|\hat{\mathcal{R}}_\omega|jm''\rangle\langle jm''|\hat{\mathcal{R}}_\omega^\dagger|jm\rangle$$

$$= \sum_{m''} \mathcal{D}^j_{m'm''}(\omega)\mathcal{D}^j_{mm''}(\omega)^\star = \delta_{m'm} \qquad (3.128)$$

where ω stands for the triplet of angles $(\alpha\beta\gamma)$.

(2) Between eigenstates of angular momentum, the operator $\hat{J}_z$ on each end can be replaced by its eigenvalue, and thus Eq. (3.126) can be re-written as

$$\mathcal{D}^j_{m'm}(\alpha\beta\gamma) = e^{i\alpha m'}\langle jm'|e^{i\beta\hat{J}_y}|jm\rangle e^{i\gamma m}$$

$$\equiv e^{i\alpha m'} d^j_{m'm}(\beta)e^{i\gamma m} \qquad (3.129)$$

Hence the heart of the rotation matrices is contained in the factor

$$d^j_{m'm}(\beta) = \langle jm'|e^{i\beta\hat{J}_y}|jm\rangle \qquad (3.130)$$

Since $\hat{J}_y = (\hat{J}_+ - \hat{J}_-)/2i$, the $d^j_{m'm}(\beta)$ are known from general principles (see Probs. 3.12–3.13).[27]

(3) The rotation matrices satisfy the following orthonormality relation [Edmonds (1974)]

$$\int_0^{2\pi} d\gamma \int_0^\pi d\beta \sin\beta \int_0^{2\pi} d\alpha\, \mathcal{D}^{j_1}_{m_1 m'_1}(\alpha\beta\gamma)^\star\, \mathcal{D}^{j_2}_{m_2 m'_2}(\alpha\beta\gamma) =$$

$$\frac{8\pi^2}{2j_1+1}\delta_{j_1 j_2}\delta_{m_1 m_2}\delta_{m'_1 m'_2} \qquad (3.131)$$

(4) In the direct-product basis for two independent angular momenta, the eigenstates of total angular momentum are determined from Eq. (3.74). This relation can be inverted by using the orthonormality properties of the C-G coefficients

$$|j_1 m_1 j_2 m_2\rangle = \sum_{j,m}\langle j_1 m_1 j_2 m_2|j_1 j_2 jm\rangle\,|j_1 j_2 jm\rangle \qquad (3.132)$$

Here the total angular momentum operator is $\hat{\mathbf{J}} = \hat{\mathbf{J}}_1 + \hat{\mathbf{J}}_2$ with $[\hat{\mathbf{J}}_1, \hat{\mathbf{J}}_2] = 0$. Now consider the matrix element of the rotation operator applied to

---

[27]A differential equation can be derived for the $d^j_{m'm}(\beta)$ relating them to Jacobi polynomials. To make contact with something familiar, we note that $\mathcal{D}^l_{0m}(\alpha\beta\gamma) = [4\pi/(2l+1)]^{1/2}Y_{lm}(\beta,\gamma)$, where the Y_{lm} are the spherical harmonics.

this expression. When operating in the direct-product basis, the rotation operator *factors* into

$$\hat{R}_{\alpha\beta\gamma} = e^{i\alpha\hat{J}_z} e^{i\beta\hat{J}_y} e^{i\gamma\hat{J}_z}$$
$$= e^{i\alpha(\hat{J}_1+\hat{J}_2)_z} e^{i\beta(\hat{J}_1+\hat{J}_2)_y} e^{i\gamma(\hat{J}_1+\hat{J}_2)_z}$$
$$= \left[e^{i\alpha\hat{J}_{1z}} e^{i\beta\hat{J}_{1y}} e^{i\gamma\hat{J}_{1z}} \right] \left[e^{i\alpha\hat{J}_{2z}} e^{i\beta\hat{J}_{2y}} e^{i\gamma\hat{J}_{2z}} \right] \qquad (3.133)$$

From our general considerations, matrix elements of the total $\hat{\mathbf{J}}$ between the states $|j_1 j_2 j m\rangle$ depend only on (j, m) and are diagonal in j. Thus matrix elements of the operator in Eq. (3.133) between the states in Eq. (3.132) lead to a composition law for the rotation matrices

$$\mathcal{D}^{j_1}_{m'_1 m_1}(\omega) \mathcal{D}^{j_2}_{m'_2 m_2}(\omega) =$$
$$\sum_{j,m,m'} \langle j_1 m'_1 j_2 m'_2 | j_1 j_2 j m' \rangle \mathcal{D}^{j}_{m'm}(\omega) \langle j_1 m_1 j_2 m_2 | j_1 j_2 j m \rangle \quad (3.134)$$

This relation can be inverted using the orthonormality of the C-G coefficients

$$\sum_{m_1,m_2} \langle j_1 m_1 j_2 m_2 | j_1 j_2 j m \rangle \mathcal{D}^{j_1}_{m'_1 m_1}(\omega) \mathcal{D}^{j_2}_{m'_2 m_2}(\omega) =$$
$$\sum_{m'} \langle j_1 m'_1 j_2 m'_2 | j_1 j_2 j m' \rangle \mathcal{D}^{j}_{m'm}(\omega) \qquad (3.135)$$

3.9.2 *Tensor Operators*

There is a fully equivalent definition of an ITO in terms of finite rotations

$$\hat{R}_{\alpha\beta\gamma} \hat{T}(\kappa, q) \hat{R}^{-1}_{\alpha\beta\gamma} = \sum_{q'} \mathcal{D}^{\kappa}_{q'q}(\alpha\beta\gamma) \hat{T}(\kappa, q') \qquad ; \text{ defines ITO} \quad (3.136)$$

Let us establish the equivalence of the definitions in Eq. (3.112) and (3.136).

First we show that Eq. (3.136) implies Eq. (3.112). Take the following three specific rotations: (i) $\hat{R}_{\varepsilon,0,0} = e^{i\varepsilon\hat{J}_z}$; (ii) $\hat{R}_{0,\varepsilon,0} = e^{i\varepsilon\hat{J}_y}$; (iii) $\hat{R}_{\pi/2,\varepsilon,-\pi/2} = e^{i\varepsilon\hat{J}_x}$, where the last relation follows since

$$e^{i(\pi/2)\hat{J}_z} \hat{J}_y e^{-i(\pi/2)\hat{J}_z} = \hat{J}_y \cos\frac{\pi}{2} + \hat{J}_x \sin\frac{\pi}{2} = \hat{J}_x \qquad (3.137)$$

Now expand both sides of Eq. (3.136) in ε

$$\hat{T}(\kappa, q) + i\varepsilon[\hat{J}_i, \hat{T}(\kappa, q)] + \cdots = \sum_{q'} \langle \kappa q' | 1 + i\varepsilon\hat{J}_i + \cdots | \kappa q \rangle \hat{T}(\kappa, q') \quad (3.138)$$

In the limit $\varepsilon \to 0$, one recovers Eq. (3.112).

To show that Eq. (3.112) implies Eq. (3.136), make use first of the following exponentiated relation (see Prob. 2.5)

$$e^{i\gamma \hat{J}_z}\hat{T}(\kappa,q)e^{-i\gamma \hat{J}_z} = \hat{T}(\kappa,q) + i\gamma[\hat{J}_z,\hat{T}(\kappa,q)] + \frac{(i\gamma)^2}{2!}[\hat{J}_z,[\hat{J}_z,\hat{T}(\kappa,q)]] + \cdots$$

$$= \hat{T}(k,q)\left[1 + i\gamma q + \frac{(i\gamma q)^2}{2!} + \cdots\right] = \hat{T}(\kappa,q)e^{i\gamma q} \quad (3.139)$$

Now make use of a second relation

$$e^{i\beta \hat{J}_y}\hat{T}(\kappa,q)e^{-i\beta \hat{J}_y} = \hat{T}(\kappa,q) + i\beta[\hat{J}_y,\hat{T}(\kappa,q)] + \frac{(i\beta)^2}{2!}[\hat{J}_y,[\hat{J}_y,\hat{T}(\kappa,q)]] + \cdots$$

$$= \hat{T}(k,q) + i\beta \sum_{q'}\langle \kappa q'|\hat{J}_y|\kappa q\rangle\hat{T}(\kappa,q') +$$

$$\frac{(i\beta)^2}{2!}\sum_{q'}\sum_{q''}\langle \kappa q'|\hat{J}_y|\kappa q''\rangle\langle \kappa q''|\hat{J}_y|\kappa q\rangle\hat{T}(\kappa,q') + \cdots$$

$$= \sum_{q'}\langle \kappa q'|e^{i\beta \hat{J}_y}|\kappa q\rangle\hat{T}(\kappa,q') = \sum_{q'}d^{\kappa}_{q'q}(\beta)\hat{T}(\kappa,q') \quad (3.140)$$

Finally, repeat the first step. In this manner, it follows from Eqs. (3.112) that

$$\hat{\mathcal{R}}_{\alpha\beta\gamma}\hat{T}(\kappa,q)\hat{\mathcal{R}}^{-1}_{\alpha\beta\gamma} = \sum_{q'}\mathcal{D}^{\kappa}_{q'q}(\alpha\beta\gamma)\,\hat{T}(\kappa,q') \quad (3.141)$$

which is just Eq. (3.136).

It is generally easiest to identify an ITO through its commutation relations in Eqs. (3.106) or (3.112), while the exponentiated form in Eq. (3.136) is more adapted to formal manipulations. To illustrate the latter point, we proceed to provide the missing step in the above derivation of the Wigner-Eckart theorem.

3.9.3 *Wigner-Eckart Theorem (Completed)*

We postponed one step in the proof of the W-E theorem, and that is the demonstration that the constructed state $|\Psi_{j'm'}\rangle$ is indeed an eigenstate of angular momentum. The proof follows from the assumed commutation relations and the definition of the C-G coefficients, but that proof is rather long and tedious. We are now in a position to give an immediate, more elegant, demonstration of this result. Recall Eq. (3.114)

$$|\Psi_{j'm'}\rangle \equiv \sum_{qm} \langle \kappa q j m | \kappa j j' m' \rangle \, \hat{T}(\kappa, q) | \gamma j m \rangle \tag{3.142}$$

Apply $\hat{\mathcal{R}}_\omega$, insert $\hat{\mathcal{R}}_\omega^{-1}\hat{\mathcal{R}}_\omega$ appropriately, and then use Eqs. (3.126) and (3.136)

$$\hat{\mathcal{R}}_\omega |\Psi_{j'm'}\rangle = \sum_{qm} \langle \kappa q j m | \kappa j j' m' \rangle \, \hat{\mathcal{R}}_\omega \hat{T}(\kappa, q) \hat{\mathcal{R}}_\omega^{-1} \hat{\mathcal{R}}_\omega | \gamma j m \rangle$$

$$= \sum_{q,m}\sum_{q',m'''} \langle \kappa q j m | \kappa j j' m' \rangle \, \mathcal{D}^\kappa_{q'q}(\omega) \mathcal{D}^j_{m'''m}(\omega) \, \hat{T}(\kappa, q') | \gamma j m''' \rangle$$

$$\tag{3.143}$$

Now use the composition law for the rotation matrices obtained in Eq. (3.135)

$$\hat{\mathcal{R}}_\omega |\Psi_{j'm'}\rangle = \sum_{q',m'',m'''} \langle \kappa q' j m''' | \kappa j j' m'' \rangle \, \mathcal{D}^{j'}_{m''m'}(\omega) \, \hat{T}(\kappa, q') | \gamma j m''' \rangle \tag{3.144}$$

And once again identify $|\Psi_{j'm''}\rangle$ from Eq. (3.142)

$$\hat{\mathcal{R}}_\omega |\Psi_{j'm'}\rangle = \sum_{m''} \mathcal{D}^{j'}_{m''m'}(\omega) |\Psi_{j'm''}\rangle \tag{3.145}$$

This was the goal — to show that the state has the proper angular momentum transformation property. The reduction to infinitesimals, as above, gives

$$\hat{J}_i |\Psi_{j'm'}\rangle = \sum_{m''} \langle j'm''|\hat{J}_i|j'm'\rangle \, |\Psi_{j'm''}\rangle \tag{3.146}$$

Specialization to $(\hat{J}_z, \hat{J}_\pm = \hat{J}_x \pm i\hat{J}_y)$ immediately demonstrates that this is an eigenstate of $\hat{J}_z$ with eigenvalue m' and of $\hat{\mathbf{J}}^2$ with eigenvalue $j'(j'+1)$. The proof of the W-E theorem is now complete.

3.10   Tensor Products

Consider two ITO $\hat{T}(\kappa_1, q_1)$, $\hat{U}(\kappa_2, q_2)$. They can act in different spaces, or in the same space. The *tensor product* of these two operators is defined by

$$\hat{X}(K,Q) \equiv \sum_{q_1,q_2} \langle \kappa_1 q_1 \kappa_2 q_2 | \kappa_1 \kappa_2 K Q \rangle \, \hat{T}(\kappa_1, q_1) \hat{U}(\kappa_2, q_2) \qquad ;$$

$$\text{tensor product} \quad (3.147)$$

We claim that $\hat{X}(K,Q)$ is *again an ITO*, of rank K. The proof follows exactly as above, through a demonstration that

$$\hat{\mathcal{R}}_\omega \hat{X}(K,Q) \hat{\mathcal{R}}_\omega^{-1} = \sum_{Q'} \mathcal{D}^K_{Q'Q}(\omega) \hat{X}(K,Q') \qquad ; \text{ defines ITO} \qquad (3.148)$$

We leave it as Prob. 3.16 to fill in the steps.

3.11   Vector Model

It remains to justify the *vector model* for the addition of angular momenta, which was used so extensively in Vol. I. Consider, for concreteness, the magnetic moment of the nucleus. The nuclear magnetic dipole operators for the proton and neutron are given in units of the nuclear magneton $\mu_N = |e|\hbar/2m_p$ by

$$\frac{\hat{\boldsymbol{\mu}}_p}{\mu_N} = \hat{\mathbf{l}} + 2\lambda_p \hat{\mathbf{s}} \qquad ; \lambda_p = +2.793$$

$$\frac{\hat{\boldsymbol{\mu}}_n}{\mu_N} = 2\lambda_n \hat{\mathbf{s}} \qquad ; \lambda_n = -1.913 \qquad (3.149)$$

where only the term in λ contributes for the neutron. The magnetic moment of the nucleus is defined as the expectation value of the magnetic dipole operator in the state where the nucleus is lined up as well as possible along the z-axis

$$\mu \equiv \langle jj|\hat{\mu}_{10}|jj \rangle = \frac{\langle jj10|j1jj \rangle}{\sqrt{2j+1}} \langle j||\mu||j \rangle \qquad (3.150)$$

The second equality follows from the Wigner-Eckart theorem. The required C-G coefficient is $\langle jj10|j1jj \rangle = \sqrt{j/(j+1)}$ from Table 12.1.

To evaluate the remaining reduced matrix element in the single-particle shell model one needs $\langle l\frac{1}{2}j|| \, l \, ||l\frac{1}{2}j \rangle$ and $\langle l\frac{1}{2}j|| \, s \, ||l\frac{1}{2}j \rangle$. These are the reduced matrix elements of an ITO acting on the first and second part of a coupled scheme respectively; they may be evaluated using the results in Prob. 13.14

$$\langle l\frac{1}{2}j\| l \| l\frac{1}{2}j\rangle = (-1)^{l+1/2+j+1}(2j+1)\left\{\begin{array}{ccc} l & j & 1/2 \\ j & l & 1 \end{array}\right\}\langle l\| l \| l\rangle \qquad (3.151)$$

$$\langle l\frac{1}{2}j\| s \| l\frac{1}{2}j\rangle = (-1)^{l+1/2+j+1}(2j+1)\left\{\begin{array}{ccc} 1/2 & j & l \\ j & 1/2 & 1 \end{array}\right\}\langle \frac{1}{2}\| s \| \frac{1}{2}\rangle$$

The required expressions can now be looked up in [Edmonds (1974)], with the results

$$\langle l\frac{1}{2}j\| l \| l\frac{1}{2}j\rangle = \frac{1}{2}\left[\frac{2j+1}{j(j+1)}\right]^{1/2}[j(j+1)+l(l+1)-s(s+1)] \quad ; \; s=\frac{1}{2}$$

$$\langle l\frac{1}{2}j\| s \| l\frac{1}{2}j\rangle = \frac{1}{2}\left[\frac{2j+1}{j(j+1)}\right]^{1/2}[j(j+1)+s(s+1)-l(l+1)] \qquad (3.152)$$

Equation (3.150) then becomes

$$\frac{\mu}{\mu_N} = \frac{1}{2(j+1)}\{[j(j+1)+l(l+1)-s(s+1)]$$
$$+2\lambda[j(j+1)+s(s+1)-l(l+1)]\} \qquad ; \; s=1/2 \quad (3.153)$$

This result for the magnetic moment will be recognized as exactly the same expression obtained in the simple vector model of angular momenta. In this model, the vectors **l** and **s** add to give the resultant $\mathbf{j}=\mathbf{l}+\mathbf{s}$ (Fig. 3.5). They then precess around this resultant so that the effective magnetic moment is only that component along **j**.

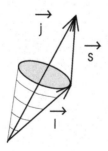

Fig. 3.5   Vector model of angular momenta.

$$\boldsymbol{\mu}_{\text{eff}} = \frac{(\boldsymbol{\mu}\cdot\mathbf{j})}{\mathbf{j}^2}\mathbf{j} \qquad (3.154)$$

The insertion of the definition of the magnetic dipole operator in Eq. (3.149) gives

$$\frac{\mu_{\text{eff}}}{\mu_N} = \frac{j}{j(j+1)}[\mathbf{l} \cdot \mathbf{j} + 2\lambda \mathbf{s} \cdot \mathbf{j}] \tag{3.155}$$

The square of the relations $\mathbf{j} - \mathbf{l} = \mathbf{s}$ and $\mathbf{j} - \mathbf{s} = \mathbf{l}$ gives

$$
\begin{aligned}
2\mathbf{l} \cdot \mathbf{j} &= \mathbf{j}^2 + \mathbf{l}^2 - \mathbf{s}^2 = j(j+1) + l(l+1) - s(s+1) \\
2\mathbf{s} \cdot \mathbf{j} &= \mathbf{j}^2 + \mathbf{s}^2 - \mathbf{l}^2 = j(j+1) + s(s+1) - l(l+1)
\end{aligned} \tag{3.156}
$$

A combination of these results indeed reproduces Eq. (3.153), thus validating the use of the vector model.

Chapter 4

Scattering Theory

Given a hamiltonian $\hat{H}$, the goal of this chapter is to solve the Schrödinger equation for a scattering problem and derive general expressions for the S-matrix, T-matrix, and transition rate, many of whose consequences have already been examined in Vol. I.[1] We work in the abstract Hilbert space.

4.1 Interaction Picture

Assume the hamiltonian can be split into two parts $\hat{H} = \hat{H}_0 + \hat{H}_1$, the first part of which leads to an exactly solvable problem, for example, free quanta with no interactions. $\hat{H}_1$ may, or may not, have an explicit time dependence; that depends on the problem at hand.[2] We then want to solve the Schrödinger equation

$$\hat{H} = \hat{H}_0 + \hat{H}_1$$
$$i\hbar \frac{\partial}{\partial t} |\Psi(t)\rangle = \hat{H} |\Psi(t)\rangle \qquad \text{; Schrödinger-equation} \qquad (4.1)$$

Define a new state vector $|\Psi_{\rm I}(t)\rangle$ by

$$|\Psi_{\rm I}(t)\rangle \equiv e^{\frac{i}{\hbar}\hat{H}_0 t} |\Psi(t)\rangle \qquad \text{; interaction picture}$$
$$|\Psi_{\rm I}(0)\rangle = |\Psi(0)\rangle \qquad \text{; coincide at } t = 0 \qquad (4.2)$$

---

[1]For a comprehensive treatment of scattering theory, see [Goldberger and Watson (2004)].

[2]Scattering in a given external field, for example, may lead to an explicitly time-dependent $\hat{H}_1(t)$.

What equation of motion does this new state satisfy? Just compute

$$i\hbar\frac{\partial}{\partial t}|\Psi_{\rm I}(t)\rangle = -\hat{H}_0\,e^{\frac{i}{\hbar}\hat{H}_0 t}|\Psi(t)\rangle + e^{\frac{i}{\hbar}\hat{H}_0 t}\,i\hbar\frac{\partial}{\partial t}|\Psi(t)\rangle$$

$$= -\hat{H}_0\,|\Psi_{\rm I}(t)\rangle + e^{\frac{i}{\hbar}\hat{H}_0 t}(\hat{H}_0 + \hat{H}_1)e^{-\frac{i}{\hbar}\hat{H}_0 t}|\Psi_{\rm I}(t)\rangle \quad (4.3)$$

The terms in $\hat{H}_0$ cancel, and thus

$$i\hbar\frac{\partial}{\partial t}|\Psi_{\rm I}(t)\rangle = \hat{H}_{\rm I}(t)|\Psi_{\rm I}(t)\rangle \qquad\qquad ;\text{ interaction picture}$$

$$\hat{H}_{\rm I}(t) \equiv e^{\frac{i}{\hbar}\hat{H}_0 t}\,\hat{H}_1\,e^{-\frac{i}{\hbar}\hat{H}_0 t} \qquad\qquad (4.4)$$

The advantage of this new formulation is that in the limit $\hat{H}_1 \to 0$, the state $|\Psi_{\rm I}(t)\rangle$ becomes time-independent; the free time variation, which can be extremely rapid, has been explicitly dealt with. Equations (4.2) and (4.4) are said to be a formulation of the problem in the *interaction picture*.

4.2 Adiabatic Approach

We will find that when we try to solve the resulting equations and generate the S-matrix, there will be infinite time integrals to carry out over oscillating integrands. In order to give the theory a well-defined mathematical meaning, we introduce an *adiabatic damping factor* $e^{-\epsilon|t|}$ with $\epsilon \geq 0$, and use the following interaction in the interaction picture

$$\hat{H}_{\rm I}^{\epsilon}(t) \equiv e^{-\epsilon|t|}\hat{H}_{\rm I}(t) \qquad\qquad ;\text{ adiabatic damping} \qquad (4.5)$$

The theory is then defined to be what is obtained in the limit as $\epsilon \to 0$.[3]

This is a somewhat archaic approach, and there are more sophisticated ways of doing formal scattering theory, which, however, can easily lead to spurious results if one is not very careful and thoughtful. The great advantage of this adiabatic approach is that it *allows one to do well-defined mathematics at each step*.

One can imagine that the interaction in Eq. (4.5) is being turned on and off very slowly ("adiabatically") as the time $t \to \pm\infty$, that is, in the infinite past and infinite future.[4] This allows us to easily specify the initial and final states in any scattering process, since now as $t \to \pm\infty$,

---

[3]There may, or may not, be other limits — we will not go there.

[4]Explicitly dealing with the scattering of wave packets can play the same role.

the hamiltonian simply reduces to $\hat{H}_0$, and we know how to solve the non-interacting problem

$$\hat{H} = \hat{H}_0 \qquad\qquad ; \ t \to \pm\infty$$

$$i\hbar\frac{\partial}{\partial t}|\Psi(t)\rangle = \hat{H}_0\,|\Psi(t)\rangle$$

$$|\Psi(t)\rangle = e^{-\frac{i}{\hbar}E_0 t}\,|\psi\rangle \tag{4.6}$$

Here $|\psi\rangle$ is simply a *solution to the free, time-independent, Schrödinger equation*

$$\hat{H}_0\,|\psi\rangle = E_0|\psi\rangle \tag{4.7}$$

The interaction-picture state vector in Eq. (4.2) is then given in this same limit by

$$|\Psi_\mathrm{I}(t)\rangle = e^{\frac{i}{\hbar}\hat{H}_0 t}\,|\Psi(t)\rangle = |\psi\rangle \qquad ; \ t \to \pm\infty$$

$$i\hbar\frac{\partial}{\partial t}|\Psi_\mathrm{I}(t)\rangle = 0 \tag{4.8}$$

Thus, in summary, with the adiabatic approach in the interaction picture, one has

$$|\Psi_\mathrm{I}(t)\rangle = |\psi\rangle \qquad\qquad ; \ t \to \pm\infty$$

$$\hat{H}_0\,|\psi\rangle = E_0|\psi\rangle \tag{4.9}$$

One starts with an initial state of this type, and then slowly turns on and off the interaction. The transition amplitude into a final state of this type is then calculated. The (transition probability)/(time interval the interaction is on) gives the transition rate,[5] and the path from the transition rate to a cross section was detailed in Vol. I.

It is then necessary to determine what happens when the interaction in Eq. (4.5) is turned on and off adiabatically. This is done through the construction of the *time-development operator* for the problem.

---

[5] We shall get more sophisticated here and actually derive a general expression for the transition rate itself.

4.3 $\hat{U}$-Operator

Let us look for an operator that develops our system in time

$$|\Psi_I(t)\rangle = \hat{U}_\epsilon(t,t_0)|\Psi_I(t_0)\rangle$$
$$i\hbar\frac{\partial}{\partial t}|\Psi_I(t)\rangle = i\hbar\frac{\partial}{\partial t}\hat{U}_\epsilon(t,t_0)|\Psi_I(t_0)\rangle = \hat{H}_I^\epsilon(t)\hat{U}_\epsilon(t,t_0)|\Psi_I(t_0)\rangle \quad (4.10)$$

If this is to hold for all $|\Psi_I(t_0)\rangle$, then $\hat{U}_\epsilon(t,t_0)$ must satisfy the operator relation

$$i\hbar\frac{\partial}{\partial t}\hat{U}_\epsilon(t,t_0) = \hat{H}_I^\epsilon(t)\hat{U}_\epsilon(t,t_0)$$
$$\hat{U}_\epsilon(t_0,t_0) = 1 \quad\quad\quad (4.11)$$

This differential equation, with its initial condition, can be rewritten as an *integral equation*

$$\hat{U}_\epsilon(t,t_0) = 1 - \frac{i}{\hbar}\int_{t_0}^t e^{-\epsilon|t'|}\,\hat{H}_I(t')\hat{U}_\epsilon(t',t_0)\,dt' \quad (4.12)$$

It is readily verified that Eqs. (4.11) are reproduced by this expression.

We will try to find a solution to this equation as a power series in $\hat{H}_I$.[6] Let us substitute this expression for $\hat{U}_\epsilon(t',t_0)$ in the integrand on the r.h.s.

$$\hat{U}_\epsilon(t,t_0) = 1 - \frac{i}{\hbar}\int_{t_0}^t e^{-\epsilon|t'|}\,\hat{H}_I(t')\,dt' +$$
$$\left(-\frac{i}{\hbar}\right)^2\int_{t_0}^t e^{-\epsilon|t'|}\,\hat{H}_I(t')\,dt'\int_{t_0}^{t'} e^{-\epsilon|t''|}\,\hat{H}_I(t'')\hat{U}_\epsilon(t'',t_0)\,dt'' \quad (4.13)$$

This expression is still exact. Repeated application of this process leads to the following infinite series in $\hat{H}_I$

$$\hat{U}_\epsilon(t,t_0) = \sum_{n=0}^\infty\left(-\frac{i}{\hbar}\right)^n\int_{t_0}^t e^{-\epsilon|t_1|}\,dt_1\int_{t_0}^{t_1} e^{-\epsilon|t_2|}\,dt_2\cdots\int_{t_0}^{t_{n-1}} e^{-\epsilon|t_n|}\,dt_n\times$$
$$\hat{H}_I(t_1)\hat{H}_I(t_2)\cdots\hat{H}_I(t_n) \quad\quad (4.14)$$

---

[6]One can only expect a power series to hold for scattering amplitudes at all energies in the absence of bound states; however, we will eventually "zip things up again" and obtain closed forms that are also valid in the presence of bound states.

By convention, the first term in this series is 1. Note that it is important to keep the *ordering* of the operators $\hat{H}_I(t)$ straight in the integrand, since they do not necessarily commute at different times. It is easy to remember the ordering since the operators are *time-ordered*, with the operator at the latest time appearing furthest to the left.

Equation (4.14) can be rewritten in the following manner

$$\hat{U}_\epsilon(t, t_0) = \sum_{n=0}^{\infty} \left(-\frac{i}{\hbar}\right)^n \frac{1}{n!} \int_{t_0}^{t} e^{-\epsilon|t_1|} dt_1 \int_{t_0}^{t} e^{-\epsilon|t_2|} dt_2 \cdots \int_{t_0}^{t} e^{-\epsilon|t_n|} dt_n \times$$

$$T\left[\hat{H}_I(t_1)\hat{H}_I(t_2)\cdots\hat{H}_I(t_n)\right] \qquad\qquad ; t \geq t_0 \qquad (4.15)$$

Here

- All the integrals are now over the full range $\int_{t_0}^{t}$;
- The "T-product" carries the instruction that the operators are to be time-ordered, with the operator at the latest time sitting to the left;
- Each term in the sum is divided by $n!$.

The proof that Eq. (4.15) reproduces Eq. (4.14) is quite simple. There are $n!$ possible orderings of the times in the multiple integral, pick one, say $t_1 > t_2 > t_3 > \cdots > t_n$. All possible time orderings of these integration variables provides a complete enumeration of the region of integration in the multiple integral. The operator in the integrand is time-ordered in each case. But now all of these contributions are *identical* by a change of dummy integration variables. Thus Eq. (4.14) is reproduced.[7]

The *scattering operator* $\hat{S}$ is now defined in the following manner

$$\hat{S} \equiv \mathrm{Lim}_{\epsilon \to 0}\, \mathrm{Lim}_{t \to +\infty}\, \mathrm{Lim}_{t_0 \to -\infty}\, \hat{U}_\epsilon(t, t_0) \qquad (4.16)$$

One lets the initial time $t_0 \to -\infty$, the final time $t \to +\infty$, and then, at the very end, the limit of the adiabatic damping factor $\epsilon \to 0$ is taken. Thus

$$\hat{S} = \mathrm{Lim}_{\epsilon \to 0}\, \hat{S}_\epsilon$$

$$= \mathrm{Lim}_{\epsilon \to 0} \sum_{n=0}^{\infty} \left(-\frac{i}{\hbar}\right)^n \frac{1}{n!} \int_{-\infty}^{\infty} e^{-\epsilon|t_1|} dt_1 \cdots \int_{-\infty}^{\infty} e^{-\epsilon|t_n|} dt_n \times$$

$$T\left[\hat{H}_I(t_1)\hat{H}_I(t_2)\cdots\hat{H}_I(t_n)\right] \qquad (4.17)$$

Everything so far has assumed $t \geq t_0$ in Eqs. (4.11) and the subsequent development; however, one can equally well write these equations for $t \leq t_0$.

---

[7]The explicit demonstration of this equality for $n = 2$ is assigned as Prob. 4.1.

How is the above analysis modified? Write Eq. (4.12) in the following fashion

$$\hat{U}_\epsilon(t, t_0) = 1 + \frac{i}{\hbar} \int_t^{t_0} e^{-\epsilon|t'|} \hat{H}_I(t') \hat{U}_\epsilon(t', t_0)\, dt' \tag{4.18}$$

It is readily verified that this expression reproduces Eqs. (4.11), and it is most convenient since the integral now runs in the positive direction if $t_0 \geq t$. A repetition of the above arguments in this case then leads to the following infinite series

$$\hat{U}_\epsilon(t, t_0) = \sum_{n=0}^{\infty} \left(\frac{i}{\hbar}\right)^n \frac{1}{n!} \int_t^{t_0} e^{-\epsilon|t_1|}\, dt_1 \int_t^{t_0} e^{-\epsilon|t_2|}\, dt_2 \cdots \int_t^{t_0} e^{-\epsilon|t_n|}\, dt_n \times$$
$$\overline{T}\left[\hat{H}_I(t_1)\hat{H}_I(t_2)\cdots\hat{H}_I(t_n)\right] \qquad ; t \leq t_0 \tag{4.19}$$

The "$\overline{T}$-product" instructs the operators to be anti-time-ordered such that the operator with the *earliest* time sits to the left. A simple reversal of the limits of integration in each integral then gives the equivalent expression

$$\hat{U}_\epsilon(t, t_0) = \sum_{n=0}^{\infty} \left(-\frac{i}{\hbar}\right)^n \frac{1}{n!} \int_{t_0}^{t} e^{-\epsilon|t_1|}\, dt_1 \int_{t_0}^{t} e^{-\epsilon|t_2|}\, dt_2 \cdots \int_{t_0}^{t} e^{-\epsilon|t_n|}\, dt_n \times$$
$$\overline{T}\left[\hat{H}_I(t_1)\hat{H}_I(t_2)\cdots\hat{H}_I(t_n)\right] \qquad ; t \leq t_0 \tag{4.20}$$

We are now in a position to exhibit some of the properties of $\hat{U}_\epsilon(t, t_0)$ from these series expansions:[8]

(1) Since the adjoint of a product is the product of the adjoints in the reverse order, it follows immediately from Eqs. (4.15) and (4.19) that

$$\hat{U}_\epsilon(t, t_0)^\dagger = \hat{U}_\epsilon(t_0, t) \tag{4.21}$$

which holds for both $t > t_0$ and $t < t_0$.

(2) If one ends up back at the start time, no matter whether $t > t_0$ or $t < t_0$, it must be true that

$$\hat{U}_\epsilon(t_0, t)\hat{U}_\epsilon(t, t_0) = 1 \tag{4.22}$$

This follows from the series expansions, and the explicit demonstration of this relation for $n = 2$ is left as Prob. 4.1.

(3) It follows from the results in (1) and (2) that

$$\hat{U}_\epsilon(t, t_0)^\dagger = \hat{U}_\epsilon(t, t_0)^{-1} \qquad ; \text{unitary} \tag{4.23}$$

---

[8]Note that relations (1)–(4) hold for finite ϵ.

The time-evolution operator is *unitary*. We know this must be true, since the Schrödinger equation preserves the *norm* of the states. This is now readily verified from the relation

$$\langle \Psi_I(t)|\Psi_I(t)\rangle = \langle \Psi_I(t_0)|\hat{U}_\epsilon(t,t_0)^\dagger \hat{U}_\epsilon(t,t_0)|\Psi_I(t_0)\rangle$$
$$= \langle \Psi_I(t_0)|\hat{U}_\epsilon(t,t_0)^{-1}\hat{U}_\epsilon(t,t_0)|\Psi_I(t_0)\rangle$$
$$= \langle \Psi_I(t_0)|\Psi_I(t_0)\rangle \tag{4.24}$$

(4) If one propagates the system from $t_0 \to t_1$, and then from $t_1 \to t_2$, the result must be the same as propagation from $t_0 \to t_2$. Thus the time-evolution operator must obey the *group property*

$$\hat{U}_\epsilon(t_2,t_1)\hat{U}_\epsilon(t_1,t_0) = \hat{U}_\epsilon(t_2,t_0) \qquad ; \text{ group property} \tag{4.25}$$

Let us demonstrate this result for $t_2 > t_1 > t_0$. The result in (2) can then be used to extend it to any relative times. For example, if $t_1 > t_2$, just write

$$\hat{U}_\epsilon(t_2,t_1)\hat{U}_\epsilon(t_1,t_0) = \hat{U}_\epsilon(t_2,t_1)\hat{U}_\epsilon(t_1,t_2)\hat{U}_\epsilon(t_2,t_0)$$
$$= \hat{U}_\epsilon(t_2,t_0) \qquad ; t_1 > t_2 \tag{4.26}$$

Write out the νth term in the sum on the r.h.s. of Eq. (4.25)

$$\hat{U}_\epsilon^{(\nu)}(t_2,t_0) = \left(-\frac{i}{\hbar}\right)^\nu \frac{1}{\nu!} \int_{t_0}^{t_2} e^{-\epsilon|t_1'|}\,dt_1' \cdots \int_{t_0}^{t_2} e^{-\epsilon|t_\nu'|}\,dt_\nu' \times$$
$$T\left[\hat{H}_I(t_1')\hat{H}_I(t_2')\cdots\hat{H}_I(t_\nu')\right] \tag{4.27}$$

Now note:

- There are $\nu!/n!m!$ ways to partition the times $t_1' \cdots t_\nu'$ so that n times are greater than the intermediate time t_1, and m times are less than t_1 — pick one;
- Now integrate over all possible relative orderings of the times within this particular partition;
- Then sum over all possible choices of the times within this particular partition. This provides a complete enumeration of the regions of integration for a given (n,m);
- The contributions in the sum are identical by a change of dummy integration variables, giving $\nu!/n!m!$ equal contributions;

- Then sum over all values of (n, m) for which $m + n = \nu$. This provides a complete evaluation of the multiple integral in Eq. (4.27)

$$\hat{U}_\epsilon^{(\nu)}(t_2, t_0) = \frac{1}{\nu!} \sum_{n+m=\nu} \left(-\frac{i}{\hbar}\right)^{n+m} \frac{\nu!}{n!m!} \times \tag{4.28}$$

$$\int_{t_1}^{t_2} e^{-\epsilon|t_1'|}\, dt_1' \cdots \int_{t_1}^{t_2} e^{-\epsilon|t_n'|}\, dt_n'\, T\left[\hat{H}_I(t_1') \cdots \hat{H}_I(t_n')\right] \times$$

$$\int_{t_0}^{t_1} e^{-\epsilon|t_{n+1}'|}\, dt_{n+1}' \cdots \int_{t_0}^{t_1} e^{-\epsilon|t_{n+m}'|}\, dt_{n+m}'\, T\left[\hat{H}_I(t_{n+1}') \cdots \hat{H}_I(t_{n+m}')\right]$$

- Finally, use $\sum_\nu \sum_{n+m=\nu} = \sum_n \sum_m$. This establishes Eq. (4.25).

4.4 $\hat{U}$-Operator for Finite Times

We started from the hamiltonian

$$\hat{H}_\epsilon = \hat{H}_0 + e^{-\epsilon|t|}\,\hat{H}_1 \tag{4.29}$$

In the end, we are to take the limit $\epsilon \to 0$, which restores the proper hamiltonian. Let us assume that we have used the preceding analysis to propagate the system from its initial state at $t_0 \to -\infty$ to a finite time such that

$$|t| \ll 1/\epsilon \qquad\qquad ; \text{ finite time} \tag{4.30}$$

Now, for this time,

$$\hat{H} = \hat{H}_0 + \hat{H}_1 \qquad ; \text{ full } \hat{H} \tag{4.31}$$

In this case, we can write a formal solution to the full Schrödinger equation as[9]

$$|\Psi_i(t)\rangle = e^{-\frac{i}{\hbar}\hat{H}t}|\Psi_i(0)\rangle \tag{4.32}$$

Here $|\Psi_i(0)\rangle = |\Psi_I^i(0)\rangle$ is the state that has propagated up to the time $t = 0$ from the initial state $|\psi_i\rangle$ prepared at $t_0 \to -\infty$ [see Eqs. (4.2)]. With the aid of the previous time-evolution operator, one can write this state as

$$|\Psi_i(0)\rangle = |\Psi_I^i(0)\rangle = \hat{U}_\epsilon(0, -\infty)|\psi_i\rangle \equiv |\psi_i^{(+)}\rangle \tag{4.33}$$

---

[9]We assume here and henceforth that $\hat{H}_1$ now has no explicit time dependence.

This relation defines $|\psi_i^{(+)}\rangle$. A combination of Eqs. (4.32) and (4.33) allows the solution to the Schrödinger equation at a finite time, which satisfies Eq. (4.30), to be expressed as

$$|\Psi_i(t)\rangle = e^{-\frac{i}{\hbar}\hat{H}t}|\psi_i^{(+)}\rangle \tag{4.34}$$

Some comments:

- This is the full Schrödinger state vector that develops from the state $|\psi_i\rangle$ at $t_0 \to -\infty$;
- One needs the adiabatic damping factor to bring that state vector up to finite time with $|\Psi_i(0)\rangle = \hat{U}_\epsilon(0, -\infty)|\psi_i\rangle \equiv |\psi_i^{(+)}\rangle$;
- From there, one can use the formal solution to the full Schrödinger equation in Eq. (4.34).

We note that under the conditions that one can indeed use the formal solution to the full Schrödinger equation, it follows that the interaction-picture state vector at the time t is given by

$$
\begin{aligned}
|\Psi_I(t)\rangle &= e^{\frac{i}{\hbar}\hat{H}_0 t}|\Psi(t)\rangle = e^{\frac{i}{\hbar}\hat{H}_0 t}e^{-\frac{i}{\hbar}\hat{H}(t-t_0)}|\Psi(t_0)\rangle \\
&= e^{\frac{i}{\hbar}\hat{H}_0 t}e^{-\frac{i}{\hbar}\hat{H}(t-t_0)}e^{-\frac{i}{\hbar}\hat{H}_0 t_0}|\Psi_I(t_0)\rangle
\end{aligned} \tag{4.35}
$$

Here $|\Psi_I(t_0)\rangle$ is the interaction-picture state vector at the time t_0. But now we can immediately identify the time development operator $\hat{U}(t, t_0)$ from the first of Eqs. (4.10)!

$$\hat{U}(t, t_0) = e^{\frac{i}{\hbar}\hat{H}_0 t}e^{-\frac{i}{\hbar}\hat{H}(t-t_0)}e^{-\frac{i}{\hbar}\hat{H}_0 t_0} \quad ; \ |t|, |t_0| \ll 1/\epsilon \tag{4.36}$$

It is only necessary to keep careful track of the ordering of the operators, and make sure that one never interchanges factors that do not commute.

Several of our previous properties of the time-development operator follow immediately from the expression in Eq. (4.36):

$$
\begin{aligned}
\hat{U}(t, t_0)^\dagger &= \hat{U}(t_0, t) \\
\hat{U}(t, t_0)^\dagger &= \hat{U}(t, t_0)^{-1} \quad ; \ \text{unitary} \\
\hat{U}(t_1, t_2)\hat{U}(t_2, t_3) &= \hat{U}(t_1, t_3) \quad ; \ \text{group property}
\end{aligned} \tag{4.37}
$$

4.5 The S-Matrix

The interaction-picture state vector in the infinite future $|\Psi_I(+\infty)\rangle$ that develops from the interaction-picture state vector in the infinite past

$|\Psi_I(-\infty)\rangle$ is obtained with the scattering operator in Eq. (4.17)

$$|\Psi_I(+\infty)\rangle = \hat{S}\,|\Psi_I(-\infty)\rangle \qquad ; \text{scattering operator} \qquad (4.38)$$

Now, with the adiabatic damping factor, the interaction state vectors in the infinite past and infinite future are simple, they are just the individual non-interacting state vectors in Eq. (4.9), or linear combinations of them. Thus, if one starts with one such prepared state $|\Psi_I^i(-\infty)\rangle = |\psi_i\rangle$, and asks for the probability for finding a particular state $|\psi_f\rangle$ in the final state $|\Psi_I^i(+\infty)\rangle$ that evolves, in the presence of all the interactions, from that initial prepared state, one has

$$P_{fi} = |\langle\psi_f|\Psi_I^i(+\infty)\rangle|^2 = |\langle\psi_f|\hat{S}\,|\Psi_I^i(-\infty)\rangle|^2 = |\langle\psi_f|\hat{S}|\psi_i\rangle|^2 \quad (4.39)$$

This is the probability of finding the initial state $|\psi_i\rangle$ in the final state $|\psi_f\rangle$ *after* the scattering has taken place. Here $|\psi_i\rangle$ and $|\psi_f\rangle$ are eigenstates of the free hamiltonian $\hat{H}_0$. The *amplitude* for this process to take place is given by the S-matrix

$$S_{fi} \equiv \langle\psi_f|\hat{S}|\psi_i\rangle \qquad ; S\text{-matrix} \qquad (4.40)$$

It was argued in Vol. I that the general form of the S-matrix for a scattering process is

$$S_{fi} = \delta_{fi} - 2\pi i\delta(E_f - E_i)\,\tilde{T}_{fi} \qquad (4.41)$$

where $\tilde{T}_{fi}$ is the T-matrix. There will always be an energy-conserving delta function here coming out of any calculation.[10]

The probability of making a *transition* to a state $f \neq i$ is therefore

$$P_{fi} = |2\pi i\delta(E_f - E_i)|^2|\tilde{T}_{fi}|^2 \qquad ; \text{probability of transition} \quad (4.42)$$

It was argued in Vol. I that the square of the energy-conserving δ-function is to be interpreted as

$$|2\pi i\delta(E_f - E_i)|^2 = 2\pi\delta(E_f - E_i)\frac{1}{\hbar}\int_{-T/2}^{T/2} dt\, e^{\frac{i}{\hbar}(E_f - E_i)t}$$

$$= \frac{2\pi}{\hbar}\delta(E_f - E_i)T \qquad ; T \to \infty \quad (4.43)$$

---

[10]Compare Eq. (4.53) and Prob. 4.8. In Vol. I we removed some additional factors in the definition of the T-matrix element T_{fi} [see EqI. (7.36) and Eq. (7.38)].

where $T \to \infty$ is the total time the interaction is turned on. The transition *rate* is then given by

$$\omega_{fi} = \frac{P_{fi}}{T}$$

$$\omega_{fi} = \frac{2\pi}{\hbar}\delta(E_f - E_i)|\tilde{T}_{fi}|^2 \qquad ; \text{ transition rate } (f \neq i) \quad (4.44)$$

This is the transition rate into *one* final state in the continuum. To get the transition rate into the *group* of states that actually get into our detectors when the states are spaced very close together, one must multiply this expression by the appropriate number of states dn_f. To get a *cross section*, one divides by the incident flux

$$d\sigma = \frac{2\pi}{\hbar}\delta(E_f - E_i)|\tilde{T}_{fi}|^2\frac{dn_f}{I_{\text{inc}}} \qquad ; \text{ cross section } \quad (4.45)$$

Some comments:

- All of these expressions were discussed and utilized frequently in Vol. I;
- Eq. (4.44) is the full expression for Fermi's Golden Rule, to all orders in the interaction;
- The derivation of the result for the transition rate involves some refinement when adiabatic switching is invoked, in contrast to the sudden turn-on and turn-off of the interaction in Vol. I; however, a proper derivation of the transition rate in this case, which we shall subsequently carry out, gives essentially the same result

$$S_{fi} = \delta_{fi} - 2\pi i\delta(E_f - E_i)\tilde{T}_{fi}$$

$$\omega_{fi} = \frac{2}{\hbar}\delta_{fi}\operatorname{Im}\tilde{T}_{ii} + \frac{2\pi}{\hbar}\delta(E_f - E_i)|\tilde{T}_{fi}|^2 \quad ; \text{ transition rate } \quad (4.46)$$

4.6 Time-Independent Analysis

We will now perform some formal manipulations on the above results. Let us try to *explicitly carry out the time integrations* in the general term in the S-matrix in Eq. (4.17), which we rewrite in its initial time-ordered form

$$\langle\psi_f|\hat{S}_\epsilon^{(n)}|\psi_i\rangle = \left(-\frac{i}{\hbar}\right)^n \int_{-\infty}^{\infty} e^{-\epsilon|t_1|}\, dt_1 \int_{-\infty}^{t_1} e^{-\epsilon|t_2|}\, dt_2 \cdots \int_{-\infty}^{t_{n-1}} e^{-\epsilon|t_n|}\, dt_n$$

$$\times\langle\psi_f|e^{\frac{i}{\hbar}\hat{H}_0 t_1}\hat{H}_1 e^{-\frac{i}{\hbar}\hat{H}_0 t_1}e^{\frac{i}{\hbar}\hat{H}_0 t_2}\hat{H}_1 e^{-\frac{i}{\hbar}\hat{H}_0 t_2}\cdots$$

$$\cdots\hat{H}_1 e^{-\frac{i}{\hbar}\hat{H}_0 t_{n-1}}e^{\frac{i}{\hbar}\hat{H}_0 t_n}\hat{H}_1 e^{-\frac{i}{\hbar}\hat{H}_0 t_n}|\psi_i\rangle \quad (4.47)$$

Here we have simply written out $\langle\psi_f|\hat{H}_I(t_1)\cdots\hat{H}_I(t_n)|\psi_i\rangle$ in detail.
We will change variables in the integrals as follows

$$
\begin{aligned}
x_1 &= t_1 & ; \quad t_1 &= x_1 \\
x_2 &= t_2 - t_1 & ; \quad t_2 &= x_1 + x_2 \\
x_3 &= t_3 - t_2 & ; \quad t_3 &= x_1 + x_2 + x_3 \\
&\;\;\vdots & &\;\;\vdots \\
x_n &= t_n - t_{n-1} & ; \quad t_n &= x_1 + x_2 + \cdots + x_n
\end{aligned}
\tag{4.48}
$$

First, let the hamiltonians $\hat{H}_0$ on either end of the operator in Eq. (4.47) act on $|\psi_i\rangle$ and $|\psi_f\rangle$, which are eigenstates of $\hat{H}_0$ with eigenvalues E_0 and E_f respectively. Equation (4.47) then can be written as

$$
\langle\psi_f|\hat{S}_\epsilon^{(n)}|\psi_i\rangle = \left(-\frac{i}{\hbar}\right)^n \int_{-\infty}^\infty e^{-\epsilon|t_1|}\,dt_1 \int_{-\infty}^{t_1} e^{-\epsilon|t_2|}\,dt_2 \cdots \int_{-\infty}^{t_{n-1}} e^{-\epsilon|t_n|}\,dt_n
$$
$$
\times\langle\psi_f|e^{\frac{i}{\hbar}(E_f-E_0)t_1}\hat{H}_1 e^{-\frac{i}{\hbar}\hat{H}_0(t_1-t_2)}e^{\frac{i}{\hbar}E_0(t_1-t_2)}\hat{H}_1 e^{-\frac{i}{\hbar}\hat{H}_0(t_2-t_3)}e^{\frac{i}{\hbar}E_0(t_2-t_3)}\cdots
$$
$$
\cdots e^{-\frac{i}{\hbar}\hat{H}_0(t_{n-1}-t_n)}e^{\frac{i}{\hbar}E_0(t_{n-1}-t_n)}\hat{H}_1|\psi_i\rangle
\tag{4.49}
$$

Next, introduce the change in variables in Eqs. (4.48), starting from the right

$$
\langle\psi_f|\hat{S}_\epsilon^{(n)}|\psi_i\rangle = \left(-\frac{i}{\hbar}\right)^n \int_{-\infty}^\infty e^{\frac{i}{\hbar}(E_f-E_0)x_1}e^{-\epsilon|x_1|}\,dx_1 \times
$$
$$
\langle\psi_f|\hat{H}_1\int_{-\infty}^0 dx_2\,e^{\{-\frac{i}{\hbar}(E_0-\hat{H}_0)x_2-\epsilon|x_1+x_2|\}}\hat{H}_1 \times
$$
$$
\int_{-\infty}^0 dx_3\,e^{\{-\frac{i}{\hbar}(E_0-\hat{H}_0)x_3-\epsilon|x_1+x_2+x_3|\}}\hat{H}_1 \times \cdots
$$
$$
\cdots\hat{H}_1\int_{-\infty}^0 dx_n\,e^{\{-\frac{i}{\hbar}(E_0-\hat{H}_0)x_n-\epsilon|x_1+\cdots+x_n|\}}\hat{H}_1|\psi_i\rangle
\tag{4.50}
$$

Now *do* all the integrals starting on the right, keeping all the other variables fixed while so doing.

Consider the first integral over dx_n at fixed $(x_1,\cdots,x_{n-1})$. What we really need is $\mathrm{Lim}_{\epsilon\to 0}\,\langle\psi_f|\hat{S}_\epsilon^{(n)}|\psi_i\rangle$. Since the damping factors are just there to cut off the oscillating exponentials, we should get the same results no matter how we go to that limit, if the theory is to make sense. We claim that *in the limit*, we can replace $e^{-\epsilon|x_1+\cdots+x_n|} \doteq e^{\epsilon x_n}$ in the integral over x_n, since it is only important for very large negative x_n. Repetition of this argument, as we do the integrals from right to left, allows us to replace

$$\text{Lim}_{\epsilon \to 0} \int \cdots \int e^{-\epsilon|x_1|} e^{-\epsilon|x_1 + x_2|} \cdots e^{-\epsilon|x_1 + \cdots x_n|} \cdots =$$

$$\text{Lim}_{\epsilon \to 0} \int \cdots \int e^{-\epsilon|x_1|} e^{\epsilon x_2} \cdots e^{\epsilon x_n} \cdots \quad (4.51)$$

The integrals now *factor*, and they can all be immediately carried out

$$\langle \psi_f | \hat{S}_\epsilon^{(n)} | \psi_i \rangle = \left(-\frac{i}{\hbar} \right)^n 2\pi\hbar \, \delta(E_f - E_0) \times$$

$$\langle \psi_f | \hat{H}_1 \frac{1}{-i(E_0 - \hat{H}_0)/\hbar + \epsilon} \hat{H}_1 \frac{1}{-i(E_0 - \hat{H}_0)/\hbar + \epsilon} \hat{H}_1 \cdots$$

$$\cdots \frac{1}{-i(E_0 - \hat{H}_0)/\hbar + \epsilon} \hat{H}_1 | \psi_i \rangle$$

$$(4.52)$$

The operator $\hat{H}_1$ appears n times in this expression. This equation has meaning in terms of a complete set of eigenstates of $\hat{H}_0$ inserted between each term. With the redefinition $\epsilon\hbar \equiv \varepsilon$, one arrives at the time-independent power series expansion of the S-matrix

$$\text{Lim}_{\varepsilon \to 0} \langle \psi_f | \hat{S}_\varepsilon | \psi_i \rangle = \langle \psi_f | \psi_i \rangle - \text{Lim}_{\varepsilon \to 0} 2\pi i \delta(E_f - E_0) \times$$

$$\langle \psi_f | \hat{H}_1 \sum_{n=0}^{\infty} \left(\frac{1}{E_0 - \hat{H}_0 + i\varepsilon} \hat{H}_1 \right)^n | \psi_i \rangle \quad (4.53)$$

Several comments:

- The $n = 0$ term is exactly Fermi's Golden Rule (see Vol. I);
- The $+i\varepsilon$ in the denominator, with the sign coming from the correct convergence factor in the integrals, just determines the correct *boundary conditions* to put in the Green's function (see later);
- We have proceeded to take the $\varepsilon \to 0$ limit in the final factor

$$\text{Lim}_{\epsilon \to 0} \int_{-\infty}^{\infty} dx_1 \, e^{\{\frac{i}{\hbar}(E_f - E_0)x_1 - \epsilon|x_1|\}} = 2\pi\hbar \, \delta(E_f - E_0) \quad (4.54)$$

- The T-matrix can now be identified from Eqs. (4.41) and (4.53)

$$\tilde{T}_{fi} \equiv \langle \psi_f | \hat{T} | \psi_i \rangle$$

$$\langle \psi_f | \hat{T} | \psi_i \rangle = \langle \psi_f | \hat{H}_1 \sum_{n=0}^{\infty} \left(\frac{1}{E_0 - \hat{H}_0 + i\varepsilon} \hat{H}_1 \right)^n | \psi_i \rangle \quad ; \text{ T-matrix} \quad (4.55)$$

- This last relation can be rewritten as

$$\langle \psi_f | \hat{T} | \psi_i \rangle = \langle \psi_f | \hat{H}_1 | \psi_i^{(+)} \rangle$$

$$|\psi_i^{(+)}\rangle \equiv \sum_{n=0}^{\infty} \left(\frac{1}{E_0 - \hat{H}_0 + i\varepsilon} \hat{H}_1 \right)^n |\psi_i\rangle \qquad (4.56)$$

We show below that this is indeed identical to the state $|\psi_i^{(+)}\rangle$ previously introduced in Eq. (4.33). If the first term is separated out in Eq. (4.56), and the series for $|\psi_i^{(+)}\rangle$ again identified in the second, this relation can be rewritten as

$$|\psi_i^{(+)}\rangle = |\psi_i\rangle + \frac{1}{E_0 - \hat{H}_0 + i\varepsilon} \hat{H}_1 |\psi_i^{(+)}\rangle \quad ; \text{ Lippmann-Schwinger} \quad (4.57)$$

In this form, when projected into the coordinate representation, one has an *integral equation* for $|\psi_i^{(+)}\rangle$. This is the *Lippmann-Schwinger equation* [Lippmann and Schwinger (1950)], which has a meaning that extends beyond the power series expansion though which it has been derived. Note that from Eq. (4.57), one observes

$$(E_0 - \hat{H}_0)|\psi_i^{(+)}\rangle = \hat{H}_1 |\psi_i^{(+)}\rangle$$

or $\qquad (E_0 - \hat{H})|\psi_i^{(+)}\rangle = 0 \qquad ; \Omega \to \infty$

$$\varepsilon \to 0 \qquad (4.58)$$

Thus the state $|\psi_i^{(+)}\rangle$, in the limits as the quantization volume $\Omega \to \infty$, and as the adiabatic damping factor $\varepsilon \to 0$, is a scattering state that is an *eigenstate of the full $\hat{H}$ with eigenvalue E_0.*[11] This is the same energy we started with at $t \to -\infty$ in the interaction picture.

- One therefore does not generate all of the eigenstates of $\hat{H}$ in this manner, if there are bound states, but only the continuum scattering states.[12]

- The terms with $n \geq 1$ in Eq. (4.56) give the *higher Born approximations* for the scattering amplitude. This is just "old-fashioned" perturbation theory, except that with the $+i\varepsilon$ in them, *we now know what to do when the denominators vanish.*

- People tried to do QED with this perturbation scheme; however, by singling out the time integration, the scattering amplitude is no longer

---

[11]Although the dependence on Ω is not explicit, we know, for example, that with a potential $V(r)$ in a big box with rigid walls there will be a finite shift in the energy levels as the interaction is turned on; this energy shift only vanishes in the limit $\Omega \to \infty$.

[12]The completeness relation is now $\sum_i |\psi_i^{(+)}\rangle\langle\psi_i^{(+)}| + \sum_{\text{bnd states}} |\psi_b\rangle\langle\psi_b| = \hat{1}$.

explicitly covariant. Infinities arise from various sources, which are not interpretable in a non-covariant approach. We will find that by leaving the time integrations in, and starting from Eq. (4.17), we are able to maintain a covariant, gauge-invariant S-matrix, which proves essential to developing a consistent renormalization scheme.[13]

4.7 Scattering State

The *Heisenberg picture* for the state vector is defined as follows

$$|\Psi_{\mathrm{H}}\rangle \equiv e^{\frac{i}{\hbar}\hat{H}t}|\Psi(t)\rangle \qquad ; \text{ Heisenberg picture} \qquad (4.59)$$

Correspondingly, an operator in the Heisenberg picture is defined by

$$\hat{O}_{\mathrm{H}} \equiv e^{\frac{i}{\hbar}\hat{H}t}\,\hat{O}\,e^{-\frac{i}{\hbar}\hat{H}t} \qquad ; \text{ Heisenberg picture} \qquad (4.60)$$

It follows from Eq. (4.32) that the Heisenberg state vector is independent of time[14]

$$i\hbar\frac{\partial}{\partial t}|\Psi_{\mathrm{H}}\rangle = 0 \qquad (4.61)$$

The interaction-picture state vector is defined in Eq. (4.2). The state vectors in all the different pictures *coincide* at $t = 0$

$$|\Psi_{\mathrm{H}}\rangle = |\Psi(0)\rangle = |\Psi_{\mathrm{I}}(0)\rangle \qquad (4.62)$$

This provides further motivation for looking at the scattering state $|\psi_i^{(+)}\rangle$ defined in Eq. (4.33) by

$$|\psi_i^{(+)}\rangle \equiv \hat{U}_\epsilon(0, -\infty)|\psi_i\rangle \qquad (4.63)$$

The *nth* order contribution to $\hat{U}_\varepsilon(0, -\infty)$ explicitly contains n powers of $\hat{H}_1$

$$\hat{U}_\epsilon^{(n)}(0, -\infty)|\psi_i\rangle = \left(-\frac{i}{\hbar}\right)^n \int_{-\infty}^0 e^{\epsilon t_1}\,dt_1 \int_{-\infty}^{t_1} e^{\epsilon t_2}\,dt_2 \cdots \int_{-\infty}^{t_{n-1}} e^{\epsilon t_n}\,dt_n \times$$
$$e^{\frac{i}{\hbar}\hat{H}_0 t_1}\hat{H}_1 e^{-\frac{i}{\hbar}\hat{H}_0(t_1-t_2)}\hat{H}_1 e^{-\frac{i}{\hbar}\hat{H}_0(t_2-t_3)}\ldots e^{-\frac{i}{\hbar}\hat{H}_0(t_{n-1}-t_n)}\hat{H}_1 e^{-\frac{i}{\hbar}\hat{H}_0 t_n}|\psi_i\rangle$$
$$(4.64)$$

---

[13]See the discussion in Vol. I.

[14]We remind the reader of the assumption, at this point, that $\hat{H}$ has no explicit time dependence.

In comparing with our starting point in Eq. (4.47) from which we proceeded to explicitly carrying out the time integrations, we note two differences:

- All the times satisfy $t \leq 0$, hence the adiabatic damping factors in all cases become $e^{-\epsilon|t|} = e^{\epsilon t}$;
- There is no eigenstate $|\psi_f\rangle$ on the left, and hence the operator $\hat{H}_0$ on the left can no longer be replaced by its eigenvalue E_f.

We may proceed to change variables as in Eqs. (4.48)–(4.50). This time, instead of $e^{-\epsilon|x_1+x_2+\cdots+x_n|}$, for example, we have $e^{\epsilon(x_1+x_2+\cdots+x_n)}$ so that all the adiabatic damping factors can simply be moved to their appropriate position in the multiple integral. Thus we arrive at

$$
\hat{U}_\epsilon^{(n)}(0,-\infty)|\psi_i\rangle = \left(-\frac{i}{\hbar}\right)^n \int_{-\infty}^0 dx_1\, e^{n\epsilon x_1} e^{\frac{i}{\hbar}(\hat{H}_0-E_0)x_1}\, \hat{H}_1 \times
$$

$$
\int_{-\infty}^0 dx_2\, e^{(n-1)\epsilon x_2} e^{\frac{i}{\hbar}(\hat{H}_0-E_0)x_2}\, \hat{H}_1 \int_{-\infty}^0 dx_3\, e^{(n-2)\epsilon x_3} e^{\frac{i}{\hbar}(\hat{H}_0-E_0)x_3}\, \hat{H}_1 \times \cdots
$$

$$
\cdots \hat{H}_1 \int_{-\infty}^0 dx_n\, e^{\epsilon x_n} e^{\frac{i}{\hbar}(\hat{H}_0-E_0)x_n}\, \hat{H}_1|\psi_i\rangle \tag{4.65}
$$

All the integrals now *explicitly factor*, and they can immediately be done just as before with the result

$$
\hat{U}_\epsilon^{(n)}(0,-\infty)|\psi_i\rangle = \frac{1}{E_0 - \hat{H}_0 + in\varepsilon}\hat{H}_1 \frac{1}{E_0 - \hat{H}_0 + i(n-1)\varepsilon}\hat{H}_1 \cdots
$$

$$
\cdots \frac{1}{E_0 - \hat{H}_0 + i\varepsilon}\hat{H}_1|\psi_i\rangle \tag{4.66}
$$

Again, we are interested in the limit as $\varepsilon \to 0$. Each of the $i\bar{n}\varepsilon$ in the denominators, where $\bar{n} = (1, 2, \cdots, n)$, simply serves to define how one treats the singularity in the individual Green's functions.[15] Hence, we can simply *replace them all by* $i\varepsilon$ *in the limit*. Thus we indeed reproduce the previously employed expression in Eq. (4.56)

$$
|\psi_i^{(+)}\rangle = \hat{U}_\varepsilon(0,-\infty)|\psi_i\rangle
$$

$$
= \sum_{n=0}^{\infty} \left(\frac{1}{E_0 - \hat{H}_0 + i\varepsilon}\hat{H}_1\right)^n |\psi_i\rangle \qquad ; \text{ scattering state} \tag{4.67}
$$

Again, by separating out the first term in the second line, and then re-

---

[15]They serve to define a contour in the evaluation of the Green's functions (see later).

identifying the series for $|\psi_i^{(+)}\rangle$, this can be rewritten as an integral equation

$$|\psi_i^{(+)}\rangle = |\psi_i\rangle + \frac{1}{E_0 - \hat{H}_0 + i\varepsilon}\hat{H}_1|\psi_i^{(+)}\rangle \qquad (4.68)$$

and the integral equation has a meaning, even when the power series solution to it does not.

Let us also consider the fully interacting state $|\psi_f^{(-)}\rangle \equiv |\Psi_I^f(0)\rangle$ that as $t \to +\infty$ reduces to the state $|\psi_f\rangle$, so that $|\psi_f^{(-)}\rangle = \hat{U}_\varepsilon(0, +\infty)|\psi_f\rangle$. If we go back to Eq. (4.19), and go through the arguments leading from Eq. (4.64) to (4.67), we see that the only changes are the replacements $E_0 \to E_f$ and $\varepsilon \to -\varepsilon$ (see Prob. 4.3). Thus

$$|\psi_f^{(-)}\rangle \equiv \hat{U}_\varepsilon(0, +\infty)|\psi_f\rangle$$
$$= \sum_{n=0}^{\infty}\left(\frac{1}{E_f - \hat{H}_0 - i\varepsilon}\hat{H}_1\right)^n |\psi_f\rangle \quad ; \text{ scattering state} \quad (4.69)$$

This can again be written as an integral equation, which has meaning even when the power series solution for it does not

$$|\psi_f^{(-)}\rangle = |\psi_f\rangle + \frac{1}{E_f - \hat{H}_0 - i\varepsilon}\hat{H}_1|\psi_f^{(-)}\rangle \qquad (4.70)$$

The state $|\psi^{(+)}\rangle$ is known as the *outgoing* scattering state, and $|\psi^{(-)}\rangle$ as the *incoming* scattering state.[16]

There are some important properties of these scattering states that follow immediately:

(1) The unitarity of the $\hat{U}_\varepsilon$ operator implies that

$$\langle\psi_{i'}^{(+)}|\psi_i^{(+)}\rangle = \langle\psi_{i'}|\hat{U}_\varepsilon(0, -\infty)^\dagger\hat{U}_\varepsilon(0, -\infty)|\psi_i\rangle = \langle\psi_{i'}|\psi_i\rangle = \delta_{i'i} \quad (4.71)$$

Similarly[17]

$$\langle\psi_{f'}^{(-)}|\psi_f^{(-)}\rangle = \delta_{f'f} \qquad (4.72)$$

---

[16]The Green's function in the former case has outgoing scattered waves, while in the latter case they are incoming [compare Eq. (4.107) and Prob. 4.9].

[17]The completeness relation can also be written $\sum_f |\psi_f^{(-)}\rangle\langle\psi_f^{(-)}| + \sum_{\text{bnd states}} |\psi_b\rangle\langle\psi_b| = \hat{1}$.

(2) Furthermore, from Eq. (4.21) and the group property of $\hat{U}_\varepsilon$, it follows that

$$
\begin{aligned}
\langle \psi_f^{(-)} | \psi_i^{(+)} \rangle &= \langle \psi_f | \hat{U}_\varepsilon(0,+\infty)^\dagger \hat{U}_\varepsilon(0,-\infty) | \psi_i \rangle \\
&= \langle \psi_f | \hat{U}_\varepsilon(+\infty,0)\hat{U}_\varepsilon(0,-\infty) | \psi_i \rangle \\
&= \langle \psi_f | \hat{U}_\varepsilon(+\infty,-\infty) | \psi_i \rangle
\end{aligned}
\tag{4.73}
$$

Thus the inner product of $|\psi_f^{(-)}\rangle$ and $|\psi_i^{(+)}\rangle$ is just the S-matrix!

$$
\langle \psi_f^{(-)} | \psi_i^{(+)} \rangle = \langle \psi_f | \hat{S} | \psi_i \rangle \qquad ; \; S\text{-matrix} \tag{4.74}
$$

(3) Since taking the adjoint merely reverses the order of the operators and changes the sign of the $i\varepsilon$, the T-matrix in Eq. (4.55) can also be written in the case $E_f = E_0$ as

$$
\begin{aligned}
\langle \psi_f | \hat{T} | \psi_i \rangle &= \langle \psi_f | \hat{H}_1 \sum_{n=0}^{\infty} \left(\frac{1}{E_0 - \hat{H}_0 + i\varepsilon} \hat{H}_1 \right)^n | \psi_i \rangle \\
&= \langle \psi_f | \left[\sum_{n=0}^{\infty} \left(\frac{1}{E_0 - \hat{H}_0 - i\varepsilon} \hat{H}_1 \right)^n \right]^\dagger \hat{H}_1 | \psi_i \rangle \\
&= \langle \psi_f^{(-)} | \hat{H}_1 | \psi_i \rangle \qquad ; \; E_f = E_0
\end{aligned}
\tag{4.75}
$$

(4) Thus, in *summary*, in addition to the explicit power-series expansions in Eqs. (4.53) and (4.55), we have expressions for the S-matrix and T-matrix in terms of the incoming and outgoing scattering states that are more general than the power-series solutions through which they were derived

$$
\begin{aligned}
\langle \psi_f | \hat{S} | \psi_i \rangle &= \langle \psi_f^{(-)} | \psi_i^{(+)} \rangle \qquad\qquad\qquad ; \; S\text{-matrix} \\
&= \langle \psi_f | \psi_i \rangle - 2\pi i \delta(E_f - E_0) \langle \psi_f | \hat{T} | \psi_i \rangle \\
\langle \psi_f | \hat{T} | \psi_i \rangle &= \langle \psi_f | \hat{H}_1 | \psi_i^{(+)} \rangle = \langle \psi_f^{(-)} | \hat{H}_1 | \psi_i \rangle \qquad ; \; T\text{-matrix} \quad (4.76)
\end{aligned}
$$

4.8   Transition Rate

We now calculate the transition rate directly, in the presence of the adiabatic switching. The derivation is from [Gell-Mann and Goldberger (1953)], in their classic paper on scattering theory. The only subtlety in the calculation is identifying those expressions that are well-defined in the limit $\varepsilon \to 0$, and knowing when to take that limit. This takes a little experience.

The Schrödinger state vector at the finite time t for a system that started as $|\psi_i\rangle$ at $t \to -\infty$ is

$$|\Psi_i(t)\rangle = e^{-\frac{i}{\hbar}\hat{H}t}|\Psi_i(0)\rangle = e^{-\frac{i}{\hbar}\hat{H}t}\hat{U}_\varepsilon(0, -\infty)|\psi_i\rangle = e^{-\frac{i}{\hbar}\hat{H}t}|\psi_i^{(+)}\rangle \quad (4.77)$$

The states that one *observes experimentally* in scattering, decays, *etc.* are the free-particle states

$$|\Phi_f(t)\rangle = e^{-\frac{i}{\hbar}E_f t}|\psi_f\rangle \quad (4.78)$$

From the general principles of quantum mechanics, the probability of finding the system in the state $|\Phi_f(t)\rangle$ at the time t, if it started in $|\psi_i\rangle$ at $t \to -\infty$, is then

$$P_{fi}(t) = |\langle \Phi_f(t)|\Psi_i(t)\rangle|^2 \equiv |M_{fi}(t)|^2 \quad (4.79)$$

This is the probability of having made a transition to the state $|\Phi_f(t)\rangle$ at the time t. The transition *rate* is the time derivative of this quantity

$$\omega_{fi} = \frac{d}{dt}P_{fi}(t) = M_{fi}^\star(t)\frac{d}{dt}M_{fi}(t) + \text{c.c.} \quad (4.80)$$

We will show that this transition rate is independent of time for times such that $|t| \ll 1/\varepsilon$. In the end, we will again let $\Omega \to \infty$, and $\varepsilon \to 0$, where Ω is the quantization volume. Let us proceed to calculate the transition rate.

From Eqs. (4.77)–(4.79) one has

$$\begin{aligned} M_{fi}(t) &= \langle \Phi_f(t)|\Psi_i(t)\rangle \\ &= \langle \psi_f|e^{\frac{i}{\hbar}E_f t}e^{-\frac{i}{\hbar}\hat{H}t}|\psi_i^{(+)}\rangle \end{aligned} \quad (4.81)$$

This relation may be differentiated with respect to time to give

$$\frac{d}{dt}M_{fi}(t) = -\frac{i}{\hbar}\langle \psi_f|(\hat{H} - E_f)e^{\frac{i}{\hbar}E_f t}e^{-\frac{i}{\hbar}\hat{H}t}|\psi_i^{(+)}\rangle \quad (4.82)$$

The observation that $(\hat{H} - E_f)|\psi_f\rangle = (\hat{H}_0 + \hat{H}_1 - E_f)|\psi_f\rangle = \hat{H}_1|\psi_f\rangle$ gives

$$\frac{d}{dt}M_{fi}(t) = -\frac{i}{\hbar}e^{\frac{i}{\hbar}E_f t}\langle \psi_f|\hat{H}_1\, e^{-\frac{i}{\hbar}\hat{H}t}|\psi_i^{(+)}\rangle \quad (4.83)$$

Now Eq. (4.58) states that in the above limit

$$(E_0 - \hat{H})|\psi_i^{(+)}\rangle = 0 \qquad ; \, \Omega \to \infty$$
$$\varepsilon \to 0 \quad (4.84)$$

Use of this relation in Eqs. (4.83) and (4.81) then gives

$$\frac{d}{dt}M_{fi}(t) = -\frac{i}{\hbar}e^{\frac{i}{\hbar}(E_f - E_0)t}\langle\psi_f|\hat{H}_1|\psi_i^{(+)}\rangle$$

$$M_{fi}(t) = e^{\frac{i}{\hbar}(E_f - E_0)t}\langle\psi_f|\psi_i^{(+)}\rangle \qquad (4.85)$$

Substitution of these relations into Eq. (4.80) then expresses the transition rate as

$$\omega_{fi} = \frac{2}{\hbar}\mathrm{Im}\,\langle\psi_f|\hat{H}_1|\psi_i^{(+)}\rangle\langle\psi_f|\psi_i^{(+)}\rangle^\star \qquad (4.86)$$

This expression now has the following properties:

- It is independent of time;
- It is *well-defined in the limit* $\Omega \to \infty$, $\varepsilon \to 0$.[18]

From our previous analysis in Eqs. (4.76) and (4.68), we have

$$\langle\psi_f|\hat{H}_1|\psi_i^{(+)}\rangle = \langle\psi_f|\hat{T}|\psi_i\rangle = \tilde{T}_{fi}$$

$$|\psi_i^{(+)}\rangle = |\psi_i\rangle + \frac{1}{E_0 - \hat{H}_0 + i\varepsilon}\hat{H}_1|\psi_i^{(+)}\rangle \qquad (4.87)$$

The inner product of the second relation with $|\psi_f\rangle$ gives

$$\langle\psi_f|\psi_i^{(+)}\rangle = \langle\psi_f|\psi_i\rangle + \frac{1}{E_0 - E_f + i\varepsilon}\tilde{T}_{fi} \qquad (4.88)$$

Substitution of this relation and the first of Eqs. (4.87) into Eq. (4.86) then gives

$$\omega_{fi} = \frac{2}{\hbar}\delta_{fi}\,\mathrm{Im}\,\tilde{T}_{ii} + \frac{2}{\hbar}\mathrm{Im}\,\frac{1}{E_0 - E_f - i\varepsilon}|\tilde{T}_{fi}|^2 \qquad (4.89)$$

Finally, we make use of the relation

$$\frac{1}{E_0 - E_f - i\varepsilon} = \mathcal{P}\frac{1}{E_0 - E_f} + i\pi\delta(E_0 - E_f) \qquad (4.90)$$

Here $\mathcal{P}$ denotes the Cauchy principal value, defined by deleting an infinitesimal symmetric region of integration through the singularity, and then letting the size of that region go to zero. Equation (4.90) is a statement on

---

[18]Here we will simply justify this observation *a posteriori*, through the many applications of the final expression. Note that by taking this limit too early in the derivation, one can arrive at spurious results [for example, try substituting the second of Eqs. (4.85) into Eq. (4.79)].

contour integration; it is derived in Prob. B.4. With the use of this relation, Eqs. (4.89) and (4.76) become

$$\omega_{fi} = \frac{2}{\hbar}\delta_{fi}\operatorname{Im}\tilde{T}_{ii} + \frac{2\pi}{\hbar}\delta(E_0 - E_f)|\tilde{T}_{fi}|^2 \quad ; \text{transition rate}$$

$$S_{fi} = \delta_{fi} - 2\pi i\delta(E_0 - E_f)\tilde{T}_{fi} \qquad ; S\text{-matrix} \qquad (4.91)$$

These expressions are exact. They are the results quoted in Eqs. (4.46) and used extensively in Vol. I.

4.9 Unitarity

The first term on the r.h.s. of ω_{fi} in Eq. (4.91) only contributes if $f = i$; it is there to take into account the depletion of the initial state. Return to Eq. (4.79). With the completeness of the states $|\Phi_f(t)\rangle$, and the normalization of the state $|\Psi_i(t)\rangle$, a sum over all final states gives[19]

$$\sum_f P_{fi}(t) = \sum_f \langle\Psi_i(t)|\Phi_f(t)\rangle\langle\Phi_f(t)|\Psi_i(t)\rangle$$

$$= \langle\Psi_i(t)|\Psi_i(t)\rangle = 1 \qquad (4.92)$$

This is the statement of conservation of probability—the initial state must end up *somewhere*. The time derivative of this sum then vanishes

$$\frac{d}{dt}\sum_f P_{fi}(t) = \sum_f \frac{d}{dt}P_{fi}(t) = \sum_f \omega_{fi} = 0 \qquad (4.93)$$

Here the transition rate has been identified from Eq. (4.80). A substitution of the expression for the transition rate in Eq. (4.91) into this relation then gives

$$-\frac{2}{\hbar}\operatorname{Im}\tilde{T}_{ii} = \sum_f \frac{2\pi}{\hbar}\delta(E_f - E_0)|\tilde{T}_{fi}|^2 \qquad ; \text{unitarity} \qquad (4.94)$$

This relation for the imaginary part of the elastic T-matrix, reflecting conservation of probability and depletion of the initial state, is known as *unitarity*.

---

[19]This sum now *includes* the state $f = i$; the reader should note that there is no sum over the repeated index i implied in Eqs. (4.91) and (4.94).

4.10   Example: Potential Scattering

To see one practical application of the preceding scattering theory, consider the elastic scattering of a non-relativistic particle of mass m from a spherically symmetric potential $\hat{H}_1 = V(|\hat{\mathbf{x}}|)$ in three dimensions. First we calculate the Green's function, or propagator.

4.10.1   *Green's Function (Propagator)*

The Green's function in this case is defined by the following matrix element taken between eigenstates of position

$$G_0(\mathbf{x} - \mathbf{y}) = \langle \mathbf{x} | \frac{1}{\hat{H}_0 - E_0 - i\varepsilon} | \mathbf{y} \rangle \qquad ; \text{ Green's function} \qquad (4.95)$$

Here

$$\hat{H}_0 = \frac{\hat{\mathbf{p}}^2}{2m} \qquad ; E_0 \equiv \frac{\hbar^2 \mathbf{k}^2}{2m} \qquad (4.96)$$

As usual, we start in a big cubical box of volume Ω where the eigenstates of momentum are plane waves satisfying periodic boundary conditions

$$\hat{\mathbf{p}} | \mathbf{t} \rangle = \hbar \mathbf{t} | \mathbf{t} \rangle$$

$$\langle \mathbf{x} | \mathbf{t} \rangle = \phi_{\mathbf{t}}(\mathbf{x}) = \frac{1}{\sqrt{\Omega}} e^{i\mathbf{t}\cdot\mathbf{x}} \qquad ; \text{ p.b.c.} \qquad (4.97)$$

The eigenstates of momentum satisfy the completeness relation

$$\sum_{\mathbf{t}} | \mathbf{t} \rangle \langle \mathbf{t} | = \hat{1} \qquad (4.98)$$

Insert this expression in Eq. (4.95), and use Eqs. (4.96) and (4.97)

$$G_0(\mathbf{x} - \mathbf{y}) = \frac{2m}{\hbar^2} \sum_{\mathbf{t}} \langle \mathbf{x} | \mathbf{t} \rangle \frac{1}{t^2 - k^2 - i\varepsilon} \langle \mathbf{t} | \mathbf{y} \rangle$$

$$= \frac{2m}{\hbar^2} \frac{1}{\Omega} \sum_{\mathbf{t}} e^{i\mathbf{t}\cdot(\mathbf{x}-\mathbf{y})} \frac{1}{t^2 - k^2 - i\varepsilon} \qquad (4.99)$$

We have redefined $(2m/\hbar^2)\varepsilon \to \varepsilon$ in this expression.

Now take the limit as the volume $\Omega \to \infty$, in which case the sum over states becomes an integral, in the familiar fashion, $\sum_{\mathbf{t}} \to \Omega(2\pi)^{-3} \int d^3t$. In this limit

$$G_0(\mathbf{x} - \mathbf{y}) = \frac{2m}{\hbar^2} \frac{1}{(2\pi)^3} \int d^3t \, e^{i\mathbf{t}\cdot(\mathbf{x}-\mathbf{y})} \frac{1}{t^2 - k^2 - i\varepsilon} \qquad (4.100)$$

It remains to do this integral. Take $\mathbf{r} \equiv \mathbf{x} - \mathbf{y}$ to define the z-axis. Then $\mathbf{t} \cdot (\mathbf{x} - \mathbf{y}) = tr \cos\theta$ and $d^3t = t^2 dt\, d\phi \sin\theta\, d\theta$. The angular integrations are then immediately performed

$$\int_0^{2\pi} d\phi \int_0^\pi \sin\theta\, d\theta\, e^{itr\cos\theta} = 2\pi \int_{-1}^1 dx\, e^{itrx} = 4\pi \frac{\sin tr}{tr} \qquad (4.101)$$

We are left with

$$G_0(\mathbf{x} - \mathbf{y}) = \frac{2m}{\hbar^2} \frac{4\pi}{(2\pi)^3} \frac{1}{r} \int_0^\infty t\, dt\, \sin tr \frac{1}{t^2 - k^2 - i\varepsilon} \qquad (4.102)$$

Now write the integral as

$$\int_0^\infty t\, dt\, \sin tr \cdots = \int_0^\infty t\, dt\, \frac{1}{2i}(e^{itr} - e^{-itr}) \cdots$$

$$= \frac{1}{2i} \int_{-\infty}^\infty t\, dt\, e^{itr} \cdots \qquad (4.103)$$

Here we have simple changed variables $t \to -t$ in the second term, and combined it with the first (the rest of the integrand is a function of t^2). The required integral is then reduced to

$$G_0(\mathbf{x} - \mathbf{y}) = \frac{2m}{\hbar^2} \frac{4\pi}{(2\pi)^3} \frac{1}{2ir} \int_{-\infty}^\infty t\, dt\, e^{itr} \frac{1}{t^2 - k^2 - i\varepsilon} \qquad ; \mathbf{r} \equiv \mathbf{x} - \mathbf{y} \quad (4.104)$$

where the integral now runs along the entire real t-axis. There is sufficient convergence in the integrand that closing the contour with a semi-circle in the upper-$1/2$ t-plane makes a vanishing contribution to the integral in the limit as the radius R of that semi-circle becomes infinite.[20] Thus the free Green's function has been reduced to a *contour integral* where the contour C is that illustrated in Fig. 4.1.

The integral is then evaluated using the complex-variable techniques summarized in appendix B. The integrand is an analytic function of t except at the poles where the denominator vanishes. That denominator can be rewritten as

$$\frac{1}{t^2 - k^2 - i\varepsilon} = \frac{1}{(t - k - i\varepsilon)(t + k + i\varepsilon)} \qquad (4.105)$$

where we have again redefined $\varepsilon \to 2k\varepsilon$ (here $k > 0$), and neglected $O(\varepsilon^2)$. The integrand thus has simple poles at $t = k + i\varepsilon$ and $t = -k - i\varepsilon$, only the first of which lies inside C.

---

[20]See Prob. 4.4.

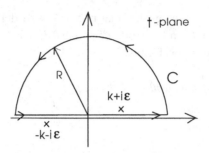

Fig. 4.1 Contour for the evaluation of the Green's function $G_0(\mathbf{x} - \mathbf{y})$ in the complex t-plane, together with the singularity structure arrived at with adiabatic damping. Here $R \to \infty$.

The integral is then given by $2\pi i \times$ (residue at k). Thus

$$G_0(\mathbf{x} - \mathbf{y}) = \frac{2m}{\hbar^2} \frac{4\pi}{(2\pi)^3} \frac{1}{2ir} 2\pi i \left(\frac{e^{ikr}}{2} \right) \tag{4.106}$$

Hence we arrive at our final result for the free Green's function in potential scattering

$$G_0(\mathbf{x} - \mathbf{y}) = \frac{2m}{\hbar^2} \frac{e^{ikr}}{4\pi r} \qquad ; \ \mathbf{r} \equiv \mathbf{x} - \mathbf{y} \tag{4.107}$$

This is recognized as the familiar Green's function for the scalar Helmholtz equation (see [Fetter and Walecka (2003)]).

4.10.2 *Scattering Wave Function*

The scattering state $|\psi_i^{(+)}\rangle$ can be similarly projected onto eigenstates of position. With the use of the completeness relation for these eigenstates, and the definition of the Green's function in Eq. (4.95), one has[21]

$$\langle \mathbf{x}|\psi_i^{(+)}\rangle = \langle \mathbf{x}|\psi_i\rangle - \int d^3 y \, \langle \mathbf{x}| \frac{1}{\hat{H}_0 - E_0 - i\varepsilon}|\mathbf{y}\rangle \, V(y) \, \langle \mathbf{y}|\psi_i^{(+)}\rangle \tag{4.108}$$

With the definition $\langle \mathbf{x}|\psi_i^{(+)}\rangle \equiv \psi_i^{(+)}(\mathbf{x})/\sqrt{\Omega}$, this becomes an integral equation for the scattering wave function

$$\langle \mathbf{x}|\psi_i^{(+)}\rangle \equiv \frac{1}{\sqrt{\Omega}} \psi_i^{(+)}(\mathbf{x})$$

$$\psi_i^{(+)}(\mathbf{x}) = e^{i\mathbf{k}\cdot\mathbf{x}} - \int d^3 y \, G_0(\mathbf{x} - \mathbf{y}) \, V(y) \, \psi_i^{(+)}(\mathbf{y}) \tag{4.109}$$

---

[21]We have used $V(|\hat{\mathbf{x}}|)\,|\mathbf{y}\rangle = V(y)\,|\mathbf{y}\rangle$ where $y \equiv |\mathbf{y}|$; note the sign of the second term.

4.10.3 *T-matrix*

The T-matrix can also be expressed in the coordinate representation as

$$\tilde{T}_{fi} = \int d^3y \, \langle \mathbf{k}_f | \mathbf{y} \rangle \, V(y) \, \langle \mathbf{y} | \psi_i^{(+)} \rangle \qquad (4.110)$$

where the final state is now written explicitly as an eigenstate of momentum $|\psi_f\rangle \equiv |\mathbf{k}_f\rangle$. With the introduction of the corresponding wave functions, one has

$$\tilde{T}_{fi} = \frac{1}{\Omega} \int d^3y \, e^{-i\mathbf{k}_f \cdot \mathbf{y}} \, V(y) \, \psi_i^{(+)}(\mathbf{y}) \qquad (4.111)$$

4.10.4 *Cross Section*

The differential cross section follows from the transition rate according to Eq. (4.45)

$$d\sigma = \frac{2\pi}{\hbar} \delta(E_f - E_0) |\tilde{T}_{fi}|^2 \frac{dn_f}{I_{\text{inc}}} \qquad (4.112)$$

In this expression:

(1) The incident wave function is $\psi_i(\mathbf{x}) = e^{i\mathbf{k}\cdot\mathbf{x}}/\sqrt{\Omega}$. This yields an incident probability flux of

$$I_{\text{inc}} = \frac{1}{\Omega} \frac{\hbar k}{m} \qquad (4.113)$$

(2) The number of final states in a big box with periodic boundary conditions is

$$dn_f = \frac{\Omega}{(2\pi)^3} d^3 k_f = \frac{\Omega}{(2\pi)^3} k_f^2 dk_f \, d\Omega_f \qquad (4.114)$$

(3) The integral over the energy-conserving delta function gives

$$\int \delta(E_f - E_0) k_f^2 dk_f = \frac{2m}{\hbar^2} \frac{k_f}{2} \qquad ; \ |\mathbf{k}_f| = |\mathbf{k}| \qquad (4.115)$$

(4) A combination of the results in Eqs. (4.112)–(4.115) gives

$$\frac{d\sigma}{d\Omega_f} = \frac{2\pi}{\hbar} \left[\frac{\Omega}{(2\pi)^3} \frac{mk}{\hbar^2} \right] \left[\frac{\Omega m}{\hbar k} \right] |\tilde{T}_{fi}|^2 \qquad (4.116)$$

The factors of Ω cancel, as they must, and the final result for the differential cross section for elastic scattering of a particle of energy

$E_0 = \hbar^2 k^2/2m$ from the potential $V(|\mathbf{x}|)$ takes the form

$$\frac{d\sigma}{d\Omega_f} = |f(k,\theta)|^2$$

$$f(k,\theta) \equiv -\frac{1}{4\pi}\frac{2m}{\hbar^2}\int d^3y\, e^{-i\mathbf{k}_f\cdot\mathbf{y}}\, V(y)\,\psi_i^{(+)}(\mathbf{y}) \qquad (4.117)$$

The minus sign is conventional.

(5) The scattering wave function $\psi_i^{(+)}(\mathbf{x})$ in this expression is the solution to the integral equation

$$\psi_i^{(+)}(\mathbf{x}) = e^{i\mathbf{k}\cdot\mathbf{x}} - \frac{2m}{\hbar^2}\int d^3y\, \frac{e^{ik|\mathbf{x}-\mathbf{y}|}}{4\pi|\mathbf{x}-\mathbf{y}|}\, V(y)\,\psi_i^{(+)}(\mathbf{y}) \qquad (4.118)$$

4.10.5 *Unitarity*

The scattering amplitude $f(k,\theta)$ and the T-matrix are related through Eqs. (4.117) and (4.111), and thus

$$-\frac{2}{\hbar}\text{Im}\,\tilde{T}_{ii} = \frac{4\pi}{\Omega}\frac{\hbar}{m}\text{Im}\,f(k,0) \qquad (4.119)$$

The unitarity relation in Eq. (4.94) states that

$$-\frac{2}{\hbar}\text{Im}\,\tilde{T}_{ii} = \sum_f \frac{2\pi}{\hbar}\delta(E_f - E_0)|\tilde{T}_{fi}|^2 \qquad (4.120)$$

Within a factor of the incident flux, the r.h.s. of this relation is just the total cross section σ_{tot}. Thus Eq. (4.120) can be rewritten as

$$-\frac{2}{\hbar}\text{Im}\,\tilde{T}_{ii} = I_{\text{inc}}\,\sigma_{\text{tot}} = \frac{1}{\Omega}\frac{\hbar k}{m}\sigma_{\text{tot}} \qquad (4.121)$$

A comparison of Eqs. (4.119) and (4.121) then leads to the *optical theorem* relating the imaginary part of the forward elastic scattering amplitude and the total cross section[22]

$$\text{Im}\,f(k,0) = \frac{k}{4\pi}\sigma_{\text{tot}} \qquad ; \text{optical theorem} \qquad (4.122)$$

The analysis of potential scattering in this section provides the underlying basis for the study of scattering in quantum mechanics, as presented, for example, in [Schiff (1968)].[23]

---

[22]So far, there is only elastic scattering in this potential model, but the optical theorem is more general and holds in the presence of additional inelastic processes.

[23]Problems 1.1–1.5 in [Walecka (2004)] take the reader through the essentials of the partial-wave analysis of the scattering problem.

Chapter 5

Lagrangian Field Theory

It was argued in Vol. I that a consistent relativistic quantum mechanics presents from the outset a many-body problem, involving the creation and destruction of particles. The appropriate framework for dealing with this problem is quantum field theory, as introduced in chapter 12 of Vol. I. In order to have a consistent *dynamical* framework for quantum field theory, we turn to the lagrangian formulation of continuum mechanics.[1] There are several reasons for this:

- One can easily pass from a lagrangian to a hamiltonian, through which the Schrödinger equation $i\hbar\partial/\partial t \, |\Psi(t)\rangle = \hat{H}|\Psi(t)\rangle$ is formulated;
- We know how to quantize with, for example, $[\hat{p}, \hat{q}] = \hbar/i$ in discrete mechanics, and the canonical momentum is obtained directly from the lagrangian;
- Lagrange's equations turn out to be manifestly covariant with a Lorentz-invariant lagrangian density;
- The stress tensor, which provides the energy and momentum of the system, is obtained directly from the lagrangian;
- Symmetry properties of the lagrangian immediately lead to conserved currents (Noether's theorem);
- This approach provides a framework for introducing new fields and new interactions.

So far, there is no other completely consistent mechanics framework for dealing with the quantum problem of many interacting relativistic particles.

To see how this all works, let us first review the results obtained in Vol. I for the classical lagrangian mechanics of a point particle of mass m.

---

[1] See, for example, [Fetter and Walecka (2003)].

5.1   Particle Mechanics

Classical lagrangian mechanics for a point particle of mass m was developed in Probls. 10.3–10.5. Here we summarize the results from those problems. Suppose one has a system with n degrees of freedom and n generalized coordinates q^i with $i = 1, 2, \cdots, n$. This can be *any* set of n linearly independent coordinates that completely specify the configuration of the system.[2] One introduces the *lagrangian* as follows

$$L(q, \dot{q}; t) \equiv T(q, \dot{q}; t) - V(q; t) \qquad \text{; lagrangian} \qquad (5.1)$$

where T is the kinetic energy, and V is the potential energy. We use the following shorthand for the coordinate dependence of the lagrangian

$$L(q, \dot{q}; t) \equiv L(q^1, \cdots, q^n, \dot{q}^1, \cdots, \dot{q}^n; t) \qquad \text{; shorthand} \qquad (5.2)$$

where $\dot{q}^i \equiv dq^i(t)/dt$, and the last t in the argument denotes a possible explicit dependence on the time.[3]

5.1.1   *Hamilton's Principle*

The *action* is defined by

$$S \equiv \int_{t_1}^{t_2} dt\, L(q, \dot{q}; t) \qquad \text{; action} \qquad (5.3)$$

Hamilton's principle states that the dynamical path the system takes is one that leaves the action stationary

$$\delta S = 0 \qquad \text{; Hamilton's principle}$$
$$\text{fixed endpoints} \qquad (5.4)$$

This expression has the following meaning:

(1) Let the coordinates undergo an infinitesimal variation from the actual path $q_0^i(t)$ to the path $q^i(t) = q_0^i(t) + \lambda \eta^i(t)$ where λ is an infinitesimal and $\eta^i(t)$ is arbitrary, except for the fact that $\eta^i(t_1) = \eta^i(t_2) = 0$;
(2) Substitute these expressions in L, and make a Taylor expansion in λ;

---

[2]We could equally well, for example, be discussing a collection of N such particles moving in three dimensions, in which case $n = 3N$.

[3]The systematic way to arrive at this lagrangian is to start in a cartesian basis, and then explicitly introduce the transformation to the generalized coordinates.

(3) Perform a partial integration on the time to produce a net change in the action of

$$S(\lambda) - S_0 = \lambda \int_{t_1}^{t_2} dt \sum_i \eta^i(t) \left[\frac{\partial L(q, \dot{q}; t)}{\partial q^i} - \frac{d}{dt} \frac{\partial L(q, \dot{q}; t)}{\partial \dot{q}^i} \right] + O(\lambda^2) \quad (5.5)$$

(4) If the action is stationary, it must be unchanged under any infinitesimal variation in the coordinates. Hence the derivative of the action with respect to λ must vanish at $\lambda = 0$

$$\left[\frac{dS}{d\lambda} \right]_{\lambda=0} = 0$$

$$= \int_{t_1}^{t_2} dt \sum_i \eta^i(t) \left[\frac{\partial L(q, \dot{q}; t)}{\partial q^i} - \frac{d}{dt} \frac{\partial L(q, \dot{q}; t)}{\partial \dot{q}^i} \right] \quad (5.6)$$

This is Hamilton's principle of stationary action.

5.1.2 *Lagrange's Equations*

If the expression in Eq. (5.6) is to vanish for arbitrary $\eta^i(t)$, then each coefficient in the integrand must vanish

$$\frac{d}{dt} \frac{\partial L(q, \dot{q}; t)}{\partial \dot{q}^i} - \frac{\partial L(q, \dot{q}; t)}{\partial q^i} = 0 \qquad ; \text{ Lagrange's equations}$$

$$i = 1, 2, \cdots, n \qquad (5.7)$$

These are *Lagrange's equations*. The partial derivatives in this expression imply that all the other variables in $L(q^1, \cdots, q^n, \dot{q}^1, \cdots, \dot{q}^n; t)$ are to be held fixed, while the time derivative in the first term is then a *total* derivative. It was shown in Probl. 10.5 that Hamilton's principle is fully equivalent to Newton's laws for a particle moving in a potential $V(\mathbf{x})$.

As an example, consider a particle moving in a radial potential $V(r)$ in two dimensions, and introduce polar coordinates. Then

$$x = r \cos \theta \qquad ; y = r \sin \theta$$
$$\dot{x}^2 + \dot{y}^2 = \dot{r}^2 + r^2 \dot{\theta}^2 \qquad (5.8)$$

Hence the lagrangian is

$$L(r, \theta, \dot{r}, \dot{\theta}; t) = \frac{1}{2} m(\dot{r}^2 + r^2 \dot{\theta}^2) - V(r) \qquad (5.9)$$

Here there is no dependence on θ, and no explicit dependence on t, so that in this case the functional form of the lagrangian simplifies to

$$L(r, \dot{r}, \dot{\theta}) = \frac{1}{2}m(\dot{r}^2 + r^2\dot{\theta}^2) - V(r) \tag{5.10}$$

Lagrange's equations are then

$$\frac{d}{dt}(m\dot{r}) = mr\dot{\theta}^2 - \frac{dV(r)}{dr} \qquad ; r \text{ eqn}$$

$$\frac{d}{dt}(mr^2\dot{\theta}) = 0 \qquad ; \theta \text{ eqn} \tag{5.11}$$

These are simply recognized as Newton's laws in polar coordinates.

5.1.3   Hamiltonian

The classical *canonical momentum* is defined by

$$p^i \equiv \frac{\partial L(q, \dot{q}; t)}{\partial \dot{q}^i} \qquad ; \text{ canonical momentum}$$

$$i = 1, 2, \cdots, n \tag{5.12}$$

Lagrange's equations can then be written as

$$\frac{dp^i}{dt} = \frac{\partial L}{\partial q^i} \qquad ; i = 1, 2, \cdots, n \tag{5.13}$$

We speak of the momentum p^i as being *conjugate* to the coordinate q^i, and a coordinate is *cyclic* if it does not appear in the lagrangian. The following theorem then follows immediately from Eq. (5.13)

- The momentum conjugate to a cyclic coordinate is a constant of the motion.

In the previous example, the angular momentum $p_\theta = mr^2\dot{\theta}$, conjugate to the cyclic polar angle θ, is a constant of the motion.

Equations (5.12) can, in principle, be inverted to give $\dot{q}^i = \dot{q}^i(p, q; t)$. The *hamiltonian* is then, in turn, defined by[4]

$$H(p, q; t) \equiv \sum_i p^i \dot{q}^i - L(q, \dot{q}; t) \qquad ; \text{ hamiltonian} \tag{5.14}$$

---

[4]This is known as a *Legendre transformation* to the new set of variables, a concept used frequently in thermodynamics.

Write out the total differential of this expression

$$dH(p,q;t) = \sum_i \left[\frac{\partial H}{\partial p^i} dp^i + \frac{\partial H}{\partial q^i} dq^i \right] + \frac{\partial H}{\partial t} dt$$

$$= \sum_i \left[\dot{q}^i dp^i + p^i d\dot{q}^i - \frac{\partial L}{\partial \dot{q}^i} d\dot{q}^i - \frac{\partial L}{\partial q^i} dq^i \right] - \frac{\partial L}{\partial t} dt \quad (5.15)$$

Here the partial derivatives in the first line imply that all the other variables in $H(p,q;t) \equiv H(p^1, \cdots, p^n, q^1, \cdots, q^n; t)$ are to be held fixed. The second and third terms in the second line cancel identically by Eqs. (5.12), and with the use of Eqs. (5.13), this relation becomes

$$dH = \sum_i \left[\frac{\partial H}{\partial p^i} dp^i + \frac{\partial H}{\partial q^i} dq^i \right] + \frac{\partial H}{\partial t} dt = \sum_i \left[\dot{q}^i dp^i - \dot{p}^i dq^i \right] - \frac{\partial L}{\partial t} dt \quad (5.16)$$

From this relation one learns two things:

- The hamiltonian is indeed a function of the indicated variables;
- The partial derivatives of the hamiltonian with respect to these variables are identified as

$$\frac{\partial H}{\partial p^i} = \frac{dq^i}{dt} \qquad ; \text{ Hamilton's equations}$$

$$\frac{\partial H}{\partial q^i} = -\frac{dp^i}{dt} \qquad i = 1, 2, \cdots, n$$

$$\frac{\partial H}{\partial t} = -\frac{\partial L}{\partial t} \qquad\qquad\qquad (5.17)$$

These are *Hamilton's equations.*[5]

Divide the first line in Eqs. (5.15) by dt to obtain the total time derivative of the hamiltonian. Substitution of Hamilton's equations then allows the result to be rewritten as follows

$$\frac{dH}{dt} = \sum_i \left[\frac{\partial H}{\partial p^i} \dot{p}^i + \frac{\partial H}{\partial q^i} \dot{q}^i \right] + \frac{\partial H}{\partial t}$$

$$= \sum_i \left[-\frac{\partial H}{\partial p^i} \frac{\partial H}{\partial q^i} + \frac{\partial H}{\partial q^i} \frac{\partial H}{\partial p^i} \right] + \frac{\partial H}{\partial t}$$

$$= \frac{\partial H}{\partial t} = -\frac{\partial L}{\partial t} \qquad\qquad (5.18)$$

---

[5]Remember that different sets of variables are held fixed in the partial derivatives in the last line.

The first of two important theorems regarding the hamiltonian follows immediately from this relation:

(1) If the lagrangian has no explicit dependence on the time, then the hamiltonian is a constant of the motion;
(2) A second theorem states that if the kinetic energy is a quadratic form in the generalized velocities so that $L = \sum_i \sum_j t_{ij}(q)\dot{q}^i\dot{q}^j - V(q)$, then the hamiltonian is the total energy.[6]

To see how this all works in the above example, the canonical momenta are calculated directly from Eq. (5.10) as

$$p_r = \frac{\partial L}{\partial \dot{r}} = m\dot{r} \qquad\qquad ; \; p_\theta = \frac{\partial L}{\partial \dot{\theta}} = mr^2\dot{\theta} \qquad (5.19)$$

The hamiltonian is then

$$H(p_r, p_\theta, r) = p_r\dot{r} + p_\theta\dot{\theta} - L = \frac{1}{2m}\left(p_r^2 + \frac{1}{r^2}p_\theta^2\right) + V(r) \qquad (5.20)$$

Hamilton's equations read

$$\frac{\partial H}{\partial p_r} = \frac{p_r}{m} = \dot{r} \qquad\qquad ; \; \frac{\partial H}{\partial p_\theta} = \frac{p_\theta}{mr^2} = \dot{\theta}$$

$$\frac{\partial H}{\partial r} = -\frac{p_\theta^2}{mr^3} + \frac{dV}{dr} = -\dot{p}_r \qquad\qquad ; \; \frac{\partial H}{\partial \theta} = 0 = -\dot{p}_\theta \qquad (5.21)$$

Since there is no explicit t-dependence in L, the hamiltonian is a constant of the motion, and since the kinetic energy is a quadratic form in $(\dot{r}, \dot{\theta})$, the hamiltonian is also the total energy

$$\frac{dH}{dt} = -\frac{\partial L}{\partial t} = 0$$

$$H = \frac{1}{2m}\left(p_r^2 + \frac{1}{r^2}p_\theta^2\right) + V(r) = E = \text{constant} \qquad (5.22)$$

With this review of classical *particle* mechanics, we turn to classical *continuum* mechanics, and review this topic by returning to the analysis in Vol. I of the transverse oscillations of the uniform continuous string.

---

[6]See Prob. 5.1.

5.2 Continuum Mechanics (String–a Review)

Recall the problem of a string of uniform mass density σ and tension τ stretched around a cylinder, and free to undergo transverse oscillations. Let $q(x,t)$ denote the transverse displacement of the string from equilibrium as illustrated in Fig. 5.1. The length of the string is now denoted by l.

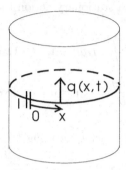

Fig. 5.1   String stretched around a cylinder free to oscillate without friction in a direction parallel to the axis of the cylinder. Illustration of periodic boundary conditions.

5.2.1 *Lagrangian Density*

The continuum mechanics of the string was analyzed in ProbI. 12.11, and we review that material here. The lagrangian for the string is[7]

$$L = T - V = \frac{\sigma}{2} \int_0^l dx \left\{ \left[\frac{\partial q(x,t)}{\partial t} \right]^2 - c^2 \left[\frac{\partial q(x,t)}{\partial x} \right]^2 \right\} \qquad (5.23)$$

This allows us to identify the *lagrangian density* $\mathcal{L}$ through[8]

$$L \equiv \int_0^l dx\, \mathcal{L}\left(\frac{\partial q}{\partial t}, \frac{\partial q}{\partial x}, q;\ t \right) \qquad \text{; lagrangian density } \mathcal{L}$$

$$\mathcal{L}\left(\frac{\partial q}{\partial t}, \frac{\partial q}{\partial x}, q;\ t \right) = \frac{\sigma}{2} \left\{ \left[\frac{\partial q(x,t)}{\partial t} \right]^2 - c^2 \left[\frac{\partial q(x,t)}{\partial x} \right]^2 \right\} \qquad (5.24)$$

Although we have made room for it in our notation, there is no *explicit* time dependence in this lagrangian density (and no explicit q dependence).

---

[7] We now use l for the length of the string so as not to confuse it with the lagrangian.
[8] Recall that here c is the sound velocity with $c^2 = \tau / \sigma$.

5.2.2 Hamilton's Principle

The *action* is the integral of the lagrangian over time

$$S = \int_{t_1}^{t_2} L \, dt = \int_{t_1}^{t_2} dt \int_0^l dx \, \mathcal{L}\left(\frac{\partial q}{\partial t}, \frac{\partial q}{\partial x}, q; \, t\right) \qquad ; \text{ action} \qquad (5.25)$$

Hamilton's principle of stationary action then takes the form

$$\delta S = \delta \int_{t_1}^{t_2} L \, dt = 0 \qquad ; \text{ Hamilton's principle}$$

$$\delta \int_{t_1}^{t_2} dt \int_0^l dx \, \mathcal{L}\left(\frac{\partial q}{\partial t}, \frac{\partial q}{\partial x}, q; \, t\right) = 0 \qquad \text{fixed endpoints in time} \quad (5.26)$$

This expression has the following meaning:[9]

(1) Let the coordinate undergo an infinitesimal variation from the actual path $q_0(x, t)$ to the path $q(x, t) = q_0(x, t) + \lambda \eta(x, t)$ where λ is an infinitesimal and $\eta(x, t)$ is arbitrary, except for the fact that it vanishes at the endpoints in time $\eta(x, t_1) = \eta(x, t_2) = 0$;

(2) Substitute these expressions in $\mathcal{L}$, and make a Taylor expansion in λ;

(3) Perform a partial integration on the *time* using the condition of fixed endpoints in time. Also perform a partial integration in *space* using the p.b.c. to eliminate the boundary contributions. These steps produce a net change in the action of

$$S(\lambda) - S_0 = \lambda \int_{t_1}^{t_2} dt \int_0^l dx \, \eta(x, t) \times$$
$$\left\{\frac{\partial \mathcal{L}}{\partial q} - \frac{\partial}{\partial t}\left[\frac{\partial \mathcal{L}}{\partial(\partial q/\partial t)}\right] - \frac{\partial}{\partial x}\left[\frac{\partial \mathcal{L}}{\partial(\partial q/\partial x)}\right]\right\} + O(\lambda^2) \quad (5.27)$$

Here the partial derivatives of $\mathcal{L}$ keep all the other variables in $\mathcal{L}$ fixed, while the final partial derivatives with respect to (x, t) keep the other variable in this pair fixed.

(4) If the action is stationary, it must be unchanged under any infinitesimal variation in the coordinates. Hence the derivative of the action with

---

[9]In going over from discrete to continuum mechanics we have effectively replaced the coordinates $q^i(t) \to q(x, t)$.

respect to λ must vanish at $\lambda = 0$

$$\left[\frac{dS}{d\lambda}\right]_{\lambda=0} = 0 \qquad (5.28)$$

$$= \int_{t_1}^{t_2} dt \int_0^l dx\, \eta(x,t) \left\{ \frac{\partial \mathcal{L}}{\partial q} - \frac{\partial}{\partial t}\left[\frac{\partial \mathcal{L}}{\partial(\partial q/\partial t)}\right] - \frac{\partial}{\partial x}\left[\frac{\partial \mathcal{L}}{\partial(\partial q/\partial x)}\right] \right\}$$

This is Hamilton's principle of stationary action for the continuum problem.

5.2.3 *Lagrange's Equation*

If Eq. (5.28) is to be satisfied for arbitrary $\eta(x,t)$, then its coefficient in the integrand must vanish

$$\frac{\partial}{\partial t}\left[\frac{\partial \mathcal{L}}{\partial(\partial q/\partial t)}\right] + \frac{\partial}{\partial x}\left[\frac{\partial \mathcal{L}}{\partial(\partial q/\partial x)}\right] - \frac{\partial \mathcal{L}}{\partial q} = 0 \;\; ; \text{ Lagrange's equation} \quad (5.29)$$

This is *Lagrange's equation* for the continuum problem. We remind the reader that the partial derivatives of $\mathcal{L}$ keep all the other variables in $\mathcal{L}(\partial q/\partial t, \partial q/\partial x, q; t)$ fixed, while the final partial derivatives with respect to (x,t) keep the other variable in this pair fixed. Lagrange's equation in the continuum case produces a partial differential equation for $q(x,t)$.

5.2.4 *Two-Vectors*

With the introduction of the "two-vector" $x_\mu = (x_1, x_2) \equiv (x, ict)$, and the convention that repeated Greek indices are summed from 1 to 2, the lagrangian density can be rewritten as

$$\mathcal{L} = -\frac{\tau}{2}\left(\frac{\partial q}{\partial x_\mu}\right)^2 \qquad ; \text{ Lagrangian density} \quad (5.30)$$

Lagrange's equation can be similarly rewritten in "invariant form" as

$$\frac{\partial}{\partial x_\mu}\left[\frac{\partial \mathcal{L}}{\partial(\partial q/\partial x_\mu)}\right] - \frac{\partial \mathcal{L}}{\partial q} = 0 \qquad ; \text{ Lagrange's equation} \quad (5.31)$$

It is readily shown that the wave equation for the string is now reproduced

$$\frac{\partial^2 q(x,t)}{\partial x^2} = \frac{1}{c^2}\frac{\partial^2 q(x,t)}{\partial t^2} \qquad ; \text{ wave equation} \quad (5.32)$$

5.2.5   *Momentum Density*

The *canonical momentum density* is defined in general by

$$\pi(x,t) \equiv \frac{\partial \mathcal{L}}{\partial(\partial q/\partial t)} \qquad ; \text{ momentum density} \qquad (5.33)$$

It follows from Eq. (5.24) that for the string

$$\pi(x,t) = \sigma \frac{\partial q(x,t)}{\partial t} \qquad ; \text{ string} \qquad (5.34)$$

Here $\pi(x,t)$ is just the *kinetic* momentum density due to the transverse motion of the string.

5.2.6   *Hamiltonian Density*

The *hamiltonian density* in continuum mechanics is defined in general by

$$\mathcal{H} \equiv \pi(x,t)\frac{\partial q(x,t)}{\partial t} - \mathcal{L} \qquad ; \text{ hamiltonian density} \qquad (5.35)$$

It follows from Eqs. (5.34) and (5.24) that the hamiltonian density for the string is simply the energy density

$$\mathcal{H} = \frac{\sigma}{2}\left[\frac{\partial q(x,t)}{\partial t}\right]^2 + \frac{\tau}{2}\left[\frac{\partial q(x,t)}{\partial x}\right]^2 = \mathcal{T} + \mathcal{V} = \mathcal{E} \qquad ; \text{ string} \qquad (5.36)$$

5.3   Quantization

We proceed to a discussion of the *quantization* of these classical mechanical systems, and we start with particle mechanics.

5.3.1   *Particle Mechanics*

In quantum mechanics, the observables become linear operators in an abstract Hilbert space that obey specified commutation relations. The commutation relations to be imposed on the previous generalized coordinates and their conjugate momenta are as follows

$$[\hat{p}_i, \hat{q}_j] = \frac{\hbar}{i}\delta_{ij} \qquad ; \text{ canonical commutation relations}$$

$$[\hat{p}_i, \hat{p}_j] = [\hat{q}_i, \hat{q}_j] = 0 \qquad (i,j) = 1,2,\cdots,n \qquad (5.37)$$

Some comments:

- To represent observables, these operators must be *hermitian*;
- These canonical commutation relations are ultimately one of the underlying assumptions of quantum mechanics;
- The *justification* of these relations is that, together with Ehrenfest's theorem and Hamilton's equations, they lead to the *correspondence principle* whereby quantum mechanics reproduces classical mechanics in the appropriate limit. This was demonstrated for a particle in a potential, while working in a cartesian basis, in Probl. 4.9. The result holds quite generally (see Probs. 5.2–5.3);
- Quantum mechanics is here formulated in the *Schrödinger picture* where the operators are time-independent, and all the time dependence is put into the state vector $|\Psi(t)\rangle$.

To see how this works, let us return to our example of a particle moving in two dimensions in a radial potential $V(r)$. The hamiltonian operator is now given by Eq. (5.20) as

$$\hat{H} = \frac{1}{2m}\left(\hat{p}_r^2 + \frac{1}{\hat{r}^2}\hat{p}_\theta^2\right) + V(\hat{r}) \tag{5.38}$$

and the commutation relations to be imposed are

$$[\hat{p}_r, \hat{r}] = [\hat{p}_\theta, \hat{\theta}] = \frac{\hbar}{i}$$
$$[\hat{p}_r, \hat{\theta}] = [\hat{p}_\theta, \hat{r}] = [\hat{p}_r, \hat{p}_\theta] = [\hat{r}, \hat{\theta}] = 0 \tag{5.39}$$

Let us again project onto a basis of eigenstates of position and work with wavefunctions and differential operators in the coordinate representation.[10] We seek a differential representation of the commutation relations. For p_θ the situation is straightforward. Take

$$p_\theta = \frac{\hbar}{i}\frac{\partial}{\partial\theta} \tag{5.40}$$

This reproduces all the results on the polar angle analyzed in Vol. I.

For p_r the situation is a little more complicated. The two-dimensional volume element in polar coordinates is

$$d^2x = rdrd\theta \equiv \rho(r)drd\theta \tag{5.41}$$

There is a radial "measure" $\rho(r) = r$. As we saw in Probls. 4.4–4.6, the establishment of the *hermiticity* of the operators involves partial integrations, and one must take into account the presence of this measure. Since

---

[10]We drop the hats on the operators in the coordinate representation.

it is p_r^2 that appears in H, which certainly must be hermitian, let us focus on this quantity. One has the general relations

$$[p_r^2, r] \equiv p_r[p_r, r] + [p_r, r]p_r = 2\frac{\hbar}{i}p_r$$

$$[[p_r^2, r], r] = -2\hbar^2 \qquad (5.42)$$

We will use the commutation relation in the second line to determine a satisfactory p_r^2, and then use the relation in the first line to determine p_r. The operator p_r^2 that does the job for us is

$$p_r^2 = -\hbar^2 \frac{1}{\rho(r)} \frac{\partial}{\partial r} \rho(r) \frac{\partial}{\partial r} \qquad (5.43)$$

This operator has the following properties (see Prob. 5.4):[11]

(1) It is hermitian with respect to the volume element in Eq. (5.41);
(2) It satisfies the commutation relation in the second line of Eq. (5.42).

The hamiltonian in differential form in polar coordinates then becomes

$$H = \frac{-\hbar^2}{2m}\left(\frac{1}{r}\frac{\partial}{\partial r}r\frac{\partial}{\partial r} + \frac{1}{r^2}\frac{\partial^2}{\partial\theta^2}\right) + V(r) \qquad (5.44)$$

As demonstrated in Probl. 4.26, this is exactly what one gets by starting in a cartesian basis and then transforming the laplacian to polar coordinates.

5.3.2 *Continuum Mechanics (String)*

The string serves as our archetypical continuum system, and the quantization of the string was discussed in Vol. I. We review that material here. In the *Schrödinger picture* where the operators are independent of time, and all the time dependence is put into the state vector $|\Psi(t)\rangle$, the canonical commutation relations for the string fields are

$$[\hat{q}(x), \hat{\pi}(x')] = i\hbar\delta(x - x') \qquad \text{; canonical commutation rules}$$

$$[\hat{q}(x), \hat{q}(x')] = [\hat{\pi}(x), \hat{\pi}(x')] = 0 \qquad (5.45)$$

---

[11]The implied p_r is determined in Prob. 5.5. The expression in Eq. (5.43) is now of the same form as the differential operator that appears in the Sturm-Liouville equation, whose properties are analyzed in [Fetter and Walecka (2003)].

The quantum fields satisfying these commutation relations can be given a representation as follows

$$\hat{q}(x) = \sum_k \left(\frac{\hbar}{2\omega_k \sigma l}\right)^{1/2} \left[a_k e^{ikx} + a_k^\dagger e^{-ikx}\right] \quad ; \text{Schrödinger picture}$$

$$\hat{\pi}(x) = \frac{1}{i} \sum_k \left(\frac{\hbar\omega_k \sigma}{2l}\right)^{1/2} \left[a_k e^{ikx} - a_k^\dagger e^{-ikx}\right] \quad (5.46)$$

Here the normal-mode wave functions, wave vectors, and angular frequencies are given by

$$\phi_k(x) = \frac{1}{\sqrt{l}} e^{ikx} \quad ; k = \frac{2\pi m}{l} \quad ; m = 0, \pm 1, \pm 2, \cdots \quad ;$$

$$\omega_k = |k|c \quad (5.47)$$

The creation and destruction operators satisfy the familiar harmonic oscillator commutation relations

$$[a_k, a_{k'}^\dagger] = \delta_{kk'}$$

$$[a_k^\dagger, a_{k'}^\dagger] = [a_k, a_{k'}] = 0 \quad (5.48)$$

The free-string hamiltonian $\hat{H}_0$ in the Schrödinger picture is obtained from Eq. (5.36), making use of Eq. (5.34),

$$\hat{H}_0 = \int_0^l dx\, \hat{\mathcal{H}} = \int_0^l dx \left\{\frac{1}{2\sigma}\hat{\pi}^2(x) + \frac{\sigma c^2}{2}\left[\frac{\partial\hat{q}(x)}{\partial x}\right]^2\right\} \quad (5.49)$$

If the quantum field representations in Eqs. (5.46) are substituted into this expression, one simply obtains the uncoupled contribution of each of the oscillator normal modes

$$\hat{H}_0 = \sum_k \hbar\omega_k(a_k^\dagger a_k + 1/2) \quad (5.50)$$

The transformation to the *interaction picture* in Eq. (4.4) is readily carried out using the following relations for the creation and destruction operators, obtained using the operator relation in Eq. (3.9) and the commutation relations in Eq. (5.48),[12]

$$e^{\frac{i}{\hbar}\hat{H}_0 t} a_k e^{-\frac{i}{\hbar}\hat{H}_0 t} = a_k e^{-i\omega_k t}$$

$$e^{\frac{i}{\hbar}\hat{H}_0 t} a_k^\dagger e^{-\frac{i}{\hbar}\hat{H}_0 t} = a_k^\dagger e^{i\omega_k t} \quad (5.51)$$

---

[12]See Prob. 5.6.

Hence, in the interaction picture,

$$\hat{q}(x,t) = \sum_k \left(\frac{\hbar}{2\omega_k \sigma l}\right)^{1/2} \left[a_k e^{i(kx - \omega_k t)} + a_k^\dagger e^{-i(kx - \omega_k t)}\right]$$

$$\hat{\pi}(x,t) = \frac{1}{i}\sum_k \left(\frac{\hbar\omega_k \sigma}{2l}\right)^{1/2} \left[a_k e^{i(kx - \omega_k t)} - a_k^\dagger e^{-i(kx - \omega_k t)}\right] \quad ;$$

interaction picture   (5.52)

These fields satisfy the canonical *equal-time* commutation relations[13]

$$[\hat{q}(x,t),\ \hat{\pi}(x',t')]_{t=t'} = i\hbar\delta(x - x')$$
$$[\hat{q}(x,t),\ \hat{q}(x',t')]_{t=t'} = [\hat{\pi}(x,t),\ \hat{\pi}(x',t')]_{t=t'} = 0 \quad ;$$

canonical commutation relations (5.53)

The fields in the interaction picture carry the free-field time dependence.

If the time-dependent fields in Eqs. (5.52) are substituted in the hamiltonian $\hat{H}_0$ obtained from Eqs. (5.36) and (5.34)

$$\hat{H}_0 = \int_0^l dx\,\hat{\mathcal{H}} = \int_0^l dx\,\left\{\frac{1}{2\sigma}\hat{\pi}^2(x,t) + \frac{\sigma c^2}{2}\left[\frac{\partial\hat{q}(x,t)}{\partial x}\right]^2\right\} \quad (5.54)$$

one obtains exactly the result in Eq. (5.50)

$$\hat{H}_0 = \sum_k \hbar\omega_k(a_k^\dagger a_k + 1/2) \quad (5.55)$$

This is how Eq. (5.55) was derived in Vol. I.

For the free fields, since $\hat{H} \equiv \hat{H}_0$ in this case, the results in the *Heisenberg picture* [see Eq. (4.60)] are identical to those in the interaction picture.

5.4   Relativistic Field Theory

To discuss physical fields in four-dimensional space-time where $x_\mu = (x_1, x_2, x_3, x_4) \equiv (\mathbf{x}, ict)$, with c now the speed of light, one can simply extend the string results by adding two more spatial dimensions. We are now working in the four-dimensional Minkowski space of special relativity.

---

[13]These equal-time commutation relations are independent of picture–see Prob. 5.23.

5.4.1 *Scalar Field*

For example, suppose $\phi(\mathbf{x}, t)$ is a massless Lorentz scalar field that at a given point in space-time is the same as viewed in any Lorentz frame.[14] The lagrangian density is then immediately obtained from Eq. (5.30) as

$$\mathcal{L}\left(\frac{\partial\phi}{\partial x_\mu}\right) = -\frac{c^2}{2}\left(\frac{\partial\phi}{\partial x_\mu}\right)^2 \qquad ; \text{ scalar field}$$

$$= \frac{1}{2}\left(\frac{\partial\phi}{\partial t}\right)^2 - \frac{c^2}{2}\left(\boldsymbol{\nabla}\phi\right)^2 \qquad (5.56)$$

Here we have written $\tau = \sigma c^2$, and then replaced $\sqrt{\sigma}\, q(x, t) \to \phi(\mathbf{x}, t)$.[15] We also here and henceforth re-identify c with the speed of light.

The action is then given by

$$S = \int_{t_1}^{t_2} dt \int_\Omega d^3x\, \mathcal{L}\left(\frac{\partial\phi}{\partial t}, \boldsymbol{\nabla}\phi\right) \qquad ; \text{ action}$$

$$= \frac{1}{c}\int_{t_1}^{t_2}\int_\Omega d^4x\, \mathcal{L}\left(\frac{\partial\phi}{\partial x_\mu}\right) \qquad ; \, d^4x \equiv d^3x(cdt) \quad (5.57)$$

Here $\Omega = l^3$ is the quantization volume, and we note the definition $d^4x \equiv d^3x(cdt)$. Now Hamilton's principle works just as before. The resulting Lagrange's equation is just that of Eq. (5.31), now in four-dimensional Minkowski space

$$\frac{\partial}{\partial x_\mu}\left[\frac{\partial\mathcal{L}}{\partial(\partial\phi/\partial x_\mu)}\right] - \frac{\partial\mathcal{L}}{\partial\phi} = 0 \qquad ; \text{ Lagrange's equation} \quad (5.58)$$

Here we assume a functional form $\mathcal{L}(\partial\phi/\partial x_\mu, \phi\,; t)$, and all the other variables are held fixed in computing the partial derivatives of $\mathcal{L}$, while the other members of the set $x_\mu = (\mathbf{x}, ict)$ are held fixed in computing the final partial derivatives with respect to the components of x_μ.

The resulting Lagrange's equation for the scalar field is the extension of the wave Eq. (5.32) to three spatial dimensions

$$\Box\,\phi(\mathbf{x}, t) = \boldsymbol{\nabla}^2\phi(\mathbf{x}, t) - \frac{1}{c^2}\frac{\partial^2\phi(\mathbf{x}, t)}{\partial t^2} = 0 \qquad (5.59)$$

---

[14]That is, $\phi(x) = \phi(x')$ where x and x' refer to the same space-time point. The pion is described by a (essentially massless) Lorentz *pseudo*scalar field.

[15]We will later also replace $\pi(x, t)/\sqrt{\sigma} \to \pi(\mathbf{x}, t)$. Repeated Greek indices are now summed from 1 to 4.

The quantum fields in the interaction picture follow as in Eqs. (5.52)

$$\hat{\phi}(\mathbf{x}, t) = \sum_{\mathbf{k}} \left(\frac{\hbar}{2\omega_k \Omega}\right)^{1/2} \left[a_{\mathbf{k}} e^{i(\mathbf{k}\cdot\mathbf{x} - \omega_k t)} + a_{\mathbf{k}}^{\dagger} e^{-i(\mathbf{k}\cdot\mathbf{x} - \omega_k t)}\right] \quad ; \; \omega_k \equiv |\mathbf{k}|c$$

$$\hat{\pi}(\mathbf{x}, t) = \frac{1}{i} \sum_{\mathbf{k}} \left(\frac{\hbar\omega_k}{2\Omega}\right)^{1/2} \left[a_{\mathbf{k}} e^{i(\mathbf{k}\cdot\mathbf{x} - \omega_k t)} - a_{\mathbf{k}}^{\dagger} e^{-i(\mathbf{k}\cdot\mathbf{x} - \omega_k t)}\right] \qquad (5.60)$$

Here $\omega_k \equiv |\mathbf{k}|c$. These fields satisfy the equal-time commutation relations

$$[\hat{\phi}(\mathbf{x}, t), \hat{\pi}(\mathbf{x}', t')]_{t=t'} = i\hbar\delta^{(3)}(\mathbf{x} - \mathbf{x}') \qquad\qquad ; \text{ C.C.R.}$$

$$[\hat{\phi}(\mathbf{x}, t), \hat{\phi}(\mathbf{x}', t')]_{t=t'} = [\hat{\pi}(\mathbf{x}, t), \hat{\pi}(\mathbf{x}', t')]_{t=t'} = 0 \qquad\qquad (5.61)$$

The hamiltonian is obtained exactly as in Eq. (5.55)

$$\hat{H}_0 = \sum_{\mathbf{k}} \hbar\omega_k (a_{\mathbf{k}}^{\dagger} a_{\mathbf{k}} + 1/2) \qquad ; \; k_i = \frac{2\pi p_i}{l} \qquad ; \; p_i = 0, \pm 1, \pm 2, \cdots \qquad ;$$

$$i = x, y, z \qquad\qquad (5.62)$$

Several comments:

(1) If ϕ is a Lorentz scalar, then the lagrangian density in the first of Eqs. (5.56) is *also* a Lorentz scalar;
(2) Since d^4x is unchanged under a Lorentz transformation (see Vol. I), the integrand in the action is also Lorentz invariant;[16]
(3) Lagrange's Eq. (5.58) is also invariant if ϕ is a Lorentz scalar;
(4) The theory is readily quantized by invoking the canonical commutation relations in Eqs. (5.61);
(5) As discussed in Vol. I and chapter 2, the eigenstates of the hamiltonian in Eq. (5.62) provide a direct-product basis within which one can describe all the effects of the quantum field;
(6) We will assume items (1)–(5) to be general properties of any theory;
(7) Lagrangian quantum field theory in Minkowski space then provides a marvelous framework for dealing with the dynamics of any interacting, relativistic, quantum many-body system.

As a simple example, the invariant lagrangian density of Eq. (5.56) can be augmented to describe a *massive* scalar field by just including a term $V_{\text{mass}}(\phi) = m_s^2 c^2 \phi^2/2$ where $m_s \equiv m_0 c/\hbar$ is the inverse Compton

---

[16]The boundaries of the integration region for the action itself can get tilted under a Lorentz transformation — we shall return to this.

wavelength[17]

$$\mathcal{L}\left(\frac{\partial\phi}{\partial x_\mu}, \phi\right) = -\frac{c^2}{2}\left[\left(\frac{\partial\phi}{\partial x_\mu}\right)^2 + m_s^2\phi^2\right] \quad ; \text{ massive scalar field} \quad (5.63)$$

The modifications of the previous analysis are as follows (see Prob. 5.8):

(1) The wave equation is changed to the Klein-Gordon equation

$$(\Box - m_s^2)\phi(\mathbf{x}, t) = 0 \qquad ; \text{ Klein-Gordon equation} \qquad (5.64)$$

(2) The angular frequency appearing in the quantum fields and hamiltonian is changed to

$$\omega_k \rightarrow c\sqrt{\mathbf{k}^2 + m_s^2} \qquad (5.65)$$

5.4.2 *Stress Tensor*

In classical continuum mechanics there is a stress tensor, or energy-momentum tensor, that provides the energy density and momentum density of the system.[18] If there are no external forces, then one has *energy conservation*, which can be written as a continuity equation for the energy density, assumed here to be identical to the hamiltonian density,

$$\frac{\partial \mathcal{H}}{\partial t} + \boldsymbol{\nabla} \cdot \boldsymbol{\mathcal{S}} = 0 \qquad ; \text{ energy conservation} \qquad (5.66)$$

Here $\boldsymbol{\mathcal{S}}$ denotes the energy flux. There will also be *momentum conservation* in this case, which can also be written as a conservation equation, this time for a vector quantity

$$\frac{\partial \mathcal{P}_k}{\partial t} + \frac{\partial}{\partial x_j}T_{jk} = 0 \qquad ; \text{ momentum conservation}$$

$$(k, j) = 1, 2, 3 \qquad (5.67)$$

In special relativity, one constructs a divergenceless second-rank tensor $T_{\mu\nu}$, whose components give these conservation laws[19]

$$\frac{\partial}{\partial x_\mu}T_{\mu\nu} = 0 \qquad ; \text{ energy-momentum conservation} \quad (5.68)$$

---

[17] In the string problem, this is the same as adding a uniform Hooke's law resorting force density $\delta\mathcal{F}_y = -\kappa q(x, t)$, or, equivalently, a potential energy density $\delta\mathcal{V} = \kappa q^2(x, t)/2$ [see Eq. (5.36)] where $\kappa \equiv \sigma m_s^2 c^2$.

[18] See [Fetter and Walecka (2003)]. We use the terms stress tensor, and energy-momentum tensor, interchangeably in this book.

[19] See, for example, [Walecka (2007)].

There is a sophisticated way of deriving the stress tensor from variations of the boundary in Hamilton's principle; however, rather than pursuing that here, we just introduce a divergenceless tensor in Minkowski space and show that it does the job for us.

Assume that $q(\mathbf{x}, t)$ is a Lorentz scalar field, and define

$$T_{\mu\nu} \equiv \mathcal{L}\,\delta_{\mu\nu} - \frac{\partial \mathcal{L}}{\partial(\partial q/\partial x_\mu)}\,\frac{\partial q}{\partial x_\nu} \qquad ;\ \text{stress tensor} \qquad (5.69)$$

Here $\mathcal{L}$ is the lagrangian density, and the partial derivative of $\mathcal{L}$ implies that all the other variables in $\mathcal{L}(\partial q/\partial x_\mu,\, q\,;\, x_\mu)$ are to be held fixed; the last argument in $\mathcal{L}$ denotes a possible *explicit* dependence on the space-time position x_μ. The partial derivatives with respect to x imply that all other variables in the set $(\mathbf{x}, ict)$ are to be held fixed.[20]

Since both $\mathcal{L}$ and q are Lorentz scalars, $T_{\mu\nu}$ is, indeed, a second-rank tensor. Let us show that it is also divergenceless. Just compute

$$\frac{\partial}{\partial x_\mu}T_{\mu\nu} = \frac{\partial \mathcal{L}}{\partial x_\nu} - \frac{\partial \mathcal{L}}{\partial(\partial q/\partial x_\mu)}\frac{\partial^2 q}{\partial x_\mu \partial x_\nu} - \left[\frac{\partial}{\partial x_\mu}\frac{\partial \mathcal{L}}{\partial(\partial q/\partial x_\mu)}\right]\frac{\partial q}{\partial x_\nu} \quad (5.70)$$

The first term on the r.h.s. is now given through the chain rule as

$$\frac{\partial \mathcal{L}}{\partial x_\nu} = \frac{\partial \mathcal{L}}{\partial q}\frac{\partial q}{\partial x_\nu} + \frac{\partial \mathcal{L}}{\partial(\partial q/\partial x_\mu)}\frac{\partial^2 q}{\partial x_\nu \partial x_\mu} + \left[\frac{\partial \mathcal{L}}{\partial x_\nu}\right]_{(q,\,\partial q/\partial x_\mu)} \quad (5.71)$$

The second term in Eq. (5.70) is cancelled by the second term in Eq. (5.71). The last term in Eq. (5.70) cancels the first term in Eq. (5.71) through the use of Lagrange's equation

$$\frac{\partial}{\partial x_\mu}\frac{\partial \mathcal{L}}{\partial(\partial q/\partial x_\mu)} = \frac{\partial \mathcal{L}}{\partial q} \qquad ;\ \text{Lagrange's equation} \qquad (5.72)$$

Thus one is left with

$$\frac{\partial T_{\mu\nu}}{\partial x_\mu} = \left[\frac{\partial \mathcal{L}}{\partial x_\nu}\right]_{(q,\,\partial q/\partial x_\mu)} \qquad (5.73)$$

It follows that if the lagrangian density has no *explicit* dependence on the space-time coordinate x_ν, then the stress tensor defined in Eq. (5.69) has a vanishing divergence in Minkowski space.

---

[20]If there are additional generalized coordinates, then there is a corresponding set of additional terms in the stress tensor in Eq. (5.69); see Prob. 5.13.

The *four-momentum* of the system is now defined by

$$P_\mu = \left(\mathbf{P}, \frac{i}{c}H\right) \equiv \frac{1}{ic}\int_\Omega d^3x\, T_{4\mu} \qquad ; \text{ four-momentum} \quad (5.74)$$

To make contact with something we know, consider the fourth component of this relation, and substitute the defining Eq. (5.69),

$$H = -\int_\Omega d^3x\, T_{44}$$

$$= -\int_\Omega d^3x\left[\mathcal{L} - \frac{\partial\mathcal{L}}{\partial(\partial q/\partial t)}\frac{\partial q}{\partial t}\right] = \int_\Omega d^3x\left[\pi\frac{\partial q}{\partial t} - \mathcal{L}\right] \quad (5.75)$$

where the canonical momentum density has been identified through Eq. (5.33). The integrand now indeed reproduces our previous definition of the hamiltonian density in Eq. (5.35).

As an example, consider massive scalar field theory where the lagrangian density is given by Eq. (5.63). The stress tensor then follows from Eq. (5.69) as

$$T_{\mu\nu} = c^2\frac{\partial\phi}{\partial x_\mu}\frac{\partial\phi}{\partial x_\nu} - \frac{c^2}{2}\left[\left(\frac{\partial\phi}{\partial x_\lambda}\right)^2 + m_s^2\phi^2\right]\delta_{\mu\nu} \quad ; \text{ scalar field} \quad (5.76)$$

The canonical momentum density follows from Eq. (5.63) as

$$\pi = \frac{\partial\mathcal{L}}{\partial(\partial\phi/\partial t)} = \frac{\partial\phi}{\partial t} \qquad ; \text{ momentum density} \qquad (5.77)$$

The total energy and momentum in the field given by Eq. (5.74) are then

$$H = \frac{1}{2}\int_\Omega d^3x\left\{\pi^2(\mathbf{x},t) + c^2[\boldsymbol{\nabla}\phi(\mathbf{x},t)]^2 + m_s^2c^2\phi^2(\mathbf{x},t)\right\}$$

$$\mathbf{P} = -\int_\Omega d^3x\,\pi(\mathbf{x},t)\boldsymbol{\nabla}\phi(\mathbf{x},t) \qquad (5.78)$$

Substitution of the quantum fields in the interaction picture in Eqs. (5.60) gives[21]

$$\hat{H} = \sum_k \hbar\omega_k\left(a_\mathbf{k}^\dagger a_\mathbf{k} + 1/2\right)$$

$$\hat{\mathbf{P}} = \sum_k \hbar\mathbf{k}\, a_\mathbf{k}^\dagger a_\mathbf{k} \qquad (5.79)$$

---

[21]See Prob. 5.12; note Eq. (5.65).

To investigate the implications of the continuity Eqs. (5.66)–(5.67), as well as the Lorentz invariance of the theory, we first modify the spatial boundary conditions, just as we did when passing from Fourier series to Fourier integrals in Vol. I. After using the discrete nature of the eigenvalues with periodic boundary conditions in the quantization volume Ω to obtain well-defined expressions for such things as the number density, energy density, transition rate, and cross section, we let $\Omega \to \infty$. Instead of p.b.c., we then go over to a *localized* boundary condition, which assumes that any disturbance we are considering is localized within that infinite volume and vanishes sufficiently fast as $|\mathbf{x}| \to \infty$.

Consider, then, the *time derivative* of the four-momentum in Eq. (5.74)

$$\frac{dP_\nu}{dt} = \frac{1}{ic}\frac{d}{dt}\int_\Omega d^3x\, T_{4\nu} = \frac{1}{ic}\int_\Omega d^3x\,\frac{\partial T_{4\nu}}{\partial t}$$

$$= \int_\Omega d^3x\,\frac{\partial T_{4\nu}}{\partial x_4} = -\int_\Omega d^3x\,\frac{\partial T_{k\nu}}{\partial x_k} \tag{5.80}$$

Here the continuity equation has been used in obtaining the last equality.[22] Now use Gauss' theorem to convert the integral of the spatial divergence over the volume Ω to a surface integral over the boundary S surrounding the volume Ω, and then let $\Omega \to \infty$

$$\int_\Omega d^3x\,\frac{\partial T_{k\nu}}{\partial x_k} = \int_S dS_k T_{k\nu} \to 0 \qquad ; \Omega \to \infty$$

$$\text{localized disturbance} \tag{5.81}$$

This integral vanishes if the disturbance is localized and the bounding surface S is far enough away. Hence we have the important result that *if the lagrangian density has no explicit dependence on x_μ, then the four-momentum is a constant of the motion*

$$\frac{dP_\nu}{dt} = 0 \qquad\qquad ; \text{constant of motion}$$

$$\nu = 1, \cdots, 4 \tag{5.82}$$

Let us further examine the four-momentum in Eq. (5.74). In three dimensions, a two-dimensional surface element is a three-vector (see Fig. 5.2).

$$d\mathbf{S} = \mathbf{n}\,dS \qquad\qquad ; \text{three-vector} \tag{5.83}$$

---

[22]Note that the continuity equation is equivalent to the vanishing of the four-divergence of the stress tensor. As usual, the repeated Latin index is summed from 1 to 3.

Here dS is the magnitude of the infinitesimal surface area and $\mathbf{n}$ is a unit vector normal to it. $d\mathbf{S}$ is a vector since if one rotates the coordinate system, the magnitude dS does not change, while the direction of the normal $\mathbf{n}$ does. Under a rotation characterized by a rotation matrix $a_{ij}(\boldsymbol{\omega})$, one has

$$dS_i' = a_{ij}(\boldsymbol{\omega})dS_j \qquad ; \text{rotation} \qquad (5.84)$$

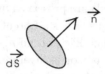

Fig. 5.2 A little element of surface area of magnitude dS with a unit vector normal $\mathbf{n}$ forms a vector $d\mathbf{S} = \mathbf{n}\,dS$ in three dimensions.

In four dimensions, one has a three-dimensional hypersurface (three-dimensional volume) in exactly the same fashion. Thus an infinitesimal amount of "surface area" forms the four-vector

$$dS_\mu = n_\mu dS \qquad ; \text{four-vector} \qquad (5.85)$$

Here dS is the "area of the surface" perpendicular to the unit normal n_μ. One particular hypersurface is characterized by a unit normal in the fourth (or time) direction $n_\mu = (0,0,0,i)$, and the orthogonal three-dimensional volume element d^3x

$$n_\mu = (0,0,0,i) \qquad ; \text{time direction}$$
$$dS_\mu = (0,0,0,id^3x) \qquad \text{surface element} \qquad (5.86)$$

With this surface element, the four-momentum in Eq. (5.74) can be written as[23]

$$P_\nu = \frac{1}{ic}\int d^3x\, T_{4\nu} = -\frac{1}{c}\int T_{\mu\nu}\, dS_\mu \qquad (5.87)$$

We use this relation to establish two results:

1) In this form, P_ν is explicitly a four-vector;

---

[23] For the remainder of this argument, we work in the limit $\Omega \to \infty$.

2) Consider what happens under a Lorentz transformation. Intervals are of three types in Minkowski space

$$x_\mu^2 > 0 \qquad ; \text{ space-like}$$
$$x_\mu^2 < 0 \qquad ; \text{ time-like}$$
$$x_\mu^2 = 0 \qquad ; \text{ light-like} \tag{5.88}$$

The second and third types represent intervals that can be connected with a light signal. Since intervals are preserved under Lorentz transformations, so are these characterizations. Thus a Lorentz transformation on dS_μ takes it into dS'_μ with another time-like normal n'_μ (see Fig. 5.3). We proceed to show that since the stress tensor is divergenceless, the surface integral in Eq. (5.87) is *independent of the particular time-like surface over which it is evaluated*.

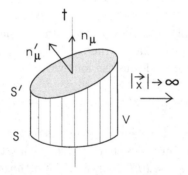

Fig. 5.3　Sketch of the four-dimensional volume V over which the four-divergence is integrated; one of the spatial coordinates is suppressed. The sides of the region are far-away space-like surfaces with $|\mathbf{x}| \to \infty$. The time-like normal n_μ is taken into n'_μ under a Lorentz transformation.

The essence of Eq. (5.87) is the integral of a four-vector with vanishing four-divergence over a time-like surface. Assume that v_μ is divergenceless and integrate it over the four-dimensional volume V sketched in Fig. 5.3.

$$\frac{\partial v_\mu}{\partial x_\mu} = 0 \qquad \Rightarrow \qquad \int_V d^4x \left(\frac{\partial v_\mu}{\partial x_\mu} \right) = 0 \tag{5.89}$$

Now use Gauss' theorem on the four-divergence to convert it into a surface integral over the entire bounding surface A, and then write out the individual contributions

$$\int_V d^4x \left(\frac{\partial v_\mu}{\partial x_\mu}\right) = \int_A v_\mu dS_\mu = 0 \tag{5.90}$$

$$= \int_{S'} v_\mu dS_\mu - \int_S v_\mu dS_\mu + \int \cdots \int_{|\mathbf{x}| \to \infty} v_\mu dS_\mu$$

The second term has a minus sign due to the choice of inward pointing normal, and the last term represents an integral over the space-like surfaces on the "sides" of the volume V. Since the disturbance is assumed to be localized, the last term *vanishes as* $|\mathbf{x}| \to \infty$. Thus

$$\int_{S'} v_\mu dS_\mu = \int_S v_\mu dS_\mu \tag{5.91}$$

and the result is established. As a consequence, one can always return to the particular time-like surface in Eq. (5.86), without penalty, after any Lorentz transformation.

5.4.3 *Dirac Field*

The quantization of the Dirac field was covered in chapter 12 of Vol. I. Here we go back to the beginning and start by considering the problem to be one in continuum mechanics. We construct a lagrangian density which, through Lagrange's equations, leads to the Dirac equation.

The first thing to note is that the Dirac field ψ is *complex*, and hence it, and its hermitian adjoint $\psi^\dagger$, are really *independent fields* since it is necessary to specify both their real and imaginary parts in order to define them. Thus we can treat their *variations* as independent quantities in applying Hamilton's principle.[24]

Try the following lagrangian density

$$\mathcal{L} = -\hbar c\, \bar\psi \left(\gamma_\mu \frac{\partial}{\partial x_\mu} + M\right) \psi \qquad ; \text{Dirac} \tag{5.92}$$

Here $M \equiv m_0 c/\hbar$ is the inverse Compton wavelength. Recall that $\bar\psi = \psi^\dagger \gamma_4$, and one can just as well take $(\psi, \bar\psi)$ as the independent fields. In exactly the same fashion as before, Hamilton's principle then leads to a pair of

---

[24]See Prob. 5.15(a).

Lagrange's equations, one for each independent field

$$\frac{\partial}{\partial x_\mu}\frac{\partial \mathcal{L}}{\partial(\partial\psi/\partial x_\mu)} - \frac{\partial \mathcal{L}}{\partial \psi} = 0 \qquad ; \psi \text{ eqn}$$

$$\frac{\partial}{\partial x_\mu}\frac{\partial \mathcal{L}}{\partial(\partial\bar\psi/\partial x_\mu)} - \frac{\partial \mathcal{L}}{\partial \bar\psi} = 0 \qquad ; \bar\psi \text{ eqn} \qquad (5.93)$$

In fact, this notation is very compact. The Dirac field ψ is a spinor, with components ψ_α where $\alpha = (1, \cdots, 4)$, and similarly for $\bar\psi$. Lagrange's equations then actually form a set of eight equations, one for each component of ψ, and one for each component of $\bar\psi$.

With the lagrangian density of Eq. (5.92), Lagrange's equations then read

$$\left(\gamma_\mu\frac{\partial}{\partial x_\mu} + M\right)\psi = 0 \qquad ; \bar\psi \text{ eqn}$$

$$\bar\psi\left(\gamma_\mu\frac{\overleftarrow{\partial}}{\partial x_\mu} - M\right) = 0 \qquad ; \psi \text{ eqn} \qquad (5.94)$$

Here the notation $\overleftarrow{\partial}$ indicates that the derivative acts on the function sitting to its left. The first is, indeed, the Dirac equation, and the second is its proper adjoint.[25]

The stress tensor is given by Eq. (5.69), where there is now one contribution for each independent field

$$T_{\mu\nu} = \mathcal{L}\,\delta_{\mu\nu} - \frac{\partial \mathcal{L}}{\partial(\partial\psi/\partial x_\mu)}\frac{\partial\psi}{\partial x_\nu} - \frac{\partial\bar\psi}{\partial x_\nu}\frac{\partial \mathcal{L}}{\partial(\partial\bar\psi/\partial x_\mu)} \qquad (5.95)$$

Since there is no dependence of $\mathcal{L}$ on $\partial\bar\psi/\partial x_\mu$, the last term vanishes, and hence

$$T_{\mu\nu} = \hbar c\left[\bar\psi\gamma_\mu\frac{\partial\psi}{\partial x_\nu} - \delta_{\mu\nu}\bar\psi\left(\gamma_\lambda\frac{\partial}{\partial x_\lambda} + M\right)\psi\right] \qquad (5.96)$$

The four-momentum is obtained from Eq. (5.74) as

$$P_\nu = \left(\mathbf{P}, \frac{i}{c}H\right) = \frac{1}{ic}\int_\Omega d^3x\, T_{4\nu} \qquad (5.97)$$

---

[25]See Prob. 5.11.

Hence

$$H = \hbar c \int_\Omega d^3x \, \bar\psi \left(\boldsymbol\gamma \cdot \boldsymbol\nabla + M \right) \psi$$

$$\mathbf{P} = -i\hbar \int_\Omega d^3x \, \bar\psi \gamma_4 \boldsymbol\nabla \psi \tag{5.98}$$

With the identification of $\boldsymbol\gamma = i\boldsymbol\alpha\beta = i\boldsymbol\alpha\gamma_4$, and $\mathbf{p} = -i\hbar\boldsymbol\nabla$, these expressions become

$$H = \int_\Omega d^3x \, \psi^\dagger (c\boldsymbol\alpha \cdot \mathbf{p} + \beta m_0 c^2)\psi$$

$$\mathbf{P} = \int_\Omega d^3x \, \psi^\dagger \mathbf{p} \, \psi \tag{5.99}$$

The first equation was used in the development of the Dirac field in Vol. I, and both expressions are familiar from the discussion of second quantization.

To *quantize* the theory, one needs the canonical momentum density conjugate to ψ. It follows from Eq. (5.92) that this is given by

$$\pi_\psi = \frac{\partial\mathcal{L}}{\partial(\partial\psi/\partial t)} = i\hbar\bar\psi\gamma_4 = i\hbar\psi^\dagger \tag{5.100}$$

With this lagrangian density one has $\pi_\psi = i\hbar\psi^\dagger$, and there is just enough flexibility in this approach to treat (ψ, π_ψ) as *independent fields*, which is a requisite for the canonical quantization procedure.

As explained in Vol. I, with fermion fields, one must impose canonical *anticommutation* relations in order to incorporate the Pauli Principle. In the interaction picture, these equal-time anticommutation relations between the components of the fields become

$$\{\hat\psi_\alpha(\mathbf{x},t), \hat\pi_\beta(\mathbf{x}',t')\}_{t=t'} = i\hbar\delta^{(3)}(\mathbf{x} - \mathbf{x}')\delta_{\alpha\beta} \quad ;$$

$$\text{all others anticommute} \tag{5.101}$$

An interaction-picture representation of the Dirac field that satisfies these canonical equal-time anticommutation relations was given in chapter 12 of Vol. I, and we reproduce that material here.

$$\hat\psi(\mathbf{x},t) = \frac{1}{\sqrt{\Omega}} \sum_\mathbf{k} \sum_\lambda \left[a_{\mathbf{k}\lambda} u_\lambda(\mathbf{k}) e^{i(\mathbf{k}\cdot\mathbf{x}-\omega_k t)} + b_{\mathbf{k}\lambda}^\dagger v_\lambda(-\mathbf{k}) e^{-i(\mathbf{k}\cdot\mathbf{x}-\omega_k t)} \right]$$

$$\omega_k = c\sqrt{\mathbf{k}^2 + M^2} \qquad ; \text{ Dirac field} \tag{5.102}$$

where $\lambda = (\uparrow, \downarrow)$ denotes the helicity. The field $\hat{\psi}(\mathbf{x}, t)$ is a 4-component spinor since (u, v) are; we suppress the underlining of these column vectors. The anticommutation relations for the particle and antiparticle creation and destruction operators for all modes are defined by

$$\{a_{\mathbf{k}\lambda}, a_{\mathbf{k}'\lambda'}^\dagger\} = \{b_{\mathbf{k}\lambda}, b_{\mathbf{k}'\lambda'}^\dagger\} = \delta_{\mathbf{k}\mathbf{k}'}\delta_{\lambda\lambda'} \qquad ;$$

all other anticommutators vanish   (5.103)

The equal-time anticommutator of the fields was then shown in Probl. 12.7 to be

$$\left\{\hat{\psi}_\alpha(\mathbf{x}, t), \hat{\psi}_\beta^\dagger(\mathbf{x}', t')\right\}_{t=t'} = \delta_{\alpha\beta}\,\delta^{(3)}(\mathbf{x} - \mathbf{x}') \qquad (5.104)$$

which reproduces Eq. (5.101).

If the hamiltonian is calculated from these interaction-picture fields, one arrives at the result obtained in Vol. I

$$\hat{H}_0 = \sum_{\mathbf{k}}\sum_\lambda \hbar\omega_k \left(a_{\mathbf{k}\lambda}^\dagger a_{\mathbf{k}\lambda} + b_{\mathbf{k}\lambda}^\dagger b_{\mathbf{k}\lambda} - 1\right) \qquad (5.105)$$

The zero-particle state now defines the *vacuum*. With respect to that state, the operator $a_{\mathbf{k}\lambda}^\dagger$ creates a one-particle state with energy increased by $\hbar\omega_k$, while the operator $b_{\mathbf{k}\lambda}^\dagger$ creates an *antiparticle* state with positive energy increase of $\hbar\omega_k$ — this is now Dirac hole theory done correctly. The momentum operator obtained with these interaction-picture fields is[26]

$$\hat{\mathbf{P}} = \sum_{\mathbf{k}}\sum_\lambda \hbar\mathbf{k} \left(a_{\mathbf{k}\lambda}^\dagger a_{\mathbf{k}\lambda} + b_{\mathbf{k}\lambda}^\dagger b_{\mathbf{k}\lambda}\right) \qquad (5.106)$$

5.4.4   *Noether's Theorem*

Noether's theorem is a very simple, yet extremely powerful and useful, result.[27] It is easily stated:

> *For every continuous symmetry of the lagrangian density, there is a corresponding conserved current.*

The proof involves a straightforward application of Lagrange's equations. Assume n generalized coordinates $q^i(\mathbf{x}, t)$ with $1, 2, \cdots, n$, and a lagrangian

---

[26]See Prob. 5.12.
[27]Due to Emmy Noether (1915).

density $\mathcal{L}(q^1, \cdots, q^n, \partial q^1/\partial x_\mu, \cdots, \partial q^n/\partial x_\mu; t)$. Make a set of infinitesimal changes in these coordinates as follows

$$q^i(\mathbf{x}, t) \rightarrow q^i(\mathbf{x}, t) + \delta q^i(\mathbf{x}, t) \qquad ; i = 1, 2, \cdots, n \qquad (5.107)$$

Compute the corresponding change in the lagrangian density by making a Taylor series expansion to first order in infinitesimals

$$\delta\mathcal{L} = \sum_{i=1}^{n} \left[\frac{\partial \mathcal{L}}{\partial q^i} \delta q^i(\mathbf{x}, t) + \frac{\partial \mathcal{L}}{\partial(\partial q^i/\partial x_\mu)} \frac{\partial}{\partial x_\mu} \delta q^i(\mathbf{x}, t) \right] \qquad (5.108)$$

Use Lagrange's equations for the generalized coordinates to rewrite the first term

$$\delta\mathcal{L} = \sum_{i=1}^{n} \left\{ \left[\frac{\partial}{\partial x_\mu} \frac{\partial \mathcal{L}}{\partial(\partial q^i/\partial x_\mu)} \right] \delta q^i(\mathbf{x}, t) + \frac{\partial \mathcal{L}}{\partial(\partial q^i/\partial x_\mu)} \frac{\partial}{\partial x_\mu} \delta q^i(\mathbf{x}, t) \right\}$$

$$= \sum_{i=1}^{n} \frac{\partial}{\partial x_\mu} \left[\frac{\partial \mathcal{L}}{\partial(\partial q^i/\partial x_\mu)} \delta q^i(\mathbf{x}, t) \right] \qquad (5.109)$$

Now suppose that $\mathcal{L}$ is *unchanged* under the coordinate transformation in Eq. (5.107), that is, suppose the transformation represents a *symmetry* of the lagrangian density. Then

$$\delta\mathcal{L} = 0 \qquad ; \text{ invariant} \qquad (5.110)$$

In this case, Eq. (5.109) immediately implies the existence of a corresponding conserved current

$$\frac{\partial J_\mu(x)}{\partial x_\mu} = 0 \qquad ; \text{ conserved current}$$

$$J_\mu(x) = \sum_{i=1}^{n} \left[\frac{\partial \mathcal{L}}{\partial(\partial q^i/\partial x_\mu)} \delta q^i(\mathbf{x}, t) \right] \qquad (5.111)$$

This is Noether's theorem.

To see an example of how this works, consider the Dirac lagrangian density in Eq. (5.92). It is clearly invariant under the following *global phase transformation*

$$\psi \rightarrow e^{i\theta} \psi \qquad ; \text{ global phase transformation}$$
$$\bar{\psi} \rightarrow e^{-i\theta} \bar{\psi} \qquad (5.112)$$

Let $\theta \equiv \epsilon$ be an infinitesimal, with $\epsilon \to 0$. In this case, the infinitesimal changes in the fields are

$$\delta\psi = i\epsilon\psi \qquad ; \ \delta\bar{\psi} = -i\epsilon\bar{\psi} \qquad (5.113)$$

The current in Eq. (5.111) is then given by

$$J_\mu(x) = \frac{\partial \mathcal{L}}{\partial(\partial\psi/\partial x_\mu)} \delta\psi + \delta\bar{\psi} \frac{\partial \mathcal{L}}{\partial(\partial\bar{\psi}/\partial x_\mu)}$$
$$= -i\epsilon\hbar c \, \bar{\psi}\gamma_\mu\psi \qquad (5.114)$$

If we remove the constant factor $-\epsilon\hbar c$ from the definition of the current, then the new current is

$$j_\mu(x) = i\bar{\psi}(x)\gamma_\mu\psi(x) \qquad ; \ \text{Dirac current} \qquad (5.115)$$

In the case of electrons, for example, this is recognized as the conserved electromagnetic current of Vol. I.

5.4.4.1 *Normal-Ordered Current*

In the interaction picture, the Dirac field of an electron is quantized according to Eq. (5.102). The corresponding electromagnetic current is obtained by substituting this field expansion into Eq. (5.115), which is then bilinear in the creation and destruction operators. We now observe that one always *measures the electromagnetic current relative to the vacuum*, and the true current operator is

$$\hat{j}_\mu(x) = i\hat{\bar{\psi}}(x)\gamma_\mu\hat{\psi}(x) - i\langle 0|\hat{\bar{\psi}}(x)\gamma_\mu\hat{\psi}(x)|0\rangle$$
$$\equiv i{:}\hat{\bar{\psi}}(x)\gamma_\mu\hat{\psi}(x){:} \qquad ; \ \text{normal-ordered} \qquad (5.116)$$

The effect of taking this difference is to simply turn around the term in $b_{k\lambda}b^\dagger_{k'\lambda'}$, with an appropriate change in sign. We say that the observed current is *normal-ordered*. This current no longer contains the c-number contribution from the filled negative-energy sea, which, after all, really should not be there since it depends on which one, e^- or e^+, is initially defined to be the particle.[28]

---

[28]The normal-ordered current has the desirable property that it now simply changes sign under the interchange of particle and antiparticle, that is, it is *odd under charge conjugation* (we leave the proof of this for the dedicated reader; see [Bjorken and Drell (1965)]). See also Prob. 5.20.

5.4.5 *Electromagnetic Field*

The lagrangian field theory of the electromagnetic field is presented in detail in appendix C. The most important results for the quantized free electromagnetic field have already been covered in chapter 12 of Vol. I. Also included in appendix C is an analysis of the quantized electromagnetic field interacting with a specified external current, a topic that provides a nice introduction to our subsequent discussion of quantum electrodynamics (QED).

5.4.6 *Interacting Fields (Dirac-Scalar)*

As an introduction to the theory of interacting fields, consider a Dirac field ψ interacting with a massive scalar field ϕ, which provides a useful model in nuclear physics.[29] The free lagrangian densities for these fields have already been examined

$$\mathcal{L}_0^{\text{Dirac}} = -\hbar c\, \bar{\psi} \left(\gamma_\mu \frac{\partial}{\partial x_\mu} + M \right) \psi$$

$$\mathcal{L}_0^{\text{scalar}} = -\frac{c^2}{2} \left[\left(\frac{\partial \phi}{\partial x_\mu} \right)^2 + m_s^2 \phi^2 \right] \tag{5.117}$$

The simplest way to include a coupling of the fields, while maintaining the Lorentz structure of the lagrangian density, is to just make the following replacement in the Dirac lagrangian

$$M \to M - \frac{g}{\hbar c} \phi \tag{5.118}$$

This gives an interaction lagrangian density

$$\mathcal{L}_1 = g \bar{\psi} \psi \phi \tag{5.119}$$

Here g is a coupling constant. This makes the problem particularly simple since the interaction *contains no derivatives*. It follows that:

- The only modification of the total stress tensor from the sum of the free stress tensors comes from the additional contribution of $\mathcal{L}_1 \delta_{\mu\nu}$;
- The only modification of the expression for the total hamiltonian density from the sum of the free hamiltonian densities $\mathcal{H}_0$ is then to add a term $\mathcal{H}_1 = -\mathcal{L}_1$;

---

[29]See Probl. 9.5.

- The expression for the total momentum is unmodified from the sum of the free expressions $\mathbf{P}_0$;
- The conjugate canonical momentum densities are unmodified from the free-field case;
- Consequently, when the theory is quantized, one can use the free-field expansions in the interaction picture as a representation of these equal-time canonical (anti)commutation relations.

As a result of these observations, the four-momentum operator $\hat{P}_\mu = (\hat{\mathbf{P}}, i\hat{H}/c)$ takes the following form in the Schrödinger picture

$$\hat{H} = \hat{H}_0^{\text{Dirac}} + \hat{H}_0^{\text{scalar}} - g \int_\Omega d^3x \,\hat{\bar{\psi}}(\mathbf{x})\hat{\psi}(\mathbf{x})\hat{\phi}(\mathbf{x})$$

$$\hat{\mathbf{P}} = \hat{\mathbf{P}}_0^{\text{Dirac}} + \hat{\mathbf{P}}_0^{\text{scalar}} \tag{5.120}$$

In the interaction picture, the additional term in the hamiltonian is expressed in terms of the Dirac and scalar fields as

$$\hat{H}_I(t) = -g \int_\Omega d^3x \,\hat{\bar{\psi}}(\mathbf{x},t)\hat{\psi}(\mathbf{x},t)\hat{\phi}(\mathbf{x},t) \tag{5.121}$$

$$\hat{\phi}(\mathbf{x},t) = \sum_{\mathbf{k}} \left(\frac{\hbar}{2\omega_k\Omega}\right)^{1/2} \left[c_{\mathbf{k}}e^{i(\mathbf{k}\cdot\mathbf{x}-\omega_k t)} + c_{\mathbf{k}}^\dagger e^{-i(\mathbf{k}\cdot\mathbf{x}-\omega_k t)}\right]$$

$$\hat{\psi}(\mathbf{x},t) = \frac{1}{\sqrt{\Omega}} \sum_{\mathbf{k}} \sum_{\lambda} \left[a_{\mathbf{k}\lambda}u_\lambda(\mathbf{k})e^{i(\mathbf{k}\cdot\mathbf{x}-\omega_k t)} + b_{\mathbf{k}\lambda}^\dagger v_\lambda(-\mathbf{k})e^{-i(\mathbf{k}\cdot\mathbf{x}-\omega_k t)}\right]$$

where we now use $(c^\dagger, c)$ for the creation and destruction operators in the scalar field.

The interacting relativistic quantum field theory is now well formulated. In subsequent chapters we will develop procedures for extracting consequences of the theory. Here, we just leave it as a problem for the reader to discuss the processes described by the interaction hamiltonian $\hat{H}_I(t)$.[30]

---

[30]See Prob. 5.22.

Chapter 6

Symmetries

We now have a formulation of quantum mechanics in the abstract Hilbert space, where there is a state vector satisfying the Schrödinger equation, scattering theory to extract physical results from the theory, and a lagrangian formulation of relativistic field theory, which provides a basis for describing the dynamics. We know how to pass from a lagrangian density to a hamiltonian density, and how to quantize the theory. We have constructed lagrangian densities for several free fields: scalar, complex scalar, Dirac, and electromagnetic. Interacting field theories have been formulated for two examples: scalar-Dirac fields, and the electromagnetic field interacting with a specified current. The issue now arises as to the construction of lagrangian densities describing more complicated interacting relativistic field theories. In order to accomplish this, one first invokes general *symmetry* properties of the theory.

6.1   Lorentz Invariance

The first general principle is *Lorentz invariance*. Special relativity implies that one must obtain the same physics in any Lorentz frame. We have seen that this will be the case if the lagrangian density is a *Lorentz scalar*. Thus all the lagrangian densities we subsequently consider will be constructed in this fashion.

Just as we described translations and rotations in the abstract Hilbert space, one can consider Lorentz transformations in that space. Since two consecutive Lorentz transformations again form a Lorentz transformation, the Lorentz transformations form a group. This is the *inhomogeneous*

Lorentz, or Poincaré, group.[1] We postpone a discussion of the full inhomogeneous Lorentz group, and here return to our previous analysis of *rotations*, which form a subgroup of the full Lorentz group.

6.2 Rotational Invariance

To the best of our current knowledge, space is isotropic. This implies that the hamiltonian must be invariant under rotations. The chapter on angular momentum investigated in detail the implications of this observation. The generators of rotations are the components of the angular momentum operator $\hat{\mathbf{J}} = (\hat{J}_1, \hat{J}_2, \hat{J}_3)$, and rotational invariance implies that the hamiltonian commutes with these generators

$$[\hat{\mathbf{J}}, \hat{H}] = 0 \qquad ; \text{ rotational invariance} \qquad (6.1)$$

The eigenstates of the hamiltonian can be labeled with the eigenvalues of $(\hat{\mathbf{J}}^2, \hat{J}_3)$, in our notation $|jm\rangle$, and the first important observation is that if the hamiltonian is invariant under rotations, then the energy eigenvalues will be *independent of m*, that is, independent of the orientation of the system in space. Thus

If the hamiltonian is invariant under rotations, the energy eigenvalues for a given j will exhibit a $(2j + 1)$-fold degeneracy in m.

We saw this m-degeneracy in all our central-field applications in Vol. I.[2]

The commutation relations between components of the angular momentum provide the basis for the angular momentum theory in chapter 3

$$[\hat{J}_i, \hat{J}_j] = i\epsilon_{ijk}\hat{J}_k \qquad ; \text{ angular momentum}$$
$$(i, j, k) = 1, 2, 3 \qquad (6.2)$$

This algebra of the generators is closed under commutation. The *fundamental basis* for angular momentum is $|\tfrac{1}{2}m\rangle$, since any other angular momentum can be obtained from direct products of these states. The finite transformations acting on this fundamental basis take the form

$$e^{i\boldsymbol{\omega}\cdot\hat{\mathbf{J}}}|\tfrac{1}{2}m\rangle = \underline{r}_{m'm}(\boldsymbol{\omega})|\tfrac{1}{2}m'\rangle \qquad ; (m', m) = \pm\tfrac{1}{2}$$
$$\underline{r}_{m'm}(\boldsymbol{\omega}) = \langle\tfrac{1}{2}m'| e^{i\boldsymbol{\omega}\cdot\hat{\mathbf{J}}}|\tfrac{1}{2}m\rangle \qquad \boldsymbol{\omega} = (\omega_1, \omega_2, \omega_3) \quad (6.3)$$

---

[1]The inhomogeneous Lorentz group includes translations.
[2]See the section on Lie groups for the general proof.

Here repeated m-indices are summed over $\pm\frac{1}{2}$. The last expression is an element of the 2×2 matrix

$$\underline{r}(\boldsymbol{\omega}) = \exp\left\{\frac{i}{2}\boldsymbol{\sigma} \cdot \boldsymbol{\omega}\right\} \tag{6.4}$$

where $\boldsymbol{\sigma} = (\sigma_1, \sigma_2, \sigma_3)$ are the Pauli matrices. This matrix is unitary

$$\underline{r}(\boldsymbol{\omega})^{-1} = \underline{r}(\boldsymbol{\omega})^{\dagger} \qquad\qquad ;\text{ unitary} \tag{6.5}$$

It was demonstrated in Probl. 7.9 that these 2×2, unitary, unimodular matrices $\underline{r}(\boldsymbol{\omega})$ satisfy the following relations

$$\begin{aligned} \underline{r}(\boldsymbol{\omega})\underline{r}(\boldsymbol{\omega}') &= \underline{r}(\boldsymbol{\omega}'') && ;\det\underline{r}(\boldsymbol{\omega}) = 1 \\ \underline{r}(\boldsymbol{\omega})^{-1} &= \underline{r}(-\boldsymbol{\omega}) && ;\underline{r}(0) = 1 \end{aligned} \tag{6.6}$$

They form a *three-parameter continuous group*, the group SU(2). As discussed in Vol. I and chapter 3, SU(2) is the group of rotations in quantum mechanics.

Any lagrangian density that is a Lorentz scalar will automatically be invariant under rotations, since rotations form a subgroup of the full Lorentz group.

6.3 Internal Symmetries

Suppose that in addition to their properties in real space-time, the quanta (particles) carry additional quantum numbers and there is also an *internal space* in the theory. The simplest example of this is isospin (see Vol. I).

6.3.1 *Isospin–SU(2)*

Let us examine isospin from the point of view of quantum field theory.

6.3.1.1 *Isovector*

Consider the free lagrangian density for three equal-mass, real (hermitian), scalar fields, which we write as a vector in an internal isospin space[3]

$$\mathcal{L} = -\frac{c^2}{2}\left[\frac{\partial\boldsymbol{\phi}}{\partial x_\mu} \cdot \frac{\partial\boldsymbol{\phi}}{\partial x_\mu} + m^2\boldsymbol{\phi}\cdot\boldsymbol{\phi}\right] \qquad ;\boldsymbol{\phi} = (\phi_1, \phi_2, \phi_3) \tag{6.7}$$

---

[3]Pions form such a *pseudo*scalar multiplet (see later).

This lagrangian density is evidently invariant under rotations in the internal space. Consider an infinitesimal rotation of the form[4]

$$\phi \to \phi + \delta\phi = \phi - \varepsilon \times \phi \qquad ; \varepsilon \text{ an infinitesimal} \qquad (6.8)$$

Noether's theorem implies the existence of a corresponding conserved current

$$\tilde{J}_\mu = \frac{\partial \mathcal{L}}{\partial(\partial\phi/\partial x_\mu)} \cdot \delta\phi = c^2 \left[\frac{\partial\phi}{\partial x_\mu} \cdot (\varepsilon \times \phi) \right]$$

$$= -c^2 \varepsilon \cdot \left(\frac{\partial\phi}{\partial x_\mu} \times \phi \right) \qquad (6.9)$$

This equation holds for any ε, and hence each isospin component of the quantity in parentheses must be individually conserved. Thus, with an appropriate choice of rescaling factor $-1/\hbar c$, we have the conserved currents

$$\mathbf{J}_\mu = \frac{c}{\hbar} \left(\frac{\partial\phi}{\partial x_\mu} \times \phi \right) \qquad ; \text{isovector currents}$$

$$\frac{\partial \mathbf{J}_\mu}{\partial x_\mu} = 0 \qquad ; \text{conserved} \qquad (6.10)$$

The bold-face notation $\mathbf{J}_\mu$ now indicates a vector in isospin space.[5]

We know from the arguments in Eqs. (5.80)–(5.82) that the integral over all space of the fourth component of a conserved current will be a constant of the motion. The momentum density conjugate to ϕ is

$$\pi = \frac{\partial\mathcal{L}}{\partial(\partial\phi/\partial t)} = \frac{\partial\phi}{\partial t} \qquad ; \text{momentum density} \qquad (6.11)$$

Hence the conserved charge can be written as

$$\mathbf{Q} = -\frac{1}{\hbar} \int_\Omega d^3x \, (\boldsymbol{\pi} \times \boldsymbol{\phi}) \qquad ; \text{conserved charge}$$

$$Q_i = -\frac{1}{\hbar} \epsilon_{ijk} \int_\Omega d^3x \, (\pi_j \phi_k) \qquad (6.12)$$

where the second line is the component form of the first. Let us write this expression another way. Introduce a three-component column vector $\underline{\phi}$, and a set of 3×3 matrices

$$[\underline{t}_i]_{jk} \equiv -i\epsilon_{ijk} \qquad ; (i,j,k) = 1,2,3 \qquad (6.13)$$

---

[4]Compare with Eq. (3.12).

[5]When required, the spatial part of this current will carry the familiar vector notation $\mathbf{J}_\mu \equiv (\vec{\mathbf{J}}, i\rho)$; note that $\mathbf{Q} = \int_\Omega d^3x \, \rho$.

Thus

$$\underline{\phi} = \begin{bmatrix} \phi_1 \\ \phi_2 \\ \phi_3 \end{bmatrix} \; ; \; \underline{t}_1 = \begin{bmatrix} 0 & 0 & 0 \\ 0 & 0 & -i \\ 0 & i & 0 \end{bmatrix} \; ; \; \underline{t}_2 = \begin{bmatrix} 0 & 0 & i \\ 0 & 0 & 0 \\ -i & & 0 \end{bmatrix} \; ; \; \underline{t}_3 = \begin{bmatrix} 0 & -i & 0 \\ i & 0 & 0 \\ 0 & 0 & 0 \end{bmatrix} \quad (6.14)$$

Write the canonical momentum density as a three-component row matrix

$$\underline{\pi} = (\pi_1, \pi_2, \pi_3) \tag{6.15}$$

Equation (6.12) can then be rewritten as a matrix relation

$$Q_i = \frac{1}{i\hbar} \int_\Omega d^3x \, \underline{\pi} \, \underline{t}_i \, \underline{\phi} \qquad ; \text{ matrix form} \tag{6.16}$$

As in the last chapter, the theory is quantized by imposing the following canonical commutation relations in the Schrödinger picture[6]

$$[\hat{\phi}_i(\mathbf{x}), \, \hat{\pi}_j(\mathbf{x}')] = i\hbar \, \delta_{ij} \delta^{(3)}(\mathbf{x} - \mathbf{x}') \qquad ; \text{ C.C.R.}$$

$$[\hat{\phi}_i(\mathbf{x}), \, \hat{\phi}_j(\mathbf{x}')] = [\hat{\pi}_i(\mathbf{x}), \, \hat{\pi}_j(\mathbf{x}')] = 0 \qquad (i, j) = 1, 2, 3 \tag{6.17}$$

The conserved charge in Eq. (6.16) becomes

$$\hat{Q}_i = \frac{1}{i\hbar} \int_\Omega d^3x \, \hat{\underline{\pi}}(\mathbf{x}) \, \underline{t}_i \, \hat{\underline{\phi}}(\mathbf{x}) \tag{6.18}$$

We are now in a position to compute the *commutation relations* of these charges[7]

$$[\hat{Q}_i, \hat{Q}_j] = \left(\frac{1}{i\hbar}\right)^2 \int_\Omega d^3x \int_\Omega d^3y \left[\hat{\pi}_a(\mathbf{x})[\underline{t}_i]_{ab}\hat{\phi}_b(\mathbf{x}), \, \hat{\pi}_r(\mathbf{y})[\underline{t}_j]_{rs}\hat{\phi}_s(\mathbf{y})\right]$$

$$= \frac{1}{i\hbar} \int_\Omega d^3x \int_\Omega d^3y \left\{ \hat{\pi}_a(\mathbf{x})[\underline{t}_i]_{ab} \, \delta_{br}\delta^{(3)}(\mathbf{x} - \mathbf{y}) \, [\underline{t}_j]_{rs}\hat{\phi}_s(\mathbf{y}) - \right.$$

$$\left. \hat{\pi}_r(\mathbf{y})[\underline{t}_j]_{rs} \, \delta_{sa}\delta^{(3)}(\mathbf{x} - \mathbf{y}) \, [\underline{t}_i]_{ab}\hat{\phi}_b(\mathbf{x}) \right\}$$

$$= \frac{1}{i\hbar} \int_\Omega d^3x \, \hat{\underline{\pi}}(\mathbf{x}) \, [\underline{t}_i, \underline{t}_j] \, \hat{\underline{\phi}}(\mathbf{x}) \tag{6.19}$$

Thus we have the lovely, general, result that *the commutator of the charges in Eq. (6.18) is directly related to the commutator of the matrices involved*

---

[6] All the subsequent quantum field operators are then constructed in the *Schrödinger picture*, where the operators are time-independent and all the time dependence is put into the state vector $|\Psi(t)\rangle$.

[7] The task of verifying that the second equality follows from the first is assigned as Prob. 6.1; as usual, repeated Latin indices are summed, here from 1 to 3.

in their definition

$$[\hat{Q}_i, \hat{Q}_j] = \frac{1}{i\hbar} \int_\Omega d^3x \, \hat{\underline{\pi}}(\mathbf{x}) \, [\underline{t}_i, \underline{t}_j] \, \hat{\underline{\phi}}(\mathbf{x}) \tag{6.20}$$

An explicit calculation establishes that the 3×3 matrices in Eqs. (6.14) provide a representation of the commutation relations for the generators of $SU(2)$[8]

$$[\underline{t}_i, \underline{t}_j] = i\epsilon_{ijk} \underline{t}_k \qquad ; \; (i,j,k) = 1,2,3 \tag{6.21}$$

Hence Eq, (6.20) becomes

$$[\hat{Q}_i, \hat{Q}_j] = i\epsilon_{ijk} \hat{Q}_k \tag{6.22}$$

If the conserved charges are now identified with the components of the isospin operator $\hat{\mathbf{T}} = (\hat{T}_1, \hat{T}_2, \hat{T}_3)$

$$\hat{Q}_i \equiv \hat{T}_i \qquad ; \; i = 1,2,3 \tag{6.23}$$

then one has a direct analogy to angular momentum in real space

$$[\hat{T}_i, \hat{T}_j] = i\epsilon_{ijk}\hat{T}_k \qquad ; \; (i,j,k) = 1,2,3 \tag{6.24}$$

The symmetry group of these isospin transformations is $SU(2)$, and

> *All of the angular momentum analysis in chapter 3 can now be taken over directly to isospin!*

In a sense the above discussion is trivial, since the lagrangian density in Eq. (6.7) is simply the sum of three free contributions; however, suppose that this symmetry is imposed on the appropriate lagrangian of an *interacting* field theory involving these fields, for example

$$\mathcal{L}_1 = -\frac{\lambda}{4!}(\boldsymbol{\phi} \cdot \boldsymbol{\phi})^2 \qquad ; \text{ example interaction} \tag{6.25}$$

One could then conclude that, *even in the presence of interactions,*

- Since $\hat{\mathcal{L}}$ is invariant under isospin rotations, the components of the isospin operator $\hat{\mathbf{T}} = (\hat{T}_1, \hat{T}_2, \hat{T}_3)$ are conserved;
- These components satisfy the angular momentum algebra in Eq. (6.24);
- The components of $\hat{\mathbf{T}}$ are the generators of rotations in the internal isospin space (verified below);

---

[8]See Prob. 6.2.

- The states in the internal space can be taken as eigenstates of $(\hat{\mathbf{T}}^2, \hat{T}_3)$, denoted by $|tm_t\rangle$;
- The energy eigenstates of given t will be $(2t + 1)$-fold degenerate in m_t.

These are powerful results, particularly if the coupling is strong, and one has no practical means of solving the theory!

6.3.1.2 *Isospinor*

Consider the free Dirac lagrangian density with two (essentially) equal-mass contributions, as would apply, for example, to the proton and neutron (p, n)

$$\mathcal{L} = -\hbar c \bar{\psi}_p \left(\gamma_\mu \frac{\partial}{\partial x_\mu} + M \right) \psi_p - \hbar c \bar{\psi}_n \left(\gamma_\mu \frac{\partial}{\partial x_\mu} + M \right) \psi_n \quad (6.26)$$

One can introduce two-component spinors to keep track of the particle *type*

$$\underline{\eta}_p = \begin{pmatrix} 1 \\ 0 \end{pmatrix} \qquad ; \underline{\eta}_n = \begin{pmatrix} 0 \\ 1 \end{pmatrix} \quad (6.27)$$

The Dirac field can then be written as

$$\underline{\psi} = \psi_p \underline{\eta}_p + \psi_n \underline{\eta}_n = \begin{pmatrix} \psi_p \\ \psi_n \end{pmatrix} \quad (6.28)$$

Here both (ψ_p, ψ_n) are four-component Dirac spinors, but, as previously, we suppress the underlining of these column vectors and of the Dirac matrices. For clarity, the isospin matrices are now underlined. The Dirac lagrangian density in Eq. (6.26) then takes the form

$$\mathcal{L} = -\hbar c \, \underline{\bar{\psi}} \left(\gamma_\mu \frac{\partial}{\partial x_\mu} + M \right) \underline{\psi} \quad (6.29)$$

The momentum density conjugate to $\underline{\psi}$ is[9]

$$\underline{\pi} = \frac{\partial \mathcal{L}}{\partial (\partial \underline{\psi} / \partial t)} = i\hbar \, \underline{\psi}^\dagger \quad (6.30)$$

We now observe that the lagrangian density in Eq. (6.29) is invariant under the transformation

$$\underline{\psi} \to \underline{\psi}' = \underline{r}(\boldsymbol{\omega}) \underline{\psi} \qquad ; \underline{r}(\boldsymbol{\omega}) = \exp \left\{ \frac{i}{2} \underline{\boldsymbol{\tau}} \cdot \boldsymbol{\omega} \right\} \quad (6.31)$$

---

[9]Explicitly, in matrix notation, $\underline{\pi} = i\hbar(\psi_p^\dagger, \psi_n^\dagger)$.

Here $\underline{r}(\boldsymbol{\omega})$ is the SU(2) matrix in Eq. (6.4), where the Pauli matrices for isospin are now denoted by $\boldsymbol{\tau} = (\tau_1, \tau_2, \tau_3)$. The infinitesimal form of this transformation with $\boldsymbol{\omega} = \boldsymbol{\varepsilon} \to 0$ gives

$$\underline{\psi} \to \underline{\psi} + \delta\underline{\psi} = \underline{\psi} + \frac{i}{2}\boldsymbol{\tau} \cdot \boldsymbol{\varepsilon}\,\underline{\psi} \qquad ; \varepsilon \text{ an infinitesimal} \qquad (6.32)$$

Noether's theorem then gives the corresponding conserved current

$$\tilde{\mathbf{J}}_\mu = = \frac{\partial\mathcal{L}}{\partial(\partial\underline{\psi}/\partial x_\mu)}\delta\underline{\psi} = -\frac{i\hbar c}{2}\boldsymbol{\varepsilon} \cdot \left(\bar{\underline{\psi}}\gamma_\mu\boldsymbol{\tau}\,\underline{\psi}\right) \qquad (6.33)$$

Since this holds for any ε, each isospin component of the quantity in parentheses must be conserved, and with the same rescaling factor as before, one has the conserved isovector currents

$$\mathbf{J}_\mu = i\bar{\underline{\psi}}\,\gamma_\mu\frac{1}{2}\boldsymbol{\tau}\,\underline{\psi} \qquad ; \text{ isovector currents}$$

$$\frac{\partial\mathbf{J}_\mu}{\partial x_\mu} = 0 \qquad ; \text{ conserved} \qquad (6.34)$$

The integral over all space of the fourth component of these conserved currents is again a constant of the motion, and these charges can be written in terms of the canonical momentum density in Eq. (6.30) as

$$\mathbf{Q} = \frac{1}{i\hbar}\int_\Omega d^3x\,\underline{\pi}\frac{1}{2}\boldsymbol{\tau}\,\underline{\psi} \qquad ; \text{ conserved charges} \qquad (6.35)$$

To quantize the theory, the following canonical *anticommutation* relations are imposed on the Dirac field in the Schrödinger picture

$$\{\hat{\psi}_{\alpha a}(\mathbf{x}),\,\hat{\pi}_{\beta b}(\mathbf{x}')\} = i\hbar\delta_{\alpha\beta}\,\delta_{ab}\delta^{(3)}(\mathbf{x} - \mathbf{x}') \qquad ; \text{ C.(A)C.R.} \quad ; (a,b) = 1,2$$

$$\{\hat{\psi}_{\alpha a}(\mathbf{x}),\,\hat{\psi}_{\beta b}(\mathbf{x}')\} = \{\hat{\pi}_{\alpha a}(\mathbf{x}),\,\hat{\pi}_{\beta b}(\mathbf{x}')\} = 0 \qquad (\alpha,\beta) = 1,2,3,4 \qquad (6.36)$$

Here the first index on the subscript refers to the component of the Dirac spinor, and the second to the isospin component. We are now in a position to compute the commutation relations of the charges

$$\hat{Q}_j = \frac{1}{i\hbar}\int_\Omega d^3x\,\hat{\underline{\pi}}(\mathbf{x})\frac{1}{2}\tau_j\,\hat{\underline{\psi}}(\mathbf{x}) \qquad ; j = 1,2,3 \qquad (6.37)$$

Despite the fact that we here quantize with *anticommutation* relations, the calculation *goes through exactly as in Eqs. (6.19)*[10]

$$[\hat{Q}_i, \hat{Q}_j] = \frac{1}{i\hbar}\int_\Omega d^3x\,\hat{\underline{\pi}}(\mathbf{x})\,[\frac{1}{2}\tau_i, \frac{1}{2}\tau_j]\,\hat{\underline{\psi}}(\mathbf{x}) \qquad (6.38)$$

---

[10]See Prob. 6.3.

We know the Pauli matrices provide a 2×2 representation of the commutation relations for the generators of SU(2)

$$[\frac{1}{2}\mathcal{I}_i, \frac{1}{2}\mathcal{I}_j] = i\epsilon_{ijk}\frac{1}{2}\mathcal{I}_k \qquad ; (i,j,k) = 1,2,3 \qquad (6.39)$$

Hence Eq. (6.38) becomes

$$[\hat{Q}_i, \hat{Q}_j] = i\epsilon_{ijk}\hat{Q}_k \qquad (6.40)$$

If the conserved charges are again identified with the components of the isospin operator $\hat{\mathbf{T}} = (\hat{T}_1, \hat{T}_2, \hat{T}_3)$

$$\hat{Q}_i \equiv \hat{T}_i \qquad ; i = 1,2,3 \qquad (6.41)$$

then one again has a direct analogy to angular momentum in real space

$$[\hat{T}_i, \hat{T}_j] = i\epsilon_{ijk}\hat{T}_k \qquad ; (i,j,k) = 1,2,3 \qquad (6.42)$$

The symmetry group of these isospin transformations is again SU(2), and all the comments made after Eq. (6.24) again apply. With the two-component isospinor transformation law in Eq. (6.31), one is working with the fundamental SU(2) matrices. With the insertion of the momentum density in Eq. (6.30), the isospin can be written in a form more familiar from second quantization

$$\hat{\mathbf{T}} = \int_\Omega d^3x\, \hat{\psi}^\dagger(\mathbf{x})\frac{1}{2}\boldsymbol{\tau}\,\hat{\psi}(\mathbf{x}) \qquad (6.43)$$

6.3.1.3 *Transformation Law*

Consider the commutator of the isospin operator with the field in the isovector case

$$[\hat{\mathbf{T}}, \hat{\phi}_j(\mathbf{x})] = \frac{1}{i\hbar}\int_\Omega d^3y\left[\hat{\pi}_a(\mathbf{y})[\mathbf{t}]_{ab}\hat{\phi}_b(\mathbf{y}), \hat{\phi}_j(\mathbf{x})\right]$$

$$= \frac{1}{i\hbar}\int_\Omega d^3y\,\{\hat{\pi}_a(\mathbf{y})[\mathbf{t}]_{ab}\hat{\phi}_b(\mathbf{y})\hat{\phi}_j(\mathbf{x}) - \hat{\phi}_j(\mathbf{x})\hat{\pi}_a(\mathbf{y})[\mathbf{t}]_{ab}\hat{\phi}_b(\mathbf{y})\}$$

$$= -\int_\Omega d^3y\,\delta_{ja}\delta^{(3)}(\mathbf{x}-\mathbf{y})[\mathbf{t}]_{ab}\hat{\phi}_b(\mathbf{y})$$

$$= -[\mathbf{t}\,\hat{\underline{\phi}}(\mathbf{x})]_j \qquad (6.44)$$

To put this back in matrix notation

$$[\hat{\mathbf{T}}, \hat{\underline{\phi}}(\mathbf{x})] = -\mathbf{t}\,\hat{\underline{\phi}}(\mathbf{x}) \qquad (6.45)$$

This is the generator relation for the following finite transformation[11]

$$e^{-i\boldsymbol{\omega}\cdot\hat{\mathbf{T}}}\,\underline{\hat{\phi}}(\mathbf{x})\,e^{i\boldsymbol{\omega}\cdot\hat{\mathbf{T}}} = \exp\left\{i\boldsymbol{\omega}\cdot\underline{\mathbf{t}}\right\}\underline{\hat{\phi}}(\mathbf{x}) \qquad (6.46)$$

This is the transformation law for an *isovector field.*

Consider the commutator of the isospin operator with the field in the isospinor case

$$[\hat{\mathbf{T}},\,\hat{\psi}_j(\mathbf{x})] = \frac{1}{i\hbar}\int_{\Omega} d^3y\left[\hat{\pi}_a(\mathbf{y})\frac{1}{2}[\underline{\boldsymbol{\tau}}]_{ab}\hat{\psi}_b(\mathbf{y}),\,\hat{\psi}_j(\mathbf{x})\right] \quad ;\,(a,b,j) = 1,2 \ (6.47)$$

Even though the quantization here is accomplished with *anticommutation* relations, the calculation again goes through in *exactly the same fashion*[12]

$$[\hat{\mathbf{T}},\,\underline{\hat{\psi}}(\mathbf{x})] = -\frac{1}{2}\underline{\boldsymbol{\tau}}\,\underline{\hat{\psi}}(\mathbf{x}) \qquad (6.48)$$

This is the generator relation for the following finite transformation

$$e^{-i\boldsymbol{\omega}\cdot\hat{\mathbf{T}}}\,\underline{\hat{\psi}}(\mathbf{x})\,e^{i\boldsymbol{\omega}\cdot\hat{\mathbf{T}}} = \exp\left\{\frac{i}{2}\boldsymbol{\omega}\cdot\underline{\boldsymbol{\tau}}\right\}\underline{\hat{\psi}}(\mathbf{x}) \qquad (6.49)$$

This is the transformation law for an *isospinor field,* which reproduces our starting point in Eq. (6.31), and confirms that $\hat{\mathbf{T}}$ is indeed the generator of rotations in isospin space.

6.3.2 *Lie Groups*

The angular momentum analysis in chapter 3 is a special case [SU(2)] of the theory of continuous, or *Lie groups.* Let us review that analysis from a group-theory point of view. This is not a book on group theory, and the reader is referred to one of the classic texts, for example [Hamermesh (1989)], to become truly informed on this topic. It is important, however, to become familiar with some of the basic concepts, and that is the goal here.

Commutation Rules. There is a set of hermitian generators $\hat{J}_i$, operators in abstract Hilbert space, satisfying an algebra that is closed under commutation

$$[\hat{J}_i,\,\hat{J}_j] = i\epsilon_{ijk}\hat{J}_k \qquad ; \text{Lie algebra}$$
$$\hat{J}_i^{\dagger} = \hat{J}_i \qquad\qquad (i,j,k) = 1,2,3 \qquad (6.50)$$

---

[11]See the beginning of chapter 3.
[12]See Prob. 6.4.

The commutator of two generators is a linear combination of the generators; they are said to form a *Lie algebra*.[13] These commutation relations are independent of any particular representation.

A *Casimir operator* is one that commutes with all the generators. Here the square of the total angular momentum is such an operator

$$\hat{\mathbf{J}}^2 = \hat{J}_1^2 + \hat{J}_2^2 + \hat{J}_3^2 \qquad ; \text{Casimir operator}$$

$$[\hat{J}_i, \hat{\mathbf{J}}^2] = 0 \qquad ; i = 1, 2, 3 \qquad (6.51)$$

Starting from these commutation relations, we were able to deduce the *spectrum*

$$\hat{\mathbf{J}}^2|jm\rangle = j(j+1)|jm\rangle \qquad ; j = 0, 1/2, 1, 3/2, 2, \cdots$$

$$\hat{J}_3|jm\rangle = m|jm\rangle \qquad ; -j \leq m \leq j \qquad ; \text{integer steps} \quad (6.52)$$

as well as the effects of $(\hat{J}_1, \hat{J}_2)$ on these states [see Eqs. (3.59)].

Group. A *Lie group* can be associated with every Lie algebra. In the case of rotations one has

$$\hat{R}(\boldsymbol{\omega}) = e^{i\boldsymbol{\omega}\cdot\hat{\mathbf{J}}} \qquad ; \boldsymbol{\omega}\cdot\hat{\mathbf{J}} = \omega_1\hat{J}_1 + \omega_2\hat{J}_2 + \omega_3\hat{J}_3 \qquad (6.53)$$

Since the product of two rotations is again a rotation, these form a *three-parameter continuous group*

$$\hat{R}(\boldsymbol{\omega})\hat{R}(\boldsymbol{\omega}') = \hat{R}(\boldsymbol{\omega}'')$$

$$\hat{R}(\boldsymbol{\omega})^{-1} = \hat{R}(-\boldsymbol{\omega})$$

$$\hat{R}(0) = \hat{1} \qquad (6.54)$$

The first relation is established through the isomorphism with the SU(2) matrices $\underline{r}(\boldsymbol{\omega})$ [see Eqs. (6.3)–(6.6)]. It is a *unitary group* in quantum mechanics, with

$$\hat{R}(\boldsymbol{\omega})^\dagger = \hat{R}(\boldsymbol{\omega})^{-1} \qquad ; \text{unitary} \qquad (6.55)$$

Irreducible Representations. The states $|jm\rangle$ form a basis for an *irreducible representation* of the group

$$\hat{R}(\boldsymbol{\omega})|jm\rangle = \sum_{m'} \mathcal{D}^j_{m'm}(\boldsymbol{\omega})|jm'\rangle \qquad (6.56)$$

There are two important elements to this relation

• Only states with a given j are involved;

---

[13]For an extended discussion of Lie algebras in physics, see [Georgi (1999)].

- All states with various m are accessed by *some* rotation.

The matrices $\mathcal{D}^j_{m'm}(\omega)$ then multiply together in the same way as do the group elements

$$\hat{R}(\omega)\hat{R}(\omega')|jm\rangle = \sum_{m''}\sum_{m'}\mathcal{D}^j_{m''m'}(\omega)\mathcal{D}^j_{m'm}(\omega')|jm''\rangle$$

$$= \hat{R}(\omega'')|jm\rangle = \sum_{m''}\mathcal{D}^j_{m''m}(\omega'')|jm''\rangle \qquad (6.57)$$

Hence the $\mathcal{D}^j_{m'm}(\omega)$ form a *matrix representation* of the continuous group

$$\sum_{m'}\mathcal{D}^j_{m''m'}(\omega)\mathcal{D}^j_{m'm}(\omega') = \mathcal{D}^j_{m''m}(\omega'') \qquad (6.58)$$

The representation is *irreducible* since the group operators connect all m-states for a given j. The matrices are given explicitly by

$$\mathcal{D}^j_{m'm}(\omega) = \langle jm'|\hat{R}(\omega)|jm\rangle \qquad (6.59)$$

As we have seen, these matrices can be computed starting only from the commutation relations of the generators.

Direct product—Clebsch–Gordon Series. The direct-product states, say for two particles, $|j_1m_1\rangle|j_2m_2\rangle$, provide a basis for a *reducible* representation. One can write

$$|j_1m_1\rangle|j_2m_2\rangle = \sum_{JM}\langle j_1m_1j_2m_2|j_1j_2JM\rangle\,|j_1j_2JM\rangle \qquad (6.60)$$

This is a unitary transformation to states that transform among themselves under $\hat{R}(\omega)$. The states $|j_1j_2JM\rangle$ again form a basis for an *irreducible* representation

$$\hat{R}(\omega)|j_1j_2JM\rangle = \sum_{M'}\mathcal{D}^J_{M'M}(\omega)|j_1j_2JM'\rangle \qquad (6.61)$$

We know that each J appears once, and only once, in the sum in Eq. (6.60), with $J = (j_1+j_2,\ j_1+j_2-1,\ \cdots,\ |j_1-j_2|)$.[14] The coeffients in Eq. (6.60) are the *Clebsch-Gordon (C-G) coefficients*, which we have constructed for SU(2).

---

[14]This is a peculiarity of SU(2).

The transformation in Eq. (6.60) can be written as a relation between the corresponding representation matrices in the following schematic fashion [compare Eq. (3.134)]

$$\mathcal{D}^{j_1}_{m_1' m_1} \otimes \mathcal{D}^{j_2}_{m_2' m_2} = \underline{\mathcal{D}}^{j_1+j_2} \oplus \underline{\mathcal{D}}^{j_1+j_2-1} \oplus \cdots \oplus \underline{\mathcal{D}}^{|j_1-j_2|} \qquad (6.62)$$

The l.h.s. represents the direct product of two square matrices of dimension $(2j_1 + 1)$ and $(2j_2 + 1)$ respectively. It is a square matrix of dimension $(2j_1 + 1) \times (2j_2 + 1)$. The r.h.s. represents a direct sum of square matrices. Pictorially, by a change of basis, the large matrix has been reduced to the block-diagonal form shown in Fig. 6.1.

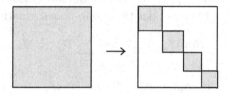

Fig. 6.1 Reduction of the direct-product representation into the direct sum of irreducible representations by a change of basis, for angular momentum, as indicated in Eq. (6.62).

As a shorthand for the matrix relation in Eq. (6.62), one can express the decomposition of the direct-product representation into the direct sum of irreducible representations, using the dimensions of each, as

$$[2j_1 + 1] \otimes [2j_2 + 1] = [2(j_1 + j_2) + 1] \oplus [2(j_1 + j_2 - 1) + 1] \oplus \cdots \oplus$$
$$[2|j_1 - j_2| + 1] \qquad (6.63)$$

Eigenvalue Spectrum. Assume that the generators $\hat{J}_i$ commute with the hamiltonian

$$[\hat{J}_i, \hat{H}] = 0 \qquad ; i = 1, 2, 3 \qquad (6.64)$$

The eigenstates of the hamiltonian can then be written as $|\gamma E j m\rangle$ where $(\hat{\Gamma}, \hat{H}, \hat{\mathbf{J}}^2, \hat{J}_3)$ form a complete set of mutually commuting hermitian operators for the problem at hand. One then has the following central *theorem*:

The degenerate eigenstates of $\hat{H}$ form a basis for an irreducible representation of SU(2); or, all states belonging to the basis for an irreducible representation of SU(2) are degenerate.

The proof is immediate. Consider an energy eigenstate

$$\hat{H}|\gamma E j m\rangle = E|\gamma E j m\rangle \tag{6.65}$$

Now let $\hat{J}_i$ act on this relation, and use Eq. (6.64),

$$\hat{H}\left[\hat{J}_i|\gamma E j m\rangle\right] = E\left[\hat{J}_i|\gamma E j m\rangle\right] \tag{6.66}$$

By repeated application of this result, one can reach all the states in the basis for the irreducible representation, that is, all the m-states for a given j. The theorem is thus established. We have seen examples of this result is all the central-field applications in Vol. I.

Irreducible Tensor Operators. An irreducible tensor operator (ITO) $\hat{T}_{\kappa q}$ is defined by its behavior under the group transformation

$$\hat{R}(\boldsymbol{\omega})\hat{T}_{\kappa q}\hat{R}(\boldsymbol{\omega})^{-1} = \sum_{q'} \mathcal{D}^{\kappa}_{q'q}(\boldsymbol{\omega})\hat{T}_{\kappa q'} \tag{6.67}$$

An ITO has $(2\kappa+1)$ components. If one lets $\boldsymbol{\omega} \to 0$, this defining equation leads to a set of commutation relations with the generators $\hat{J}_i$.

Wigner-Eckart Theorem. The proof of the Wigner-Eckart theorem uses only the group properties reviewed above. It states that the matrix element of an ITO taken between states belonging to the basis for an irreducible representation of the group take the form

$$\langle\mu'\nu'|\hat{T}_{\kappa q}|\mu\nu\rangle = \langle\mu\nu\kappa q|\mu\kappa\mu'\nu'\rangle \frac{1}{(2\mu'+1)^{1/2}} \langle\mu'||T_\kappa||\mu\rangle \tag{6.68}$$

Here (μ, μ') label the irreducible representations, and (ν, ν') are the *components* within those representations. $\langle\mu\nu\kappa q|\mu\kappa\mu'\nu'\rangle$ is a C-G coefficient. The Wigner-Eckart theorem

- Explicitly exhibits the dependence on (ν, ν', q);
- Manifests the selection rules on κ through the C-G coefficient.

6.3.3 Sakata Model–SU(3)

The Sakata model adds one degenerate isoscalar Dirac field to the previous discussion of the isospinor Dirac field. The model is presented here with the following motivation:

- It is the simplest extension of SU(2) to a higher rank group, here SU(3);
- It illustrates the methods;

- Despite the fact that when applied to baryons it does not accurately describe the observed baryon multiplets, it does give the correct classification of mesons within the framework of the Fermi-Yang model of mesons as bound baryon-antibaryon pairs;
- With some messaging in terms of selective retention of irreducible representations obtained from direct products of the fundamental basis, it does reproduce the "Eightfold-Way" of [Gell-Mann and Ne'eman (1963)], and the quark model of Gell-Mann and Zweig.

6.3.3.1 *Dirac Triplet*

The Sakata model, as presented here, adds one degenerate baryon with vanishing isospin, the Λ particle, to the nucleon (p, n) isodoublet considered previously. A three-component spinor now keeps track of the particle type

$$\underline{\eta}_p = \begin{pmatrix} 1 \\ 0 \\ 0 \end{pmatrix} \qquad ; \underline{\eta}_n = \begin{pmatrix} 0 \\ 1 \\ 0 \end{pmatrix} \qquad ; \underline{\eta}_\Lambda = \begin{pmatrix} 0 \\ 0 \\ 1 \end{pmatrix} \qquad (6.69)$$

The Dirac field can then be written as

$$\underline{\psi} = \psi_p \underline{\eta}_p + \psi_n \underline{\eta}_n + \psi_\Lambda \underline{\eta}_\Lambda = \begin{pmatrix} \psi_p \\ \psi_n \\ \psi_\Lambda \end{pmatrix} \qquad (6.70)$$

Here all three $(\psi_p, \psi_n, \psi_\Lambda)$ are four-component Dirac spinors, but, as previously, we suppress the underlining of these column vectors. For clarity, the internal matrices are now underlined. The Dirac lagrangian density then takes the form

$$\mathcal{L} = -\hbar c \, \underline{\bar{\psi}} \left(\gamma_\mu \frac{\partial}{\partial x_\mu} + M \right) \underline{\psi} \qquad (6.71)$$

We now observe that the lagrangian density in Eq. (6.71) is invariant under the transformation

$$\underline{\psi} \rightarrow \underline{\psi}' = \underline{r}(\omega)\underline{\psi} \qquad (6.72)$$

where $\underline{r}(\omega)$ is a unitary 3×3 matrix of the form

$$\underline{r}(\omega) = \exp\left\{ \frac{i}{2} \underline{\lambda}^a \omega^a \right\} \qquad ; \omega = (\omega^1, \omega^2, \cdots, \omega^8) \qquad (6.73)$$

The repeated Latin index is here summed from 1 to 8.

As discussed in Vol. I, the eight hermitian, traceless, 3×3 Gell-Mann matrices $\underline{\lambda}^a$ for $a = 1, \cdots, 8$ are the analogues of the Pauli matrices; they are given in order by

$$\begin{pmatrix} & 1 & \\ 1 & & \\ & & \end{pmatrix} \begin{pmatrix} & -i & \\ i & & \\ & & \end{pmatrix} \begin{pmatrix} 1 & & \\ & -1 & \\ & & \end{pmatrix} \begin{pmatrix} & & 1 \\ & & \\ 1 & & \end{pmatrix} \begin{pmatrix} & & -i \\ & & \\ i & & \end{pmatrix}$$

$$\begin{pmatrix} & & \\ & & 1 \\ & 1 & \end{pmatrix} \begin{pmatrix} & & \\ & & -i \\ & i & \end{pmatrix} \begin{pmatrix} 1/\sqrt{3} & & \\ & 1/\sqrt{3} & \\ & & -2/\sqrt{3} \end{pmatrix} \qquad (6.74)$$

These matrices satisfy the following Lie algebra

$$[\frac{1}{2}\lambda^a, \frac{1}{2}\lambda^b] = i f^{abc} \frac{1}{2}\lambda^c \qquad ; (a,b,c) = 1, 2, \cdots, 8 \qquad (6.75)$$

Here the f^{abc} are the *structure constants*; they are real and antisymmetric in the indices (abc).

The 3×3, unitary, unimodular matrices $\underline{r}(\omega)$ satisfy the following relations[15]

$$\begin{aligned} \underline{r}(\omega)\underline{r}(\omega') &= \underline{r}(\omega'') & ; \det \underline{r}(\omega) &= 1 \\ \underline{r}(\omega)^{-1} &= \underline{r}(-\omega) & ; \underline{r}(0) &= 1 \end{aligned} \qquad (6.76)$$

The matrices $\underline{r}(\omega)$ form an eight-parameter continuous Lie group, the group SU(3), with the Lie algebra of Eq. (6.75).

There is a corresponding set of generators in the abstract Hilbert space

$$[\hat{G}^a, \hat{G}^b] = i f^{abc} \hat{G}^c \qquad ; (a,b,c) = 1, 2, \cdots, 8 \qquad (6.77)$$

The connection to the above is made through the basis for the fundamental representation, denoted with its dimension as $|(3)m\rangle$, with $m = (p, n, \Lambda)$. Thus, in direct analogy to Eq. (6.3),

$$\begin{aligned} e^{i\omega^a \hat{G}^a} |(3)m\rangle &= \underline{r}_{m'm}(\omega) |(3)m'\rangle & ; (m', m) &= (p, n, \Lambda) \\ \underline{r}_{m'm}(\omega) &= \langle (3)m'| e^{i\omega^a \hat{G}^a} |(3)m\rangle & \omega &= (\omega^1, \cdots, \omega^8) \end{aligned} \quad (6.78)$$

The generators can be constructed exactly as in Eq. (6.43)

$$\hat{G}^a = \int_\Omega d^3x \, \hat{\underline{\psi}}^\dagger(\mathbf{x}) \frac{1}{2}\underline{\lambda}^a \, \hat{\underline{\psi}}(\mathbf{x}) \qquad (6.79)$$

---

[15]See Prob. 6.14.

The Dirac triplet field transforms according to the fundamental representation as in Eq. (6.49)

$$e^{-i\omega^a \hat{G}^a} \, \hat{\psi}(\mathbf{x}) \, e^{i\omega^a \hat{G}^a} = \exp\left\{\frac{i}{2}\omega^a \underline{\lambda}^a\right\} \hat{\psi}(\mathbf{x}) \qquad (6.80)$$

6.3.3.2 *Scalar Octet*

Suppose one has eight degenerate, real (hermitian), scalar fields ϕ^a with $a = 1, 2, \cdots, 8$, and a lagrangian density of the form

$$\mathcal{L} = -c^2 \left[\frac{\partial \phi^a}{\partial x_\mu} \frac{\partial \phi^a}{\partial x_\mu} + m^2 \phi^a \phi^a\right] \qquad ; \, a = 1, 2, \cdots, 8 \qquad (6.81)$$

Let ε^a be a set of infinitesimals. In analogy to Eq. (6.8), this lagrangian density is invariant under the transformation

$$\phi^a \rightarrow \phi^a + \delta\phi^a = \phi^a + f^{abc}\phi^b\varepsilon^c \qquad (6.82)$$

The previous analysis of the isovector scalar multiplet can be carried out just as before. Define the eight component column vector field $\underline{\phi}$ as

$$[\underline{\phi}]_a \equiv \phi^a \qquad ; \, a = 1, 2, \cdots, 8 \qquad (6.83)$$

In analogous fashion, define the eight 8×8 matrices $\underline{t}^a$ by

$$[\underline{t}^a]_{bc} \equiv -if^{abc} \qquad ; \, (a, b, c) = 1, 2, \cdots, 8 \qquad (6.84)$$

The generators are then constructed exactly as in Eq. (6.18)[16]

$$\hat{G}^a = \frac{1}{i\hbar} \int_\Omega d^3x \, \underline{\hat{\pi}}(\mathbf{x}) \, \underline{t}^a \, \underline{\hat{\phi}}(\mathbf{x}) \qquad (6.85)$$

The field $\underline{\hat{\phi}}$ transforms under the group just as in Eq. (6.46)

$$e^{-i\omega^a \hat{G}^a} \, \underline{\hat{\phi}}(\mathbf{x}) \, e^{i\omega^a \hat{G}^a} = \exp\left\{i\omega^a \underline{t}^a\right\} \underline{\hat{\phi}}(\mathbf{x}) \qquad (6.86)$$

---

[16]We leave it for the dedicated reader to demonstrate that the matrices defined in Eq. (6.84) provide an 8×8 matrix representation, the *regular representation*, of the SU(3) algebra; they satisfy $[\underline{t}^a, \underline{t}^b] = if^{abc}\underline{t}^c$ [compare with Eq. (6.21)].

6.3.3.3 *Interacting Fields (Dirac-Scalar)*

Consider a theory composed of a Dirac triplet and a scalar octet. A possible interaction that is invariant under $SU(3)$ transformations can be constructed as[17]

$$\mathcal{L}_1 = g\,\underline{\bar{\psi}}\,\frac{1}{2}\lambda^a\,\underline{\psi}\,\phi^a \tag{6.87}$$

6.4   Phase Invariance

In addition to their internal symmetries, the model lagrangian densities discussed above exhibit a *global phase invariance*.

6.4.1   *Global Phase Invariance*

The lagrangian densities in Eqs. (6.29) and (6.71) are invariant under the following global phase transformation

$$\underline{\psi} \to e^{i\theta}\,\underline{\psi} \qquad ; \; \underline{\bar{\psi}} \to e^{-i\theta}\,\underline{\bar{\psi}} \tag{6.88}$$

It is clear that any Lorentz-invariant lagrangian density constructed in the form $\bar{\psi}(\cdots)\psi$ in the baryon sector will exhibit such an invariance. The arguments in Eqs. (5.112)–(5.115) then lead, through Noether's theorem, to the existence of a conserved baryon current

$$j_\mu = i\underline{\bar{\psi}}\gamma_\mu\underline{\psi} \qquad ; \; \text{conserved baryon current} \tag{6.89}$$

The integral over all space of the fourth component of this current is a constant of the motion. This is the *baryon number*, which played an important role in Vol. I. When quantized, the baryon number operator takes the form

$$\hat{B} = \int_\Omega d^3x\,\underline{\hat{\psi}}^\dagger(\mathbf{x})\,\underline{\hat{\psi}}(\mathbf{x}) \qquad ; \; \text{baryon number} \tag{6.90}$$

It follows from the arguments in Eqs. (6.48)–(6.49) that the baryon number operator is the generator of the phase transformation

$$[\hat{B},\,\underline{\hat{\psi}}(\mathbf{x})] = -\underline{\hat{\psi}}(\mathbf{x})$$
$$e^{-i\theta\hat{B}}\,\underline{\hat{\psi}}(\mathbf{x})\,e^{i\theta\hat{B}} = e^{i\theta}\,\underline{\hat{\psi}}(\mathbf{x}) \tag{6.91}$$

---

[17]Compare Probs. 6.8–6.9.

Since the phase transformations form a simple abelian one-parameter group,[18] one says that the symmetry group of the above models is $SU(n) \otimes U(1)$ with $n = 2, 3$. It follows from the arguments in Eqs. (6.43) and (6.38) that the baryon number operator commutes with all the generators of the $SU(n)$ symmetry group.

6.4.2 Local Phase Invariance

Let us return to the problem of a charged Dirac field ψ interacting with the electromagnetic field through the vector potential A_μ. It follows from Probl. 9.4 and appendix C that the starting lagrangian density for this interacting field theory (QED) takes the form

$$\mathcal{L}_{\text{QED}} = -\hbar c\, \bar{\psi} \left[\gamma_\mu \left(\frac{\partial}{\partial x_\mu} - \frac{ie}{\hbar} A_\mu \right) + M \right] \psi - \frac{\varepsilon_0}{4} F_{\mu\nu} F_{\mu\nu}$$

$$\frac{1}{c} F_{\mu\nu} = \frac{\partial A_\nu}{\partial x_\mu} - \frac{\partial A_\mu}{\partial x_\nu} \tag{6.92}$$

This lagrangian density is invariant under the following transformation

$$A_\mu \rightarrow A_\mu + \frac{\partial \Lambda(x)}{\partial x_\mu}$$

$$\psi \rightarrow \exp\left\{ \frac{ie}{\hbar} \Lambda(x) \right\} \psi \tag{6.93}$$

Here $\Lambda(x)$ is some differentiable scalar function of the space-time coordinate x. While the first relation is a change in gauge, the second is simply a *local phase transformation* on the charged Dirac field ψ. A local phase transformation on the Dirac field ψ is thus equivalent to a change in gauge, which cannot affect the physics.[19]

Let us turn the argument around. Suppose one starts with the lagrangian density for the free Dirac field

$$\mathcal{L} = -\hbar c\, \bar{\psi} \left(\gamma_\mu \frac{\partial}{\partial x_\mu} + M \right) \psi \tag{6.94}$$

Now *impose the requirement that the theory be invariant under local phase transformations*. How is this to be accomplished?

---

[18] "Abelian" implies that the group elements commute with each other.

[19] The *global* phase invariance of the lagrangian density in Eq. (6.92) again leads to the conserved electromagnetic current of Eq. (5.115).

(1) First, introduce a photon field A_μ and couple it in the following fashion ("minimal coupling")

$$\frac{\partial}{\partial x_\mu} \rightarrow \left(\frac{\partial}{\partial x_\mu} - \frac{ie}{\hbar} A_\mu\right) \tag{6.95}$$

This *covariant derivative* acting on the Dirac field then transforms under Eqs. (6.93) as if it were a *global* phase transformation

$$\left(\frac{\partial}{\partial x_\mu} - \frac{ie}{\hbar} A_\mu\right) \psi \rightarrow \exp\left\{\frac{ie}{\hbar} \Lambda(x)\right\} \left(\frac{\partial}{\partial x_\mu} - \frac{ie}{\hbar} A_\mu\right) \psi \tag{6.96}$$

The lagrangian density arising from Eq. (6.94) is then left invariant.

(2) In addition to terms coupling the photon (gauge) field to the charged Dirac field, there must be a kinetic energy term for the photon itself. To be gauge invariant, it must be constructed from the combination

$$\frac{1}{c} F_{\mu\nu} \equiv \frac{\partial A_\nu}{\partial x_\mu} - \frac{\partial A_\mu}{\partial x_\nu} \tag{6.97}$$

The leading Lorentz-invariant term is just

$$\mathcal{L}_{\text{EM}} = -\frac{\varepsilon_0}{4} F_{\mu\nu} F_{\mu\nu} \tag{6.98}$$

We then know from appendix C that the field equations are just Maxwell's equations with the charge $e = e_{\text{SI}}$.[20]

(3) A quadratic mass term for the photon must *not* be included because it is *not gauge invariant*

$$\mathcal{L}_{\text{mass}} = -\frac{1}{2} m_\gamma^2 c^2 A_\mu A_\mu \qquad ; \underline{\text{not}} \text{ included} \tag{6.99}$$

These arguments lead to the lagrangian density in Eq. (6.92), and remarkably enough

Imposing local phase invariance leads to the theory of QED, the most accurate physical theory known!

---

[20]The coupling constant e is now defined through the lagrangian. For example, by rescaling $c\sqrt{\varepsilon_0} A_\mu \rightarrow A_\mu$, one can write $\mathcal{L}_{\text{EM}} = -(1/4) F_{\mu\nu} F_{\mu\nu}$, where now $F_{\mu\nu} \equiv \partial A_\nu/\partial x_\mu - \partial A_\mu/\partial x_\nu$, and $e = e_{\text{SI}}/c\sqrt{\varepsilon_0}$ in Eq. (6.95). This latter form of the lagrangian provides the point of departure in the next section. Note that upon conversion from SI to H-L units (see appendix K in Vol. I), $e_{\text{SI}}/c\sqrt{\varepsilon_0} \rightarrow e_{\text{HL}}/c$.

6.5 Yang-Mills Theories

In analogy to the above, Yang and Mills asked the question, "How does one construct a theory that is *invariant under a local non-abelian internal symmetry transformation?*" A local internal symmetry transformation is one where the parameters in the transformation, instead of being global constants, are functions of the space-time position x. We illustrate their method in the case of isospin–SU(2), as they did in their original paper [Yang and Mills (1954)], although the method is applicable to any Lie group. We indicate the extension to SU(3) at the end of this section.

Start with the lagrangian density for an isospinor Dirac field in Eq. (6.29)

$$\mathcal{L} = -\hbar c\, \underline{\bar{\psi}} \left(\gamma_\mu \frac{\partial}{\partial x_\mu} + M \right) \underline{\psi} \tag{6.100}$$

This possesses a global internal SU(2) symmetry, and is invariant under the transformation

$$\underline{\psi} \rightarrow \underline{\psi}' = \underline{r}(\boldsymbol{\omega})\underline{\psi} \qquad ; \ \underline{r}(\boldsymbol{\omega}) = \exp\left\{ \frac{i}{2}\boldsymbol{\tau}\cdot\boldsymbol{\omega} \right\} \tag{6.101}$$

Here $\boldsymbol{\omega} = (\omega_1, \omega_2, \omega_3)$ is a set of three real parameters, and $\boldsymbol{\tau} = (\underline{\tau}_1, \underline{\tau}_2, \underline{\tau}_3)$ are the Pauli matrices. Now consider a *local* symmetry transformation where $\boldsymbol{\omega} = \boldsymbol{\omega}(x)$

$$\underline{\psi} \rightarrow \underline{\psi}' = \underline{r}[\boldsymbol{\omega}(x)]\,\underline{\psi} \qquad ; \ \underline{r}[\boldsymbol{\omega}(x)] = \exp\left\{ \frac{i}{2}\boldsymbol{\tau}\cdot\boldsymbol{\omega}(x) \right\} \tag{6.102}$$

Those terms in the lagrangian where the fields appear in the combination $\bar{\psi}\psi$ will be invariant under this transformation since it is still true that

$$\underline{\bar{\psi}}'\underline{\psi}' = \underline{\bar{\psi}}\,\underline{r}^\dagger\underline{r}\,\underline{\psi} = \underline{\bar{\psi}}\,\underline{\psi} \tag{6.103}$$

But what about the *gradient* terms in $\mathcal{L}$? If they appeared in the form of a covariant derivative that, when applied to the field, again transformed as if the transformation were global, then the lagrangian density would again remain invariant by the same argument. Thus one seeks to use a covariant derivative in $\mathcal{L}$ for which

$$\frac{D'}{Dx_\mu}\underline{\psi}' = \underline{r}[\boldsymbol{\omega}(x)]\,\frac{D}{Dx_\mu}\underline{\psi} \tag{6.104}$$

Working in analogy to the QED argument given above, but now taking into account the fact that one has a non-abelian internal symmetry group where the group elements do not commute, Yang and Mills accomplished this through the following series of steps:

(1) Introduce a set of vector fields A_μ^i, with $i = 1, 2, 3$, one for each generator of the symmetry group,

$$A_\mu^i(x) \qquad ; i = 1, 2, 3 \qquad ; \text{ gauge bosons} \qquad (6.105)$$

These are referred to as *gauge bosons*;

(2) Introduce a covariant derivative in the form

$$\frac{D}{Dx_\mu}\psi \equiv \left[\frac{\partial}{\partial x_\mu} - \frac{ig}{2}\boldsymbol{\tau} \cdot \mathbf{A}_\mu(x)\right]\psi \qquad (6.106)$$

where g is a coupling constant, defined through the lagrangian;

(3) Ensure that the gauge bosons transform in such a way that Eq. (6.104) is satisfied.

We thus ask how the gauge boson field must transform in order to satisfy

$$\left[\frac{\partial}{\partial x_\mu} - \frac{ig}{2}\boldsymbol{\tau} \cdot \mathbf{A}_\mu'(x)\right]\psi' = \underline{r}[\boldsymbol{\omega}(x)]\left[\frac{\partial}{\partial x_\mu} - \frac{ig}{2}\boldsymbol{\tau} \cdot \mathbf{A}_\mu(x)\right]\psi \qquad (6.107)$$

Substitution of Eq. (6.102), and cancellation of the term in $\partial\psi/\partial x_\mu$, gives

$$\left\{\frac{\partial}{\partial x_\mu}\underline{r}[\boldsymbol{\omega}(x)] - \frac{ig}{2}\boldsymbol{\tau} \cdot \mathbf{A}_\mu'(x)\,\underline{r}[\boldsymbol{\omega}(x)]\right\}\psi =$$
$$\underline{r}[\boldsymbol{\omega}(x)]\left\{-\frac{ig}{2}\boldsymbol{\tau} \cdot \mathbf{A}_\mu(x)\right\}\psi \qquad (6.108)$$

This relation will hold provided

$$\frac{\partial}{\partial x_\mu}\underline{r}[\boldsymbol{\omega}(x)] - \frac{ig}{2}\boldsymbol{\tau} \cdot \mathbf{A}_\mu'(x)\,\underline{r}[\boldsymbol{\omega}(x)] = \underline{r}[\boldsymbol{\omega}(x)]\left\{-\frac{ig}{2}\boldsymbol{\tau} \cdot \mathbf{A}_\mu(x)\right\} \qquad (6.109)$$

This relation, in turn, will hold if

$$-\frac{ig}{2}\boldsymbol{\tau} \cdot \mathbf{A}_\mu'(x) = \underline{r}[\boldsymbol{\omega}(x)]\left\{-\frac{ig}{2}\boldsymbol{\tau} \cdot \mathbf{A}_\mu(x)\right\}\underline{r}[\boldsymbol{\omega}(x)]^{-1}$$
$$-\left\{\frac{\partial}{\partial x_\mu}\underline{r}[\boldsymbol{\omega}(x)]\right\}\underline{r}[\boldsymbol{\omega}(x)]^{-1} \qquad (6.110)$$

Thus, Eq. (6.107) will hold provided the gauge bosons transform according to

$$\frac{1}{2}\boldsymbol{\tau} \cdot \mathbf{A}'_\mu(x) = \underline{r}[\boldsymbol{\omega}(x)] \left\{ \frac{1}{2}\boldsymbol{\tau} \cdot \mathbf{A}_\mu(x) \right\} \underline{r}[\boldsymbol{\omega}(x)]^{-1}$$
$$-\frac{i}{g} \left\{ \frac{\partial}{\partial x_\mu} \underline{r}[\boldsymbol{\omega}(x)] \right\} \underline{r}[\boldsymbol{\omega}(x)]^{-1} \qquad (6.111)$$

The condition in Eq. (6.111) appears to depend on the particular representation of the generators of the symmetry, in this case the Pauli matrices. A key insight of Yang and Mills was that, in fact, *it only depends on the commutation relations satisfied by these generators, that is, on the Lie algebra of the relevant symmetry group.* To see this, go to infinitesimals with $\boldsymbol{\omega}(x) \equiv \boldsymbol{\varepsilon}(x) \to 0.$[21] Retaining terms exact through $O(\varepsilon)$, Eq. (6.111) then becomes

$$\frac{1}{2}\boldsymbol{\tau} \cdot \mathbf{A}'_\mu(x) = \left(1 + \frac{i}{2}\boldsymbol{\tau} \cdot \boldsymbol{\varepsilon}\right) \left\{ \frac{1}{2}\boldsymbol{\tau} \cdot \mathbf{A}_\mu(x) \right\} \left(1 - \frac{i}{2}\boldsymbol{\tau} \cdot \boldsymbol{\varepsilon}\right) - \frac{i}{2g} \left\{ i\boldsymbol{\tau} \cdot \frac{\partial \boldsymbol{\varepsilon}}{\partial x_\mu} \right\}$$
$$= \frac{1}{2}\boldsymbol{\tau} \cdot \mathbf{A}_\mu(x) + i[\frac{1}{2}\tau^i, \frac{1}{2}\tau^j]\,\varepsilon^i A_\mu^j + \frac{1}{2g}\boldsymbol{\tau} \cdot \frac{\partial \boldsymbol{\varepsilon}}{\partial x_\mu} \qquad (6.112)$$

Now use the fact that the Pauli matrices satisfy the Lie algebra of SU(2)

$$[\frac{1}{2}\tau^i, \frac{1}{2}\tau^j] = i\epsilon_{ijk}\frac{1}{2}\tau^k \qquad ; \ (i,j,k) = 1,2,3 \qquad (6.113)$$

Since the Pauli matrices are linearly independent, their coefficients can then be equated to give

$$A_\mu^{i\,\prime} = A_\mu^i - \epsilon_{ijk}\,\varepsilon^j A_\mu^k + \frac{1}{g}\frac{\partial \varepsilon^i}{\partial x_\mu} \qquad (6.114)$$

The *change* in the gauge fields is then given to first-order in the local transformation parameter $\varepsilon(x)$ by

$$\delta A_\mu^i = -\epsilon_{ijk}\,\varepsilon^j A_\mu^k + \frac{1}{g}\frac{\partial \varepsilon^i}{\partial x_\mu} \qquad (6.115)$$

We make some observations on this result:

- It only depends on the *structure constants* ϵ_{ijk}, in this case for SU(2), as advertised;

---

[21] Recall that the application of Noether's theorem requires only invariance under infinitesimal transformations.

- With an abelian symmetry group, as is the case of QED, the structure constant term is absent and this result reduces to the familiar gauge transformation of Eq. (6.93);
- In the case of SU(2), this relation can be written in the form of an ordinary vector relation in the internal space

$$\delta \mathbf{A}_\mu = -\boldsymbol{\varepsilon} \times \mathbf{A}_\mu + \frac{1}{g} \frac{\partial \boldsymbol{\varepsilon}}{\partial x_\mu} \qquad (6.116)$$

- Consider the isovector triplet of scalar mesons in SU(2). Define the covariant derivative as

$$\frac{D}{Dx_\mu} \underline{\phi} \equiv \left[\frac{\partial}{\partial x_\mu} - ig \, \underline{\mathbf{t}} \cdot \mathbf{A}_\mu(x) \right] \underline{\phi} \qquad (6.117)$$

where $\underline{\mathbf{t}}$ is the set of matrices in Eqs. (6.13) and (6.14), and $\underline{\phi}$ transforms according to Eq. (6.46). Then the calculation of the required δA_μ^i goes through *exactly the same way as above, since the matrices $\underline{\mathbf{t}}$ satisfy the same Lie algebra* [see Eq. (6.21)]. Equation (6.117) can also be rewritten as a vector relation

$$\frac{D}{Dx_\mu} \phi = \left(\frac{\partial}{\partial x_\mu} + g\mathbf{A}_\mu \times \right) \phi \qquad (6.118)$$

In line with the discussion of QED, it is now appropriate to include a kinetic energy term for the gauge bosons themselves. It must be included in such a way as to preserve the local gauge invariance, thus it must be invariant under the transformation in Eq. (6.115). A second key insight of Yang and Mills was to introduce the following *field tensor*

$$\mathcal{F}_{\mu\nu}^i \equiv \frac{\partial A_\nu^i}{\partial x_\mu} - \frac{\partial A_\mu^i}{\partial x_\nu} + g\epsilon_{ijk} A_\mu^j A_\nu^k \qquad ; \text{ field tensor}$$

$$(i,j,k) = 1,2,3 \quad (6.119)$$

This again depends only on the structure constants and transforms in the following manner under the infinitesimal gauge transformation in Eq. (6.115)

$$\delta \mathcal{F}_{\mu\nu}^i = -\epsilon_{ijk} \, \varepsilon^j \, \mathcal{F}_{\mu\nu}^k \qquad (6.120)$$

For SU(2), these relations can again be written in a vector notation

$$\boldsymbol{\mathcal{F}}_{\mu\nu} = \frac{\partial \mathbf{A}_\nu}{\partial x_\mu} - \frac{\partial \mathbf{A}_\mu}{\partial x_\nu} + g\mathbf{A}_\mu \times \mathbf{A}_\nu$$

$$\delta \boldsymbol{\mathcal{F}}_{\mu\nu} = -\boldsymbol{\varepsilon} \times \boldsymbol{\mathcal{F}}_{\mu\nu} \qquad (6.121)$$

Since the last is simply a rotation in the internal space, it is evident that it leaves the following kinetic-energy lagrangian density for the gauge field unchanged

$$\mathcal{L}_0 = -\frac{1}{4}\mathcal{F}_{\mu\nu} \cdot \mathcal{F}_{\mu\nu} \tag{6.122}$$

Three comments:

- In the abelian case of QED, where the structure-constant term is absent, this reduces to the electromagnetic field tensor;
- In the non-abelian case, here SU(2), there is an additional *intrinsic non-linearity* in the field tensor
- There is a single coupling constant g in the theory.

It remains to establish Eq. (6.120). Just compute the first-order change $\delta\mathcal{F}_{\mu\nu}$ of Eq. (6.121) in terms of the first-order change $\delta\mathbf{A}_\mu$ in Eq. (6.116)

$$\delta\mathcal{F}_{\mu\nu} = \frac{\partial}{\partial x_\mu}\left(-\varepsilon \times \mathbf{A}_\nu + \frac{1}{g}\frac{\partial\varepsilon}{\partial x_\nu}\right) - \frac{\partial}{\partial x_\nu}\left(-\varepsilon \times \mathbf{A}_\mu + \frac{1}{g}\frac{\partial\varepsilon}{\partial x_\mu}\right) +$$
$$g\mathbf{A}_\mu \times \left(-\varepsilon \times \mathbf{A}_\nu + \frac{1}{g}\frac{\partial\varepsilon}{\partial x_\nu}\right) + g\left(-\varepsilon \times \mathbf{A}_\mu + \frac{1}{g}\frac{\partial\varepsilon}{\partial x_\mu}\right) \times \mathbf{A}_\nu \tag{6.123}$$

The second and fourth terms in the first line cancel, and the terms with derivatives of ε from the first and third terms are cancelled by the second and fourth terms in the second line. Thus

$$\delta\mathcal{F}_{\mu\nu} = -\varepsilon \times \left(\frac{\partial\mathbf{A}_\nu}{\partial x_\mu} - \frac{\partial\mathbf{A}_\mu}{\partial x_\nu}\right) - g\left[\mathbf{A}_\mu \times (\varepsilon \times \mathbf{A}_\nu) + (\varepsilon \times \mathbf{A}_\mu) \times \mathbf{A}_\nu\right] \tag{6.124}$$

Judicious use of a standard relation for the vector triple product reduces the last term to

$$\mathbf{A}_\mu \times (\varepsilon \times \mathbf{A}_\nu) + (\varepsilon \times \mathbf{A}_\mu) \times \mathbf{A}_\nu$$
$$= \varepsilon (\mathbf{A}_\mu \cdot \mathbf{A}_\nu) - (\varepsilon \cdot \mathbf{A}_\mu)\mathbf{A}_\nu + \mathbf{A}_\mu(\varepsilon \cdot \mathbf{A}_\nu) - \varepsilon(\mathbf{A}_\mu \cdot \mathbf{A}_\nu)$$
$$= \varepsilon \times (\mathbf{A}_\mu \times \mathbf{A}_\nu) \tag{6.125}$$

Hence Eq. (6.124) becomes

$$\delta\mathcal{F}_{\mu\nu} = -\varepsilon \times \mathcal{F}_{\mu\nu} \tag{6.126}$$

which was the result to be established.

The final step would be to include a mass term for the gauge bosons

$$\mathcal{L}_{\text{mass}} = -\frac{1}{2}m_g^2 c^2 \mathbf{A}_\mu \cdot \mathbf{A}_\mu \qquad ; \ \underline{\text{not}} \text{ included} \qquad (6.127)$$

However, as this term is *not gauge invariant*, it *cannot be included.*

In *summary*, the lagrangian density for the Yang-Mills relativistic inter-acting field theory with local SU(2) symmetry built on the Dirac isodoublet field in Eq. (6.29) is then

$$\mathcal{L} = -\hbar c \underline{\bar{\psi}} \left\{ \gamma_\mu \left[\frac{\partial}{\partial x_\mu} - \frac{ig}{2} \boldsymbol{\tau} \cdot \mathbf{A}_\mu(x) \right] + M \right\} \underline{\psi} - \frac{1}{4} \boldsymbol{\mathcal{F}}_{\mu\nu} \cdot \boldsymbol{\mathcal{F}}_{\mu\nu} \qquad ;$$

$$\text{Yang-Mills–SU(2)} \quad (6.128)$$

The inclusion of an isovector triplet of scalars would add a term

$$\mathcal{L}_{\text{scalar}} = -\frac{c^2}{2} \left\{ \left[\left(\frac{\partial}{\partial x_\mu} + g\mathbf{A}_\mu \times \right) \boldsymbol{\phi} \right] \cdot \left[\left(\frac{\partial}{\partial x_\mu} + g\mathbf{A}_\mu \times \right) \boldsymbol{\phi} \right] + m^2 \boldsymbol{\phi} \cdot \boldsymbol{\phi} \right\} \ ;$$

$$\text{isovector scalars} \quad (6.129)$$

to the lagrangian density.

As stated before, Yang and Mills developed this as a gauge theory based on local isospin invariance. As there is no triplet of massless gauge bosons corresponding to this symmetry observed in nature, they presented their work primarily as an academic exercise. Indeed, their paper sat in the literature for almost two decades before the deep implications of imposing such a local symmetry on a theory were realized. In fact

> *The very successful theories of both the strong interactions (QCD) and the electroweak interactions (Standard Model) start from the assumption of invariance under local symmetry transformations, the former for an internal $SU(3)_C$ color symmetry, and the latter for a weak $SU(2)_W \otimes U(1)_W$.*

In order for this to happen, however, it was first necessary to face the fact that the only massless gauge boson we do observe in nature is the photon, corresponding to QED. To deal with this dilemma, it is necessary to have some idea of how mass is generated in relativistic quantum field theories, and it is to that topic that we turn in the next section.

To conclude this section, we summarize the required modifications of the Dirac triplet lagrangian density in Eq. (6.71), when the global SU(3) internal symmetry is made into a local one:

- *Eight* massless gauge boson fields $G_\mu^a(x)$ with $a = 1, \cdots, 8$ are now introduced, one for each generator;
- Correspondingly, all repeated Latin indices are now summed from $1, \cdots, 8$;
- Instead of the Pauli matrices $\underline{\tau}_i$, one uses the Gell-Mann matrices $\underline{\lambda}^a$;
- instead of ϵ_{ijk}, the structure constants f^{abc} of SU(3) are now employed.

The covariant derivative D/Dx_μ acting on the triplet Dirac field, the infinitesimal gauge transformation δG_μ^a, the field tensor $\mathcal{G}_{\mu\nu}^a$, and the infinitesimal change in the field tensor $\delta \mathcal{G}_{\mu\nu}^a$ then take the form

$$\frac{D}{Dx_\mu}\underline{\psi} \equiv \left[\frac{\partial}{\partial x_\mu} - \frac{ig}{2}\underline{\lambda}^a G_\mu^a(x)\right]\underline{\psi} \qquad ; \text{SU(3)}$$

$$\delta G_\mu^a = -f^{abc}\varepsilon^b G_\mu^c + \frac{1}{g}\frac{\partial \varepsilon^a}{\partial x_\mu} \qquad (a,b,c) = 1,\cdots,8$$

$$\mathcal{G}_{\mu\nu}^a = \frac{\partial G_\nu^a}{\partial x_\mu} - \frac{\partial G_\mu^a}{\partial x_\nu} + g f^{abc} G_\mu^b G_\nu^c$$

$$\delta \mathcal{G}_{\mu\nu}^a = -f^{abc}\varepsilon^b \mathcal{G}_{\mu\nu}^c \qquad (6.130)$$

The lagrangian density for the Yang-Mills relativistic interacting field theory with local SU(3) symmetry built on the Dirac triplet field in Eq. (6.71) is then

$$\mathcal{L} = -\hbar c\underline{\bar{\psi}}\left\{\gamma_\mu\left[\frac{\partial}{\partial x_\mu} - \frac{ig}{2}\underline{\lambda}^a G_\mu^a(x)\right] + M\right\}\underline{\psi} - \frac{1}{4}\mathcal{G}_{\mu\nu}^a \mathcal{G}_{\mu\nu}^a \qquad ;$$

$$\text{Yang-Mills–SU(3)} \quad (6.131)$$

6.6 Chiral Symmetry

Let us return to the isospinor Dirac field in Eq. (6.28) and the Dirac lagrangian density for that field in Eq. (6.29). Suppose the mass term is absent, then Eq. (6.29) takes the form

$$\mathcal{L} = -\hbar c\,\underline{\bar{\psi}}\left(\gamma_\mu \frac{\partial}{\partial x_\mu}\right)\underline{\psi} \qquad (6.132)$$

This lagrangian density is invariant under a wider class of transformations than just the isospin transformation of Eq. (6.31). In that transformation, the Dirac fields ψ_p and ψ_n are treated as units, and the Dirac spinors remain unchanged through all the subsequent manipulations. Consider, more generally, transformations where *the components of the Dirac spinors are also mixed during the transformation*. In particular, consider the transformation where they are mixed according to

$$\psi \to \psi' = \underline{r}_5(\boldsymbol{\omega})\psi$$

$$\underline{r}_5(\boldsymbol{\omega}) \equiv \exp\left\{\frac{i}{2}\gamma_5\,\boldsymbol{\tau}\cdot\boldsymbol{\omega}\right\} \tag{6.133}$$

Here γ_5 is the Dirac matrix introduced in Vol. I, with the properties

$$\gamma_5^2 = 1 \qquad ; \gamma_5^\dagger = \gamma_5$$

$$\gamma_5\gamma_\mu + \gamma_\mu\gamma_5 = 0 \qquad ; \mu = 1, 2, 3, 4 \tag{6.134}$$

The matrix $\underline{r}_5(\boldsymbol{\omega})$ in Eq. (6.133) has a well-defined meaning in terms of its power series expansion

$$\exp\left\{\frac{i}{2}\gamma_5\,\boldsymbol{\tau}\cdot\boldsymbol{\omega}\right\} = 1 + \frac{i}{2}\gamma_5\,\boldsymbol{\tau}\cdot\boldsymbol{\omega} + \frac{1}{2!}\left(\frac{i}{2}\gamma_5\,\boldsymbol{\tau}\cdot\boldsymbol{\omega}\right)^2 + \frac{1}{3!}\left(\frac{i}{2}\gamma_5\,\boldsymbol{\tau}\cdot\boldsymbol{\omega}\right)^3 + ..$$

$$= \cos\left(\frac{\omega}{2}\right) + i\gamma_5\,\boldsymbol{\tau}\cdot\mathbf{n}\,\sin\left(\frac{\omega}{2}\right) \qquad ; \mathbf{n} \equiv \frac{\boldsymbol{\omega}}{\omega} \tag{6.135}$$

Since γ_5 is hermitian, the matrix $\underline{r}_5(\boldsymbol{\omega})$ is unitary

$$\underline{r}_5(\boldsymbol{\omega})^\dagger = \underline{r}_5(\boldsymbol{\omega})^{-1} \qquad ; \text{unitary} \tag{6.136}$$

Recall that $\bar{\psi} = \psi^\dagger\gamma_4$. It then follows from Eqs. (6.133)–(6.136) that

$$\bar{\psi}'\gamma_\mu = \psi^\dagger\underline{r}_5(\boldsymbol{\omega})^\dagger\gamma_4\gamma_\mu = \psi^\dagger\gamma_4\gamma_\mu\,\underline{r}_5(\boldsymbol{\omega})^\dagger = \bar{\psi}\gamma_\mu\,\underline{r}_5(\boldsymbol{\omega})^{-1} \tag{6.137}$$

Hence the isospinor Dirac lagrangian in Eq. (6.132) is *invariant* under the *chiral transformation* in Eq. (6.133).[22]

If one now goes to infinitesimals where $\boldsymbol{\omega} \equiv \boldsymbol{\varepsilon} \to 0$, then Eq. (6.133) becomes

$$\psi \to \psi + \delta\psi = \psi + \frac{i}{2}\gamma_5\,\boldsymbol{\tau}\cdot\boldsymbol{\varepsilon}\,\psi \qquad ; \boldsymbol{\varepsilon} \text{ an infinitesimal} \tag{6.138}$$

Now Noether's theorem can be invoked, and all the arguments in the section on isospin–SU(2) follow through just as before. Just as there is a set of

---

[22]Note that the Dirac mass term $\mathcal{L}_{\text{mass}} = -\hbar c\bar{\psi}M\psi$ is *not* invariant under this transformation (see Prob. 6.10).

isovector, Lorentz vector, conserved currents

$$\mathbf{J}_\mu = i\bar{\psi}\,\gamma_\mu \frac{1}{2}\boldsymbol{\tau}\,\underline{\psi} \qquad ; \text{ vector currents}$$

$$\frac{\partial \mathbf{J}_\mu}{\partial x_\mu} = 0 \qquad ; \text{ conserved} \qquad (6.139)$$

there will now be a *additional* set of isovector, Lorentz *axial*-vector, conserved currents

$$\mathbf{J}_{5\mu} = i\bar{\psi}\,\gamma_\mu \gamma_5 \frac{1}{2}\boldsymbol{\tau}\,\underline{\psi} \qquad ; \text{ axial-vector currents}$$

$$\frac{\partial \mathbf{J}_{5\mu}}{\partial x_\mu} = 0 \qquad ; \text{ conserved} \qquad (6.140)$$

The integral over all space of the fourth component of these new currents will again be a constant of the motion, and just as with the generators of the isospin transformations,

$$\hat{\mathbf{T}} = \int_\Omega d^3x\,\hat{\underline{\psi}}^\dagger(\mathbf{x})\,\frac{1}{2}\boldsymbol{\tau}\,\hat{\underline{\psi}}(\mathbf{x}) \qquad (6.141)$$

there will now be an additional set of generators of the chiral transformation

$$\hat{\mathbf{T}}^5 = \int_\Omega d^3x\,\hat{\underline{\psi}}^\dagger(\mathbf{x})\,\gamma_5 \frac{1}{2}\boldsymbol{\tau}\,\hat{\underline{\psi}}(\mathbf{x}) \qquad (6.142)$$

The arguments on the commutator of these generators follows through exactly as in Eq. (6.38),[23] for example

$$[\hat{T}_i^5, \hat{T}_j^5] = \int_\Omega d^3x\,\hat{\underline{\psi}}^\dagger(\mathbf{x})\left[\gamma_5 \frac{1}{2}\tau_i,\,\gamma_5 \frac{1}{2}\tau_j\right]\hat{\underline{\psi}}(\mathbf{x}) \qquad (6.143)$$

One can then just read off the following commutation relations from the matrices involved in the above expressions

$$[\hat{T}_i, \hat{T}_j] = i\epsilon_{ijk}\hat{T}_k \qquad ; (i,j,k) = 1,2,3$$
$$[\hat{T}_i, \hat{T}_j^5] = i\epsilon_{ijk}\hat{T}_k^5$$
$$[\hat{T}_i^5, \hat{T}_j^5] = i\epsilon_{ijk}\hat{T}_k \qquad (6.144)$$

This is a Lie algebra, and the set of isospin and chiral transformations then forms a Lie group of transformations under which the massless isospinor Dirac lagrangian in Eq. (6.132) is left invariant. Just which Lie group it

---

[23] Note Eq. (6.30).

is can be evidenced more clearly in the following fashion. Make use of the following properties of γ_5[24]

$$P_L \equiv \frac{1}{2}(1 + \gamma_5) \qquad ; \; P_R \equiv \frac{1}{2}(1 - \gamma_5)$$
$$P_L^2 = P_L \qquad\qquad ; \; P_R^2 = P_R$$
$$P_L P_R = P_R P_L = 0 \qquad\qquad\qquad (6.145)$$

Now define new linear combinations of the generators

$$\hat{\mathbf{T}}^L = \int_\Omega d^3x \, \underline{\hat{\psi}}^\dagger(\mathbf{x}) \, P_L \frac{1}{2}\underline{\boldsymbol{\tau}} \, \underline{\hat{\psi}}(\mathbf{x})$$
$$\hat{\mathbf{T}}^R = \int_\Omega d^3x \, \underline{\hat{\psi}}^\dagger(\mathbf{x}) \, P_R \frac{1}{2}\underline{\boldsymbol{\tau}} \, \underline{\hat{\psi}}(\mathbf{x}) \qquad (6.146)$$

It again follows simply from the properties of the matrices involved that

$$[\hat{T}_i^L \, \hat{T}_j^L] = i\epsilon_{ijk}\hat{T}_k^L \qquad ; \; (i,j,k) = 1,2,3$$
$$[\hat{T}_i^R \, \hat{T}_j^R] = i\epsilon_{ijk}\hat{T}_k^R$$
$$[\hat{T}_i^L, \hat{T}_j^R] = 0 \qquad\qquad\qquad (6.147)$$

Thus the full Lie group decomposes into the direct product of two SU(2) subgroups

$$SU(2)_L \otimes SU(2)_R \qquad ; \; \text{symmetry group} \qquad (6.148)$$

All the angular momentum analysis of chapter 3 can now be applied to each SU(2) subgroup!

This is more that just an academic exercise. We know from Vol. I that the leptons in β-decay, and in the other semi-leptonic weak interactions, couple to both isovector Lorentz vector, and isovector Lorentz axial-vector, currents of the form in Eqs. (6.139) and (6.140). The vector current is conserved (CVC), and the axial-vector current is "almost" conserved (PCAC). Here PCAC stands for "partially conserved axial-vector current"; it appears that in these weak interactions the axial-vector current is strictly conserved

---

[24]It was shown in Vol. I that P_L projects left-handed particles, and right-handed antiparticles, from the spinors for massless Dirac particles, and P_R does just the opposite.

only in the limit that the pion mass vanishes

$$\frac{\partial \mathbf{J}_\mu}{\partial x_\mu} = 0 \qquad\qquad ; \text{CVC}$$

$$\frac{\partial \mathbf{J}_{5\mu}}{\partial x_\mu} \propto m_\pi^2\, \boldsymbol{\pi} \qquad\qquad ; \text{PCAC}$$

$$\to 0 \qquad\qquad m_\pi^2 \to 0 \qquad\qquad (6.149)$$

Here $\boldsymbol{\pi}$ is the pion field, and $m_\pi \equiv m_{0\pi}c/\hbar$ is the pion inverse Compton wavelength. Thus, it is very important to build this chiral symmetry into any more realistic lagrangian densities constructed to describe strongly interacting hadronic systems. In attempting to do this, however, one is faced with two glaring problems:

- Although the lagrangian density in Eq. (6.132) provides a fine basis for introducing the concept of chiral invariance, it is not clear that it has any relevance to physics since *the observed isospinor Dirac multiplet, in this case the nucleon, is not massless!*
- It is the pion that is responsible for the long-range part of the nuclear force, and that forms the basis for the above statement of PCAC. The most obvious coupling of the isovector, pseudoscalar pion to the Dirac isodoublet is[25]

$$\mathcal{L}_{\pi N} = i g_\pi \underline{\bar{\psi}} \gamma_5 \underline{\boldsymbol{\tau}}\, \underline{\psi} \cdot \boldsymbol{\pi} \qquad\qquad ; \text{pion-Dirac coupling} \quad (6.150)$$

This interaction is *not* chiral invariant.

The σ-model addresses both of these issues [Schwinger (1957); Gell-Mann and Levy (1960)]. Although it is a simple model, it has had a most profound influence on the development of modern physics, for it both

- Provides a paradigm for the *generation of mass* in relativistic quantum field theory;
- Yields a concrete illustration of a *spontaneously broken symmetry* whereby the observed eigenstates do not reflect the underlying symmetry of the lagrangian.

---

[25]The pion is represented by an isovector field $\boldsymbol{\pi}$, a standard notation. To avoid confusion, the canonical momentum density will here be denoted by $\boldsymbol{\Pi}$. The pion field $\boldsymbol{\pi}$ behaves exactly as our previously discussed real (hermitian) isovector scalar field $\boldsymbol{\phi}$, except that it changes sign under spatial reflections. To construct a lagrangian density for the strong interactions which conserves parity, one must therefore use the hermitian pseudoscalar Dirac density $i\bar{\psi}\gamma_5\psi$ (see Probl. 9.10).

6.6.1   σ-Model

Let us augment the lagrangian density for the massless, isospinor Dirac field ψ in Eq. (6.132) with a massless, isovector, pseudoscalar pion field $\boldsymbol{\pi}$ and the interaction of Eq. (6.150). At the same time, we anticipate and include a massless, isoscalar, scalar field σ that couples to the Dirac scalar density $\bar{\psi}\psi$ with the same coupling strength

$$\mathcal{L}_{\text{Dirac}} = -\hbar c\, \bar{\psi} \left[\gamma_\mu \frac{\partial}{\partial x_\mu} - g\left(i\gamma_5\, \boldsymbol{\tau} \cdot \boldsymbol{\pi} + \sigma\right) \right] \psi \qquad (6.151)$$

Consider the infinitesimal chiral transformation of the Dirac field in Eq. (6.138), and let the pion and scalar fields simultaneously undergo the infinitesimal transformations $(\delta\boldsymbol{\pi}, \delta\sigma)$. The change in this lagrangian density, exact through first order in infinitesimals, is then given by[26]

$$\delta\mathcal{L}_{\text{Dirac}} = \hbar c g\, \bar{\psi} \left[i\gamma_5 \frac{1}{2}\boldsymbol{\tau} \cdot \boldsymbol{\varepsilon} \left(i\gamma_5\, \boldsymbol{\tau} \cdot \boldsymbol{\pi} + \sigma\right) + \left(i\gamma_5\, \boldsymbol{\tau} \cdot \boldsymbol{\pi} + \sigma\right) i\gamma_5 \frac{1}{2}\boldsymbol{\tau} \cdot \boldsymbol{\varepsilon} + \right.$$

$$\left. \left(i\gamma_5\, \boldsymbol{\tau} \cdot \delta\boldsymbol{\pi} + \delta\sigma\right) \right]\psi$$

$$= \hbar c g\, \bar{\psi} \left[(\delta\sigma - \boldsymbol{\varepsilon} \cdot \boldsymbol{\pi}) + i\gamma_5\, \boldsymbol{\tau} \cdot (\delta\boldsymbol{\pi} + \boldsymbol{\varepsilon}\sigma) \right]\psi \qquad (6.152)$$

This will vanish, provided the pion and scalar fields transform according to

$$\delta\boldsymbol{\pi} = -\boldsymbol{\varepsilon}\sigma \qquad ;\ \boldsymbol{\varepsilon}\ \text{an infinitesimal}$$

$$\delta\sigma = \boldsymbol{\varepsilon} \cdot \boldsymbol{\pi} \qquad\qquad\qquad\qquad (6.153)$$

We note that this transformation leaves the quantity $\boldsymbol{\pi}^2 + \sigma^2$ invariant to this order

$$\delta(\boldsymbol{\pi}^2 + \sigma^2) = 2\boldsymbol{\pi} \cdot \delta\boldsymbol{\pi} + 2\sigma\delta\sigma = 2\boldsymbol{\pi} \cdot (-\boldsymbol{\varepsilon}\sigma) + 2\sigma(\boldsymbol{\varepsilon} \cdot \boldsymbol{\pi}) = 0 \quad (6.154)$$

It is appropriate to also include a kinetic energy term for the massless fields $(\boldsymbol{\pi}, \sigma)$, and since $\boldsymbol{\varepsilon}$ is a spatial constant here, that term will also be invariant by the same argument

$$\mathcal{L}^0_{\pi\sigma} = -\frac{c^2}{2} \left(\frac{\partial\boldsymbol{\pi}}{\partial x_\mu} \cdot \frac{\partial\boldsymbol{\pi}}{\partial x_\mu} + \frac{\partial\sigma}{\partial x_\mu} \cdot \frac{\partial\sigma}{\partial x_\mu} \right)$$

$$\delta\mathcal{L}^0_{\pi\sigma} = 0 \qquad\qquad\qquad\qquad (6.155)$$

Finally, any potential energy term of the form $\mathcal{V}(\boldsymbol{\pi}^2 + \sigma^2)$ can be included in the lagrangian density since this is chirally invariant by Eq. (6.154). Thus,

---

[26]Recall that $\delta\bar{\psi} = +i\bar{\psi}\gamma_5 \frac{1}{2}\boldsymbol{\tau} \cdot \boldsymbol{\varepsilon}$.

we arrive at the σ-model lagrangian

$$\mathcal{L} = -\hbar c \bar{\psi} \left[\gamma_\mu \frac{\partial}{\partial x_\mu} - g\left(i\gamma_5 \,\underline{\boldsymbol{\tau}} \cdot \boldsymbol{\pi} + \sigma\right) \right] \psi - \frac{c^2}{2} \left(\frac{\partial \boldsymbol{\pi}}{\partial x_\mu} \cdot \frac{\partial \boldsymbol{\pi}}{\partial x_\mu} + \frac{\partial \sigma}{\partial x_\mu} \cdot \frac{\partial \sigma}{\partial x_\mu} \right)$$
$$-\mathcal{V}(\boldsymbol{\pi}^2 + \sigma^2) \qquad\qquad\qquad ; \ \sigma\text{-model} \qquad (6.156)$$

This lagrangian density is invariant under both the infinitesimal isospin and chiral transformations of the fields $(\psi, \boldsymbol{\pi}, \sigma)$. Noether's theorem then leads to the following conserved isovector vector, and isovector axial-vector, currents for the interacting system[27]

$$\mathbf{J}_\mu = i\bar{\psi}\, \gamma_\mu \frac{1}{2}\underline{\boldsymbol{\tau}}\, \psi + \frac{c}{\hbar}\left(\frac{\partial \boldsymbol{\pi}}{\partial x_\mu} \times \boldsymbol{\pi} \right)$$
$$\mathbf{J}_{5\mu} = i\bar{\psi}\, \gamma_\mu \gamma_5 \frac{1}{2}\underline{\boldsymbol{\tau}}\, \psi + \frac{c}{\hbar}\left(\boldsymbol{\pi}\frac{\partial \sigma}{\partial x_\mu} - \sigma\frac{\partial \boldsymbol{\pi}}{\partial x_\mu} \right) \qquad (6.157)$$

The fields $(\boldsymbol{\pi}, \sigma)$ can now be given mass in a chirally symmetric fashion by choosing the familiar quadratic form for the potential

$$\mathcal{V}(\boldsymbol{\pi}^2 + \sigma^2) = \frac{1}{2}m^2 c^2(\boldsymbol{\pi}^2 + \sigma^2) \qquad ; \ \text{usual mass term} \quad (6.158)$$

This potential has a minimum at $\boldsymbol{\pi}^2 + \sigma^2 = 0$, as illustrated in Fig. 6.2(a).

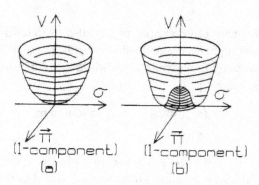

Fig. 6.2 Meson potential surfaces $\mathcal{V}$ for (a) usual mass terms; (b) spontaneous symmetry breaking.

---

[27]See Prob. 6.11; note the first current is simply the sum of the isospin currents for the isospinor Dirac and isovector pion fields, while the second now contains an additional contribution arising from the kinetic energy term for $(\boldsymbol{\pi}, \sigma)$.

Suppose, instead, that the potential has the form introduced by [Gell-Mann and Levy (1960)][28]

$$\mathcal{V}(\pi^2 + \sigma^2) = \frac{\lambda}{4}\left[(\pi^2 + \sigma^2) - v^2\right]^2 \qquad ; \sigma\text{-model} \qquad (6.159)$$

The lagrangian density remains chirally invariant, one still has the conserved currents of Eqs. (6.157), but now the minimum of the potential is no longer at $\pi^2 + \sigma^2 = 0$, and remarkable things happen [see Fig. 6.2(b)]!

6.6.2 *Spontaneous Symmetry Breaking*

Readers are familiar with at least one example of spontaneous symmetry breaking. The intrinsic hamiltonian for a ferromagnet is invariant under overall spatial rotations, yet the ground state of a ferromagnet has the magnetization pointing in some direction in space. One way to analyze this problem is to apply a small holding field, and then let the strength of the holding field go to zero. The ferromagnet then retains the memory of the direction of the holding field.

A similar phenomenon occurs in the σ-model. A small *intrinsic* chiral-symmetry-breaking term $\delta\mathcal{V}_{\text{csb}} \equiv \epsilon\sigma$ is first included in the lagrangian density, with the idea that one is interested in the limit $\epsilon \to 0$.[29] Consider therefore the σ-model with the potential

$$\mathcal{V}(\pi^2 + \sigma^2) = \frac{\lambda}{4}\left[(\pi^2 + \sigma^2) - v^2\right]^2 + \epsilon\sigma \qquad ; \epsilon \to 0 \qquad (6.160)$$

The extra term tilts the potential in Fig. 6.2 so that the minimum now occurs along the negative σ axis is illustrated in Fig. 6.3. In the minimum of $\mathcal{V}$, that is, in the ground state, there will be an additional constant field σ_0, with vanishing π,

$$\langle\sigma\rangle_{\text{gs}} = \sigma_0 \qquad ; \text{at minimum}$$
$$\langle\pi\rangle_{\text{gs}} = 0 \qquad\qquad\qquad\qquad (6.161)$$

Note that here $\sigma_0 < 0$. The ground state no longer manifests chiral symmetry, which treats (π, σ) on an equal footing, and thus chiral symmetry is *spontaneously broken* in the observed ground state. In the quantized field theory, the ground state is the *vacuum*.

---

[28]The reason for stopping with quartic powers of the meson fields is that the model in then *renormalizable*.

[29]It will turn out that here $\epsilon \propto m_\pi^2$, where m_π is the pion inverse Compton wavelength.

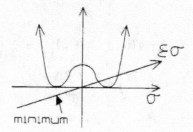

Fig. 6.3 Contribution of an additional chiral-symmetry-breaking term $\varepsilon\sigma$ to the potential $\mathcal{V}$ as viewed along the σ axis.

Now note that an additional constant scalar field is *equivalent to a mass term for the Dirac field in the lagrangian density*

$$\mathcal{L}_{\text{mass}} = -\hbar c\,\bar{\psi}(-g\sigma_0)\psi \equiv -\hbar c\,\bar{\psi}M\psi \qquad \text{; Dirac mass term} \qquad (6.162)$$

Thus the spontaneous chiral symmetry breaking gives rise to a mass M for the Dirac isodoublet, a term that was ruled out by chiral symmetry in the starting lagrangian density itself!

The minimum of the potential $\mathcal{V}(\sigma,\boldsymbol{\pi})$ in Eq. (6.160) occurs at

$$\left[\frac{d\mathcal{V}(\sigma,0)}{d\sigma}\right]_{\sigma_0} = 0$$

$$\Rightarrow \qquad \lambda\sigma_0(\sigma_0^2 - v^2) = -\epsilon \qquad \text{; minimum} \qquad (6.163)$$

The lagrangian density can be expanded about the minimum by writing

$$\sigma \equiv \sigma_0 + \varphi \qquad \text{; } \boldsymbol{\pi} = \boldsymbol{\pi} \qquad (6.164)$$

The expression of the lagrangian density in terms of the new fields $(\boldsymbol{\pi},\varphi)$ is straightforward algebra, and is assigned as Prob. 6.11. With the definitions

$$\sigma_0 \equiv -\frac{M}{g} \qquad \text{; } \epsilon \equiv \frac{M}{g}m_\pi^2 c^2 \qquad \text{; } \lambda \equiv \frac{c^2}{2}\left(\frac{g}{M}\right)^2 (m_s^2 - m_\pi^2) \qquad (6.165)$$

the result is

$$\mathcal{L} = -\hbar c \bar{\psi} \left[\gamma_\mu \frac{\partial}{\partial x_\mu} + M - g \left(i\gamma_5 \underline{\tau} \cdot \boldsymbol{\pi} + \varphi \right) \right] \psi - \frac{c^2}{2} \left(\frac{\partial \boldsymbol{\pi}}{\partial x_\mu} \cdot \frac{\partial \boldsymbol{\pi}}{\partial x_\mu} + m_\pi^2 \boldsymbol{\pi}^2 \right)$$

$$- \frac{c^2}{2} \left(\frac{\partial \varphi}{\partial x_\mu} \cdot \frac{\partial \varphi}{\partial x_\mu} + m_s^2 \varphi^2 \right) + \frac{c^2}{2} \left(\frac{g}{M} \right) (m_s^2 - m_\pi^2) \varphi(\varphi^2 + \boldsymbol{\pi}^2)$$

$$- \frac{c^2}{8} \left(\frac{g}{M} \right)^2 (m_s^2 - m_\pi^2)(\boldsymbol{\pi}^2 + \varphi^2)^2 + \text{constant} \qquad ; \sigma\text{-model} \quad (6.166)$$

We make several comments on these results for this remarkable model:

- The intrinsic chiral-symmetry-breaking term in the lagrangian density is $\epsilon\sigma$, and in the limit as $\epsilon = (M/g)m_\pi^2 c^2 \to 0$, the underlying lagrangian density exhibits exact chiral invariance. In this limit, the currents in Eqs. (6.157) are strictly conserved;
- At the minimum of the potential, in the ground state, the scalar field develops a constant value $\sigma_0 \equiv -M/g$ throughout space. Chiral symmetry, which treats $(\boldsymbol{\pi}, \sigma)$ on the same footing, is "spontaneously broken";
- In that constant scalar field, the Dirac isodoublet acquires a mass $M = -g\sigma_0$, which can be arbitrarily large;
- When expanded about the minimum in the potential as in Eq. (6.164), the isovector pion field $\boldsymbol{\pi}$ acquires a mass $m_\pi = m_{\pi 0} c/\hbar$, and the isoscalar field φ develops a mass $m_s = m_{0s} c/\hbar$, which can again be arbitrarily large;
- When expanded about that minimum, there are cubic and quartic self-couplings of the fields $(\boldsymbol{\pi}, \varphi)$ in the lagrangian density;
- There remains a linear coupling $g(i\gamma_5 \underline{\tau} \cdot \boldsymbol{\pi} + \varphi)$ of the fields $(\boldsymbol{\pi}, \varphi)$ to the Dirac field;
- It follows from the currents in Eq. (6.157), and the field equations including $\delta\mathcal{V}_{\text{csb}} \equiv \epsilon\sigma$, that the model yields both CVC and PCAC[30]

$$\frac{\partial \mathbf{J}_\mu}{\partial x_\mu} = 0 \qquad\qquad\qquad ; \text{CVC}$$

$$\frac{\partial \mathbf{J}_{5\mu}}{\partial x_\mu} = \frac{\epsilon\boldsymbol{\pi}}{\hbar c} = \frac{c}{\hbar} \left(\frac{M}{g} \right) m_\pi^2 \boldsymbol{\pi} \qquad ; \text{PCAC} \qquad (6.167)$$

The excitation spectrum in the σ-model is easily understood by looking at Fig. 6.4. The excitations of the scalar field φ take place about $\sigma = \sigma_0$

---

[30]See Prob. 6.11. The scalar field φ of the σ-model is distinct from the scalar field ϕ of nuclear physics (see [Walecka (2004)]).

in the direction of the σ-axis, and the mass m_s depends on the curvature of the potential at the minimum. To the extent that the potential is not strictly quadratic about the minimum, there will be additional non-linear interactions in φ. The excitations of the pion field π take place in a direction orthogonal to the σ-axis, and if the intrinsic chiral-symmetry-breaking parameter $\epsilon = 0$, the potential is flat along the trough and the pion is massless.[31] A non-zero ϵ tilts the potential slightly, provides a restoring force, and gives the pion field a non-zero mass.

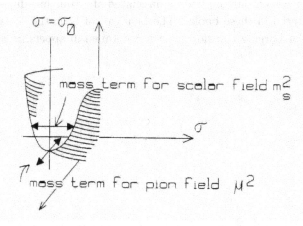

Fig. 6.4 Sketch of the meson potential in which the excitations about the minimum take place. Here $\mu^2 \equiv m_\pi^2$.

Note that as $\epsilon \to 0$ the full potential in Fig. 6.4 is still *chiral symmetric*. The unsymmetric form of the lagrangian density in Eq. (6.166) is obtained by expanding about one particular point in the trough of that potential and using fields in the radial and tangential directions. An improvement here is to use a pion field that still exhibits the full chiral symmetry of the trough in the potential, but we leave that for a more advanced course.[32]

---

[31]This is the archetypical illustration of *Goldstone's theorem*, which, roughly stated, says that there is a zero-mass boson associated with every spontaneously broken symmetry.

[32]This development is detailed, for example, in [Donoghue, Golowich, and Holstein (1993); Walecka (2004)]. Problem 10.10 points the reader in the right direction.

6.7 Lorentz Transformations

The most fundamental symmetry in nature is that of special relativity: *physics must be invariant under Lorentz transformations between inertial frames.* Lorentz transformations are technically more complicated, basically because one has to work in terms of basis states with covariant norm in the abstract Hilbert space. In contrast, quantization in a big box with periodic boundary conditions has provided an essential framework for all our previous calculations. Although many texts will start here, in the author's opinion, the formulation of Lorentz transformations in the abstract Hilbert space can only be truly appreciated after one has digested the previous material in these books. The behavior of both the scalar and Dirac fields under Lorentz transformations is detailed in appendix E.

Chapter 7

Feynman Rules

Now that we have developed procedures for obtaining interaction lagrangians, and the corresponding hamiltonians, we turn to the analysis of the S-matrix for an interacting relativistic quantum field theory. The starting point will be the expression for the scattering operator in Eq. (4.17)

$$\hat{S} = \sum_{n=0}^{\infty} \left(-\frac{i}{\hbar}\right)^n \frac{1}{n!} \int_{-\infty}^{\infty} dt_1 \cdots \int_{-\infty}^{\infty} dt_n \, T\left[\hat{H}_I(t_1)\hat{H}_I(t_2)\cdots\hat{H}_I(t_n)\right] \quad (7.1)$$

Here $\hat{H}_I(t)$ is the interaction expressed in the interaction picture, and we have incorporated the adiabatic damping factor into it. The analysis will eventually lead to the Feynman diagrams and Feynman rules that played such a key role in Vol. I.

To be concrete, we shall focus on the simplest such theory, that of a Dirac particle interacting with a massive neutral scalar field. The analysis is immediately extended to other theories, such as QED, and we shall subsequently do so.[1] To start with, however, we take the interaction hamiltonian in Eq. (5.121)[2]

$$\hat{H}_I(t) = -g \int_\Omega d^3x \, \hat{\bar{\psi}}(\mathbf{x}, t)\hat{\psi}(\mathbf{x}, t)\hat{\phi}(\mathbf{x}, t) \quad (7.2)$$

$$\hat{\phi}(\mathbf{x}, t) = \sum_{\mathbf{k}} \left(\frac{\hbar}{2\omega_k \Omega}\right)^{1/2} \left[c_{\mathbf{k}} e^{i(\mathbf{k}\cdot\mathbf{x} - \omega_k t)} + c_{\mathbf{k}}^\dagger e^{-i(\mathbf{k}\cdot\mathbf{x} - \omega_k t)}\right]$$

$$\hat{\psi}(\mathbf{x}, t) = \frac{1}{\sqrt{\Omega}} \sum_{\mathbf{k}} \sum_{\lambda} \left[a_{\mathbf{k}\lambda} u(\mathbf{k}\lambda) e^{i(\mathbf{k}\cdot\mathbf{x} - \omega_k t)} + b_{\mathbf{k}\lambda}^\dagger v(-\mathbf{k}\lambda) e^{-i(\mathbf{k}\cdot\mathbf{x} - \omega_k t)}\right]$$

---

[1]The Feynman rules for non-linear Yang-Mills theories are more immediately arrived at using path-integral techniques (see later).

[2]We relabel the Dirac spinors as $u(\mathbf{k}\lambda)$, $v(-\mathbf{k}\lambda)$ for ease in what follows.

It is important for what follows to separate these interaction-picture fields into their *destruction* parts, labeled with a superscript $(+)$ corresponding to positive frequency, and their *creation* parts, labeled with a superscript $(-)$ corresponding to negative frequency. Thus[3]

$$\hat{\psi}(\mathbf{x}, t) = \hat{\psi}^{(+)}(\mathbf{x}, t) + \hat{\psi}^{(-)}(\mathbf{x}, t) \quad ; \ \text{destruction}^{(+)} + \text{creation}^{(-)}$$
$$\hat{\bar{\psi}}(\mathbf{x}, t) = \hat{\bar{\psi}}^{(+)}(\mathbf{x}, t) + \hat{\bar{\psi}}^{(-)}(\mathbf{x}, t)$$
$$\hat{\phi}(\mathbf{x}, t) = \hat{\phi}^{(+)}(\mathbf{x}, t) + \hat{\phi}^{(-)}(\mathbf{x}, t) \tag{7.3}$$

The scattering operator now contains an infinite series of *time-ordered* products of fields. If each term only involved *normal-ordered* products of fields, where the destruction operators sit to the right of the creation operators for all times, then it is relatively simple to compute the required matrix elements of these expressions; the annihilation operators just have to annihilate the quanta in the initial state, and the creation operators create those in the final state. The key to rearranging the time-ordered products into normal-ordered products is provided by *Wick's theorem* [Wick (1950)], and we proceed to the proof of this important result.[4]

7.1 Wick's Theorem

P-products and N-Products. We begin by defining two types of interaction-picture operator products:

- The *P-product*, or "time-ordered product with fermions", orders the operators in a descending sequence in time, with the operator at the latest time to the left, and affixes a sign appropriate to the number of interchanges of fermion operators required to obtain this ordering;
- The *N-product*, or "normal-ordered product with fermions", orders the operators with all the annihilation operators to the right of the creation operators, and affixes a sign appropriate to the number of interchanges of fermion operators required to obtain this ordering. The N-product is denoted with a pair of colons.

---

[3]Note that $\hat{\bar{\psi}}^{(+)}(\mathbf{x}, t) = \hat{\psi}^{(-)}(\mathbf{x}, t)^{\dagger}\gamma_4$, and $\hat{\bar{\psi}}^{(-)}(\mathbf{x}, t) = \hat{\psi}^{(+)}(\mathbf{x}, t)^{\dagger}\gamma_4$.

[4]See also the detailing of this proof in [Fetter and Walecka (2003a)], where Wick's theorem, Feynman diagrams, and Feynman rules provide the framework for analyzing non-relativistic quantum many-body systems.

We give three examples:

$$:\hat{\phi}^{(+)}(x)\hat{\phi}^{(-)}(y): \ = \hat{\phi}^{(-)}(y)\hat{\phi}^{(+)}(x)$$
$$:\hat{\psi}^{(+)}(x)\hat{\psi}^{(-)}(y): \ = -\hat{\psi}^{(-)}(y)\hat{\psi}^{(+)}(x)$$
$$:\hat{\psi}^{(+)}(x)\hat{\bar{\psi}}^{(-)}(y): \ = -\hat{\bar{\psi}}^{(-)}(y)\hat{\psi}^{(+)}(x) \qquad (7.4)$$

If δ_P is the signature of the number of permutations of fermion operators required to reorder the operators, then one has

$$P(\hat{A}\hat{B}\hat{C}\hat{D}\cdots) = \delta_P \, P(\hat{C}\hat{A}\hat{D}\hat{B}\cdots)$$
$$:\hat{A}\hat{B}\hat{C}\hat{D}\cdots: \ = \delta_P:\hat{C}\hat{A}\hat{D}\hat{B}\cdots: \qquad (7.5)$$

As defined, one can reorder the operators *within* the P- and N-products as if all appropriate commutators or anticommutators vanished.

Both operations are *distributive*. For example

$$:(\hat{A}+\hat{B})(\hat{C}+\hat{D})\cdots: \ = \ :\hat{A}\hat{C}\cdots: \ + \ :\hat{A}\hat{D}\cdots: \ +$$
$$:\hat{B}\hat{C}\cdots: \ + \ :\hat{B}\hat{D}\cdots: \qquad (7.6)$$

One can therefore limit the proofs to expressions where each operator is either a creation or a destruction part. A general theorem on such terms is then related back to the full fields through the distributive law.

Contractions. The symbol $\hat{U}\cdot\hat{V}\cdot$ stands for the *contraction* between the operators $\hat{U}$ and $\hat{V}$. It is defined by

$$P(\hat{U}\hat{V}) \equiv \ :\hat{U}\hat{V}: \ + \ \hat{U}\cdot\hat{V}\cdot \qquad ; \text{ contraction} \qquad (7.7)$$

We note the following:

- The contraction is what is *left over* when two time-ordered operators are converted to normal-ordered form;
- In the interaction picture, the contraction between two field operators is a *c-number*;
- The contraction is *time-dependent*; it depends on the time-ordering of the operators;
- Since the vacuum expectation value of the normal-ordered product vanishes,[5] the contraction is just the *vacuum expectation value of the time-ordered product*

$$\langle 0|:\hat{U}\hat{V}:|0\rangle = 0$$
$$\Rightarrow \qquad \hat{U}\cdot\hat{V}\cdot = \langle 0|P(\hat{U}\hat{V})|0\rangle \qquad (7.8)$$

---

[5]The annihilation operators simply annihilate the vacuum.

- From the definition

$$\hat{U} \cdot \hat{V}^{\cdot} = +\hat{V} \cdot \hat{U}^{\cdot} \qquad ; \text{ bosons}$$
$$= -\hat{V} \cdot \hat{U}^{\cdot} \qquad ; \text{ fermions} \qquad (7.9)$$

We give some examples for the fields in Eqs. (7.2)–(7.3)

$$\hat{\phi}^{(+)\cdot} \hat{\phi}^{(+)\cdot} = \hat{\phi}^{(-)\cdot} \hat{\phi}^{(-)\cdot} = 0$$
$$\hat{\psi}^{(+)\cdot} \hat{\psi}^{(+)\cdot} = \hat{\psi}^{(-)\cdot} \hat{\psi}^{(-)\cdot} = \hat{\bar{\psi}}^{(+)\cdot} \hat{\psi}^{(+)\cdot} = \hat{\bar{\psi}}^{(-)\cdot} \hat{\psi}^{(-)\cdot} =$$
$$\hat{\psi}^{(+)\cdot} \hat{\psi}^{(-)\cdot} = \hat{\bar{\psi}}^{(+)\cdot} \hat{\bar{\psi}}^{(-)\cdot} = 0 \qquad (7.10)$$

These relations hold since all factors either appropriately commute or anticommute with each other. Take one in detail

$$P[\hat{\psi}^{(+)}(x)\hat{\psi}^{(-)}(y)] = \hat{\psi}^{(+)}(x)\hat{\psi}^{(-)}(y) = -\hat{\psi}^{(-)}(y)\hat{\psi}^{(+)}(x) \qquad ; t_x > t_y$$
$$= -\hat{\psi}^{(-)}(y)\hat{\psi}^{(+)}(x) \qquad ; t_y > t_x$$
$$:\hat{\psi}^{(+)}(x)\hat{\psi}^{(-)}(y): \; = -\hat{\psi}^{(-)}(y)\hat{\psi}^{(+)}(x)$$
$$\Rightarrow \qquad \hat{\psi}^{(+)}(x)^{\cdot} \hat{\psi}^{(-)}(y)^{\cdot} = 0 \qquad (7.11)$$

It is useful for purposes of orientation to realize that *most contractions are zero*. The only non-zero contractions for the scalar field are

$$\hat{\phi}^{(+)}(x)^{\cdot} \hat{\phi}^{(-)}(y)^{\cdot} = \langle 0|\hat{\phi}^{(+)}(x)\hat{\phi}^{(-)}(y)|0\rangle = \frac{\hbar}{ic}\Delta_F(x-y) \qquad ; t_x > t_y$$
$$= 0 \qquad ; t_y > t_x$$
$$\hat{\phi}^{(-)}(x)^{\cdot} \hat{\phi}^{(+)}(y)^{\cdot} = 0 \qquad ; t_x > t_y$$
$$= \langle 0|\hat{\phi}^{(+)}(y)\hat{\phi}^{(-)}(x)|0\rangle = \frac{\hbar}{ic}\Delta_F(x-y) \qquad ; t_y > t_x$$
$$(7.12)$$

Here $\Delta_F(x-y)$ is the Feynman propagator for the scalar field, studied in appendix F, and expressed as a sum over modes in Eqs. (F.43) and (F.39).[6] The contraction of the fields themselves follows from the distributive law

$$\hat{\phi}(x)^{\cdot} \hat{\phi}(y)^{\cdot} = \langle 0|T[\hat{\phi}(x)\,\hat{\phi}(y)]|0\rangle = \frac{\hbar}{ic}\Delta_F(x-y;\, m^2) \qquad ;$$
$$\text{Feynman propagator} \quad (7.13)$$

---

[6] Note that each of these contractions is only non-zero for the indicated time-orderings. The evaluation of the matrix elements is verified in Prob. 7.2.

which is just by Eq. (F.44). The four-dimensional Fourier transform in Minkowski space of $\Delta_F(x - y; m^2)$ is given by the first of Eqs. (F.50)

$$\Delta_F(x - y; m^2) = \int \frac{d^4k}{(2\pi)^4} \frac{e^{ik\cdot(x-y)}}{k^2 + m^2} \qquad ; \; m \to m - i\eta \qquad (7.14)$$

A similar calculation gives the only non-zero contractions in the case of the Dirac field as

$$
\begin{aligned}
\hat{\psi}_\alpha^{(+)}(x)^\bullet \, \hat{\bar{\psi}}_\beta^{(-)}(y)^\bullet &= \langle 0|\hat{\psi}_\alpha^{(+)}(x)\hat{\bar{\psi}}_\beta^{(-)}(y)|0\rangle = iS_{\alpha\beta}^F(x - y) &&; \; t_x > t_y \\
&= 0 &&; \; t_y > t_x \\
\hat{\psi}_\alpha^{(-)}(x)^\bullet \, \hat{\bar{\psi}}_\beta^{(+)}(y)^\bullet &= 0 &&; \; t_x > t_y \\
&= -\langle 0|\hat{\bar{\psi}}_\beta^{(+)}(y)\hat{\psi}_\alpha^{(-)}(x)|0\rangle = iS_{\alpha\beta}^F(x - y) &&; \; t_y > t_x
\end{aligned}
$$

$$(7.15)$$

Here $S_{\alpha\beta}^F(x - y)$ is the Feynman propagator for the Dirac field derived in appendix F, and given in Eq. (F.47)–(F.49). For the fields themselves

$$\hat{\psi}_\alpha(x)^\bullet \, \hat{\bar{\psi}}_\beta(y)^\bullet = \langle 0|P[\hat{\psi}_\alpha(x)\,\hat{\bar{\psi}}_\beta(y)]|0\rangle = iS_{\alpha\beta}^F(x - y) \qquad ;$$

$$\text{Feynman propagator} \qquad (7.16)$$

Equation (F.49) expresses this propagator in terms of the previous propagator $\Delta_F(x - y; M^2)$ in Eq. (7.14)

$$S_{\alpha\beta}^F(x - y) = \left[\gamma_\mu \frac{\partial}{\partial x_\mu} - M\right]_{\alpha\beta} \Delta_F(x - y; M^2) \qquad (7.17)$$

We shall often use the Feynman propagator for the Dirac field in its unrationalized, matrix form

$$S_F(x - x') = -\int \frac{d^4k}{(2\pi)^4} \left[\frac{1}{ik_\mu\gamma_\mu + M}\right] e^{ik\cdot(x-x')} \qquad (7.18)$$

Sign Convention. We adopt the following sign convention:

N- and P-products with more than one contraction will have the contractions denoted by single dots, double dots, etc. As soon as the two factors of a given contraction sit together, that contraction may be replaced by the appropriate c-number. The number of sign changes to get the factors to sit together, that is, the number of interchanges of fermion factors, whether or not they are in another contraction, must be counted.

Wick's Theorem. Wick's theorem then reads as follows:

$$P(\hat{U}\hat{V}\cdots\hat{X}\hat{Y}\hat{Z}) = :\hat{U}\hat{V}\cdots\hat{X}\hat{Y}\hat{Z}: + :\hat{U}\cdot\hat{V}\cdot\cdots\hat{X}\hat{Y}\hat{Z}: + \cdots +$$
$$:\hat{U}\cdot\hat{V}\cdot\cdot\hat{W}\cdot\cdot\cdots\hat{X}\cdots\hat{Y}\cdot\hat{Z}\cdots: + \cdots$$

$$= :(\text{sum over all possible pairs of contractions}): \quad ;$$

$$\text{Wick's theorem} \qquad (7.19)$$

Before we proceed to a proof, we give the basic idea of the theorem. Take a given time ordering. Now start moving the creation operators to the left. Every time the creation operator fails to simply commute (or anticommute) with its neighbor, there will be something left over in this process. The term left-over is just the contraction. One can include all possible contractions, since if the creation operator already sits to the left of the destruction operator, the contraction is zero.[7] The above expression then simply enumerates all the extra terms left over in going from a P-product to an N-product. Although a little reflection will convince the reader that this argument actually justifies the result, we proceed to the formal proof given by Wick.

Basic Lemma. We first prove the basic lemma:

If $:\hat{U}\hat{V}\cdots\hat{X}\hat{Y}:$ is an N-product and $\hat{Z}$ is a factor labeled with a time earlier than any of the times for $(\hat{U},\hat{V},\cdots,\hat{X},\hat{Y})$, then

$$:\hat{U}\hat{V}\cdots\hat{X}\hat{Y}: \hat{Z} = :\hat{U}\hat{V}\cdots\hat{X}\hat{Y}\cdot\hat{Z}\cdot: + :\hat{U}\hat{V}\cdots\hat{X}\cdot\hat{Y}\hat{Z}\cdot: + \cdots +$$
$$:\hat{U}\cdot\hat{V}\cdots\hat{X}\hat{Y}\hat{Z}\cdot: + :\hat{U}\hat{V}\cdots\hat{X}\hat{Y}\hat{Z}: \qquad (7.20)$$

This is proven through the following series of steps:

(1) If $\hat{Z}$ is a destruction operator, then all the contractions vanish. Therefore only the last term contributes, and the lemma holds;

(2) It can be assumed that $(\hat{U},\hat{V},\cdots,\hat{X},\hat{Y})$ are already normal-ordered. If not, simply change the order of these factors *on both sides* of the equation. This gives a common sign change, which cancels;

(3) It can then be assumed that $\hat{Z}$ is a creation operator and $(\hat{U},\hat{V},\cdots,\hat{X},\hat{Y})$ all destruction operators. Additional creation operators can subsequently be added on the left, for all further contractions vanish and the creation operators can be immediately taken inside all the normal-ordering symbols;

(4) *Hence it is sufficient to prove the lemma for $\hat{Z}$ a creation operator and $(\hat{U},\hat{V},\cdots,\hat{X},\hat{Y})$ all destruction operators.*

---

[7]Most contractions are zero!

The lemma is then proven by induction. It is true for two terms, since in that case

$$\hat{Y}\hat{Z} = P(\hat{Y}\hat{Z}) = \hat{Y}\cdot\hat{Z}\cdot + :\hat{Y}\hat{Z}: \qquad (7.21)$$

Now assume it holds for n terms, and show it then holds for $n + 1$ terms. To do this, multiply Eq. (7.20) on the left by another destruction operator $\hat{D}$ having a time later than $\hat{Z}$. This operator can be taken right inside the normal-ordering on the l.h.s. As long as $\hat{Z}$ sits in a contraction (a c-number), the operator $\hat{D}$ can also be taken inside the normal-ordering on the r.h.s. Thus

$$:\hat{D}\hat{U}\hat{V}\cdots\hat{X}\hat{Y}:\hat{Z} = :\hat{D}\hat{U}\hat{V}\cdots\hat{X}\hat{Y}\cdot\hat{Z}\cdot: + :\hat{D}\hat{U}\hat{V}\cdots\hat{X}\cdot\hat{Y}\hat{Z}\cdot: + \cdots +$$
$$:\hat{D}\hat{U}\cdot\hat{V}\cdots\hat{X}\hat{Y}\hat{Z}\cdot: + \hat{D}:\hat{U}\hat{V}\cdots\hat{X}\hat{Y}\hat{Z}: \qquad (7.22)$$

It is only the last term that creates any problem. We claim that this term can be expressed as

$$\hat{D}:\hat{U}\hat{V}\cdots\hat{X}\hat{Y}\hat{Z}: = :\hat{D}\cdot\hat{U}\hat{V}\cdots\hat{X}\hat{Y}\hat{Z}\cdot: + :\hat{D}\hat{U}\hat{V}\cdots\hat{X}\hat{Y}\hat{Z}: \qquad (7.23)$$

This proves the lemma for $n + 1$ terms.

To establish Eq. (7.23), move the factor $\hat{Z}$ to the left *within* the normal-ordering symbol on the l.h.s. This results in a sign change δ_p. Then[8]

$$\hat{D}:\hat{U}\hat{V}\cdots\hat{X}\hat{Y}\hat{Z}: = \delta_p \hat{D}\hat{Z}\hat{U}\hat{V}\cdots\hat{X}\hat{Y}$$
$$= \delta_p P(\hat{D}\hat{Z})\hat{U}\hat{V}\cdots\hat{X}\hat{Y}$$
$$= \delta_p \hat{D}\cdot\hat{Z}\cdot\hat{U}\hat{V}\cdots\hat{X}\hat{Y} + \delta_p\delta_q \hat{Z}\hat{D}\hat{U}\hat{V}\cdots\hat{X}\hat{Y} \qquad (7.24)$$

Now reorder the terms in this expression using the previous sign convention

$$\hat{D}:\hat{U}\hat{V}\cdots\hat{X}\hat{Y}\hat{Z}: = \delta_p^2 \hat{D}\cdot\hat{U}\hat{V}\cdots\hat{X}\hat{Y}\hat{Z}\cdot + \delta_p^2\delta_q^2 :\hat{D}\hat{U}\hat{V}\cdots\hat{X}\hat{Y}\hat{Z}:$$
$$= :\hat{D}\cdot\hat{U}\hat{V}\cdots\hat{X}\hat{Y}\hat{Z}\cdot: + :\hat{D}\hat{U}\hat{V}\cdots\hat{X}\hat{Y}\hat{Z}: \qquad (7.25)$$

which is just Eq. (7.23).

The lemma can be extended to the case where the N-products in Eq. (7.20) already have contractions in them. Multiply both sides by, say, $\hat{R}\cdot\cdot\hat{S}\cdot\cdot$, and then interchange factors *on both sides* as you which. This is simply the multiplication of Eq. (7.20) by a common factor and sign, which

---

[8]The last term simply express $:\hat{D}\hat{Z}:$.

changes nothing. Thus the basic lemma takes an equivalent form as[9]

$$:\hat{U}\hat{V}\cdots\cdots\hat{W}\hat{X}\cdot\cdot\hat{Y}:\hat{Z} = :\hat{U}\hat{V}\cdots\cdots\hat{W}\hat{X}\cdot\cdot\hat{Y}\cdot\hat{Z}\cdot: +$$
$$:\hat{U}\hat{V}\cdots\cdots\hat{W}\cdot\hat{X}\cdot\cdot\hat{Y}\hat{Z}\cdot: + \cdots + :\hat{U}\cdot\hat{V}\cdots\cdots\hat{W}\hat{X}\cdot\cdot\hat{Y}\hat{Z}\cdot: +$$
$$:\hat{U}\hat{V}\cdots\cdots\hat{W}\hat{X}\cdot\cdot\hat{Y}\hat{Z}: \qquad\qquad (7.26)$$

Proof of Wick's Theorem. The proof again follows by induction. Wick's theorem holds for two terms

$$P(\hat{U}\hat{V}) = :\hat{U}\hat{V}: + \hat{U}\cdot\hat{V}\cdot \qquad\qquad (7.27)$$

Assume Eq. (7.19) holds for n terms. Multiply it on the right by another operator $\hat{\Omega}$ with a time *earlier* than that of any other factor so that

$$P(\hat{U}\hat{V}\cdots\hat{X}\hat{Y}\hat{Z})\hat{\Omega} = P(\hat{U}\hat{V}\cdots\hat{X}\hat{Y}\hat{Z}\hat{\Omega}) \qquad\qquad (7.28)$$

Then

$$P(\hat{U}\hat{V}\cdots\hat{X}\hat{Y}\hat{Z}\hat{\Omega}) = :\hat{U}\hat{V}\cdots\hat{X}\hat{Y}\hat{Z}:\hat{\Omega} + :\hat{U}\cdot\hat{V}\cdot\cdots\hat{X}\hat{Y}\hat{Z}:\hat{\Omega} + \cdots +$$
$$:\hat{U}\cdot\hat{V}\cdot\cdot\hat{W}\cdots\cdots\hat{X}\cdots\hat{Y}\cdot\hat{Z}\cdots:\hat{\Omega} + \cdots \qquad (7.29)$$

Now use the basic lemma on the r.h.s.

$$P(\hat{U}\hat{V}\cdots\hat{X}\hat{Y}\hat{Z}\hat{\Omega}) = :\hat{U}\hat{V}\cdots\hat{X}\hat{Y}\hat{Z}\hat{\Omega}: + :\hat{U}\hat{V}\cdots\hat{X}\hat{Y}\hat{Z}\cdot\hat{\Omega}\cdot: + \cdots +$$
$$:\hat{U}\cdot\hat{V}\cdot\cdots\hat{X}\hat{Y}\hat{Z}\hat{\Omega}: + :\hat{U}\cdot\hat{V}\cdot\cdots\hat{X}\hat{Y}\hat{Z}\cdot\cdot\hat{\Omega}\cdot\cdot: + \cdots +$$
$$:\hat{U}\cdot\hat{V}\cdot\cdot\hat{W}\cdots\cdots\hat{X}\cdots\hat{Y}\cdot\hat{Z}\cdots\hat{\Omega}: + \cdots$$
$$= :(\text{sum over all possible pairs of contractions}): \qquad (7.30)$$

The restriction on the time of $\hat{\Omega}$ can now be removed by rearranging the order of the terms on *both sides* of this relation; the result is an overall sign, which cancels. Thus Wick's theorem also holds for $n+1$ terms, and we have the desired proof.

The distributive law can now be used to return to the full field, and thus Wick's theorem holds in terms of the fields themselves.

Corollary. A *mixed P*-product $P(\hat{U}\hat{V}:\hat{W}\hat{X}:\cdots\hat{Z})$ containing within it equal-time normal-ordered products of fields $:\hat{W}\hat{X}:$ can be decomposed in the same manner as above, but *omitting* contractions $\hat{W}\cdot\hat{X}\cdot$ between any factors already N-ordered. This follows since it is never necessary to turn those terms around to get them into normal-ordered form.

---

[9]Here we relabel $\hat{R}\cdot\cdot\hat{S}\cdot\cdot \rightarrow \hat{V}\cdot\cdot\hat{X}\cdot\cdot$.

7.2   Example (Dirac-Scalar)

As an example, consider the Dirac-scalar theory of Eq. (7.2). To simplify the discussion, we assume, in line with our previous analysis of the electromagnetic current in Eq. (5.116), that the scalar density is *already* normal-ordered.[10] We then proceed to expand the scattering operator in powers of the coupling constant g. We assume the (N, ϕ) are both massive and stable particles.[11] The first-order $\hat{S}$-operator then vanishes by energy-momentum conservation (see Prob. 7.3)

$$\hat{S}^{(1)} = 0 \qquad ; (N, \phi) \text{ stable} \qquad (7.31)$$

The second-order $\hat{S}$ now follows from Eqs. (7.1)–(7.2) as

$$\hat{S}^{(2)} = \left(\frac{ig}{\hbar c}\right)^2 \frac{1}{2!} \int d^4x_1 \int d^4x_2 \times$$
$$P[:\hat{\bar{\psi}}(x_1)\hat{\psi}(x_1)::\hat{\bar{\psi}}(x_2)\hat{\psi}(x_2):]P[\hat{\phi}(x_1)\hat{\phi}(x_2)] \qquad (7.32)$$

We have used the fact that the interaction-picture Dirac and scalar fields commute, and we again write $d^4x = d^3x\,cdt$. Application of Wick's theorem, and its corollary, to the P-product of the Dirac fields then gives

$$P[:\hat{\bar{\psi}}(x_1)\hat{\psi}(x_1)::\hat{\bar{\psi}}(x_2)\hat{\psi}(x_2):] = :\hat{\bar{\psi}}(x_1)\hat{\psi}(x_1)\hat{\bar{\psi}}(x_2)\hat{\psi}(x_2): +$$
$$:\hat{\bar{\psi}}(x_1)\hat{\psi}(x_1)^{\boldsymbol{\cdot}}\hat{\bar{\psi}}(x_2)^{\boldsymbol{\cdot}}\hat{\psi}(x_2): + :\hat{\bar{\psi}}(x_1)^{\boldsymbol{\cdot}}\hat{\psi}(x_1)\hat{\bar{\psi}}(x_2)\hat{\psi}(x_2)^{\boldsymbol{\cdot}}: +$$
$$:\hat{\bar{\psi}}(x_1)^{\boldsymbol{\cdot\cdot}}\hat{\psi}(x_1)^{\boldsymbol{\cdot}}\hat{\bar{\psi}}(x_2)^{\boldsymbol{\cdot}}\hat{\psi}^{\boldsymbol{\cdot\cdot}}(x_2): \qquad (7.33)$$

where only the non-zero contractions have been retained. These are obtained from Eq. (7.16). Hence, with careful attention to our sign convention, and with the Dirac indices explicitly displayed,

$$P[:\hat{\bar{\psi}}_\alpha(x_1)\hat{\psi}_\alpha(x_1)::\hat{\bar{\psi}}_\beta(x_2)\hat{\psi}_\beta(x_2):] = :\hat{\bar{\psi}}_\alpha(x_1)\hat{\psi}_\alpha(x_1)\hat{\bar{\psi}}_\beta(x_2)\hat{\psi}_\beta(x_2): +$$
$$:\hat{\bar{\psi}}_\alpha(x_1)iS^F_{\alpha\beta}(x_1 - x_2)\hat{\psi}_\beta(x_2): + :\hat{\bar{\psi}}_\beta(x_2)iS^F_{\beta\alpha}(x_2 - x_1)\hat{\psi}_\alpha(x_1): -$$
$$iS^F_{\alpha\beta}(x_1 - x_2)iS^F_{\beta\alpha}(x_2 - x_1) \qquad (7.34)$$

Note, in particular, the sign of the last term.

A similar application of Wick's theorem to the scalar fields, and the corresponding expression for a contraction in Eq. (7.13), then reduces the

---

[10]For the relaxation of this assumption, see Prob. 7.11. Note that the scalar density is *bilinear* in the fermion fields, in which case its T- and P-products coincide.

[11]We use (N, ϕ) to denote the particles.

scattering operator in Eq. (7.32) to the expression

$$\hat{S}^{(2)} = \left(\frac{ig}{\hbar c}\right)^2 \frac{1}{2!} \int d^4x_1 \int d^4x_2 \left[:\hat{\bar{\psi}}_\alpha(x_1)\hat{\psi}_\alpha(x_1)\hat{\bar{\psi}}_\beta(x_2)\hat{\psi}_\beta(x_2): + \right.$$

$$:\hat{\bar{\psi}}_\alpha(x_1)iS^F_{\alpha\beta}(x_1-x_2)\hat{\psi}_\beta(x_2): + :\hat{\bar{\psi}}_\beta(x_2)iS^F_{\beta\alpha}(x_2-x_1)\hat{\psi}_\alpha(x_1): -$$

$$\left. iS^F_{\alpha\beta}(x_1-x_2)iS^F_{\beta\alpha}(x_2-x_1)\right]\left[:\hat{\phi}(x_1)\hat{\phi}(x_2): + \frac{\hbar}{ic}\Delta_F(x_1-x_2)\right]$$

$$(7.35)$$

This scattering operator describes a variety of processes, and individual amplitudes are obtained by taking appropriate matrix elements of this expression. We discuss various examples.

7.2.1 *Scattering Amplitudes*

Consider the Dirac-scalar scattering process $\phi + N \to \phi + N$. The initial and final states, assumed to be eigenstates of momentum, are

$$|i\rangle = c^\dagger_{\mathbf{k}_1} a^\dagger_{\mathbf{k}_2\lambda}|0\rangle \qquad\qquad ; |f\rangle = c^\dagger_{\mathbf{k}_3} a^\dagger_{\mathbf{k}_4\lambda'}|0\rangle \qquad (7.36)$$

In this case, only the second and third Dirac operators in $\hat{S}^{(2)}$, and the first scalar operator, contribute to the matrix element. A change of dummy variables $(x_1, \alpha) \rightleftharpoons (x_2, \beta)$ shows that these two contributions are identical. It is then sufficient to keep just the second Dirac operator and cancel the $1/2!$ in front. The creation and destruction operators in the fields must then annihilate and create the states in Eq. (7.36).[12] One is then simply left with the corresponding wave functions in the fields. Thus

$$S_{fi} = \left(\frac{ig}{\hbar c}\right)^2 \frac{1}{\Omega^2} \frac{\hbar}{\sqrt{4\omega_1\omega_3}} \int d^4x_1 \int d^4x_2$$

$$\left[\bar{u}_\alpha(\mathbf{k}_4\lambda')iS^F_{\alpha\beta}(x_1-x_2)u_\beta(\mathbf{k}_2\lambda)e^{i(k_2+k_1)\cdot x_2}e^{-i(k_4+k_3)\cdot x_1} + \right.$$

$$\left. \bar{u}_\alpha(\mathbf{k}_4\lambda')iS^F_{\alpha\beta}(x_1-x_2)u_\beta(\mathbf{k}_2\lambda)e^{i(k_2-k_3)\cdot x_2}e^{-i(k_4-k_1)\cdot x_1}\right] \quad (7.37)$$

where we explicitly exhibit the Dirac indices on the spinors.

We were very careful about passing from the discrete to the continuum limit in Vol. I. Here we go directly to the continuum limit and identify the matrix elements T_{fi}, through which all the rates and cross sections were

---

[12]See Prob. 7.16; note that there are two possibilities for $:\hat{\phi}(x_1)\hat{\phi}(x_2):$.

calculated, from the expression in Probl. 7.2[13]

$$S_{fi} = (2\pi)^4 \delta^{(4)}(K_f - K_i) \frac{1}{\Omega^{(n+2)/2}} \left[\frac{-iT_{fi}}{\hbar c} \right] \tag{7.38}$$

The four-dimensional Fourier transform of the Feynman propagator for the Dirac field is given in Eq. (7.18)

$$S_{\alpha\beta}^F(x_1 - x_2) = - \int \frac{d^4q}{(2\pi)^4} \left[\frac{1}{iq_\mu\gamma_\mu + M} \right]_{\alpha\beta} e^{iq\cdot(x_1-x_2)} \tag{7.39}$$

This expression can now be substituted in Eq. (7.37) and the integrals over all space-time performed. For the first term this gives

$$\int d^4x_1 \int d^4x_2\, e^{i(k_2+k_1)\cdot x_2} e^{iq\cdot(x_1-x_2)} e^{-i(k_4+k_3)\cdot x_1}$$

$$= (2\pi)^4 \delta^{(4)}(k_1 + k_2 - q)(2\pi)^4 \delta^{(4)}(q - k_3 - k_4)$$

$$= (2\pi)^8 \delta^{(4)}(k_1 + k_2 - k_3 - k_4)\delta^{(4)}(k_1 + k_2 - q) \tag{7.40}$$

For the second term one has

$$\int d^4x_1 \int d^4x_2\, e^{i(k_2-k_3)\cdot x_2} e^{iq\cdot(x_1-x_2)} e^{-i(k_4-k_1)\cdot x_1}$$

$$= (2\pi)^4 \delta^{(4)}(k_2 - k_3 - q)(2\pi)^4 \delta^{(4)}(q - k_4 + k_1)$$

$$= (2\pi)^8 \delta^{(4)}(k_1 + k_2 - k_3 - k_4)\delta^{(4)}(k_2 - k_3 - q) \tag{7.41}$$

The integral over $d^4q/(2\pi)^4$ from Eq. (7.39) can now be performed, and one obtains the following second-order S-matrix element for the Dirac-scalar scattering process $\phi(k_1) + N(k_2) \rightarrow \phi(k_3) + N(k_4)$

$$S_{fi}^{(2)} = (2\pi)^4 \delta^{(4)}(k_1 + k_2 - k_3 - k_4)\frac{1}{\Omega^2}\frac{\hbar}{\sqrt{4\omega_1\omega_3}}\left(\frac{ig}{\hbar c}\right)^2\left(\frac{1}{i}\right) \times$$

$$\bar{u}(\mathbf{k_4}\lambda')\left[\frac{1}{i\gamma_\mu(k_2+k_1)_\mu + M} + \frac{1}{i\gamma_\mu(k_2-k_3)_\mu + M}\right] u(\mathbf{k_2}\lambda) \tag{7.42}$$

Here we revert to a matrix notation and again suppress the Dirac indices.

As a second example, consider the scattering of two Dirac particles $N + N \rightarrow N + N$. In this case, only the normal-ordered product of the four Dirac fields contributes in Eq. (7.35), along with the scalar propagator $(\hbar/ic)\Delta_F$. The initial and final states in this case are

$$|i\rangle = a_{\mathbf{k_1}\lambda_1}^\dagger a_{\mathbf{k_2}\lambda_2}^\dagger |0\rangle \qquad ; \qquad |f\rangle = a_{\mathbf{k_3}\lambda_3}^\dagger a_{\mathbf{k_4}\lambda_4}^\dagger |0\rangle \tag{7.43}$$

---

[13] Here $(\hbar K_i, \hbar K_f)$ are the total initial and final four-momenta, and n is the number of final particles; in general, we extract a factor $\Omega^{-(n_f+n_i)/2}$ in the definition of T_{fi}.

One again simply has to pair off the creation and annihilation operators in the fields, leaving the corresponding wave functions in the matrix element. There are four possibilities, and the result is[14]

$$
S_{fi} = \left(\frac{ig}{\hbar c}\right)^2 \frac{1}{\Omega^2}\frac{1}{2!} \int d^4x_1 \int d^4x_2 \, \frac{\hbar}{ic}\Delta_F(x_1 - x_2) \times
$$

$$
\Big\{ [\bar{u}(\mathbf{k}_3\lambda_3)u(\mathbf{k}_1\lambda_1)]\,[\bar{u}(\mathbf{k}_4\lambda_4)u(\mathbf{k}_2\lambda_2)]\, e^{i(k_1-k_3)\cdot x_2}e^{i(k_2-k_4)\cdot x_1} -
$$

$$
[\bar{u}(\mathbf{k}_4\lambda_4)u(\mathbf{k}_1\lambda_1)]\,[\bar{u}(\mathbf{k}_3\lambda_3)u(\mathbf{k}_2\lambda_2)]\, e^{i(k_1-k_4)\cdot x_2}e^{i(k_2-k_3)\cdot x_1} +
$$

$$
[\bar{u}(\mathbf{k}_3\lambda_3)u(\mathbf{k}_1\lambda_1)]\,[\bar{u}(\mathbf{k}_4\lambda_4)u(\mathbf{k}_2\lambda_2)]\, e^{i(k_1-k_3)\cdot x_1}e^{i(k_2-k_4)\cdot x_2} -
$$

$$
[\bar{u}(\mathbf{k}_4\lambda_4)u(\mathbf{k}_1\lambda_1)]\,[\bar{u}(\mathbf{k}_3\lambda_3)u(\mathbf{k}_2\lambda_2)]\, e^{i(k_1-k_4)\cdot x_1}e^{i(k_2-k_3)\cdot x_2} \Big\} \quad (7.44)
$$

The last two terms differ from the first two only by the interchange ($x_1 \rightleftharpoons x_2$), and since $\Delta_F(x_1 - x_2) = \Delta_F(x_2 - x_1)$ is symmetric in its argument, a change of dummy integration variables demonstrates that one can keep only the first two, provided the $1/2!$ is deleted in front. Hence

$$
S_{fi} = \left(\frac{ig}{\hbar c}\right)^2 \frac{1}{\Omega^2} \int d^4x_1 \int d^4x_2 \, \frac{\hbar}{ic}\Delta_F(x_1 - x_2) \times
$$

$$
\Big\{ [\bar{u}(\mathbf{k}_3\lambda_3)u(\mathbf{k}_1\lambda_1)]\,[\bar{u}(\mathbf{k}_4\lambda_4)u(\mathbf{k}_2\lambda_2)]\, e^{i(k_1-k_3)\cdot x_2}e^{i(k_2-k_4)\cdot x_1} -
$$

$$
[\bar{u}(\mathbf{k}_4\lambda_4)u(\mathbf{k}_1\lambda_1)]\,[\bar{u}(\mathbf{k}_3\lambda_3)u(\mathbf{k}_2\lambda_2)]\, e^{i(k_1-k_4)\cdot x_2}e^{i(k_2-k_3)\cdot x_1} \Big\} \quad (7.45)
$$

The substitution of the four-dimensional Fourier transform in Eq. (7.14) allows one to carry out the integrations over all space-time. An argument exactly analogous to that in Eqs. (7.40)–(7.42) then leads to the following second-order S-matrix element for the Dirac-Dirac scattering process $N(k_1) + N(k_2) \to N(k_3) + N(k_4)$

$$
S_{fi}^{(2)} = (2\pi)^4\delta^{(4)}(k_1 + k_2 - k_3 - k_4)\frac{1}{\Omega^2}\left(\frac{ig}{\hbar c}\right)^2\left(\frac{\hbar}{ic}\right) \times
$$

$$
\Big[\bar{u}(\mathbf{k}_3\lambda_3)u(\mathbf{k}_1\lambda_1)\frac{1}{(k_1 - k_3)^2 + m^2}\bar{u}(\mathbf{k}_4\lambda_4)u(\mathbf{k}_2\lambda_2) -
$$

$$
\bar{u}(\mathbf{k}_4\lambda_4)u(\mathbf{k}_1\lambda_1)\frac{1}{(k_1 - k_4)^2 + m^2}\bar{u}(\mathbf{k}_3\lambda_3)u(\mathbf{k}_2\lambda_2) \Big] \quad (7.46)
$$

The second term is the "exchange amplitude", present because this theory has a single type of Dirac particle. The Dirac indices have again been

---

[14]See Prob. 7.16; it is easy to see that the second amplitude involves one additional fermion interchange, and hence has a minus sign relative to the first.

suppressed in this expression.

7.2.2 Self-Energies

There is a non-vanishing matrix element of the scattering operator $\hat{S}^{(2)}$ where there is a single Dirac particle in and out, with initial and final states

$$|i\rangle = a^{\dagger}_{\mathbf{k}_1\lambda_1}|0\rangle \qquad\qquad ; |f\rangle = a^{\dagger}_{\mathbf{k}_2\lambda_2}|0\rangle \qquad (7.47)$$

In this case, it is the second and third Dirac operators in Eq. (7.35) that contribute, along with the scalar propagator. Interchange of dummy integration variables ($x_1 \rightleftharpoons x_2$) again shows that the contributions of the two terms are identical, and it is sufficient to keep just one of them, while cancelling the $1/2!$ in front. The S-matrix element then follows as

$$S_{fi} = \left(\frac{ig}{\hbar c}\right)^2 \frac{1}{\Omega} \int d^4x_1 \int d^4x_2 \left(\frac{\hbar}{ic}\right) \Delta_F(x_1 - x_2) \times$$
$$\bar{u}(\mathbf{k}_2\lambda_2)e^{-ik_2 \cdot x_2} iS_F(x_2 - x_1)u(\mathbf{k}_1\lambda_1)e^{ik_1 \cdot x_1} \qquad (7.48)$$

Now substitute the four-dimensional Fourier transforms of the propagators

$$S_{fi} = \left(\frac{ig}{\hbar c}\right)^2 \frac{1}{\Omega} \left(\frac{\hbar}{ic}\right) \frac{1}{i} \frac{1}{(2\pi)^8} \int d^4x_1 \int d^4x_2 \int d^4q \int d^4t \frac{1}{q^2 + m^2} \times$$
$$\bar{u}(\mathbf{k}_2\lambda_2)\frac{1}{i\gamma_\mu t_\mu + M}u(\mathbf{k}_1\lambda_1)e^{-ik_2 \cdot x_2}e^{ik_1 \cdot x_1}e^{iq \cdot (x_1 - x_2)}e^{it \cdot (x_2 - x_1)} \qquad (7.49)$$

The two integrals over space-time then give

$$\int d^4x_1 \int d^4x_2\, e^{-ik_2 \cdot x_2}e^{ik_1 \cdot x_1}e^{iq \cdot (x_1 - x_2)}e^{it \cdot (x_2 - x_1)}$$
$$= (2\pi)^8 \delta^{(4)}(k_1 + q - t)\delta^{(4)}(t - q - k_2)$$
$$= (2\pi)^8 \delta^{(4)}(k_1 - k_2)\delta^{(4)}(k_1 + q - t) \qquad (7.50)$$

The integral over d^4t can now be carried out, and Eq. (7.48) then takes the form

$$S_{fi} = -(2\pi)^4 \delta^{(4)}(k_1 - k_2)\frac{i}{\Omega}\bar{u}(\mathbf{k}_2\lambda_2)\Sigma(k_1)u(\mathbf{k}_1\lambda_1)$$
$$\Sigma(k_1) \equiv \frac{ig^2}{\hbar c^3} \int \frac{d^4q}{(2\pi)^4} \frac{1}{i\gamma_\mu(k_1 + q)_\mu + M} \frac{1}{q^2 + m^2} \qquad (7.51)$$

The diagonal matrix element here corresponds to a *self-mass* of the Dirac particle, whose effects we shall subsequently discuss.

There is a similar self-energy term for the scalar particle where the initial and final states are

$$|i\rangle = c_{\mathbf{k}_1}^\dagger |0\rangle \qquad\qquad ; \; |f\rangle = c_{\mathbf{k}_2}^\dagger |0\rangle \qquad (7.52)$$

It is the fully contracted term in the Dirac operator (a c-number), and the first term in the scalar operator in $\hat{S}^{(2)}$ that contribute in this case. Again, there are two equal terms that differ merely by the interchange of the dummy integration variables $(x_1 \rightleftharpoons x_2)$, and the S-matrix element is

$$S_{fi} = \left(\frac{ig}{\hbar c}\right)^2 \frac{1}{\Omega} \frac{\hbar}{\sqrt{4\omega_1\omega_2}} \int d^4x_1 \int d^4x_2 \times$$
$$(-1)\mathrm{Tr}\left[iS_F(x_1 - x_2)iS_F(x_2 - x_1)\right] e^{-ik_2 \cdot x_2} e^{ik_1 \cdot x_1} \qquad (7.53)$$

Here "Tr" indicates the trace (the sum of the diagonal elements) of the Dirac matrix product. Substitution of the Fourier transforms of the propagators, and a repetition of the above arguments gives

$$S_{fi} = (2\pi)^4\delta^{(4)}(k_1 - k_2)\frac{ic}{\Omega}\frac{1}{\sqrt{4\omega_1\omega_2}}\Pi(k_1)$$

$$\Pi(k_1) \equiv \frac{ig^2}{\hbar c^3} \int \frac{d^4t}{(2\pi)^4}\mathrm{Tr}\left[\frac{1}{i\gamma_\mu(t + k_1)_\mu + M} \frac{1}{i\gamma_\mu t_\mu + M}\right] \qquad (7.54)$$

This term contributes to the *self-mass* of the scalar particle.

7.2.3   *Vacuum Amplitude*

There is a most unusual matrix element of the scattering operator $\hat{S}^{(2)}$, one that involves *no incoming or outgoing particles whatsoever!* This arises from the product of the c-number terms in Eq. (7.35) and gives rise to a vacuum-vacuum S-matrix element of

$$S_{00}^{(2)} = \left(\frac{ig}{\hbar c}\right)^2 \frac{1}{2!} \int d^4x_1 \int d^4x_2 \frac{\hbar}{ic}\Delta_F(x_1 - x_2) \times$$
$$(-1)\mathrm{Tr}\left[iS_F(x_1 - x_2)iS_F(x_2 - x_1)\right] \qquad (7.55)$$

It is difficult to believe that the process described by this amplitude can have a physical effect here in the laboratory, since it is taking place uniformly everywhere in space-time. In fact, we shall show that when treated properly, this vacuum-vacuum amplitude disappears from the theory.

It is important to realize that all the processes and matrix elements discussed above in this example *are inherent features of any relativistic quantum field theory*.

7.3 Feynman Diagrams

After the dust clears, the resulting S-matrix elements take a relatively simple form, and each contributing term can be viewed pictorially in terms of a diagram. This is the analysis developed by Feynman [Feynman (1949)], and *Feynman diagrams* today form an integral part of the language of physics.

Consider the S-matrix element in coordinate space for the scattering process $\phi(k_1) + N(k_2) \rightarrow \phi(k_3) + N(k_4)$. To order g^2 there are 2! identical contributions corresponding to the permutations of the *labeling* (x_1, x_2) of the space-time points at which the interactions take place. Any one labeling can be employed, provided the $1/2!$ in front of the scattering operator is deleted. There are then just two independent contributions to S_{fi} in Eq. (7.37), as illustrated in Fig. (7.1). These can be viewed as representing two distinct *paths* in space-time that the system can take in propagating from the infinite past to the infinite future.

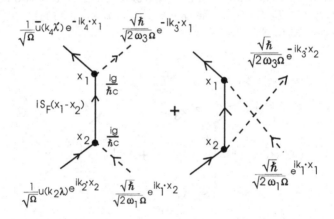

Fig. 7.1 Feynman diagrams in coordinate space for the S-matrix to second-order in g for the process $\phi(k_1) + N(k_2) \rightarrow \phi(k_3) + N(k_4)$. One must still perform $\int d^4x_1 \int d^4x_2$. The Dirac indices are suppressed along the fermion line, and there is a suppressed Dirac unit matrix at each vertex. The second diagram shows only the relevant changes.

The wave functions appearing in these amplitudes are those corresponding to the initial and final particle configurations. They are the coefficients of the creation and destruction operators in the fields. The behavior of the fields between the initial and final interactions, those connected to the external legs, is governed by the *Feynman propagators*. The basic interaction in this case involves the emission or absorption of a ϕ by the Dirac N, with a strength governed by the coupling constant g. All possible relative orderings of these interactions in space-time are included in these Feynman diagrams. In this case, there are two diagrams corresponding to the ordering of the emission and absorption of the ϕ.

The wave functions for particles with definite momenta are plane-waves. The insertion of the four-dimensional Fourier transform of the propagators in Minkowski space, allows one to carry out the integral over each d^4x. This gives rise to a four-dimensional Dirac delta function arising from each factor containing that coordinate x, that is, from each *vertex* in the coordinate-space Feynman diagram. The resulting delta functions imply that *four-momentum is conserved at each vertex*. One four-dimensional delta function expressing *overall* energy and momentum conservation can always be factored from the amplitude, and the S-matrix always reduces to the expression in Eq. (7.38). If the volume factors in $1/\sqrt{\Omega}$ arising from each external leg are also extracted, a T-matrix element $-iT_{fi}/\hbar c$ can then be identified as shown in that equation, and used in Vol. I. The resulting two Feynman diagrams for $-iT_{fi}/\hbar c$ in momentum space for $\phi(k_1)+N(k_2) \rightarrow \phi(k_3)+N(k_4)$, as obtained in Eq. (7.42), are shown in Fig. 7.2.

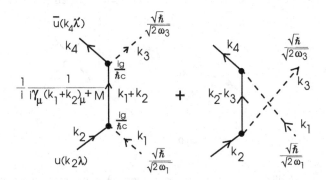

Fig. 7.2 Feynman diagrams in momentum space for $-iT_{fi}/\hbar c$ to second-order in g for the process $\phi(k_1) + N(k_2) \rightarrow \phi(k_3) + N(k_4)$. The Dirac indices are suppressed along the fermion line, and there is a suppressed Dirac unit matrix at each vertex.

These same arguments apply to the scattering process $N(k_1)+N(k_2) \rightarrow N(k_3) + N(k_4)$. The two coordinate-space Feynman diagrams remaining, after a labeling of the space-time points (x_1, x_2) is chosen, are given in Eq. (7.45) and shown in Fig. 7.3. The interaction involves the transfer of a quantum of the scalar field. Since there is only one type of Dirac particle, *either* of the final particles can be created at the vertex labeled by x_1. Hence there is an *exchange* diagram, as well as a *direct* diagram in this case, the two contributions entering with a relative minus sign.[15]

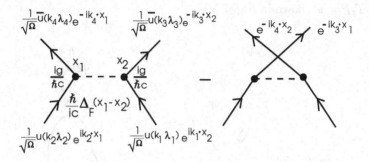

Fig. 7.3 Feynman diagrams in coordinate space for the S-matrix to second-order in g for the process $N(k_1) + N(k_2) \rightarrow N(k_3) + N(k_4)$. One must still perform $\int d^4x_1 \int d^4x_2$. The Dirac indices are suppressed along the fermion line, and there is a suppressed Dirac unit matrix at each vertex. The diagram on the right is known as the *exchange* diagram.

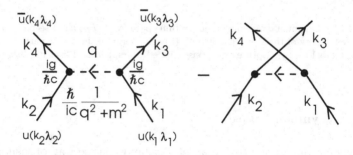

Fig. 7.4 Feynman diagrams in momentum space for $-iT_{fi}/\hbar c$ to second-order in g for the process $N(k_1) + N(k_2) \rightarrow N(k_3) + N(k_4)$. The Dirac indices are suppressed along the fermion line, and there is a suppressed Dirac unit matrix at each vertex.

The insertion of the four-dimensional Fourier transform of the propagator, and subsequent integration over both space-time coordinates, leads to the S-matrix element in Eq. (7.46), and the two momentum-space Feynman

---

[15]With two *distinct* Dirac particles, only the direct diagram would be present.

diagrams shown in Fig. 7.4.

The momentum-space S-matrix element to order g^2 for the self-energy of the N is given in Eq. (7.51), and the Feynman diagram for the corresponding $-iT_{fi}/\hbar c$ is shown in Fig. 7.5(a). The analogous expression for the ϕ is given in Eq. (7.54), and the Feynman diagram for the corresponding $-iT_{fi}/\hbar c$ is shown in Fig. 7.5(b).[16]

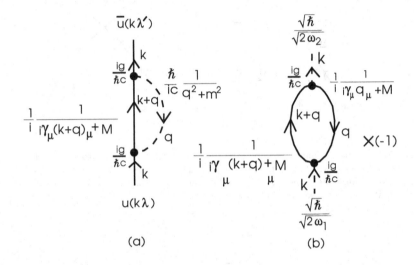

(a)   (b)

Fig. 7.5   Feynman diagrams in momentum space for $-iT_{fi}/\hbar c$ to second-order in g for: (a) self-energy of the Dirac particle N; (b) self-energy of the scalar particle ϕ (where we here relabel $t \to q$). Both expressions must still be integrated over $d^4q/(2\pi)^4$.

7.4   Feynman Rules

Feynman diagrams provide an elegant and concise way of visualizing any physical process in quantum field theory. In addition, as we have seen in the above examples, by associating factors with each element of a diagram, and summing the contributions, one arrives at *an analytic expression for each appropriate S-matrix element*. The *Feynman rules* consist of an enumeration of the appropriate factors to associate with each element of the Feynman diagrams in any quantum field theory.

Before stating the Feynman rules in the Dirac-scalar example given

---

[16]The process in (b) involves a virtual $(N\bar{N})$ pair; it is known as *vacuum polarization*.

above, let us deal with the vacuum-vacuum amplitude in Eq. (7.55). Through order g^2, the vacuum-vacuum amplitude S_{00}, and S-matrix for each of the processes with external legs discussed above, can be written as

$$S_{00} = 1 + S_{00}^{(2)}$$
$$S_{fi} = [S_{fi}]_C \times S_{00} \qquad ; \text{through } O(g^2) \qquad (7.56)$$

where the subscript C denotes the sum of all diagrams connected to the external lines. Now the *physical* S-matrix elements should always be defined relative to the amplitude that the vacuum remains the vacuum. Hence

$$[S_{fi}]_{\text{phys}} \equiv \frac{S_{fi}}{S_{00}} = [S_{fi}]_C \qquad ; \text{through } O(g^2) \qquad (7.57)$$

Thus for physics, one can concentrate on the connected diagrams. This is known as the *cancellation of the disconnected diagrams*. We shall show shortly that

- The cancellation of disconnected diagrams holds to all orders;
- The vacuum-vacuum amplitude is, in fact, simply a *phase*

$$S_{00} = e^{i\Phi} \qquad (7.58)$$

whose absolute square disappears from any physical quantity.

Let us then summarize the Feynman rules seen so far in Figs. 7.1–7.5, and Eqs. (7.37)–(7.54), for the Dirac-scalar theory. *The Feynman rules in coordinate space for the contribution to $[S_{fi}]_C$ that is nth-order in the coupling constant g are as follows:*[17]

(1) Place down the coordinates $(x_1, x_2, \cdots, x_n)$. These mark the vertices. Draw all topologically-distinct connected diagrams with one incoming N-line, one outgoing N-line, and one ϕ-line at each vertex.[18] These are the Feynman diagrams, and $[S_{fi}^{(n)}]_C$ is the sum of the contributions from these diagrams;
(2) Include a factor of $(ig/\hbar c)$ for each vertex;
(3) Include a factor of $(\hbar/ic)\Delta_F(x_j - x_i)$ for each internal ϕ-line;
(4) Include a factor of $iS_F(x_f - x_i)$ for each internal directed N-line;
(5) Include a factor of $(\hbar/2\omega\Omega)^{1/2} e^{-ik\cdot x}$ for each outgoing ϕ-line;

---

[17]Note $[S_{fi}]_C = \sum_{n=0}^{\infty} [S_{fi}^{(n)}]_C$.
[18]If there is any confusion over what "topologically distinct" means, one goes back to Wick's theorem. Note that an external fermion line must run all the way through a diagram; it cannot end.

(6) include a factor of $(\hbar/2\omega\Omega)^{1/2} e^{ik\cdot x}$ for each incoming ϕ-line;

(7) include a factor of $(1/\Omega)^{1/2} \bar{u}(k\lambda)e^{-ik\cdot x}$ for each outgoing N line;

(8) include a factor of $(1/\Omega)^{1/2} u(k\lambda)e^{ik\cdot x}$ for each incoming N-line;

(9) Take the Dirac matrix product along each fermion line;

(10) Include a factor of (-1) for each closed fermion loop;

(11) Integrate over $d^4x_1 \cdots d^4x_n$.

Insertion of the four-dimensional Fourier transforms of the Feynman propagators, and the integration over $d^4x_1 \cdots d^4x_n$, then leads to the set of Feynman rules in momentum space. *The Feynman rules in momentum space for the contribution to $[S_{fi}]_C$ that is nth-order in the coupling constant g are as follows*:

(1) Place down n vertices, and draw all topologically-distinct connected diagrams, with one incoming N-line, one outgoing N-line, and one ϕ-line at each vertex. These form the Feynman diagrams, and $[S_{fi}^{(n)}]_C$ is the sum of the contributions from these diagrams;

(2) Include a factor of $(ig/\hbar c)$ for each vertex;

(3) Assign a directed four-momentum k to each line;[19]

(4) Include a factor of $(\hbar/ic)(q^2 + m^2)^{-1}$ for each internal ϕ-line;

(5) Include a factor of $(1/i)[i\gamma_\mu k_\mu + M]^{-1}$ for each internal N-line;

(6) Conserve four-momentum with a factor of $(2\pi)^4\delta^{(4)}\left(\sum_i k_i\right)$ at each vertex;

(7) Include a factor of $(\hbar/2\omega\Omega)^{1/2}$ for each external ϕ-line;

(8) Include a factor of $(1/\Omega)^{1/2}u(k\lambda)$ for each incoming N-line;

(9) Include a factor of $(1/\Omega)^{1/2}\bar{u}(k\lambda)$ for each outgoing N-line;

(10) Take the Dirac matrix product along each fermion line;

(11) Include a factor of (-1) for each closed fermion loop;

(12) Integrate $d^4q/(2\pi)^4$ over all internal four-momenta.

The extension of these rules to cover incoming and outgoing antiparticle $\bar{N}$-lines is presented in Prob. 7.4. The application of these rules to fourth-order processes in the Dirac-scalar theory is assigned as Prob. 7.5.

7.5   Cancellation of Disconnected Diagrams

Let us establish the claims made above concerning the disconnected diagrams. First, we observe that the application of Wick's theorem in an

---

[19]Here k is a four-vector wavenumber, and the four-momentum is actually $\hbar k$.

arbitrary order in g leads to diagrams with pieces that are both connected to, and disconnected from, the external lines. As we saw in the previous examples, permutations of the coordinates $(x_1, \cdots, x_n)$ will lead to distinct *connected* diagrams, that is, distinct contributions arising from Wick's theorem.[20] Thus the claim that the $n!$ permutation of the dummy integration labels $(x_1, \cdots, x_n)$ cancels the $1/n!$ in front of the scattering operator $\hat{S}$ is correct as far as the connected diagrams are concerned.

This claim is *not* correct for the *disconnected* diagrams— witness the $1/2!$ remaining in Eq. (7.55). The reason is that for such diagrams, the permutation of the vertex labels does *not* lead to a new diagram, that is, to a new set of contractions in Wick's theorem.

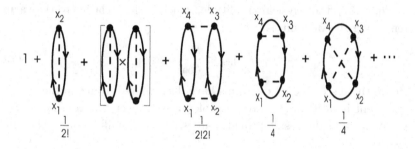

Fig. 7.6 Feynman diagrams in coordinate space for the vacuum-vacuum amplitude S_{00} up through fourth-order in the Dirac-scalar theory. The correction factor which must be applied for the overcounting when the $n!$ permutations of the coordinates $(x_1, \cdots, x_n)$ are taken as making independent, identical contributions is shown below each diagram.

This complication can be handled in the following fashion: A factor is assigned to each disconnected piece that corrects for the overcounting arising when each of the permutations of $(x_1, \cdots, x_n)$ is taken to lead to a distinct diagram that makes an identical contribution to the S-matrix. The key feature here is that this correction factor *depends only on the topology of the disconnected piece, and not on the overall order in perturbation theory in which the disconnected piece contributes*. The vacuum-vacuum S-matrix, by definition, consists of the contribution of all the disconnected pieces. The diagrammatic structure of the vacuum-vacuum amplitude S_{00} up through fourth-order in the Dirac-scalar theory is shown in Fig. 7.6, along with the

---

[20] As long as there are external lines to orient the contribution, different permutations of the vertex labels along the fermion line correspond to distinct contractions in Wick's theorem.

correction factor that must be applied to each diagram.

Now to determine the appropriate factor to be applied to a disconnected piece with a complicated topology is a formidable problem. The crucial point here is that *it does not matter what that factor is*, it is enough to know that it exists!

To understand this point, we observe that as one goes to higher and higher order, any set of connected diagrams for a given process, *will occur multiplied by the series whose first few terms are illustrated in Fig. 7.6.* Thus the S-matrix S_{fi} for any process factors into

$$S_{fi} = [S_{fi}]_C \, S_{00} \qquad (7.59)$$

where the subscript C denotes all connected diagrams. Then, exactly as in Eq. (7.57), the physical amplitude, which is that relative to the vacuum remaining the vacuum, is

$$[S_{fi}]_{\text{phys}} \equiv \frac{S_{fi}}{S_{00}} = [S_{fi}]_C \qquad (7.60)$$

Hence the disconnected diagrams have disappeared from the theory! This argument is ultimately quite a simple one; however, the consequences are profound, and the result provides an essential simplification of the analysis.[21]

Let us verify the second claim made in Eq. (7.58) concerning the vacuum-vacuum amplitude S_{00}. The scattering operator is unitary so that

$$\hat{S}^\dagger \hat{S} = \hat{1} \qquad ; \text{ unitary} \qquad (7.61)$$

Take the vacuum-vacuum matrix element of this expression and insert a complete set of states

$$\sum_n \langle 0|\hat{S}^\dagger|n\rangle\langle n|\hat{S}|0\rangle = \sum_n |\langle n|\hat{S}|0\rangle|^2 = 1 \qquad (7.62)$$

As observed many times, the scattering operator conserves four-momentum. Since in our example both (N, ϕ) are massive, the only state the scattering operator can connect to in this expression is the vacuum itself. Hence

$$|\langle 0|\hat{S}|0\rangle|^2 = 1$$
$$\Rightarrow \qquad \langle 0|\hat{S}|0\rangle = e^{i\Phi} \qquad (7.63)$$

and the complete vacuum-vacuum amplitude must indeed be a pure phase.

---

[21]The corresponding result in quantum many-body theory is another of Goldstone's theorems (see [Fetter and Walecka (2003a)]).

7.6 Mass Renormalization

We have seen that the self-energy terms in Fig. 7.5 shift the masses of the (N, ϕ), as evidenced through Eqs. (7.51) and (7.54). How do we deal with this? Consider, for example, the N. Let us go back to the starting Dirac hamiltonian density with no interactions. Relabel the mass in that expression by M_0, and the mass term $\tilde{\mathcal{H}}_0^{\text{mass}}$ in that starting hamiltonian density then takes the form

$$\tilde{\mathcal{H}}_0^{\text{mass}} = \hbar c M_0 \bar{\psi} \psi \qquad ; \text{ starting hamiltonian} \qquad (7.64)$$

Now we would like the mass here, in terms of which everything is subsequently expressed, to be the *experimental* mass M_{exp} and not the "bare mass" M_0, which is not an observable. This can be achieved by adding and subtracting a term in δM where

$$M_{\text{exp}} \equiv M = M_0 + \delta M \qquad (7.65)$$

Correspondingly, we identify a *new* mass term $\mathcal{H}_0^{\text{mass}}$ in the starting hamiltonian density through

$$\tilde{\mathcal{H}}_0^{\text{mass}} = \mathcal{H}_0^{\text{mass}} + \mathcal{H}_1^{\text{mass}}$$
$$\mathcal{H}_0^{\text{mass}} = \hbar c M \bar{\psi} \psi$$
$$\mathcal{H}_1^{\text{mass}} = -\hbar c\, \delta M \bar{\psi} \psi \qquad (7.66)$$

Now $\mathcal{H}_0^{\text{mass}}$, expressed in terms of M, can be included in the starting hamiltonian, just as we have been doing. There is, however, a price to be paid. There is now an additional *interaction density* $\mathcal{H}_1^{\text{mass}}$ that must be included, and the interaction hamiltonian in Eq. (7.2) gets extended to read

$$\hat{H}_{\text{I}}(t) = -g \int_\Omega d^3x\, \hat{\bar{\psi}}(\mathbf{x}, t) \hat{\psi}(\mathbf{x}, t) \hat{\phi}(\mathbf{x}, t) - \hbar c\, \delta M \int_\Omega d^3x\, \hat{\bar{\psi}}(\mathbf{x}, t) \hat{\psi}(\mathbf{x}, t) \quad (7.67)$$

The scattering operator $\hat{S}$ is constructed just as before. One simply has to deal with the effects arising from the additional term in the interaction. As indicated by our example, we assume that the "mass counter-term" δM has a power-series expansion in the coupling constant of the form

$$\delta M = \delta M^{(2)} + \delta M^{(4)} + \cdots$$
$$= c_2 g^2 + c_4 g^4 + \cdots \qquad (7.68)$$

In this case, there will be a contribution of order g^2 arising from the term $\hat{S}^{(1)}$ in Eq. (7.1), where the notation here indicates the order in $\hat{H}_I(t)$.[22] The contribution of the mass counter-term will then be

$$\hat{S}^{(1)}_{\text{mass}} = \left(\frac{i\hbar c \delta M}{\hbar c} \right) \int_\Omega d^4x : \hat{\bar{\psi}}(\mathbf{x}, t) \hat{\psi}(\mathbf{x}, t): \qquad (7.69)$$

where we now assume, just as in Eq. (7.32), that the scalar density is normal-ordered. The matrix element of this operator between the states in Eq. (7.47) leads to an additional contribution in Eq. (7.51), so that it now takes the form

$$S_{fi} = -(2\pi)^4 \delta^{(4)}(k_1 - k_2) \frac{i}{\Omega} \bar{u}(\mathbf{k}_2\lambda_2)[\Sigma(k_1) - \delta M]u(\mathbf{k}_1\lambda_1)$$

$$\Sigma(k_1) \equiv \frac{ig^2}{\hbar c^3} \int \frac{d^4q}{(2\pi)^4} \frac{1}{i\gamma_\mu(k_1 + q)_\mu + M} \frac{1}{q^2 + m^2} \qquad (7.70)$$

For a particle "on its mass shell", $\Sigma(k)$ is just a number since $(i\gamma_\mu k_\mu + M)u(\mathbf{k}\lambda) = 0$, and $k^2 = -M^2$. The mass counter-term will then *exactly* *cancel* the term in $\Sigma(k)$ for a real particle provided it is chosen to satisfy

$$\bar{u}(\mathbf{k}\lambda)[\Sigma(k) - \delta M^{(2)}]u(\mathbf{k}\lambda) = 0 \qquad ; \text{ mass renormalization} \qquad (7.71)$$

As a result, there is no mass shift of the N as the interaction is turned on, and the mass M in the starting hamiltonian remains the mass in the interacting theory. Thus, indeed, $M = M_{\text{exp}}$. This analysis is known as *mass renormalization*. Note that mass renormalization is not a luxury, but is an essential aspect of any relativistic quantum field theory.[23]

It is in *quantum electrodynamics* (QED) that one has a controlled perturbation expansion, where the dimensionless fine-structure constant $e^2/4\pi\varepsilon_0\hbar c = 1/137.04$ provides the small parameter. QED was developed in a very formal fashion by Schwinger (see the collected papers in [Schwinger (1958)]), and in a most intuitive fashion by Feynman [Feynman (1949)]. It was Dyson who showed the connection between the two approaches [Dyson (1949)], using an analysis whose foundation is that detailed in the present chapter. The theory of quantum electrodynamics is one of the great intellectual achievements of the 20th century. It is the most accurate physical theory we have, and every physicist should be familiar with its essentials. We now have the tools, and QED is our next topic.

---

[22] Of course, this mass term must also be included in the higher-order terms in $\hat{S}$.

[23] Recall from chapter 4 that the energy eigenvalues for the Lippmann-Schwinger equation are identical to those arising from the unperturbed hamiltonian. Thus, it is *essential* to perform mass renormalization if one wants to employ that scattering theory.

Chapter 8

Quantum Electrodynamics (QED)

For our purposes here, quantum electrodynamics (QED) is the relativistic quantum field theory of Dirac electrons interacting with the electromagnetic field. Let us start with the classical theory.

8.1 Classical Theory

The lagrangian density has already been presented in Eq. (6.92)

$$\mathcal{L} = -\frac{\varepsilon_0}{4} F_{\mu\nu} F_{\mu\nu} - \hbar c \, \bar{\psi} \left[\gamma_\mu \left(\frac{\partial}{\partial x_\mu} - \frac{ie}{\hbar} A_\mu \right) + M \right] \psi$$

$$\frac{1}{c} F_{\mu\nu} = \frac{\partial A_\nu}{\partial x_\mu} - \frac{\partial A_\mu}{\partial x_\nu} \tag{8.1}$$

It consists of the lagrangian density for the electromagnetic field, studied in appendix C, plus that for the Dirac field, where the *interaction* with the electromagnetic field has been included in a minimal, gauge-invariant fashion. This lagrangian density is invariant under the transformation

$$A_\mu \to A_\mu + \frac{\partial \Lambda(x)}{\partial x_\mu}$$

$$\psi \to \exp \left\{ \frac{ie}{\hbar} \Lambda(x) \right\} \psi \tag{8.2}$$

where $\Lambda(x)$ is some differentiable scalar function of the space-time coordinate x. While the first relation is a change in gauge, the second is simply a local phase transformation on the charged Dirac field ψ (see chapter 6).

We can rewrite $\mathcal{L}$ in the following fashion

$$\mathcal{L} = \mathcal{L}_{\text{Dirac}}^0 + \mathcal{L}_{\text{EM}}$$

$$\mathcal{L}_{\text{Dirac}}^0 = -\hbar c \, \bar{\psi} \left(\gamma_\mu \frac{\partial}{\partial x_\mu} + M \right) \psi$$

$$\mathcal{L}_{\text{EM}} = -\frac{\varepsilon_0}{4} F_{\mu\nu} F_{\mu\nu} + e c j_\mu A_\mu \qquad (8.3)$$

where $j_\mu(x)$ is the Dirac four-vector electromagnetic current of Eq. (5.115)

$$j_\mu(x) = i\bar{\psi}(x)\gamma_\mu\psi(x) = \left[\frac{1}{c}\mathbf{j}(x), \, i\rho(x) \right] \qquad (8.4)$$

The Euler-Lagrange equations for the fields following from this lagrangian density are those derived in appendix C and chapter 5

$$\frac{\partial}{\partial x_\mu} F_{\lambda\mu}(x) = \frac{e}{\varepsilon_0} j_\lambda(x) \qquad ; \lambda = 1, 2, 3, 4$$

$$\left[\gamma_\mu \left(\frac{\partial}{\partial x_\mu} - \frac{ie}{\hbar} A_\mu \right) + M \right] \psi(x) = 0$$

$$\bar{\psi}(x) \left[\gamma_\mu \left(\frac{\overleftarrow{\partial}}{\partial x_\mu} + \frac{ie}{\hbar} A_\mu \right) - M \right] = 0 \qquad (8.5)$$

These are the coupled equations of motion in four-dimensional Minkowski space for continuum electrodynamics.

8.2 Hamiltonian

Given $\mathcal{L}$, one can proceed to the hamiltonian. For $\mathcal{L}_{\text{Dirac}}^0$, this leads to the Dirac hamiltonian H_{Dirac}^0 of Eq. (5.99). The passage to the hamiltonian for the electromagnetic field in the Coulomb gauge, with a specified external current $j_\mu^{\text{ext}}(x)$, has been presented in detail in appendix C. In the present case, the current is *dynamical*, arising from the interacting Dirac field; however, the reader can easily convince himself or herself that the analysis in appendix C goes through exactly as before with the *following exception*: The electrostatic potential $\Phi(x)$, instead of being a c-number obtained from the instantaneous external charge density, is now a dynamical quantity obtained from the instantaneous Dirac charge density

$$\Phi(\mathbf{x}, t) = \frac{e}{4\pi\varepsilon_0} \int_\Omega d^3x' \frac{1}{|\mathbf{x} - \mathbf{x}'|} \rho(\mathbf{x}', t)$$

$$\rho(x) = \psi^\dagger(x)\psi(x) \tag{8.6}$$

The hamiltonian for the electromagnetic field is then just that of Eq. (C.62)

$$H = H_{\text{Dirac}}^0 + \frac{\varepsilon_0}{2} \int_\Omega d^3x \left\{ \left[\frac{\partial \mathbf{A}(\mathbf{x}, t)}{\partial t} \right]^2 + c^2 [\boldsymbol{\nabla} \times \mathbf{A}(\mathbf{x}, t)]^2 \right\} \tag{8.7}$$

$$- e \int_\Omega d^3x\, \mathbf{j}(\mathbf{x}, t) \cdot \mathbf{A}(\mathbf{x}, t) + \frac{e^2}{8\pi\varepsilon_0} \int_\Omega d^3x \int_\Omega d^3x' \frac{\rho(\mathbf{x}, t)\rho(\mathbf{x}', t)}{|\mathbf{x} - \mathbf{x}'|}$$

where the Dirac current is that of Eq. (8.4). We have chosen to work in a particular gauge, the Coulomb gauge with $\boldsymbol{\nabla} \cdot \mathbf{A} = 0$, since when quantized, there is then a one-to-one correspondence between the quanta of the vector potential and real transverse photons.

8.3 Quantization

The canonical momenta derived from the lagrangian density in Eq. (8.1) are given in Eqs. (5.100) and (C.64)

$$\pi_\psi = \frac{\partial \mathcal{L}}{\partial(\partial\psi/\partial t)} = i\hbar\bar{\psi}\gamma_4 = i\hbar\psi^\dagger$$

$$(\pi_A)_i = \frac{\partial \mathcal{L}}{\partial(\partial A_i/\partial t)} = \varepsilon_0 \left(\frac{\partial A_i}{\partial t} + \frac{\partial \Phi}{\partial x_i} \right) \qquad ; i = 1, 2, 3 \tag{8.8}$$

To quantize the theory, the canonical equal-time (anti)commutation relations must be imposed. For the Dirac field, the analysis proceeds exactly as in chapter 5. For the electromagnetic field, we note that even though the electrostatic potential is now dynamical, it depends only on the Dirac field, and the Dirac field and electromagnetic field *commute* at a given time. Hence, the canonical equal-time (anti)commutation relations in QED are *just those of the free fields* in Eqs. (5.104) and (C.66)

$$\left\{ \hat{\psi}_\alpha(\mathbf{x}, t), \hat{\psi}_\beta^\dagger(\mathbf{x}', t') \right\}_{t=t'} = \delta_{\alpha\beta}\, \delta^{(3)}(\mathbf{x} - \mathbf{x}') \qquad ; (\alpha, \beta) = (1, 2, 3, 4)$$

$$\left[\hat{A}_i(\mathbf{x}, t), \hat{A}_j(\mathbf{x}', t') \right]_{t=t'} = \frac{i\hbar}{\varepsilon_0} \delta_{ij}^{\text{T}}(\mathbf{x} - \mathbf{x}') \qquad ; (i, j) = (1, 2, 3) \tag{8.9}$$

where we use the notation $\dot{\hat{\mathbf{A}}}$ for the operator $\partial\hat{\mathbf{A}}/\partial t$, the time derivative of the field at a fixed position. These are satisfied by the interaction-picture field expansions in Eqs. (5.102) and (C.67)

$$\hat{\psi}(\mathbf{x},t) = \frac{1}{\sqrt{\Omega}}\sum_{\mathbf{k}}\sum_{\lambda}\left[a_{\mathbf{k}\lambda}u(\mathbf{k}\lambda)e^{i(\mathbf{k}\cdot\mathbf{x}-\omega_k t)} + b^{\dagger}_{\mathbf{k}\lambda}v(-\mathbf{k}\lambda)e^{-i(\mathbf{k}\cdot\mathbf{x}-\omega_k t)}\right]$$

$$\hat{\mathbf{A}}(\mathbf{x},t) = \sum_{\mathbf{k}}\sum_{s=1}^{2}\left(\frac{\hbar}{2\omega_k\varepsilon_0\Omega}\right)^{1/2}\left[c_{\mathbf{k}s}\mathbf{e}_{\mathbf{k}s}e^{i(\mathbf{k}\cdot\mathbf{x}-\omega_k t)} + c^{\dagger}_{\mathbf{k}s}\mathbf{e}_{\mathbf{k}s}e^{-i(\mathbf{k}\cdot\mathbf{x}-\omega_k t)}\right]$$

$$\frac{\partial\hat{\mathbf{A}}(\mathbf{x},t)}{\partial t} = \frac{1}{i}\sum_{\mathbf{k}}\sum_{s=1}^{2}\left(\frac{\hbar\omega_k}{2\varepsilon_0\Omega}\right)^{1/2}\left[c_{\mathbf{k}s}\mathbf{e}_{\mathbf{k}s}e^{i(\mathbf{k}\cdot\mathbf{x}-\omega_k t)} - c^{\dagger}_{\mathbf{k}s}\mathbf{e}_{\mathbf{k}s}e^{-i(\mathbf{k}\cdot\mathbf{x}-\omega_k t)}\right]$$

$$(8.10)$$

Here and henceforth in this chapter, to avoid confusion, we use $(c^{\dagger}_{\mathbf{k}s}, c_{\mathbf{k}s})$ for the photon creation and destruction operators.

Furthermore, in line with the discussion of Eq. (5.116), in defining the quantum field theory we assume the electromagnetic current to be *normal-ordered*.[1] The hamiltonian of QED is then given by Eq. (8.7) as

$$\hat{H}_{\text{QED}} = \int_{\Omega} d^3x\,\hat{\psi}^{\dagger}(x)(c\boldsymbol{\alpha}\cdot\mathbf{p} + \beta m_0 c^2)\hat{\psi}(x) \tag{8.11}$$

$$+\frac{\varepsilon_0}{2}\int_{\Omega}d^3x\left\{\left[\frac{\partial\hat{\mathbf{A}}(x)}{\partial t}\right]^2 + c^2[\boldsymbol{\nabla}\times\hat{\mathbf{A}}(x)]^2\right\}$$

$$-e\int_{\Omega}d^3x\,\hat{\mathbf{j}}(x)\cdot\hat{\mathbf{A}}(x) + \frac{e^2}{8\pi\varepsilon_0}\int_{\Omega}d^3x\int_{\Omega}d^3x'\,\frac{\hat{\rho}(\mathbf{x},t)\hat{\rho}(\mathbf{x}',t)}{|\mathbf{x}-\mathbf{x}'|}$$

where the Dirac current is

$$\hat{j}_{\mu}(x) = \left[\frac{1}{c}\hat{\mathbf{j}}(x),\,i\hat{\rho}(x)\right] = i:\hat{\bar{\psi}}(x)\gamma_{\mu}\hat{\psi}(x): \tag{8.12}$$

and the interaction-picture fields, with $(x) = (\mathbf{x},t)$, are those in Eqs. (8.10).

---

[1]The relaxation of this assumption is discussed in appendix G.

Substitution of the field expansions in Eqs. (8.10) into $\hat{H}_{\text{QED}}$ in Eq. (8.11) yields

$$
\hat{H}_{\text{QED}} = \sum_{\mathbf{k}} \sum_{\lambda} \hbar\omega_k \left(a_{\mathbf{k}\lambda}^{\dagger} a_{\mathbf{k}\lambda} + b_{\mathbf{k}\lambda}^{\dagger} b_{\mathbf{k}\lambda} - 1 \right) + \sum_{\mathbf{k}} \sum_{s=1}^{2} \hbar\omega_k \left(c_{\mathbf{k}s}^{\dagger} c_{\mathbf{k}s} + 1/2 \right)
$$

$$
- e \int_{\Omega} d^3x\, \hat{\mathbf{j}}(x) \cdot \hat{\mathbf{A}}(x) + \frac{e^2}{8\pi\varepsilon_0} \int_{\Omega} d^3x \int_{\Omega} d^3x' \, \frac{\hat{\rho}(\mathbf{x},t)\hat{\rho}(\mathbf{x}',t)}{|\mathbf{x}-\mathbf{x}'|}
$$

$$
\equiv \hat{H}_0 + \hat{H}_I(t) \tag{8.13}
$$

The relativistic field theory of quantum electrodynamics (QED) is now well-formulated.

8.4 Photon Propagator

Let us separate the interaction term in Eq. (8.13) into two parts

$$
\hat{H}_I(t) = \hat{H}_I^{\gamma}(t) + \hat{H}_I^{C}(t)
$$

$$
\hat{H}_I^{\gamma}(t) = -e \int_{\Omega} d^3x\, \hat{\mathbf{j}}(x) \cdot \hat{\mathbf{A}}(x)
$$

$$
\hat{H}_I^{C}(t) = \frac{e^2}{8\pi\varepsilon_0} \int_{\Omega} d^3x \int_{\Omega} d^3x' \, \frac{\hat{\rho}(\mathbf{x},t)\hat{\rho}(\mathbf{x}',t)}{|\mathbf{x}-\mathbf{x}'|} \tag{8.14}
$$

A slight difficulty is that these two terms are of different order in e. To calculate the scattering operator $\hat{S}^{(2)}$ that is second order in e, the term H_I^{γ}, which is of order e, must be treated to second order, while the term H_I^{C}, which is of order e^2, contributes in *first* order. Hence $\hat{S}^{(2)}$ is given by

$$
\hat{S}^{(2)} = \left(\frac{-i}{\hbar c} \right)^2 \frac{1}{2!} \int d^4x_1 \int d^4x_2 \, P[\hat{\mathcal{H}}_I^{\gamma}(x_1)\hat{\mathcal{H}}_I^{\gamma}(x_2)] +
$$

$$
\left(\frac{-i}{\hbar c} \right) \int d^4x_1 \, \hat{\mathcal{H}}_I^{C}(x_1) \tag{8.15}
$$

Substitution of the above expressions gives

$$
\hat{S}^{(2)} = \left(\frac{ie}{\hbar c} \right)^2 \frac{c^2}{2!} \int d^4x_1 \int d^4x_2 \, P[\hat{j}_i(x_1)\hat{j}_j(x_2)] P[\hat{A}_i(x_1)\hat{A}_j(x_2)] +
$$

$$
\left(\frac{ie}{\hbar c} \right)^2 \left(\frac{\hbar c}{i\varepsilon_0} \right) \frac{1}{2!} \int d^4x_1 \int d^3x_2 \, \hat{j}_4(\mathbf{x}_1 t_1)\hat{j}_4(\mathbf{x}_2 t_1) \frac{1}{4\pi|\mathbf{x}_1-\mathbf{x}_2|} \tag{8.16}
$$

where $\hat{j}_i$ is now a cartesian spatial component of the four-vector current $\hat{j}_\mu$ in Eq. (8.12), and $\hat{j}_4 = i\hat{\rho}$ is its fourth component. The first term in Eq. (8.16) is analyzed directly using Wick's theorem. Let us do a little messaging of the Coulomb interaction. One can add an integration over t_2, provided one also inserts the appropriate delta-function in the time. Time-ordering the charge densities then changes nothing, since they only contribute at equal times.[2] Thus

$$\int d^4x_1 \int d^3x_2 \, \hat{j}_4(\mathbf{x}_1 t_1)\hat{j}_4(\mathbf{x}_2 t_1)\frac{1}{4\pi|\mathbf{x}_1 - \mathbf{x}_2|} =$$
$$\int d^4x_1 \int d^4x_2 \, P[\hat{j}_4(x_1)\hat{j}_4(x_2)]\frac{\delta(ct_1 - ct_2)}{4\pi|\mathbf{x}_1 - \mathbf{x}_2|} \quad (8.17)$$

Now use the following four-dimensional Fourier transform in Minkowski space[3]

$$\frac{\delta(ct_1 - ct_2)}{4\pi|\mathbf{x}_1 - \mathbf{x}_2|} = \int \frac{d^4k}{(2\pi)^4}\frac{1}{k^2}e^{ik\cdot(x_1-x_2)}$$
$$\equiv D^C(x_1 - x_2) \quad (8.18)$$

Equation (8.16) then takes the form

$$\hat{S}^{(2)} = \left(\frac{ie}{\hbar c}\right)^2 \frac{1}{2!} \int d^4x_1 \int d^4x_2 \left\{ P[\hat{j}_4(x_1)\hat{j}_4(x_2)]\frac{\hbar c}{i\varepsilon_0}D^C(x_1 - x_2) + \right.$$
$$\left. c^2 P[\hat{j}_i(x_1)\hat{j}_j(x_2)]P[\hat{A}_i(x_1)\hat{A}_j(x_2)] \right\} \quad (8.19)$$

Use of Wick's theorem in the last line, and identification of the transverse propagator from Eq. (F.45), gives

$$\hat{S}^{(2)} = \left(\frac{ie}{\hbar c}\right)^2 \frac{1}{2!} \int d^4x_1 \int d^4x_2 \left\{ P[\hat{j}_4(x_1)\hat{j}_4(x_2)]\frac{\hbar c}{i\varepsilon_0}D^C(x_1 - x_2) + \right.$$
$$\left. P[\hat{j}_i(x_1)\hat{j}_j(x_2)]\left[c^2 {:}\hat{A}_i(x_1)\hat{A}_j(x_2){:} + \frac{\hbar c}{i\varepsilon_0}D_F^T(x_1 - x_2)_{ij}\right] \right\} \quad (8.20)$$

Here

$$D_F^T(x_1 - x_2)_{ij} = \int \frac{d^4k}{(2\pi)^4}\frac{1}{k^2}\left(\delta_{ij} - \frac{k_i k_j}{\mathbf{k}^2}\right)e^{ik\cdot(x_1-x_2)} \quad (8.21)$$

The transverse and Coulomb propagators in Eq. (8.20) can be combined in the following fashion. Define a unit four-vector in the fourth direction

---

[2] And they commute at equal times [see Eq. (8.38)].
[3] See Prob. 8.2; we shall refer to $D^C(x_1 - x_2)$ as the "Coulomb propagator".

in Minkowski space by

$$\eta_\mu \equiv (0,0,0,i) \qquad ; \text{ unit four-vector} \qquad (8.22)$$

It follows that

$$k \cdot \eta = -k_0$$
$$\mathbf{k}^2 = k^2 + (k \cdot \eta)^2$$
$$k_\mu + (k \cdot \eta)\eta_\mu = (\mathbf{k}, 0) \qquad (8.23)$$

Equation (8.20) can then be rewritten as

$$\hat{S}^{(2)} = \left(\frac{ie}{\hbar c}\right)^2 \frac{1}{2!} \int d^4x_1 \int d^4x_2 \left\{ P[\hat{j}_\mu(x_1)\hat{j}_\nu(x_2)] \right.$$
$$\left. \left[c^2 : \hat{A}_\mu(x_1)\hat{A}_\nu(x_2) : + \frac{\hbar c}{i\varepsilon_0} D_{\mu\nu}^F(x_1 - x_2) \right] \right\} \qquad (8.24)$$

Here we first introduce the *convention* that the $\hat{A}_\mu(x)$ remaining in this equation has no fourth component in the Coulomb gauge

$$\hat{A}_\mu(x) \equiv [\hat{\mathbf{A}}(x), 0] \qquad ; \text{ convention} \qquad (8.25)$$

We then write

$$D_{\mu\nu}^F(x_1 - x_2) = \int \frac{d^4k}{(2\pi)^4} \tilde{D}_{\mu\nu}^F(k) e^{ik \cdot (x_1 - x_2)} \qquad (8.26)$$

and identify the Fourier transform from Eqs. (8.18), (8.21), and (8.23) as[4]

$$k^2 \tilde{D}_{\mu\nu}^F(k) = -\eta_\mu\eta_\nu \left[1 - \frac{(k \cdot \eta)^2}{k^2 + (k \cdot \eta)^2} \right] +$$
$$\delta_{\mu\nu} + \eta_\mu\eta_\nu - \frac{1}{k^2 + (k \cdot \eta)^2} [k_\mu + (k \cdot \eta)\eta_\mu][k_\nu + (k \cdot \eta)\eta_\nu] \qquad (8.27)$$

The first term on the r.h.s. comes from the Coulomb propagator, and the second line from that of the transverse photons. This expression (quite miraculously!) simplifies to

$$\tilde{D}_{\mu\nu}^F(k) = \frac{1}{k^2}\delta_{\mu\nu} - \frac{1}{k^2}\frac{1}{k^2 + (k \cdot \eta)^2} [k_\mu k_\nu + (k \cdot \eta)(k_\mu\eta_\nu + k_\nu\eta_\mu)] \qquad ;$$
$$\text{photon propagator} \qquad (8.28)$$

This is the *photon propagator* in momentum space in the Coulomb gauge.

---

[4]Note $1/\mathbf{k}^2 = (1/k^2)(1 - k_0^2/k^2)$.

The second term on the r.h.s. of Eq. (8.28) is both frame-dependent, because of the presence of η, and gauge-dependent, because of the terms in k_μ and k_ν. We shall now prove the remarkable result that

Because of current conservation, any term proportional to k_μ or k_ν in the photon propagator does not contribute to the scattering operator.

The implication is that in the scattering operator, the *effective photon propagator* is

$$\tilde{D}^F_{\mu\nu}(k) \doteq \frac{1}{k^2}\delta_{\mu\nu} \qquad ; \text{ effective photon propagator} \qquad (8.29)$$

Some comments:

- This result is *remarkably simple*;
- This photon propagator is *covariant*;
- This photon propagator is *gauge-invariant*;
- The simplicity is somewhat misleading, since the theory *does contain all the electric and magnetic interactions of electrodynamics.*

We proceed to a proof of the above result. Write a term in $D^F_{\mu\nu}(x)$ proportional to k_μ, for example, as

$$\int \frac{d^4k}{(2\pi)^4} k_\mu F_\nu(k,\eta) e^{ik\cdot x} = \frac{1}{i}\frac{\partial}{\partial x_\mu} \int \frac{d^4k}{(2\pi)^4} F_\nu(k,\eta) e^{ik\cdot x} \qquad (8.30)$$

In the scattering operator in Eq. (8.24), this term contributes as

$$\delta\hat{S}^{(2)} = \left(\frac{ie}{\hbar c}\right)^2 \left(\frac{\hbar c}{i\varepsilon_0}\right)\frac{1}{2!}\int d^4x_1 \int d^4x_2\, P[\hat{j}_\mu(x_1)\hat{j}_\nu(x_2)]\times$$
$$\frac{1}{i}\frac{\partial}{\partial x_{1\mu}} \int \frac{d^4k}{(2\pi)^4} F_\nu(k,\eta) e^{ik\cdot(x_1-x_2)} \qquad (8.31)$$

Now carry out a partial integration on the four-dimensional gradient, getting it over onto the current. This partial integration is justified as follows:

- Before taking the continuum limit, the spatial integration on $\mathbf{x}_1$ is really over $\int_\Omega d^3x_1$. As we have seen many times, in a big box with p.b.c., this partial integration on ∇_1 is allowed;
- There is a suppressed adiabatic damping factor $e^{-\epsilon|t_1|}$ in the integrand. This eliminates any contribution from the time endpoints, and the derivative of this factor with respect to time gives a term proportional to ϵ, which vanishes in the adiabatic limit.

Thus, provided one can take the time derivative through the time-ordering symbol, the above expression becomes

$$\delta \hat{S}^{(2)} = \left(\frac{ie}{\hbar c}\right)^2 \left(\frac{\hbar c}{i\varepsilon_0}\right) \frac{1}{2!} \int d^4x_1 \int d^4x_2\, P\left[\frac{\partial \hat{j}_\mu(x_1)}{\partial x_{1\mu}}\, \hat{j}_\nu(x_2)\right] \times$$

$$i \int \frac{d^4k}{(2\pi)^4} F_\nu(k,\eta) e^{ik\cdot(x_1-x_2)} \tag{8.32}$$

Now the interaction-picture Dirac current is conserved

$$\frac{\partial \hat{j}_\mu(x)}{\partial x_\mu} = 0 \qquad ; \text{ current conservation}$$

$$\Rightarrow \qquad \delta \hat{S}^{(2)} = 0 \tag{8.33}$$

This establishes the result.

It remains to show that the time derivative can be taken through the time-ordering symbol. Write the analytic expression[5]

$$P[\hat{\rho}(x)\,\hat{j}_\lambda(y)] = \hat{\rho}(x)\,\hat{j}_\lambda(y)\theta(t_x-t_y) + \hat{j}_\lambda(y)\,\hat{\rho}(x)\theta(t_y-t_x) \tag{8.34}$$

where

$$\theta(t) = 1 \qquad ; t > 0$$

$$= 0 \qquad ; t < 0 \tag{8.35}$$

The theta function can be expressed as

$$\theta(t) = \int_{-\infty}^{t} dt'\, \delta(t')$$

$$\text{hence;} \quad \frac{d}{dt}\theta(t) = \delta(t) \tag{8.36}$$

Differentiation of Eq. (8.34) with respect to t_x then gives

$$\frac{\partial}{\partial t_x} P[\hat{\rho}(x)\,\hat{j}_\lambda(y)] = P\left[\frac{\partial \hat{\rho}(x)}{\partial t_x}\,\hat{j}_\lambda(y)\right] + \delta(t_x-t_y)[\hat{\rho}(x),\hat{j}_\lambda(y)]_{t_x=t_y} \tag{8.37}$$

The arguments in Eqs. (6.19) and (6.38) imply that the additional commutator on the r.h.s. *vanishes*[6]

$$[\hat{\rho}(x),\hat{j}_\lambda(y)]_{t_x=t_y} = i[\hat{\psi}^\dagger(\mathbf{x},t)\hat{\psi}(\mathbf{x},t),\,\hat{\psi}^\dagger(\mathbf{y},t)\gamma_4\gamma_\lambda\hat{\psi}(\mathbf{y},t)]_{t=t_x} = 0 \tag{8.38}$$

---

[5]Recall that both $\hat{\rho}(x)$ and $\hat{j}_\lambda(y)$ are bilinear in the fermion fields.
[6]See Prob. 8.1. (Note also Prob. 8.15.)

Thus the time derivative can indeed be taken through the time-ordering symbol in this case, and the proof is complete.

8.5   Second-Order Processes

The previous arguments have shown that the second-order scattering operator in QED takes the form

$$\hat{S}^{(2)} = \left(\frac{ie}{\hbar c}\right)^2 \frac{1}{2!} \int d^4x_1 \int d^4x_2 \, P[\hat{j}_\mu(x_1)\,\hat{j}_\nu(x_2)] \times$$
$$\left[c^2 : \hat{A}_\mu(x_1)\hat{A}_\nu(x_2): + \left(\frac{\hbar c}{i\varepsilon_0}\right) D^F_{\mu\nu}(x_1 - x_2)\right] \qquad (8.39)$$

Here

$$\hat{A}_4(x) \equiv 0 \qquad\qquad\qquad\qquad ; \text{ convention} \qquad (8.40)$$
$$D^F_{\mu\nu}(x_1 - x_2) \doteq \delta_{\mu\nu} \int \frac{d^4k}{(2\pi)^4}\frac{1}{k^2}e^{ik\cdot(x_1-x_2)} \qquad ; \text{ photon propagator}$$

where $\doteq$ denotes the *effective* photon propagator. The Dirac current is

$$\hat{j}_\mu(x) = i : \hat{\bar{\psi}}(x)\gamma_\mu\hat{\psi}(x): \qquad\qquad ; \text{ Dirac current} \qquad (8.41)$$

The fields $\hat{\psi}(x)$ and $\hat{\mathbf{A}}(x)$ are in the interaction picture, and they are given in Eqs. (8.10). The insertion of the Dirac current into Eq. (8.39) gives

$$\hat{S}^{(2)} = \left(\frac{-e}{\hbar c}\right)^2 \frac{1}{2!} \int d^4x_1 \int d^4x_2 \, P[:\hat{\bar{\psi}}(x_1)\gamma_\mu\hat{\psi}(x_1)::\hat{\bar{\psi}}(x_2)\gamma_\nu\hat{\psi}(x_2):] \times$$
$$\left[c^2 : \hat{A}_\mu(x_1)\hat{A}_\nu(x_2): + \left(\frac{\hbar c}{i\varepsilon_0}\right) D^F_{\mu\nu}(x_1 - x_2)\right] \qquad (8.42)$$

The problem of quantum electrodynamics has now been reduced to that studied in the previous chapter, and we proceed to some applications of these results.

8.5.1   *Scattering Amplitudes*

Consider the *Compton scattering* process $\gamma(k_1s) + e^-(k_2\lambda) \rightarrow \gamma(k_3s') + e^-(k_4\lambda')$. The initial and final states are

$$|i\rangle = c^\dagger_{\mathbf{k}_1 s} a^\dagger_{\mathbf{k}_2 \lambda}|0\rangle \qquad\qquad ; |f\rangle = c^\dagger_{\mathbf{k}_3 s'} a^\dagger_{\mathbf{k}_4 \lambda'}|0\rangle \qquad (8.43)$$

We can make immediate use of the previous analysis of Dirac-scalar scattering, and we need only take note of the modifications required in the present case. Here we proceed directly to the matrix element S_{fi} in momentum space, and Eq. (7.42) becomes

$$S_{fi}^{(2)} = (2\pi)^4 \delta^{(4)}(k_1 + k_2 - k_3 - k_4)\frac{1}{\Omega^2}\frac{\hbar c^2}{\sqrt{4\omega_1\omega_3}}\left(\frac{-e}{\hbar c\sqrt{\varepsilon_0}}\right)^2\left(\frac{1}{i}\right) \times$$

$$\bar{u}(\mathbf{k}_4\lambda')\varepsilon_\nu^f\left[\gamma_\nu\frac{1}{i\gamma_\lambda(k_2 + k_1)_\lambda + M}\gamma_\mu + \gamma_\mu\frac{1}{i\gamma_\lambda(k_2 - k_3)_\lambda + M}\gamma_\nu\right]\varepsilon_\mu^i u(\mathbf{k}_2\lambda)$$

$$(8.44)$$

where we have defined the polarization four-vectors for the initial and final photons as

$$\varepsilon_\mu^i \equiv (\mathbf{e}_{\mathbf{k}_1 s},\, 0) \qquad ; \quad \varepsilon_\nu^f \equiv (\mathbf{e}_{\mathbf{k}_3 s'},\, 0) \tag{8.45}$$

The Dirac indices are again suppressed in Eq. (8.44), and there is an implied Dirac matrix product along the electron line. This Compton amplitude can be summarized in terms of Feynman diagrams just as in Fig. 7.2, and the direct analog of that figure is shown as Fig. 8.1. Note there are now

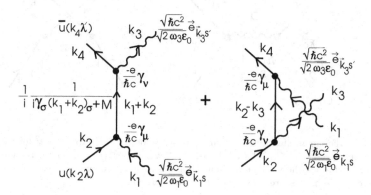

Fig. 8.1 Feynman diagrams in momentum space for $-iT_{fi}/\hbar c$ to order e^2 for the Compton scattering process $\gamma(k_1 s) + e^-(k_2\lambda) \rightarrow \gamma(k_3 s') + e^-(k_4\lambda')$. The Dirac indices are suppressed along the electron line, and there are now Dirac matrices $\gamma\cdot\varepsilon^f$ and $\gamma\cdot\varepsilon^i$ at the vertices, where the four-vectors $(\varepsilon^f, \varepsilon^i)$ are defined in Eqs. (8.45). Only the relevant changes are shown in the second diagram.

Dirac matrices (γ_ν, γ_μ) from the current at the vertices, which get dotted into the photon polarization vectors. The coupling-constant factor in front is now $(-e/\hbar c)^2$. The factor multiplying the polarization vector for an external photon, arising from $c^2 \colon \hat{A}_\mu(x_1)\hat{A}_\nu(x_2)$: in the scattering operator, is $(\hbar c^2/2\omega\varepsilon_0)^{1/2}$.

Similarly, consider the *Møller scattering* process $e^-(k_1\lambda_1) + e^-(k_2\lambda_2) \to e^-(k_3\lambda_3) + e^-(k_4\lambda_4)$. The amplitude S_{fi} for Dirac-Dirac scattering in Eq. (7.46) now becomes

$$S_{fi}^{(2)} = (2\pi)^4\delta^{(4)}(k_1 + k_2 - k_3 - k_4)\frac{1}{\Omega^2}\left(\frac{-e}{\hbar c\sqrt{\varepsilon_0}}\right)^2\left(\frac{\hbar c}{i}\right) \times$$

$$\left[\bar{u}(\mathbf{k}_3\lambda_3)\gamma_\mu u(\mathbf{k}_1\lambda_1)\frac{1}{(k_1 - k_3)^2}\bar{u}(\mathbf{k}_4\lambda_4)\gamma_\mu u(\mathbf{k}_2\lambda_2) - \right.$$

$$\left. \bar{u}(\mathbf{k}_4\lambda_4)\gamma_\mu u(\mathbf{k}_1\lambda_1)\frac{1}{(k_1 - k_4)^2}\bar{u}(\mathbf{k}_3\lambda_3)\gamma_\mu u(\mathbf{k}_2\lambda_2)\right] \qquad (8.46)$$

The Feynman diagrams in momentum space in Fig. 7.4 now become those shown in Fig. 8.2. There is now a Dirac matrix γ_μ from the current at each vertex, and the photon propagator contribution is $(\hbar c/i\varepsilon_0)(1/q^2)$.[7]

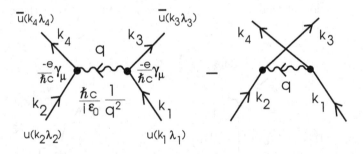

Fig. 8.2   Feynman diagrams in momentum space for $-iT_{fi}/\hbar c$ to order e^2 for the Møller scattering process $e^-(k_1\lambda_1) + e^-(k_2\lambda_2) \to e^-(k_3\lambda_3) + e^-(k_4\lambda_4)$. The Dirac indices are suppressed along the electron lines, and there is a Dirac matrix γ_μ at each vertex. Only the relevant changes are shown in the second diagram.

8.5.2   *Self-Energies*

The self-energy of the electron is immediately obtained in analogy to the self-energy of the N in the Dirac-scalar theory. The required modification

---

[7]The $\delta_{\mu\nu}$ in the photon propagator has already been incorporated.

of Eq. (7.51) is as follows

$$S_{fi} = -(2\pi)^4 \delta^{(4)}(k_1 - k_2) \frac{i}{\Omega} \bar{u}(\mathbf{k}_2 \lambda_2) \Sigma(k_1) u(\mathbf{k}_1 \lambda_1)$$

$$\Sigma(k_1) \equiv \frac{e^2}{i\hbar c\varepsilon_0} \int \frac{d^4 q}{(2\pi)^4} \gamma_\mu \frac{1}{i\gamma_\lambda(k_1 + q)_\lambda + M} \gamma_\mu \frac{1}{q^2} \tag{8.47}$$

The associated Feynman diagram in momentum space is immediately obtained from Fig. 7.5(a); it is shown in Fig. 8.3(a).

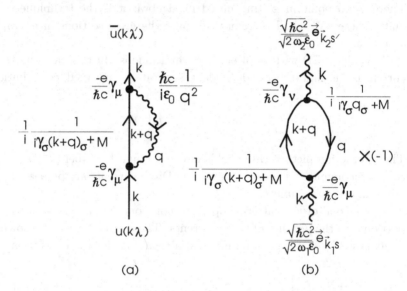

(a)   (b)

Fig. 8.3   Feynman diagrams in momentum space for $-iT_{fi}/\hbar c$ to order e^2 for: (a) self-energy of the electron ; (b) self-energy of the photon. Here $k_1 \equiv k$. Both expressions must still be integrated over $d^4 q/(2\pi)^4$.

The virtual $(e^- e^+)$ contribution to the self-energy of the photon is analogous to that of the $(N\bar{N})$ contribution to the self-energy of the ϕ in the Dirac-scalar theory. The required modification of Eq. (7.54) is

$$S_{fi} = (2\pi)^4 \delta^{(4)}(k_1 - k_2) \frac{ic}{\Omega} \frac{1}{\sqrt{4\omega_1\omega_2}} \varepsilon_\nu^f \Pi_{\nu\mu}(k_1) \varepsilon_\mu^i \tag{8.48}$$

$$\Pi_{\nu\mu}(k_1) \equiv \frac{e^2}{i\hbar c\varepsilon_0} \int \frac{d^4 q}{(2\pi)^4} \mathrm{Tr}\left[\gamma_\nu \frac{1}{i\gamma_\rho(q + k_1)_\rho + M} \gamma_\mu \frac{1}{i\gamma_\sigma q_\sigma + M} \right]$$

The *vacuum-polarization* response $\Pi_{\nu\mu}(k_1)$ is now a second-rank tensor because of the Dirac matrices (γ_ν, γ_μ) at the vertices. The associated

Feynman diagram in momentum space is again immediately obtained from Fig. 7.5(b), and it is shown in Fig. 8.3(b).

8.6 QED With Two Leptons

Once the S-matrix elements are determined, one can proceed to the cross section. The goal is to present the details of just how that is accomplished. There is no point in getting mired in algebra; it is the techniques that matter here. We shall therefore calculate the cross sections in a simpler situation.

Consider QED with both electrons and muons. In this case the Dirac current is the sum of two independent, separately conserved, contributions

$$j_\mu(x) = j_\mu(x)_{e^-} + j_\mu(x)_{\mu^-} \tag{8.49}$$

The interaction-picture Dirac fields now carry another label (e^-, μ^-), and they anticommute with each other. The Dirac currents are bilinear in the fields, and they commute.

The second-order scattering operator now contains two cross terms proportional to the product of these currents. These cross terms are shown to be identical by a change of dummy variables. Thus for the cross terms

$$\hat{S}^{(2)} = \left(\frac{-e}{\hbar c}\right)^2 \int d^4x_1 \int d^4x_2 \left[: \hat{\bar{\psi}}(x_1)\gamma_\mu \hat{\psi}(x_1) :\right]_{e^-} \left[: \hat{\bar{\psi}}(x_2)\gamma_\nu \hat{\psi}(x_2) :\right]_{\mu^-} \times$$
$$\left(\frac{\hbar c}{i\varepsilon_0}\right) D_{\mu\nu}^F(x_1 - x_2) \tag{8.50}$$

Only the non-zero contribution has been retained, and we have dropped the P-ordering symbol since the currents commute. This scattering operator governs various (e, μ) processes.

8.7 Cross Sections

8.7.1 $e^- + \mu^- \to e^- + \mu^-$

Consider the scattering process $e^-(k_1) + \mu^-(k_2) \to e^-(k_3) + \mu^-(k_4)$.

8.7.1.1 Scattering Amplitude

The matrix element $S_{fi}^{(2)}$ is calculated exactly as in Eq. (8.46), only now, since the (e, μ) are distinct particles, there is no exchange contribution

$$S_{fi}^{(2)} = (2\pi)^4 \delta^{(4)}(k_1 + k_2 - k_3 - k_4) \frac{1}{\Omega^2} \left(\frac{-e}{\hbar c \sqrt{\varepsilon_0}} \right)^2 \left(\frac{\hbar c}{i} \right) \times$$

$$[\bar{u}(\mathbf{k}_3 \lambda_3) \gamma_\mu u(\mathbf{k}_1 \lambda_1)]_{e^-} \frac{1}{q^2} [\bar{u}(\mathbf{k}_4 \lambda_4) \gamma_\mu u(\mathbf{k}_2 \lambda_2)]_{\mu^-} \quad ; \quad q = k_1 - k_3 \quad (8.51)$$

There is now just the one Feynman diagram shown in Fig. 8.4; this is what simplifies the calculation. The T-matrix is identified as

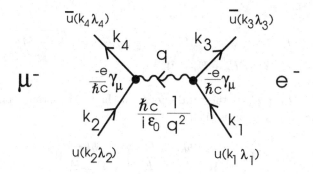

Fig. 8.4 Feynman diagram in momentum space for $-iT_{fi}/\hbar c$ to order e^2 for the scattering process $e^-(k_1 \lambda_1) + \mu^-(k_2 \lambda_2) \to e^-(k_3 \lambda_3) + \mu^-(k_4 \lambda_4)$. The Dirac indices are suppressed along the electron lines, and there is a Dirac matrix γ_μ at each vertex.

$$\frac{-iT_{fi}^{(2)}}{\hbar c} = \left(\frac{-e}{\hbar c \sqrt{\varepsilon_0}} \right)^2 \left(\frac{\hbar c}{i} \right) [\bar{u}(\mathbf{k}_3 \lambda_3) \gamma_\mu u(\mathbf{k}_1 \lambda_1)]_{e^-} \frac{1}{q^2} [\bar{u}(\mathbf{k}_4 \lambda_4) \gamma_\mu u(\mathbf{k}_2 \lambda_2)]_{\mu^-}$$

$$(8.52)$$

8.7.1.2 Cross Section

The passage from $-iT_{fi}/\hbar c$ to the cross section was carefully detailed in Vol. I, leading to the results in EqsI. (7.46) and (7.49)

$$d\sigma_{fi} = \frac{2\pi}{\hbar} \delta(W_f - W_i) \frac{1}{\Omega^2} |T_{fi}|^2 \frac{\Omega d^3 k_3}{(2\pi)^3} \frac{1}{I_{\text{inc}}} \quad (8.53)$$

where I_{inc} is the incident flux, determined from the initial relative velocities by (see Probl. 7.3)

$$I_{inc} = \frac{1}{\Omega} |\mathbf{v}(1) - \mathbf{v}(2)| \qquad (8.54)$$

We shall calculate the differential cross section in the center-of-momentum (C-M) frame where

$$\mathbf{k}_1 = -\mathbf{k}_2 \equiv \mathbf{k} \qquad ; \text{ C-M frame}$$
$$\mathbf{k}_3 = -\mathbf{k}_4 \equiv \mathbf{k}' \qquad (8.55)$$

The last equality follows from momentum conservation.[8] The incident energies are then

$$E_e/\hbar c = \sqrt{\mathbf{k}^2 + M_e^2} \qquad ; E_\mu/\hbar c = \sqrt{\mathbf{k}^2 + M_\mu^2}$$
$$W_i = E_e + E_\mu \equiv W \qquad (8.56)$$

where $W = E_e + E_\mu$ is the total energy in the C-M system. The incident flux in the C-M system is then

$$I_{inc} = \frac{c}{\Omega}\left(\frac{\hbar kc}{E_e} + \frac{\hbar kc}{E_\mu}\right) = \frac{\hbar kc^2}{\Omega E_e E_\mu} W \qquad (8.57)$$

Consider the integral over the energy-conserving δ-function, where we sum over all *lengths* of the vector $\mathbf{k}'$ that get into the detector

$$\int \delta(W_f - W)dk' = \int \delta(W_f - W)\left[\frac{dk'}{dW_f}\right]_{k'=k} dW_f \qquad (8.58)$$

Now the total final energy, and its derivative with respect to k' evaluated at energy conservation, are

$$W_f = \hbar c\left(\sqrt{\mathbf{k}'^2 + M_e^2} + \sqrt{\mathbf{k}'^2 + M_\mu^2}\right)$$
$$\left[\frac{dW_f}{dk'}\right]_{k'=k} = (\hbar c)^2 k\left(\frac{1}{E_e} + \frac{1}{E_\mu}\right) = \frac{(\hbar c)^2 k}{E_e E_\mu} W \qquad (8.59)$$

A combination of these results gives the following expression for the differ-

---

[8] Enforced by the factor $\delta_{\mathbf{k}_1+\mathbf{k}_2,\mathbf{k}_3+\mathbf{k}_4}$ in the scattering amplitude (see Vol. I).

ential cross section in the C-M system[9]

$$\frac{d\sigma}{d\Omega} = \frac{4\alpha^2}{q^4} \left(\frac{E_e E_\mu}{\hbar c W} \right)^2 \left| [\bar{u}(\mathbf{k}_3\lambda_3)\gamma_\mu u(\mathbf{k}_1\lambda_1)]_{e^-} [\bar{u}(\mathbf{k}_4\lambda_4)\gamma_\mu u(\mathbf{k}_2\lambda_2)]_{\mu^-} \right|^2 \quad ;$$

$$\alpha = \frac{e^2}{4\pi\hbar c\varepsilon_0} \tag{8.60}$$

Here we have identified the dimensionless fine-structure constant α, and the cross section evidently now has the correct dimensions of $[L^2]$.

If the initial leptons are unpolarized, we must *average* over their spins, and if the spins of the final leptons are not observed, we must *sum* over everything that gets into the detector. In this case the quantity of interest is

$$\mathcal{S}^2 = \frac{1}{4} \sum_{\lambda_1} \cdots \sum_{\lambda_4} \left| [\bar{u}(\mathbf{k}_3\lambda_3)\gamma_\mu u(\mathbf{k}_1\lambda_1)]_{e^-} [\bar{u}(\mathbf{k}_4\lambda_4)\gamma_\mu u(\mathbf{k}_2\lambda_2)]_{\mu^-} \right|^2 \tag{8.61}$$

Consider the complex conjugate of one of the currents, and make the Dirac indices explicit

$$\{u_\alpha^\star(\mathbf{k}_3\lambda_3)[\gamma_4]_{\alpha\beta}[\gamma_\nu]_{\beta\delta}u_\delta(\mathbf{k}_1\lambda_1)\}^\star = u_\alpha(\mathbf{k}_3\lambda_3)[\gamma_4]_{\alpha\beta}^\star[\gamma_\nu]_{\beta\delta}^\star u_\delta^\star(\mathbf{k}_1\lambda_1)$$
$$= u_\delta^\star(\mathbf{k}_1\lambda_1)[\gamma_\nu]_{\delta\beta}[\gamma_4]_{\beta\alpha}u_\alpha(\mathbf{k}_3\lambda_3) \tag{8.62}$$

where we have used the fact that the γ-matrices are hermitian. Hence

$$[\bar{u}(\mathbf{k}_3\lambda_3)\gamma_\nu u(\mathbf{k}_1\lambda_1)]^\star = u^\dagger(\mathbf{k}_1\lambda_1)\gamma_\nu\gamma_4 u(\mathbf{k}_3\lambda_3) \tag{8.63}$$

We observe that *the complex conjugate of the Dirac matrix product is the product of the adjoint matrices in the reverse order*. Thus the quantity of interest in Eq. (8.61) takes the form

$$\mathcal{S}^2 = \frac{1}{4} \sum_{\lambda_1} \cdots \sum_{\lambda_4} [\bar{u}(\mathbf{k}_3\lambda_3)\gamma_\mu u(\mathbf{k}_1\lambda_1)]_{e^-} [\bar{u}(\mathbf{k}_1\lambda_1)\gamma_\nu u(\mathbf{k}_3\lambda_3)]_{e^-} \times$$
$$[\bar{u}(\mathbf{k}_4\lambda_4)\gamma_\mu u(\mathbf{k}_2\lambda_2)]_{\mu^-} [\bar{u}(\mathbf{k}_2\lambda_2)\gamma_\nu u(\mathbf{k}_4\lambda_4)]_{\mu^-} \tag{8.64}$$

Here the γ_4 has been moved back through the γ_ν in *both terms*, with no resultant change in sign.

Now consider one set of spin sums, and again make the Dirac indices explicit

$$\sum_{(1)} = \sum_{\lambda_1} \sum_{\lambda_3} \bar{u}_\alpha(\mathbf{k}_3\lambda_3)[\gamma_\mu]_{\alpha\beta}u_\beta(\mathbf{k}_1\lambda_1)\bar{u}_\rho(\mathbf{k}_1\lambda_1)[\gamma_\nu]_{\rho\sigma}u_\sigma(\mathbf{k}_3\lambda_3) \tag{8.65}$$

---

[9]Note $dk'/dW_f = [dW_f/dk']^{-1}$.

Make use of the positive-energy projection in Prob. F.2

$$\sum_\lambda u_\beta(\mathbf{k}\lambda)\bar{u}_\rho(\mathbf{k}\lambda) = \left[\frac{M - i\gamma_\mu k_\mu}{2E_k/\hbar c}\right]_{\beta\rho} \qquad ; k_\mu = \left(\mathbf{k}, \frac{iE_k}{\hbar c}\right) \quad (8.66)$$

To simplify the resulting expressions, we shall make use of the *Feynman notation* for the scalar product of γ_μ and the four-vector $a_\mu = (\mathbf{a}, ia_0)$

$$\not{a} \equiv \gamma_\mu a_\mu \qquad ; \text{ Feynman notation} \qquad (8.67)$$

With this notation, Eq. (8.66) reads

$$\sum_\lambda u_\beta(\mathbf{k}\lambda)\bar{u}_\rho(\mathbf{k}\lambda) = \left[\frac{M - i\not{k}}{2E_k/\hbar c}\right]_{\beta\rho} \qquad (8.68)$$

Equation (8.65) then becomes

$$\begin{aligned}
\sum_{(1)} &= \frac{(\hbar c)^2}{4E_1 E_3}[M - i\not{k}_3]_{\sigma\alpha}[\gamma_\mu]_{\alpha\beta}[M - i\not{k}_1]_{\beta\rho}[\gamma_\nu]_{\rho\sigma} \\
&= \frac{(\hbar c)^2}{4E_1 E_3}\text{Tr}\,[M - i\not{k}_3]\gamma_\mu[M - i\not{k}_1]\gamma_\nu \qquad (8.69)
\end{aligned}$$

Here "Tr" stands for *trace*, the sum of the diagonal elements of the Dirac matrix product.[10] Thus Eq. (8.64) becomes

$$\begin{aligned}
\mathcal{S}^2 = \frac{1}{4}\frac{(\hbar c)^4}{(4E_e E_\mu)^2}&\text{Tr}\,[m - i\not{k}_3]\gamma_\mu[m - i\not{k}_1]\gamma_\nu \times \\
&\text{Tr}\,[M - i\not{k}_4]\gamma_\mu[M - i\not{k}_2]\gamma_\nu \qquad (8.70)
\end{aligned}$$

where, during the algebra, we write $M_e \equiv m$ and $M_\mu \equiv M$. Putting this all together, the cross section in the C-M system becomes

$$\frac{d\sigma}{d\Omega} = \frac{4\alpha^2}{q^4}\left(\frac{\hbar c}{W}\right)^2 \mathcal{M}^2$$

$$\mathcal{M}^2 = \frac{1}{64}\text{Tr}\,[m - i\not{k}_3]\gamma_\mu[m - i\not{k}_1]\gamma_\nu\,\text{Tr}\,[M - i\not{k}_4]\gamma_\mu[M - i\not{k}_2]\gamma_\nu \qquad ;$$

$$M_e \equiv m,\ M_\mu \equiv M \qquad (8.71)$$

It remains to evaluate the traces of the Dirac matrix products.

---

[10]Note that the initial and final indices in the matrix product in the first line of Eq. (8.69) are both σ, and that index is summed over.

8.7.1.3   Traces

Since the sum of two matrices is just the sum of the individual elements, the trace operation is distributive. Furthermore, it is clear from Eq. (8.69) that the trace is invariant under a cyclic permutation of the matrices involved. Thus the trace has two general properties

$$\text{Tr}(\underline{A} + \underline{B}) = \text{Tr}\,\underline{A} + \text{Tr}\,\underline{B}$$

$$\text{Tr}\,\underline{A}\,\underline{B}\,\underline{C} = \text{Tr}\,\underline{B}\,\underline{C}\,\underline{A} = \text{Tr}\,\underline{C}\,\underline{A}\,\underline{B} \tag{8.72}$$

The standard representation of the Dirac matrices in 2×2 form is

$$\boldsymbol{\gamma} = \begin{pmatrix} 0 & -i\boldsymbol{\sigma} \\ i\boldsymbol{\sigma} & 0 \end{pmatrix} \; ; \; \gamma_4 = \begin{pmatrix} 1 & 0 \\ 0 & -1 \end{pmatrix} \; ; \; \gamma_5 = \gamma_1\gamma_2\gamma_3\gamma_4 = \begin{pmatrix} 0 & -1 \\ -1 & 0 \end{pmatrix} \tag{8.73}$$

It follows that

$$\text{Tr}\,\gamma_\mu = \text{Tr}\,\gamma_5 = 0 \tag{8.74}$$

Now consider the relation defining the algebra of the gamma matrices

$$\gamma_\mu\gamma_\nu + \gamma_\nu\gamma_\mu = 2\delta_{\mu\nu} \times 1 \tag{8.75}$$

Here the "1" on the r.h.s. reminds the reader that there is always a suppressed unit Dirac matrix in these expressions, and consistent with our notation, the underlining of the Dirac matrices is also suppressed. Take $1/2$ of the trace of this relation, using the properties in Eqs. (8.72)

$$\frac{1}{2}\text{Tr}\,(\gamma_\mu\gamma_\nu + \gamma_\nu\gamma_\mu) = \text{Tr}\,\gamma_\mu\gamma_\nu = 4\delta_{\mu\nu} \tag{8.76}$$

Thus, for example, with the Feynman notation

$$\text{Tr}\,\not{a}\not{b} = 4(a \cdot b) \tag{8.77}$$

where $(a \cdot b)$ is the dot product of the four-vectors.

Consider the trace of the product of three gamma matrices. If two of the indices are the same, this reduces to the trace of a single gamma matrix, since

$$\gamma_1^2 = \gamma_2^2 = \gamma_3^2 = \gamma_4^2 = 1 \tag{8.78}$$

The trace of a single gamma matrix vanishes by the first of Eqs. (8.74). Suppose all the indices on the three gamma matrices are distinct. Take the trace of the following relation

$$\gamma_5\gamma_\mu + \gamma_\mu\gamma_5 = 0$$

$$\Rightarrow \quad \mathrm{Tr}\,\gamma_5\gamma_\mu = 0 \tag{8.79}$$

From the definition of γ_5, this covers all possibilities for the three distinct indices. Thus the trace of the product of three Dirac matrices always vanishes

$$\mathrm{Tr}\,\gamma_\mu\gamma_\nu\gamma_\lambda = 0 \tag{8.80}$$

With the product of four gamma matrices, the trace will vanish unless the indices are equal in pairs. The possibilities are

$$\mathrm{Tr}\,\gamma_\mu\gamma_\nu\gamma_\rho\gamma_\sigma = 4\,(\delta_{\mu\nu}\delta_{\rho\sigma} + \delta_{\mu\sigma}\delta_{\nu\rho} - \delta_{\mu\rho}\delta_{\nu\sigma}) \tag{8.81}$$

With the Feynman notation

$$\mathrm{Tr}\,\not{k}_1\not{k}_2\not{k}_3\not{k}_4 = 4[(k_1 \cdot k_2)(k_3 \cdot k_4) + (k_1 \cdot k_4)(k_2 \cdot k_3) - (k_1 \cdot k_3)(k_2 \cdot k_4)] \tag{8.82}$$

8.7.1.4 *Cross Section (Continued)*

We now have the tools to evaluate $\mathcal{M}^2$ in Eq. (8.71)

$$64\mathcal{M}^2 = 16\{m^2\delta_{\mu\nu} - [k_{3\mu}k_{1\nu} + k_{3\nu}k_{1\mu} - (k_1 \cdot k_3)\delta_{\mu\nu}]\} \times$$
$$\{M^2\delta_{\mu\nu} - [k_{4\mu}k_{2\nu} + k_{4\nu}k_{2\mu} - (k_4 \cdot k_2)\delta_{\mu\nu}]\} \tag{8.83}$$

The sum over the indices of the two tensors can be carried out to give[11]

$$4\mathcal{M}^2 = 4m^2M^2 + 2M^2(k_1 \cdot k_3) + 2m^2(k_2 \cdot k_4) +$$
$$(k_3 \cdot k_4)(k_1 \cdot k_2) + (k_1 \cdot k_4)(k_2 \cdot k_3) - (k_2 \cdot k_4)(k_1 \cdot k_3) +$$
$$(k_2 \cdot k_3)(k_1 \cdot k_4) + (k_1 \cdot k_2)(k_3 \cdot k_4) - (k_1 \cdot k_3)(k_2 \cdot k_4) +$$
$$4(k_1 \cdot k_3)(k_2 \cdot k_4) - (k_1 \cdot k_3)(k_2 \cdot k_4) - (k_1 \cdot k_3)(k_2 \cdot k_4) \tag{8.84}$$

A combination of terms then yields

$$4\mathcal{M}^2 = 4m^2M^2 + 2M^2(k_1 \cdot k_3) + 2m^2(k_2 \cdot k_4) +$$
$$2(k_3 \cdot k_4)(k_1 \cdot k_2) + 2(k_1 \cdot k_4)(k_2 \cdot k_3) \tag{8.85}$$

---

[11] Note that $\delta_{\mu\nu}\delta_{\mu\nu} = \delta_{\mu\mu} = 4$.

Note that this is a nice Lorentz-invariant expression composed entirely of Lorentz scalars.

Now specify these results to the C-M system as illustrated In Fig. 8.5.

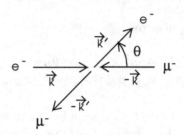

Fig. 8.5   Center of momentum (C-M) configuration for the scattering process $e^-(k_1) + \mu^-(k_2) \rightarrow e^-(k_3) + \mu^-(k_4)$. θ is the scattering angle, and $|\mathbf{k}'| = |\mathbf{k}|$.

In the C-M system, the four-vectors take the following form

$$
\begin{aligned}
k_1 &= (\mathbf{k}, iE_e/\hbar c) && ; E_e/\hbar c = \sqrt{m^2 + \mathbf{k}^2} \\
k_2 &= (-\mathbf{k}, iE_\mu/\hbar c) && ; E_\mu/\hbar c = \sqrt{M^2 + \mathbf{k}^2} \\
k_3 &= (\mathbf{k}', iE_e/\hbar c) \\
k_4 &= (-\mathbf{k}', iE_\mu/\hbar c)
\end{aligned}
\tag{8.86}
$$

where

$$
\begin{aligned}
\mathbf{k} \cdot \mathbf{k}' &= \mathbf{k}^2 \cos\theta \\
q^2 &= (k_1 - k_3)^2 = (\mathbf{k} - \mathbf{k}')^2 = 2\mathbf{k}^2(1 - \cos\theta)
\end{aligned}
\tag{8.87}
$$

In this system, the scalar products required in Eq. (8.85) are

$$
\begin{aligned}
(k_1 \cdot k_3) &= \mathbf{k} \cdot \mathbf{k}' - (E_e/\hbar c)^2 = -m^2 - \mathbf{k}^2(1 - \cos\theta) \\
(k_2 \cdot k_4) &= \mathbf{k} \cdot \mathbf{k}' - (E_\mu/\hbar c)^2 = -M^2 - \mathbf{k}^2(1 - \cos\theta) \\
(k_1 \cdot k_2) &= (k_3 \cdot k_4) \\
&= -\mathbf{k}^2 - E_e E_\mu/(\hbar c)^2 \\
(k_1 \cdot k_4) &= (k_2 \cdot k_3) \\
&= -\mathbf{k} \cdot \mathbf{k}' - E_e E_\mu/(\hbar c)^2 = -\mathbf{k}^2 \cos\theta - E_e E_\mu/(\hbar c)^2
\end{aligned}
\tag{8.88}
$$

This gives

$$4\mathcal{M}^2 = 4m^2 M^2 - 2M^2[m^2 + \mathbf{k}^2(1 - \cos\theta)] - 2m^2[M^2 + \mathbf{k}^2(1 - \cos\theta)] +$$
$$2[\mathbf{k}^2 + E_e E_\mu/(\hbar c)^2]^2 + 2[\mathbf{k}^2 \cos\theta + E_e E_\mu/(\hbar c)^2]^2 \qquad (8.89)$$

This expression simplifies to

$$4\mathcal{M}^2 = -2(M^2 + m^2)\mathbf{k}^2(1 - \cos\theta) +$$
$$2[\mathbf{k}^2 + E_e E_\mu/(\hbar c)^2]^2 + 2[\mathbf{k}^2 \cos\theta + E_e E_\mu/(\hbar c)^2]^2 \qquad (8.90)$$

The cross section in the C-M system for the process $e^-(\mathbf{k}) + \mu^-(-\mathbf{k}) \to e^-(\mathbf{k}') + \mu^-(-\mathbf{k}')$ is then given to $O(\alpha^2)$ by Eqs. (8.71) as

$$\frac{d\sigma}{d\Omega} = \frac{\alpha^2}{4\mathbf{k}^4 \sin^4(\theta/2)} \left(\frac{\hbar c}{E_e + E_\mu}\right)^2 \frac{1}{2} \times$$
$$\left\{[\mathbf{k}^2 + E_e E_\mu/(\hbar c)^2]^2 + [\mathbf{k}^2 \cos\theta + E_e E_\mu/(\hbar c)^2]^2 - 2(M_\mu^2 + M_e^2)\mathbf{k}^2 \sin^2(\theta/2)\right\} \qquad ; \text{C-M system} \quad (8.91)$$

8.7.1.5 *Limiting Cases*

Consider two limiting cases of this result:

The *extreme relativistic limit* (ERL) is obtained by letting $|\mathbf{k}| \to \infty$

$$\frac{d\sigma}{d\Omega} = \frac{\alpha^2}{4\mathbf{k}^2 \sin^4(\theta/2)} \frac{1}{8} \{4 + [1 + \cos\theta]^2\}$$
$$= \frac{\alpha^2}{4\mathbf{k}^2 \sin^4(\theta/2)} \frac{1}{2} [1 + \cos^4(\theta/2)] \qquad ; \text{ERL} \quad (8.92)$$

The *non-relativistic limit* (NRL) is obtained by letting $|\mathbf{k}| \to 0$[12]

$$\frac{d\sigma}{d\Omega} = \frac{\alpha^2}{4\mathbf{k}^4 \sin^4(\theta/2)} \left(\frac{1}{M_\mu + M_e}\right)^2 \frac{1}{2} [(M_e M_\mu)^2 + (M_e M_\mu)^2]$$
$$= \frac{\alpha^2}{4\mathbf{k}^4 \sin^4(\theta/2)} \left(\frac{M_e M_\mu}{M_\mu + M_e}\right)^2 \qquad ; \text{NRL} \quad (8.93)$$

This is recognized as the Rutherford cross section (see Prob. 8.10).

---

[12]More precisely, the ERL is $|\mathbf{k}| \gg (M_e, M_\mu)$ and the NRL here is $|\mathbf{k}| \ll (M_e, M_\mu)$.

8.7.1.6 *Møller Scattering*

The cross section for the process $e^- + e^- \rightarrow e^- + e^-$ follows in exactly the same fashion from the expression for S_{fi} in Eq. (8.46); however, this calculation involves more algebra because of the presence of the exchange term in the amplitude and two Feynman diagrams.[13] In the C-M system in the ERL, the cross section for this Møller scattering process is given by

$$\frac{d\sigma}{d\Omega} = \frac{\alpha^2}{8k^2} \left\{ \frac{1 + \cos^4(\theta/2)}{\sin^4(\theta/2)} + \frac{1 + \sin^4(\theta/2)}{\cos^4(\theta/2)} + \frac{2}{\sin^2(\theta/2)\cos^2(\theta/2)} \right\} \quad ;$$

$$\text{Møller scattering in C-M ; ERL} \qquad (8.94)$$

The original Stanford colliding-beam experiment of O'Neil, Richter, Ritson, and others, was designed to test this formula at high momentum transfer. The 500 MeV electron beam from the Mark III accelerator at the Stanford High Energy Physics Laboratory (HEPL) was inserted into two storage rings in such a fashion that the collision region in the laboratory was actually the C-M for the $e^- + e^-$ system (see Fig. 8.6). The outgoing

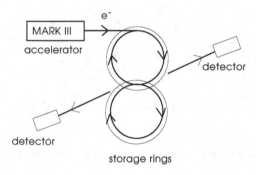

Fig. 8.6 Sketch of original Stanford colliding-beam experiment at HEPL to measure the Møller scattering process $e^-(\mathbf{k}) + e^-(-\mathbf{k}) \rightarrow e^-(\mathbf{k}') + e^-(-\mathbf{k}')$ at high momentum transfer.

scattered electrons were then detected in a back-to-back coincidence. The Møller formula in Eq. (8.92) was found to work to a few percent. More importantly, this experiment demonstrated the feasibility of using storage rings to carry out colliding-beam experiments with electrons, experiments which later found their culmination in the ground-breaking $(e^+ e^-)$ work at SLAC and CERN.

---

[13] We leave this as an exercise for our old friend, the dedicated reader.

8.7.2 $e^+ + e^- \rightarrow \mu^+ + \mu^-$

It is quite remarkable that the same analysis used to compute the cross section for the scattering of the two charged leptons $(e^- \mu^-)$ can be used to compute the cross section for an apparently unrelated process. An electron and positron pair $(e^- e^+)$ can *annihilate* to produce pairs of other particles, in this case a muon and its antiparticle $(\mu^- \mu^+)$. As this provides another nice illustration of just how one goes from the S-matrix element to the cross section, we carry out that calculation here.

8.7.2.1 *Scattering Amplitude*

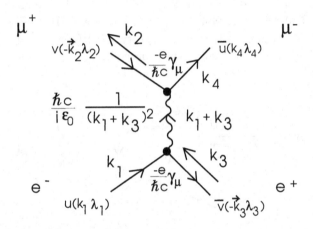

Fig. 8.7 Feynman diagram in momentum space for $-iT_{fi}/\hbar c$ to order e^2 for the scattering process $e^-(k_1\lambda_1) + e^+(k_3\lambda_3) \rightarrow \mu^-(k_4\lambda_4) + \mu^+(k_2\lambda_2)$. The Dirac indices are suppressed along the lepton lines, and there is a Dirac matrix γ_μ at each vertex.

The matrix element $S_{fi}^{(2)}$ for the process $e^-(k_1\lambda_1) + e^+(k_3\lambda_3) \rightarrow \mu^-(k_4\lambda_4) + \mu^+(k_2\lambda_2)$ is obtained from Eq. (8.50) as[14]

$$S_{fi}^{(2)} = (2\pi)^4 \delta^{(4)}(k_1 + k_3 - k_2 - k_4)\frac{1}{\Omega^2}\left(\frac{-e}{\hbar c\sqrt{\varepsilon_0}}\right)^2\left(\frac{\hbar c}{i}\right) \times$$

$$[\bar{u}(\mathbf{k}_4\lambda_4)\gamma_\mu v(-\mathbf{k}_2\lambda_2)]_{\mu^-}\frac{1}{(k_1+k_3)^2}[\bar{v}(-\mathbf{k}_3\lambda_3)\gamma_\mu u(\mathbf{k}_1\lambda_1)]_{e^-} \quad (8.95)$$

---

[14]The overall sign depends on just how the states $|i\rangle$ and $|f\rangle$ are defined.

The T-matrix is identified as

$$\frac{-iT_{fi}^{(2)}}{\hbar c} = \left(\frac{-e}{\hbar c\sqrt{\varepsilon_0}}\right)^2 \left(\frac{\hbar c}{i}\right) \times$$

$$[\bar{u}(\mathbf{k}_4\lambda_4)\gamma_\mu v(-\mathbf{k}_2\lambda_2)]_{\mu^-} \frac{1}{(k_1 + k_3)^2} [\bar{v}(-\mathbf{k}_3\lambda_3)\gamma_\mu u(\mathbf{k}_1\lambda_1)]_{e^-} \quad (8.96)$$

There is again only one Feynman diagram for this process, that depicting the *annihilation channel*, and it is shown in Fig. 8.7. The labeling of the four-momenta for this process has been chosen so that we can make immediate use of the previous results.

8.7.2.2 *Cross Section*

We shall again compute the cross section in the center of momentum (C-M) frame. The cross section is given in general by the expression in Eq. (8.53)

$$d\sigma_{fi} = \frac{2\pi}{\hbar}\delta(W_f - W_i)\frac{1}{\Omega^2}|T_{fi}|^2\frac{\Omega d^3 k_4}{(2\pi)^3}\frac{1}{I_{\text{inc}}} \quad (8.97)$$

In the C-M system, we label the variables as shown in Fig. 8.8 The

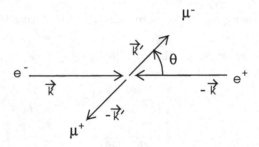

Fig. 8.8 Center of momentum (C-M) configuration for the scattering process $e^-(k_1) + e^+(k_3) \rightarrow \mu^-(k_4) + \mu^+(k_2)$. θ is the scattering angle, and $|\mathbf{k}'| < |\mathbf{k}|$.

incident flux for the $(e^- e^+)$ pair in this system is

$$I_{\text{inc}} = \frac{c}{\Omega}\left(\frac{\hbar k c}{E_e} + \frac{\hbar k c}{E_e}\right) = \frac{\hbar k c^2}{\Omega(E_e)^2}W \quad (8.98)$$

where $W \equiv W_i = E_e + E_e$ is the total initial energy.

The integral over the energy-conserving delta-function is performed as in Eqs. (8.58)–(8.59), making use of the relations

$$W_f = \hbar c \left(\sqrt{\mathbf{k}'^2 + M_\mu^2} + \sqrt{\mathbf{k}'^2 + M_\mu^2} \right)$$

$$\left[\frac{\partial W_f}{\partial k'} \right]_{W_f = W} = (\hbar c)^2 k' \left(\frac{1}{E_\mu} + \frac{1}{E_\mu} \right) = \frac{(\hbar c)^2 k'}{(E_\mu)^2} W \qquad (8.99)$$

A combination of these results expresses the cross section as

$$\frac{d\sigma}{d\Omega} = \frac{4\alpha^2}{(k_1 + k_3)^4} \left(\frac{E_e E_\mu}{\hbar c W} \right)^2 \frac{k'}{k} \times$$

$$\left| [\bar{v}(-\mathbf{k}_3\lambda_3)\gamma_\mu u(\mathbf{k}_1\lambda_1)]_{e^-} [\bar{u}(\mathbf{k}_4\lambda_4)\gamma_\mu v(-\mathbf{k}_2\lambda_2)]_{\mu^-} \right|^2 \qquad ;$$

$$\alpha = \frac{e^2}{4\pi\hbar c \varepsilon_0} \qquad (8.100)$$

The sum and average over the lepton spins is now carried out exactly as in Eqs. (8.61)–(8.70). The only new feature is the use of the alternate relation in Prob. F.2 for the sum over spins of the antiparticle spinors

$$\sum_\lambda v_\beta(-\mathbf{k}\lambda)\bar{v}_\rho(-\mathbf{k}\lambda) = \left[\frac{-M - i\not{k}}{2E_k/\hbar c} \right]_{\beta\rho} \quad ; \quad k_\mu = \left(\mathbf{k}, \frac{iE_k}{\hbar c} \right) \qquad (8.101)$$

Thus, where we again use $M_e \equiv m$ and $M_\mu \equiv M$ during the algebra,

$$\frac{1}{4} \sum_{\lambda_1, \cdots, \lambda_4} \left| [\bar{v}(-\mathbf{k}_3\lambda_3)\gamma_\mu u(\mathbf{k}_1\lambda_1)]_{e^-} [\bar{u}(\mathbf{k}_4\lambda_4)\gamma_\mu v(-\mathbf{k}_2\lambda_2)]_{\mu^-} \right|^2 = \frac{1}{4} \times$$

$$\mathrm{Tr}\left\{ \left[\frac{-m - i\not{k}_3}{2E_e/\hbar c} \right] \gamma_\mu \left[\frac{m - i\not{k}_1}{2E_e/\hbar c} \right] \gamma_\nu \right\} \times$$

$$\mathrm{Tr}\left\{ \left[\frac{M - i\not{k}_4}{2E_\mu/\hbar c} \right] \gamma_\mu \left[\frac{-M - i\not{k}_2}{2E_\mu/\hbar c} \right] \gamma_\nu \right\} \qquad (8.102)$$

The cross section then becomes

$$\frac{d\sigma}{d\Omega} = \frac{4\alpha^2}{(k_1 + k_3)^4} \left(\frac{\hbar c}{W} \right)^2 \frac{k'}{k} \mathcal{M}^2 \qquad (8.103)$$

$$\mathcal{M}^2 = \frac{1}{64} \mathrm{Tr}[-m - i\not{k}_3]\gamma_\mu[m - i\not{k}_1]\gamma_\nu \, \mathrm{Tr}[M - i\not{k}_4]\gamma_\mu[-M - i\not{k}_2]\gamma_\nu$$

The advantage of the present labeling of the four-momenta is now apparent, for we can just take over the results in Eqs. (8.83)–(8.85), replacing

$(m^2, M^2) \to (-m^2, -M^2)$ in the final result

$$4\mathcal{M}^2 = 4m^2M^2 - 2M^2(k_1 \cdot k_3) - 2m^2(k_2 \cdot k_4) +$$
$$2(k_3 \cdot k_4)(k_1 \cdot k_2) + 2(k_1 \cdot k_4)(k_2 \cdot k_3) \qquad (8.104)$$

Energy conservation in the C-M system says that $E_\mu = E_e$. Since the process only takes place above the muon threshold, one has the relations

$$E_e/\hbar c = E_\mu/\hbar c \geq M_\mu \gg M_e \qquad (8.105)$$

Above the muon threshold, the electrons are highly relativistic. We can simplify our expressions by assuming from the outset that

$$E_e/\hbar c = k \qquad ; \text{relativistic } (e^- e^+) \qquad (8.106)$$

The four-vectors in the C-M frame then take the form

$$k_1 = (\mathbf{k}, ik) \qquad ; E_e/\hbar c = k$$
$$k_3 = (-\mathbf{k}, ik)$$
$$k_4 = (\mathbf{k}', ik) \qquad ; E_\mu/\hbar c = \sqrt{M_\mu^2 + \mathbf{k}'^2} = k$$
$$k_2 = (-\mathbf{k}', ik)$$
$$\mathbf{k} \cdot \mathbf{k}' = kk' \cos\theta$$
$$(k_1 + k_3)^2 = -4\mathbf{k}^2 = -W^2/(\hbar c)^2 \qquad (8.107)$$

where $k \equiv |\mathbf{k}|$ and $k' \equiv |\mathbf{k}'|$. The required scalar products of these four-vectors are then

$$(k_1 \cdot k_3) = -2\mathbf{k}^2$$
$$(k_2 \cdot k_4) = -\mathbf{k}^2 - \mathbf{k}'^2$$
$$(k_1 \cdot k_2) = (k_3 \cdot k_4)$$
$$= -\mathbf{k}^2 - kk' \cos\theta$$
$$(k_1 \cdot k_4) = (k_2 \cdot k_3)$$
$$= -\mathbf{k}^2 + kk' \cos\theta \qquad (8.108)$$

Substitution of these relations into Eq. (8.104) yields[15]

$$4\mathcal{M}^2 = -2M_\mu^2[-2\mathbf{k}^2] + 2[\mathbf{k}^2 + kk' \cos\theta]^2 + 2[\mathbf{k}^2 - kk' \cos\theta]^2$$
$$= 4\mathbf{k}^2 M_\mu^2 + 4\mathbf{k}^4 + 4\mathbf{k}^2\mathbf{k}'^2 \cos^2\theta \qquad (8.109)$$

---

[15]With $m = 0$, and restoring $M = M_\mu$

Hence the cross section in the C-M system for $e^-(\mathbf{k}) + e^+(-\mathbf{k}) \rightarrow \mu^-(\mathbf{k}') + \mu^+(-\mathbf{k}')$, with relativistic electrons, is to $O(\alpha^2)$

$$\frac{d\sigma}{d\Omega} = \frac{4\alpha^2}{(4\mathbf{k}^2)^3}\left\{\mathbf{k}^4 + \mathbf{k}^2\mathbf{k}'^2\cos^2\theta + \mathbf{k}^2 M_\mu^2\right\}\frac{k'}{k} \qquad ; k = |\mathbf{k}| \ , k' = |\mathbf{k}'|$$

$$\text{C-M} \qquad (8.110)$$

8.7.2.3 *Limiting Cases*

There are two relevant limiting cases of this result:[16]

In the *extreme relativistic limit* (ERL), one has $k \approx k' \rightarrow \infty$, and thus

$$\frac{d\sigma}{d\Omega} = \frac{\alpha^2}{16\mathbf{k}^2}(1 + \cos^2\theta) \qquad ; \text{ERL} \qquad (8.111)$$

At *threshold*, one has $\mathbf{k}'^2 = \mathbf{k}^2 - M_\mu^2 \rightarrow 0$. In that case

$$\frac{d\sigma}{d\Omega} = \frac{\alpha^2}{8M_\mu^2}\frac{k'}{M_\mu} \qquad ; \text{Threshold} \qquad (8.112)$$

8.7.2.4 *Colliding Beams*

Colliding beam experiments with counter-rotating (e^-e^+) beams in a single ring,[17] where the beams are steered into various collision regions around the ring, have proven to be a rich source of information in particle physics. Such experiments have two great advantages:

- The laboratory system is the C-M frame in this case, and thus the full initial laboratory energy can be used to create new particles;
- Any pair of particles coupled to the photon, in particular $(q\bar{q})$ pairs, will be produced in this process (unless forbidden by selection rules).

Since the cross section in Eq. (8.110) is exactly calculated in QED, the process $e^- + e^+ \rightarrow \mu^- + \mu^+$ serves as a valuable calibration and beam monitoring tool in such experiments.

8.8   QED in External Field

Suppose that a specified, classical *external field* is also present in QED. Examples include the nuclear Coulomb field, a static magnetic field, or

---

[16]More precisely, these limits are $\mathbf{k}'^2 \gg M_\mu^2$ and $\mathbf{k}'^2 \ll M_\mu^2$.
[17]As opposed to the pair of rings in Fig. 8.6.

the electromagnetic field of a laser. The important point is that the field is both classical and specified. The interaction of the electrons with this field can be included in H_{QED} of Eq. (8.7) by making the gauge-invariant replacements $\mathbf{p} \to \mathbf{p} - e\mathbf{A}^{\text{ext}}$, $H \to H + e\Phi^{\text{ext}}$ in the Dirac hamiltonian of Eq. (5.99). This leads to an additional interaction in Eq. (8.13) of the form[18]

$$\hat{H}_I^{\text{ext}}(t) = -ec \int_\Omega d^3x \, \hat{j}_\mu(x) A_\mu^{\text{ext}}(x) \tag{8.113}$$

Here the external four-vector potential is a given function of space-time

$$A_\mu^{\text{ext}}(x) = \left[\mathbf{A}^{\text{ext}}(x), \frac{i}{c}\Phi^{\text{ext}}(x) \right] \qquad ; \text{ specified} \tag{8.114}$$

Since there are no derivatives in this interaction, the canonical quantization is unchanged, and the four-vector current in the interaction picture is the familiar expression

$$\hat{j}_\mu(x) = i : \hat{\bar{\psi}}(x) \gamma_\mu \hat{\psi}(x) : \tag{8.115}$$

The interaction in Eq. (8.14) now has *three* terms in it

$$\hat{H}_I(t) = \hat{H}_I^\gamma(t) + \hat{H}_I^C(t) + \hat{H}_I^{\text{ext}}(t) \tag{8.116}$$

The scattering operator can correspondingly be organized according to the number of powers of A_μ^{ext} occurring in it

$$\hat{S} = \hat{S}_0 + \hat{S}_1^{\text{ext}} + \hat{S}_2^{\text{ext}} + \cdots \tag{8.117}$$

where $\hat{S}_0$ is the scattering operator studied previously in this chapter. The first-order contribution in the external field can correspondingly be expanded in powers of the electric charge e appearing in Eqs. (8.113) and (8.14) as

$$\hat{S}_1^{\text{ext}} = e\hat{S}_{11}^{\text{ext}} + e^2 \hat{S}_{12}^{\text{ext}} + e^3 \hat{S}_{13}^{\text{ext}} \cdots \tag{8.118}$$

The first term in this expression is just

$$e\hat{S}_{11}^{\text{ext}} = \left(\frac{-i}{\hbar c} \right) \int d^4x \, [-ec\hat{j}_\mu(x) A_\mu^{\text{ext}}(x)]$$

$$= \left(\frac{-e}{\hbar c} \right) \int d^4x : \hat{\bar{\psi}}(x) \gamma_\mu \hat{\psi}(x) : cA_\mu^{\text{ext}}(x) \tag{8.119}$$

---

[18] Compare with Eq. (C.51).

This describes the scattering of the Dirac particle in the external field, examples of which were presented in Vol. I.

Consider the second term in Eq. (8.118). This comes from the two cross terms between $\hat{H}_I^\gamma$ and $\hat{H}_I^{\text{ext}}$, which are shown to make identical contributions by a change of dummy variables. Since $A_\mu^{\text{ext}}(x)$ is a c-number which commutes with everything, this term in the scattering operator takes the form

$$e^2 \hat{S}_{12}^{\text{ext}} = \left(\frac{-e}{\hbar c}\right)^2 \int d^4x_1 \int d^4x_2 \, P\left[:\hat{\bar{\psi}}(x_1)\gamma_\mu\hat{\psi}(x_1)::\hat{\bar{\psi}}(x_2)\gamma_\nu\hat{\psi}(x_2):\right] \times$$
$$c^2 \hat{A}_\mu(x_1) A_\nu^{\text{ext}}(x_2) \qquad\qquad ; \, \hat{A}_4(x) \equiv 0 \qquad (8.120)$$

We will discuss two applications of this result for QED processes taking place in the Coulomb field of the nucleus.

8.8.1   Nuclear Coulomb Field

In the case of the nuclear Coulomb field, there is only the electrostatic Coulomb potential and

$$cA_\mu^{\text{ext}}(x) = i\delta_{\mu 4}\left(\frac{Ze_p}{4\pi\varepsilon_0|\mathbf{x}|}\right) \qquad\qquad (8.121)$$

We will need the four-dimensional Fourier transform of this potential defined by

$$cA_\mu^{\text{ext}}(x) \equiv \int \frac{d^4q}{(2\pi)^4}\left[\frac{1}{\sqrt{\varepsilon_0}}a_\mu(q)\right]e^{iq\cdot x} \qquad\qquad (8.122)$$

Inversion of this relation, utilizing previous results, gives

$$\frac{1}{\sqrt{\varepsilon_0}}a_\mu(q) = \int d^4x \, cA_\mu^{\text{ext}}(x)e^{-iq\cdot x}$$
$$= \left(\frac{iZe_p}{4\pi\varepsilon_0}\right)\delta_{\mu 4}\int d^4x \, \frac{e^{-iq\cdot x}}{|\mathbf{x}|}$$
$$= \left(\frac{iZe_p}{4\pi\varepsilon_0}\right)\delta_{\mu 4}\, 2\pi\delta(q_0)\frac{4\pi}{\mathbf{q}^2} \qquad\qquad (8.123)$$

Hence

$$a_\mu(q) = i\delta_{\mu 4}\,\delta(q_0)\left(\frac{2\pi Ze_p}{\sqrt{\varepsilon_0}\,\mathbf{q}^2}\right) \qquad\qquad (8.124)$$

Note that the Fourier transform of this static potential contains a delta-function in the frequency.

8.8.2 Bremsstrahlung

Consider the process $e^-(k_1) \to e^-(k_2) + \gamma(l)$ taking place in the external Coulomb field of the nucleus.[19] The initial and final states in this case are

$$|i\rangle = a^\dagger_{k_1\lambda_1}|0\rangle \qquad ; |f\rangle = c^\dagger_{1s}a^\dagger_{k_2\lambda_2}|0\rangle \qquad (8.125)$$

The P-product of the Dirac currents in Eq. (8.120) is analyzed with Wick's theorem, and the appropriate matrix element of the scattering operator in Eq. (8.120) in momentum space follows as in our previous examples[20]

$$\left[e^2\hat{S}^{\text{ext}}_{12}\right]_{fi} = \frac{-i}{\sqrt{\Omega^3}}\left(\frac{-e}{\hbar c\sqrt{\varepsilon_0}}\right)^2\left(\frac{\hbar c^2}{2\omega_l}\right)^{1/2}\bar{u}(\mathbf{k}_2\lambda_2) \times \qquad (8.126)$$

$$\left[\not{\epsilon}\frac{1}{i(\not{k}_2+\not{l})+M}\not{a}(k_2+l-k_1) + \not{a}(k_2+l-k_1)\frac{1}{i(\not{k}_1-\not{l})+M}\not{\epsilon}\right]u(\mathbf{k}_1\lambda_1)$$

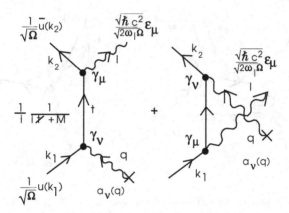

Fig. 8.9   Feynman diagrams in momentum space for $[e^2\hat{S}^{\text{ext}}_{12}]_{fi}$ for the bremmstrahlung process $e^-(k_1\lambda_1) \to \gamma(ls) + e^-(k_2\lambda_2)$ in the nuclear Coulomb field. The coupling constant at each vertex is now $(-e/\hbar c\sqrt{\varepsilon_0})$. Only the relevant changes are shown in the second diagram. Four-momentum is conserved at the vertices, in particular $q = k_2+l-k_1$.

Some comments:

---

[19]The external field is necessary to conserve energy and momentum.
[20]See Prob. 8.11.

- The photon polarization four-vector is defined as

$$\varepsilon_\mu \equiv (\mathbf{e}_{1s}, 0) \tag{8.127}$$

- Instead of an overall $\delta^{(4)}$-function, it is the Fourier transform of the external potential $a_\mu(q)/\sqrt{\varepsilon_0}$ that supplies the missing four-momentum;
- Since the external field is static, it can, in fact, only supply three-momentum [see the energy-conserving delta-function in Eq. (8.124)];
- The structure of this matrix element is clearly demonstrated through the Feynman diagrams in momentum space shown in Fig. 8.9;
- The coupling constant at each vertex is now $(-e/\hbar c\sqrt{\varepsilon_0})$;[21]
- The momentum of *both* particles can now be freely varied in the final state;
- This amplitude gives the exact cross section to order e^4 for the accelerating particle to emit a photon in QED.

8.8.3 *Pair Production*

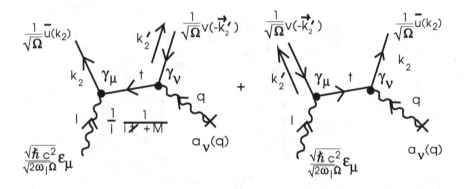

Fig. 8.10 Feynman diagrams in momentum space for $[e^2 \hat{S}_{12}^{\text{ext}}]_{fi}$ for the pair-production process $\gamma(ls) \rightarrow e^-(k_2\lambda_2) + e^+(k_2'\lambda_2')$ in the Coulomb field of the nucleus. The coupling constant at each vertex is now $(-e/\hbar c\sqrt{\varepsilon_0})$. Only the relevant changes are shown in the second diagram. Four-momentum is conserved at the vertices, in particular $q = k_2 + k_2' - l$.

Just as in the previous chapter, one obtains the amplitude for the pair-production process $\gamma(l) \rightarrow e^-(k_2) + e^+(k_2')$ in the nuclear Coulomb field by turning the incident electron line into an outgoing positron line. The

---

[21] Recall that in S-I units, $\alpha = e^2/4\pi\hbar c\varepsilon_0$ is the fine-structure constant.

appropriate matrix element of the scattering operator in Eq. (8.120) in momentum space follows as above[22]

$$\left[e^2 \hat{S}_{12}^{\text{ext}}\right]_{fi} = \frac{-i}{\sqrt{\Omega^3}} \left(\frac{-e}{\hbar c \sqrt{\varepsilon_0}}\right)^2 \left(\frac{\hbar c^2}{2\omega_l}\right)^{1/2} \bar{u}(\mathbf{k}_2 \lambda_2) \times \tag{8.128}$$

$$\left[\not{\epsilon} \frac{1}{i(\not{k}_2 - \not{l}) + M} \not{q}(k_2 + k_2' - l) + \not{q}(k_2 + k_2' - l) \frac{1}{i(\not{l} - \not{k}_2') + M} \not{\epsilon}\right] v(-\mathbf{k}_2' \lambda_2')$$

where the photon polarization four-vector is again defined as in Eq. (8.127). The corresponding Feynman diagrams in momentum space are shown in Fig. 8.10. This amplitude gives rise to the Bethe-Heitler cross section for pair production.[23]

8.9 Scattering Operator $\hat{S}_1^{\text{ext}}$ in Order e^3

Let us extend the analysis to one higher order, and examine the term $e^3 \hat{S}_{13}^{\text{ext}}$ in Eq. (8.118). In the scattering operator $\hat{S}$, in third-order in the interaction $\hat{H}_I$ of Eq. (8.116), there are three terms of $O[(\hat{H}_I^\gamma)^2 \hat{H}_I^{\text{ext}}]$, and hence of $O(e^3)$. These three terms are shown to be identical by a change in dummy variables. Similarly, in second-order in $\hat{H}_I$, there are two identical terms of $O(\hat{H}_I^C \hat{H}^{\text{ext}})$, which are *also* of $O(e^3)$. Thus

$$e^3 \hat{S}_{13}^{\text{ext}} = \frac{3}{3!} \left(\frac{ie}{\hbar c}\right)^3 c^3 \int d^4 x_1 \int d^4 x_2 \int d^4 x_3 \times \tag{8.129}$$

$$P[\hat{j}_i(x_1) \hat{j}_j(x_2) \hat{j}_\lambda(x_3)] P[\hat{A}_i(x_1) \hat{A}_j(x_2)] A_\lambda^{\text{ext}}(x_3) +$$

$$\frac{2}{2!} \left(\frac{-ie}{\hbar c}\right)^2 \left(\frac{-ec}{8\pi\varepsilon_0}\right) \int d^4 x_1 \int d^3 x' \int d^4 x_3 P\left[\frac{\hat{\rho}(x_1)\hat{\rho}(\mathbf{x}', t_1)}{|\mathbf{x}_1 - \mathbf{x}'|} \hat{j}_\lambda(x_3)\right] A_\lambda^{\text{ext}}(x_3)$$

This last term can be rewritten as

$$\frac{1}{2!} \left(\frac{ie}{\hbar c}\right)^3 \int d^4 x_1 \int d^4 x_2 \int d^4 x_3 P[\hat{j}_4(x_1) \hat{j}_4(x_2) \hat{j}_\lambda(x_3)] \times$$

$$\left(\frac{\hbar c}{i\varepsilon_0}\right) \frac{\delta(ct_1 - ct_2)}{4\pi|\mathbf{x}_1 - \mathbf{x}_2|} c A_\lambda^{\text{ext}}(x_3) \tag{8.130}$$

where $\hat{j}_4 = i\hat{\rho}$. The use of Wick's theorem on the photon fields in the first term in Eq. (8.129), and the introduction of the Coulomb propagator from

---

[22]See Prob. 8.11.

[23]The path to the cross sections, in both cases, is detailed in Vol. I.

Eq. (8.18) into the above, now allow us to rewrite Eq. (8.129) as

$$
e^3 \hat{S}_{13}^{\text{ext}} = \left(\frac{ie}{\hbar c}\right)^3 \frac{1}{2!} \int d^4x_1 \int d^4x_2 \int d^4x_3 \times
$$

$$
\left\{ P[\hat{j}_i(x_1)\hat{j}_j(x_2)\hat{j}_\lambda(x_3)] \left[c^2 : \hat{A}_i(x_1)\hat{A}_j(x_2) : + \frac{\hbar c}{i\varepsilon_0} D_F^T(x_1 - x_2)_{ij} \right] + \right.
$$

$$
\left. P[\hat{j}_4(x_1)\hat{j}_4(x_2)\hat{j}_\lambda(x_3)] \frac{\hbar c}{i\varepsilon_0} D^C(x_1 - x_2) \right\} c A_\lambda^{\text{ext}}(x_3) \qquad (8.131)
$$

The Coulomb and transverse photon propagators can now be combined *exactly as in Eqs. (8.20)–(8.24)* to give

$$
e^3 \hat{S}_{13}^{\text{ext}} = \left(\frac{ie}{\hbar c}\right)^3 \frac{1}{2!} \int d^4x_1 \int d^4x_2 \int d^4x_3 \, P[\hat{j}_\mu(x_1)\hat{j}_\nu(x_2)\hat{j}_\lambda(x_3)] \times
$$

$$
\left[c^2 : \hat{A}_\mu(x_1)\hat{A}_\nu(x_2) : + \frac{\hbar c}{i\varepsilon_0} D_{\mu\nu}^F(x_1 - x_2) \right] c A_\lambda^{\text{ext}}(x_3) \qquad ;
$$

$$
\hat{A}_4(x) \equiv 0 \qquad (8.132)
$$

where the photon propagator in the Coulomb gauge is again given by Eqs. (8.26)–(8.28)

$$
D_{\mu\nu}^F(x_1 - x_2) = \int \frac{d^4k}{(2\pi)^4} \tilde{D}_{\mu\nu}^F(k) \, e^{ik \cdot (x_1 - x_2)}
$$

$$
\tilde{D}_{\mu\nu}^F(k) = \frac{1}{k^2}\delta_{\mu\nu} - \frac{1}{k^2} \frac{1}{k^2 + (k \cdot \eta)^2} \left[k_\mu k_\nu + (k \cdot \eta)(k_\mu \eta_\nu + k_\nu \eta_\mu) \right] \qquad ;
$$

$$
\text{photon propagator} \qquad (8.133)
$$

We again adopt the convention that $\hat{A}_4(x) \equiv 0$ in writing Eq. (8.132).

Since the interaction-picture Dirac current is conserved, a repetition of the arguments in Eqs. (8.28)–(8.38) shows that one can again use the *effective photon propagator* with Fourier transform given in Eq. (8.29)[24]

$$
\tilde{D}_{\mu\nu}^F(k) \doteq \frac{1}{k^2}\delta_{\mu\nu} \qquad ; \text{ effective photon propagator} \quad (8.134)
$$

At this point, we see a return to simplicity, elegance, covariance, and gauge invariance for the photon propagator.

---

[24]For alternate derivations of this result, see Prob. 9.5 and appendix H. (Note also Prob. 9.18.)

8.10 Feynman Rules for QED

8.10.1 *General Scattering Operator*

Although we have only proven it for $e^2 \hat{S}_{02}$ and $e^3 \hat{S}_{13}^{\text{ext}}$, the general forms of the effective scattering operators $\hat{S}_0$ and $\hat{S}_1^{\text{ext}}$ are

$$
\hat{S}_0 = \sum_{n=0}^{\infty} \left(\frac{ie}{\hbar c}\right)^n \frac{1}{n!} \int d^4 x_1 \cdots \int d^4 x_n \, P[\hat{j}_\mu(x_1) \cdots \hat{j}_\nu(x_n)] \times
$$

$$
c^n P[\hat{A}_\mu(x_1) \cdots \hat{A}_\nu(x_n)]
$$

$$
\hat{S}_1^{\text{ext}} = \sum_{n=0}^{\infty} \left(\frac{ie}{\hbar c}\right)^{n+1} \frac{1}{n!} \int d^4 x_1 \cdots \int d^4 x_n \int d^4 y \, P[\hat{j}_\mu(x_1) \cdots \hat{j}_\nu(x_n)\hat{j}_\lambda(y)] \times
$$

$$
c^{n+1} P[\hat{A}_\mu(x_1) \cdots \hat{A}_\nu(x_n)] A_\lambda^{\text{ext}}(y) \tag{8.135}
$$

where

- These expressions are to be analyzed using Wick's theorem;
- The contraction of the photon fields to be used in Wick's theorem is the *effective photon propagator*

$$
c^2 \hat{A}_\mu(x)^{\cdot} \hat{A}_\nu(y)^{\cdot} = \frac{\hbar c}{i\varepsilon_0} D_{\mu\nu}^F(x-y)
$$

$$
D_{\mu\nu}^F(x-y) = \delta_{\mu\nu} \int \frac{d^4 q}{(2\pi)^4} \frac{1}{q^2} e^{iq\cdot(x-y)} \quad ; \text{ effective} \tag{8.136}
$$

- The remaining photon fields to be used in Wick's theorem create and destroy real, transverse photons, and for these remaining fields

$$
\hat{A}_4(x) \equiv 0 \qquad ; \text{ convention} \tag{8.137}
$$

- The choice of gauge for $A_\lambda^{\text{ext}}(x)$, since it is treated differently, is arbitrary.

The proof of the above statements is not difficult; we have all the tools. The proof involves

(1) Combining the Coulomb and transverse photon propagators to a given order in e to show that one can use the photon propagator in the Coulomb gauge;

(2) Showing that the terms proportional to k_μ or k_ν in that propagator do not contribute to the effective scattering operator because of the conservation of the interaction-picture current.

The first is simply a careful counting problem, and the second involves partial integration and the use of the equal-time commutation relations in the interaction picture in Eq. (8.38).[25] The general proof is, however, somewhat tedious, and we are content at this point to leave it for a future course.

8.10.2   *Feynman Diagrams*

The scattering operator now describes so many processes that it is most convenient to take specific matrix elements and work in momentum space. Our several examples have shown that only distinct diagrams with a different topology, or flow of momentum, contribute in this case.

To see how this comes about, start with a given process in coordinate space, for example, fourth-order Compton scattering, one contribution to which is illustrated in Fig. 8.11.

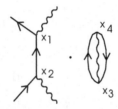

Fig. 8.11   Example of a fourth-order coordinate-space contribution to Compton scattering.

All four vertex locations $(x_1, \cdots, x_4)$ here are equivalent. Thus there are 4! equal contributions, where, as discussed previously, we assign a factor of 1/2! to the disconnected piece to correct for the overcounting. In general, this factor of $n!$ from the permutations of the vertices $(x_1, \cdots, x_n)$ cancels the factor of $1/n!$ in front of the scattering operator. There remains a single diagram with a specific set of vertex labels.

As demonstrated in chapter 7, the disconnected diagrams (those not connected to any external line) now disappear from physical processes, and one can concentrate on the *connected diagrams*.

When matrix elements are taken between eigenstates of momentum, the Feynman rules for the contribution of the connected diagrams are just those

---

[25] Alternatively, one establishes the second through the generalization of the analysis in Prob. 9.5, as given in [Feynman (1949)] and [Bjorken and Drell (1964)].

we have seen in our several examples.

8.10.3 *Feynman Rules*

The Feynman rules for the connected diagrams in coordinate space for QED are summarized as follows:

8.10.3.1 *Coordinate Space*

(1) Place down the coordinates $(x_1, x_2, \cdots, x_n)$. These mark the vertices. Draw all topologically-distinct connected diagrams with one incoming electron line, outgoing electron line, and one photon line at each vertex.[26] These are the Feynman diagrams, and $[S_{fi}^{(n)}]_C$ is the sum of the contributions from these diagrams;

(2) Include a factor of $(-e/\hbar c\sqrt{\varepsilon_0}\,)\gamma_\mu$ for each vertex;

(3) Include a factor of $(\hbar c/i)D_{\mu\nu}^F(x_j - x_i)$ for each internal photon line;

(4) Include a factor of $iS_F(x_f - x_i)$ for each internal directed electron line;

(5) Include a factor of $(\hbar c^2/2\omega_k\Omega)^{1/2}\mathbf{e}_{ks}e^{-ik\cdot x}$ for each outgoing photon line;

(6) Include a factor of $(\hbar c^2/2\omega_k\Omega)^{1/2}\mathbf{e}_{ks}e^{ik\cdot x}$ for each incoming photon line;

(7) Include a factor of $(1/\Omega)^{1/2}\,\bar{u}(k\lambda)e^{-ik\cdot x}$ for each outgoing e^- line;

(8) Include a factor of $(1/\Omega)^{1/2}\,u(k\lambda)e^{ik\cdot x}$ for each incoming e^- line;

(9) Take the Dirac matrix product along each fermion line;

(10) Include a factor of (-1) for each closed fermion loop;

(11) Integrate over $d^4x_1 \cdots d^4x_n$;

(12) With an external field at the y-vertex, include a factor $cA_\lambda^{\text{ext}}(y)$ and integrate over d^4y.

8.10.3.2 *Momentum Space*

Insertion of the four-dimensional Fourier transforms, and integration over all coordinates, then leads to the set of Feynman rules in momentum space. *The Feynman rules in momentum space for the contribution to $[S_{fi}]_C$ that is nth-order in the coupling constant e are as follows:*

(1) Place down n vertices, and draw all topologically-distinct connected diagrams, with one incoming electron line, one outgoing electron line,

---

[26] If there is any confusion over what "topologically-distinct" means, one goes back to Wick's theorem. Note that an external fermion line must run all the way through a diagram; it cannot end.

and one photon line at each vertex. These form the Feynman dia-grams, and $[S^{(n)}_{fi}]_C$ is the sum of the contributions from these dia-grams;

(2) Include a factor of $(-e/\hbar c\sqrt{\varepsilon_0})\gamma_\mu$ for each vertex;

(3) Assign a directed four-momentum k to each line;[27]

(4) Include a factor of $(\hbar c/i)\delta_{\mu\nu}\, q^{-2}$ for each internal photon line;

(5) Include a factor of $(1/i)[i\not k + M]^{-1}$ for each internal electron line;

(6) Conserve four-momentum with a factor of $(2\pi)^4\delta^{(4)}\left(\sum_i k_i\right)$ at each vertex;

(7) Include a factor of $(\hbar c^2/2\omega_k\Omega)^{1/2}\mathbf{e}_{ks}$ for each external photon line;

(8) Include a factor of $(1/\Omega)^{1/2}u(k\lambda)$ for each incoming electron line;

(9) Include a factor of $(1/\Omega)^{1/2}\bar{u}(k\lambda)$ for each outgoing electron line;[28]

(10) Take the Dirac matrix product along each fermion line;

(11) Include a factor of (-1) for each closed fermion loop;

(12) With an external field bringing q into a vertex, include a factor $a_\mu(q)$;

(13) Integrate $d^4q/(2\pi)^4$ over all internal four-momenta.

---

[27] Here k is a four-vector wavenumber, and the four-momentum is actually $\hbar k$.

[28] See Prob. 7.4 for the rules with incoming or outgoing positrons.

Chapter 9

Higher-Order Processes

We have constructed the scattering operator in second-order for the Dirac-scalar theory and for QED, and we have shown with examples how one proceeds to cross sections. It is of interest to also examine the basic *higher-order* processes described by the scattering operator in a relativistic quantum field theory, and to keep close contact with physics, we shall continue to focus on QED.

9.1 Example–Scattering in External Field

As an example, consider the scattering of an electron in an external field through $O(e^3)$, where the relevant parts of the scattering operator are given by Eqs. (8.119) and (8.132). The initial and final states are

$$|i\rangle = a_{\mathbf{k}\lambda}^\dagger |0\rangle \qquad ; \quad |f\rangle = a_{\mathbf{k}'\lambda'}^\dagger |0\rangle \qquad (9.1)$$

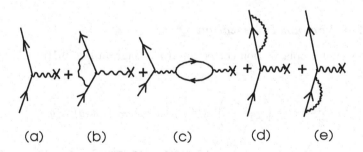

Fig. 9.1   Connected Feynman diagrams contributing to the scattering of an electron in an external field through $O(e^3)$.

9.1.1 *Feynman Diagrams*

The connected Feynman diagrams contributing through this order are shown in Fig. 9.1. We use the Feynman rules to write the momentum-space contributions of each of these diagrams.

9.1.2 *First-Order*

The first-order S-matrix element for the scattering of an electron in an external field from diagram Fig. 9.1(a) is

$$\left[e\hat{S}_{11}^{\text{ext}}\right]_{fi} = \frac{1}{\Omega}\left(\frac{-e}{\hbar c\sqrt{\varepsilon_0}}\right)\bar{u}(\mathbf{k}'\lambda')\gamma_\mu u(\mathbf{k}\lambda)a_\mu(q) \qquad ; \; q = k' - k \quad (9.2)$$

9.1.3 *Vertex Insertion*

The vertex correction from diagram Fig. 9.1(b) is, with l the virtual photon four-momentum,

$$\left[e^3\hat{S}_{13}^{\text{ext}}\right]_{fi} = \frac{1}{\Omega}\left(\frac{-e}{\hbar c\sqrt{\varepsilon_0}}\right)\bar{u}(\mathbf{k}'\lambda')\Lambda_\mu(k',k)u(\mathbf{k}\lambda)a_\mu(q) \qquad (9.3)$$

$$\Lambda_\mu(k',k) \equiv \frac{ie^2}{\hbar c\varepsilon_0}\int\frac{d^4l}{(2\pi)^4}\frac{1}{l^2}\,\gamma_\lambda\frac{1}{i(\not{k}'-\not{l})+M}\gamma_\mu\frac{1}{i(\not{k}-\not{l})+M}\gamma_\lambda$$

9.1.4 *Vacuum Polarization*

The vacuum polarization correction from diagram Fig. 9.1(c) is[1]

$$\left[e^3\hat{S}_{13}^{\text{ext}}\right]_{fi} = \frac{1}{\Omega}\left(\frac{-e}{\hbar c\sqrt{\varepsilon_0}}\right)\bar{u}(\mathbf{k}'\lambda')\gamma_\mu u(\mathbf{k}\lambda)\frac{1}{q^2}\Pi_{\mu\nu}(q)a_\nu(q) \qquad (9.4)$$

$$\Pi_{\mu\nu}(q) \equiv \frac{e^2}{i\hbar c\varepsilon_0}\int\frac{d^4k}{(2\pi)^4}\text{Tr}\,\frac{1}{i(\not{k}-\not{q}/2)+M}\gamma_\mu\frac{1}{i(\not{k}+\not{q}/2)+M}\gamma_\nu$$

---

[1]Compare Eqs. (8.48) and (G.29).

9.1.5 Self-Energy Insertions

The sum of the contributions from the self-energy insertions in diagrams Fig. 9.1(d) and (e) is[2]

$$\left[e^3 \hat{S}_{13}^{\text{ext}}\right]_{fi} = \frac{1}{\Omega} \left(\frac{-e}{\hbar c \sqrt{\varepsilon_0}}\right) \bar{u}(\mathbf{k}'\lambda') \times$$

$$\left[-\gamma_\mu \frac{1}{i\not{k} + M} \Sigma(k) - \Sigma(k') \frac{1}{i\not{k}' + M} \gamma_\mu\right] u(\mathbf{k}\lambda) a_\mu(q)$$

$$\Sigma(k) \equiv \frac{e^2}{i\hbar c \varepsilon_0} \int \frac{d^4 l}{(2\pi)^4} \frac{1}{l^2} \gamma_\lambda \frac{1}{i(\not{k} - \not{l}) + M} \gamma_\lambda \qquad (9.5)$$

So far we have not taken into account *mass renormalization* in QED. As discussed at the end of chapter 7, the effect is to add a term

$$\hat{H}_I^{\text{mass}}(t) = -\hbar c \, \delta M \int d^3 x : \hat{\bar{\psi}}(x) \hat{\psi}(x) : \qquad (9.6)$$

to the interaction in Eq. (8.116), where it is assumed that the mass shift δM is a power series in e^2. We leave it as a problem to show that the modification of the first of Eqs. (9.5) is then[3]

$$\left[e^3 \hat{S}_{13}^{\text{ext}}\right]_{fi} = \frac{1}{\Omega} \left(\frac{-e}{\hbar c \sqrt{\varepsilon_0}}\right) \bar{u}(\mathbf{k}'\lambda') \times \qquad (9.7)$$

$$\left\{\gamma_\mu \frac{1}{i\not{k} + M} [\delta M - \Sigma(k)] + [\delta M - \Sigma(k')] \frac{1}{i\not{k}' + M} \gamma_\mu\right\} u(\mathbf{k}\lambda) a_\mu(q)$$

We shall subsequently discuss each of the above contributions in some detail, as they form the basic building blocks of the theory. We first note the following:

- These third-order corrections are obtained by *insertion* of the second-order self-energies and vertex correction into the first-order diagram;
- Although simple power counting shows the integrals in these correction terms are ill-defined as they stand, we shall assume that they are *defined with dimensional regularization* as discussed in appendix G; hence, we are *free to manipulate these expressions algebraically*;
- As one immediate consequence, there is a powerful and useful relation between the vertex and electron self-energy known as *Ward's identity*

---

[2] Compare with Eq. (8.47). Note that both $\Sigma(k)$ and $\Lambda_\mu(k', k)$ are 4×4 Dirac matrices.

[3] See Prob. 9.1; as with the current, we assume that this mass counter-term is normal-ordered. The contribution of this term can be included diagrammatically by augmenting Fig. 9.1(d,e) with two diagrams with crosses on the external legs—see Fig. 9.2.

$$\frac{\partial \Sigma(k)}{\partial k_\mu} = i\Lambda_\mu(k, k) \qquad ; \text{ Ward's identity} \qquad (9.8)$$

9.2 Ward's Identity

In order to prove the identity, it is necessary to use the following relation to differentiate the inverse of an operator (in this case, merely a matrix)

$$\frac{\partial}{\partial x}\frac{1}{\Omega(x)} = -\frac{1}{\Omega(x)}\frac{\partial \Omega(x)}{\partial x}\frac{1}{\Omega(x)} \qquad (9.9)$$

To establish this relation, start from $\Omega(x)\Omega(x)^{-1} = 1$, so that

$$\frac{\partial}{\partial x}\Omega(x)\Omega(x)^{-1} = 0$$

$$= \Omega(x)\frac{\partial}{\partial x}\Omega(x)^{-1} + \frac{\partial \Omega(x)}{\partial x}\Omega(x)^{-1} \qquad (9.10)$$

Multiplication of this relation on the left by $\Omega(x)^{-1}$ then gives Eq. (9.9).

Now identify

$$\Omega(k) = i(\not{k} - \not{l}) + M \qquad ; \frac{1}{\Omega(k)} = \frac{1}{i(\not{k} - \not{l}) + M} \qquad (9.11)$$

Hence

$$\frac{\partial}{\partial k_\mu}\frac{1}{i(\not{k} - \not{l}) + M} = -\frac{1}{i(\not{k} - \not{l}) + M}i\gamma_\mu\frac{1}{i(\not{k} - \not{l}) + M} \qquad (9.12)$$

The derivative of $\Sigma(k)$ in Eq. (9.5) then gives

$$\frac{\partial \Sigma(k)}{\partial k_\mu} = i\Lambda_\mu(k, k) \qquad (9.13)$$

where $\Lambda_\mu(k, k)$ is identified from Eq. (9.3). This is Ward's identity.

9.3 Electron Self-Energy

Let us begin the study of the various components with an examination of the electron self-energy $\Sigma(k)$ defined in Eq. (9.5).

9.3.1   *General Form*

First, some general comments on integrals of this type. In order to have an integral that converges for large l, from simple power counting, it is necessary to have at least five powers of l in the denominator.[4] In the present case, we have three. Suppose we write the integral as

$$\Sigma(k) = \int d^4l \, R(k, l) \tag{9.14}$$

Then the following two *subtractions of the integrand* lead to a convergent integral

$$R(k, l) - R(0, l) - k_\mu \left[\frac{\partial}{\partial k_\mu} R(k, l) \right]_{k=0} = R_c(k, l)$$

$$\int d^4l \, R_c(k, l) < \infty \qquad \text{; convergent} \quad (9.15)$$

This is easy to see, since for large l,

$$\frac{1}{i(\not{k} - \not{l}) + M} - \frac{1}{i(-\not{l}) + M} + \frac{1}{i(-\not{l}) + M} i\not{k} \frac{1}{i(-\not{l}) + M} = O\left(\frac{1}{l^3}\right) \quad ;$$
$$l \to \infty \tag{9.16}$$

Integration of Eq. (9.15) then expresses $\Sigma(k)$ as

$$\Sigma(k) = I_0 + k_\mu I_{1\mu} + I_{2c}(k) \tag{9.17}$$

Some comments:[5]

- With dimensional regularization, $(I_0, I_{1\mu})$ are finite constants independent of k, which diverge in the limit $n \to 4$;
- By covariance $I_{1\mu} = I_1 \gamma_\mu$, since γ_μ is the only four-vector around;
- For $n = 4$, the last term in Eq. (9.17) is a finite function of k;
- We claim that $\Sigma(k)$ can now be recast in the form

$$\Sigma(k) = A + (i\not{k} + M)B + (i\not{k} + M)\Sigma_f(k)(i\not{k} + M) \tag{9.18}$$

where (A, B) share the properties of (I_0, I_1), and $\Sigma_f(k)$ is a finite matrix function of k for $n = 4$.

---

[4]The $\int dl/l$ diverges logarithmically, while the $\int dl/l^2$ converges.
[5]Recall that $\Sigma(k)$ is a 4×4 Dirac matrix, and the unit Dirac matrix is suppressed.

Let us verify this last assertion. The result holds immediately for the first two terms in Eq. (9.17). The problem is to find a construction to rewrite the last term Eq. (9.17) in this fashion

$$I_{2c}(k) = \mathcal{I}_0(-M^2) + (i\not{k} + M)\mathcal{I}_1(-M^2) + (i\not{k} + M)\mathcal{I}_2(k)(i\not{k} + M) \quad (9.19)$$

This is accomplished through the following steps:

(1) Since the only Lorentz scalars around are k^2 and $\not{k}$, and since $\not{k}\not{k} = k^2$, one has by covariance

$$I_{2c}(k) = (i\not{k} + M)J_1(k^2) + J_2(k^2) \quad (9.20)$$

(2) Write

$$\begin{aligned} J(k^2) &\equiv J(k^2) - J(-M^2) + J(-M^2) \\ &\equiv (k^2 + M^2)L(k^2) + J(-M^2) \end{aligned} \quad (9.21)$$

(3) Use

$$k^2 + M^2 \equiv 2M(i\not{k} + M) - (i\not{k} + M)^2 \quad (9.22)$$

(4) Repeat.

This establishes Eq. (9.18).

9.3.2   *Evaluation*

Let us then proceed to evaluate $\Sigma(k)$ assuming the dimensional regularization in Minkowski space of appendix G

$$\Sigma(k) \equiv \frac{e^2}{i\hbar c \varepsilon_0} \int \frac{d^n l}{(2\pi)^4} \frac{1}{l^2} \gamma_\lambda \frac{1}{i(\not{k} - \not{l}) + M} \gamma_\lambda \quad (9.23)$$

Rationalization of the denominator, and the gamma-matrix algebra, give

$$\begin{aligned} \Sigma(k) &= \frac{i\alpha}{4\pi^3} \int \frac{d^n l}{l^2} \gamma_\lambda \frac{i(\not{k} - \not{l}) - M}{(k - l)^2 + M^2} \gamma_\lambda \\ &= \frac{i\alpha}{4\pi^3} \int \frac{d^n l}{l^2} \frac{1}{(k - l)^2 + M^2} [(2 - n)i(\not{k} - \not{l}) - nM] \end{aligned} \quad (9.24)$$

where we use $\gamma_\lambda \gamma_\lambda = n$. Here $\alpha = e^2/4\pi\varepsilon_0 \hbar c$ is the fine-structure constant.

We now want to carry out the angular integrals, and the term in $k \cdot l$ in the denominator complicates matters. The following simple *Feynman parameterization* allows us to complete the square in the denominator and immediately perform those angular integrals[6]

$$\frac{1}{ab} = \int_0^1 \frac{dx}{[ax + b(1-x)]^2} \qquad ; \text{ Feynman parameterization} \quad (9.25)$$

Hence

$$\Sigma(k) = \frac{i\alpha}{4\pi^3} \int_0^1 dx \int d^n l \frac{1}{\{[(k-l)^2 + M^2]x + l^2(1-x)\}^2} \times$$
$$[(2-n)i(\not{k} - \not{l}) - nM] \qquad (9.26)$$

Write the denominator as

$$[(k-l)^2 + M^2]x + l^2(1-x) = l^2 - 2k \cdot lx + (k^2 + M^2)x \qquad (9.27)$$
$$= (l - kx)^2 + (k^2 + M^2)x(1-x) + M^2x^2$$

and then change variables in the integral from l to $t = l - kx$ with $d^n l = d^n t$

$$\Sigma(k) = \frac{i\alpha}{4\pi^3} \int_0^1 dx \int \frac{d^n t}{[t^2 + a^2 - i\eta]^2} \{(2-n)i[\not{k}(1-x) - \not{t}] - nM\} \quad ;$$
$$a^2 \equiv (k^2 + M^2)x(1-x) + M^2x^2 \qquad (9.28)$$

Note that since the coefficient of M^2 in the denominator is positive, the prescription $M \to M - i\eta$ gives us the expected Feynman singularities in the integral. This integral can now be evaluated using the results in appendix G.

We first have by symmetric integration

$$\int \frac{d^n t}{[t^2 + a^2 - i\eta]^2} t_\mu = 0 \qquad ; \text{ symmetric integration} \quad (9.29)$$

Then from Eq. (G.24)

$$\int \frac{d^n t}{[t^2 + a^2 - i\eta]^2} = \frac{i}{a^4} \frac{n(\pi a^2)^{n/2}}{\Gamma(1 + n/2)} \frac{\Gamma(2 - n/2)\Gamma(n/2)}{2\Gamma(2)}$$
$$= \frac{i}{a^4} (\pi a^2)^{n/2} \Gamma(2 - n/2) \qquad (9.30)$$

Thus from Eq. (9.28), one has the general result

$$\Sigma(k) = -\frac{\alpha}{4\pi^3} \Gamma(2 - n/2) \int_0^1 dx \frac{(\pi a^2)^{n/2}}{a^4} [(2-n)i\not{k}(1-x) - nM] \quad ;$$
$$a^2 \equiv (k^2 + M^2)x(1-x) + M^2x^2 \qquad (9.31)$$

---

[6]See Prob. 9.2.

The first term A in Eq. (9.18) can be isolated by taking diagonal matrix elements of $\Sigma(k)$ between Dirac spinors $u(k\lambda)$, where $(i\not{k} + M)u(k\lambda) = 0$ and $k^2 + M^2 = 0$. Thus

$$A = \frac{\alpha M}{4\pi^3}\Gamma(2 - n/2)\int_0^1 dx \frac{(\pi a^2)^{n/2}}{a^4}[2(1 - x) + nx] \;\; ; \; a^2 \equiv M^2 x^2 \quad (9.32)$$

The singularity as $n \to 4$ arises from the factor of $\Gamma(2-n/2)$ in front. Write $n = 4 - \epsilon$ and use $\Gamma(\epsilon/2) = (2/\epsilon)\Gamma(1 + \epsilon/2)$. The singularity as $\epsilon \to 0$ is then isolated as[7]

$$\begin{aligned}
A &= \frac{\alpha M}{\pi\epsilon}\int_0^1 dx(1 + x) \\
&= \frac{3\alpha}{2\pi\epsilon}M \qquad\qquad ; \; n = 4 - \epsilon \to 4 \qquad (9.33)
\end{aligned}$$

9.3.3 *Mass Renormalization*

It is evident from Eq. (9.7) that with mass renormalization, the self-energy insertion always enters in the combination

$$\Sigma(k) - \delta M = A - \delta M + (i\not{k} + M)B + (i\not{k} + M)\Sigma_f(k)(i\not{k} + M) \quad (9.34)$$

Now *choose*

$$\delta M = A \qquad\qquad ; \; \text{mass renormalization} \qquad (9.35)$$

This implies

$$\Sigma(k) - \delta M = (i\not{k} + M)B + (i\not{k} + M)\Sigma_f(k)(i\not{k} + M) \qquad (9.36)$$

As a consequence of mass renormalization, *the constant A, with its singularity at $n = 4$, is eliminated from the theory*. The constant B also has within it a singularity as $n \to 4$; however, we postpone a discussion of its role until after we have discussed the vertex, for we recall that the electron self-energy and vertex are related by Ward's identity.

Although it is *completely unnecessary* to do so at this point, we investigate the divergence in A in more detail to get some insight as to just where things are breaking down. From Eq. (9.28), the constant A can be written as

$$A = -\frac{i\alpha M}{4\pi^3}\int_0^1 dx \int \frac{d^n t}{[t^2 + a^2 - i\eta]^2}[2(1 - x) + nx] \;\; ; \; a^2 = M^2 x^2 \quad (9.37)$$

---

[7]There are additional finite contributions as $\epsilon \to 0$, obtained from a consistent expansion in ϵ (see Prob. 9.3).

Let us again rotate the contour from the real axis to the imaginary axis so that the integral is in euclidean space.[8] Simple power counting now shows that the resulting integral is logarithmically divergent at large t in four dimensions. We note that the superficial *linear* divergence of the self-energy has disappeared due to the symmetric integration in Eq. (9.29). This logarithmic divergence can be investigated by just cutting off the momentum integral at some large value Λ. The remaining radial measure in euclidian space in $n = 4$ dimensions, after doing all the angular integrals, is $2\pi^2 t^3 dt$ [see Eqs. (G.8) and (G.12)]. Thus the cut-off integral in euclidean space is

$$A = M\frac{\alpha}{\pi} \int_0^1 (1+x)dx \int^{\Lambda} \frac{t^3 dt}{(t^2+a^2)^2} \tag{9.38}$$

As $\Lambda \to \infty$, the a^2 can be neglected, and hence[9]

$$A = M\frac{3\alpha}{2\pi} \ln \frac{\Lambda}{M} \qquad ; \Lambda \to \infty \tag{9.39}$$

The mass shift now follows from Eq. (9.35) as

$$\frac{\delta M}{M} = \frac{3\alpha}{2\pi} \ln \frac{\Lambda}{M} \qquad ; \text{ mass renormalization} \tag{9.40}$$

Let us discuss this result. The classical electron radius r_0 is defined by equating the Coulomb energy $e^2/4\pi\varepsilon_0 r_0$ with the rest energy of the electron

$$m_e c^2 = \frac{e^2}{4\pi\varepsilon_0 r_0} \qquad ; \text{ classical electron radius} \tag{9.41}$$

This classical electromagnetic mass diverges *linearly* with r_0 as $r_0 \to 0$. In QED, the mass diverges only *logarithmically* with the cut-off $r_c \equiv 1/\Lambda$. As pointed out, with dimensional regularization, this follows immediately from the symmetric integration in Eq. (9.29). The result that the divergence is only logarithmic in QED is due to Weisskopf (1934).

Let us put in some numbers for the electron where $M = m_e c/\hbar$:

(1) For a cut-off of the nucleon mass $\Lambda = M_p$

$$\frac{\delta M}{M} = 2.6 \times 10^{-2} \qquad ; \Lambda = M_p \tag{9.42}$$

This is a very small correction.

---

[8] See appendix G; then $\int_{-i\infty}^{i\infty} dt_0 = i \int_{-\infty}^{\infty} d\bar{t}_0$ where $t_0 = i\bar{t}_0$.

[9] We have used M in the denominator for dimensional reasons, but it does not really matter. Since $\ln(\Lambda/M) = \ln\Lambda - \ln M$, one has $\ln(\Lambda/M) \to \ln\Lambda$ for any M, as $\Lambda \to \infty$.

(2) For $\delta M = M$, so that all the mass is electromagnetic,

$$\frac{\Lambda}{M} = e^{2\pi/3\alpha} = e^{287} \qquad\qquad ; \; \delta M = M$$

$$\frac{\hbar\Lambda}{c} \equiv m_\Lambda \approx 10^{125} m_e \approx 10^{98} \, \text{gm} \qquad (9.43)$$

For comparison, the mass of the sun is $m_\odot = 2 \times 10^{33}$ gm.

This logarithmic divergence is incredibly weak. There is a whole physical universe that must be included before one reaches this scale![10] There is no reason to expect that the electron-photon system of QED can be treated in an isolated subspace. Indeed, it is truly remarkable that one can get so far by assuming this to be the case!

9.4   Vertex

We next turn to the vertex in Eq. (9.3), where the integral is again defined through dimensional regularization

$$\Lambda_\mu(k',k) = \frac{ie^2}{\hbar c \varepsilon_0} \int \frac{d^n l}{(2\pi)^4} \frac{1}{l^2} \gamma_\lambda \frac{1}{i(\not{k}' - \not{l}) + M} \gamma_\mu \frac{1}{i(\not{k} - \not{l}) + M} \gamma_\lambda \quad (9.44)$$

9.4.1   General Form

Simple power counting shows the integral is logarithmically divergent in four dimensions. Write the integral as

$$\Lambda_\mu(k',k) = \int d^4 l \, R_\mu(k',k,l) \qquad (9.45)$$

Then, as in Eq. (9.15), a single subtraction of the integrand will lead to a convergent integral when $n = 4$

$$R_\mu(k',k,l) - R_\mu(0,0,l) = R_{c\mu}(k',k,l) \qquad (9.46)$$

Integration of this relation expresses the vertex as

$$\Lambda_\mu(k',k) = L' \gamma_\mu + \Lambda'_{c\mu}(k',k) \qquad (9.47)$$

where L' is a constant independent of (k',k). If one takes the diagonal matrix element of the convergent piece between Dirac spinors, the result

---

[10]The above gives $1/\Lambda \approx 10^{-125} (\hbar/m_e c) \approx 4 \times 10^{-136}$ cm.

must be of the same form as the first term on the r.h.s.[11]

$$\bar{u}(k)\Lambda'_{c\mu}(k,k)u(k) = \bar{u}(k)L''\gamma_\mu u(k) \qquad (9.48)$$

Addition and subtraction of $L''\gamma_\mu$, then expresses the vertex in the general form

$$\Lambda_\mu(k',k) = L\gamma_\mu + \Lambda_{c\mu}(k',k) \qquad ; \text{ general form}$$
$$\bar{u}(k)\Lambda_{c\mu}(k,k)u(k) = 0 \qquad (9.49)$$

With dimensional regularization, L is a finite constant, independent of (k',k), which diverges as $n \to 4$.

9.4.2 *Ward's Identity*

We are now in a position to implement Ward's identity in Eq. (9.8), which relates the electron self-energy and the vertex. Substitute the general form of the electron self-energy in Eq. (9.36) into the l.h.s. of Eq. (9.8) to give

$$\frac{\partial \Sigma(k)}{\partial k_\mu} = iB\gamma_\mu + i\gamma_\mu \Sigma_f(k)(i\not{k} + M) + (i\not{k} + M)\Sigma_f(k)i\gamma_\mu +$$
$$(i\not{k} + M)\frac{\partial \Sigma_f(k)}{\partial k_\mu}(i\not{k} + M) \qquad (9.50)$$

Now take the diagonal matrix element of this relation between Dirac spinors[12]

$$\bar{u}(k)\frac{\partial \Sigma(k)}{\partial k_\mu}u(k) = iB\bar{u}(k)\gamma_\mu u(k) \qquad (9.51)$$

Take this same matrix element of the r.h.s. of Eq. (9.8), using Eqs. (9.49)

$$i\bar{u}(k)\Lambda_\mu(k,k)u(k) = iL\bar{u}(k)\gamma_\mu u(k) \qquad (9.52)$$

Hence, with our particular choice of the general forms for the electron self-energy and vertex, Ward's identity implies that

$$B = L \qquad ; \text{ Ward's identity} \qquad (9.53)$$

This relation has significant implications for QED, as we shall see.

---

[11] See Prob. 9.4.

[12] Recall the Dirac equation is $(i\not{k} + M)u(k) = \bar{u}(k)(i\not{k} + M) = 0$.

9.4.3 *Evaluation*

Let us proceed to the evaluation of the vertex in Eq. (9.44). Obtaining the general expression involves substantial algebra, none of which is new to us, and we shall be content here to evaluate the *matrix element* $\bar{u}(k')\Lambda_\mu(k',k)u(k)$ so that the Dirac equation can be used on both sides, together with the mass-shell values $k'^2 = k^2 = -M^2$.

Rationalization of the denominators in Eq. (9.44) gives

$$\Lambda_\mu(k',k) = \frac{i\alpha}{4\pi^3} \int \frac{d^n l}{l^2} \gamma_\lambda \frac{i(\not{k'} - \not{l}) - M}{(k' - l)^2 + M^2} \gamma_\mu \frac{i(\not{k} - \not{l}) - M}{(k - l)^2 + M^2} \gamma_\lambda \quad (9.54)$$

We want to again complete the square in the denominator so that the angular integrals can be performed. For this, we use the Feynman parameterization for the product of three factors (see Prob. 9.2)

$$\frac{1}{abc} = 2! \int_0^1 dx \int_0^x dy \frac{1}{[ay + b(x - y) + c(1 - x)]^3} \quad ;$$

Feynman parameterization   (9.55)

With $k'^2 = k^2 = -M^2$, the denominators in the matrix element of $\Lambda_\mu(k',k)$ are then combined into

$$(l^2 - 2l \cdot k')y + (l^2 - 2l \cdot k)(x - y) + l^2(1 - x) = l^2 - 2l \cdot kx - 2l \cdot qy \quad (9.56)$$

where $k' = k + q$. Now complete the square in this denominator

$$l^2 - 2l \cdot k\,x - 2l \cdot q\,y = (l - qy - kx)^2 - q^2 y^2 - 2q \cdot k\,xy + M^2 x^2 \quad (9.57)$$

Use the following kinematics [remember these are n-vectors (k'_μ, k_μ)]

$$q^2 = (k' - k)^2 = -2M^2 - 2k' \cdot k \qquad\qquad ; q = k' - k$$
$$(k' + k) \cdot q = 0$$
$$k' \cdot q = -k \cdot q = \frac{1}{2}q^2 \quad (9.58)$$

Between Dirac spinors, the vertex in Eq. (9.54) then takes the form

$$\bar{u}(k')\Lambda_\mu(k',k)u(k) = \frac{i\alpha}{2\pi^3} \int_0^1 dx \int_0^x dy \int \frac{d^n l}{(t^2 + a^2 - i\eta)^3} N_\mu \quad (9.59)$$
$$N_\mu = \bar{u}(k') \gamma_\lambda [i(\not{k'} - \not{l}) - M]\gamma_\mu[i(\not{k} - \not{l}) - M]\gamma_\lambda\, u(k)$$

where

$$t = l - qy - kx \qquad ; a^2 = q^2 y(x - y) + M^2 x^2 \quad (9.60)$$

Here we have noted that the prescription $M \to M - i\eta$ again produces the appropriate Feynman singularities in the integral.

Now change variables from l to t, with $d^n l = d^n t$, and use the fact that any term linear in t goes out by symmetric integration. This gives

$$N_\mu = \bar{u}(k')\{\gamma_\lambda[i(\not{k}' - \not{q}y - \not{k}x) - M]\gamma_\mu[i(\not{k} - \not{q}y - \not{k}x) - M]\gamma_\lambda - \gamma_\lambda \not{t} \gamma_\mu \not{t} \gamma_\lambda\}u(k) \qquad (9.61)$$

Symmetric integration on the last term gives

$$\gamma_\lambda \not{t} \gamma_\mu \not{t} \gamma_\lambda \to \frac{t^2}{n}\gamma_\lambda\gamma_\rho\gamma_\mu\gamma_\rho\gamma_\lambda = \frac{(2-n)^2}{n}t^2\gamma_\mu \qquad (9.62)$$

From power counting, it is apparent that this is the term that leads to the singularity at $n = 4$ in the integral in Eq. (9.59); the rest of N_μ yields a finite expression there. Hence, we shall evaluate everything else in N_μ at $n = 4$. One can then use the following γ-matrix algebra on the rest of N_μ[13]

$$\gamma_\mu \not{d} \gamma_\mu = -2\not{d}$$
$$\gamma_\mu \not{d}\not{b} \gamma_\mu = 2(\not{d}\not{b} + \not{b}\not{d}) = 4a \cdot b$$
$$\gamma_\mu \not{d}\not{b}\not{c}\gamma_\mu = -2\not{c}\not{b}\not{d} \qquad (9.63)$$

With the subsequent use of $k' = k + q$ to get all the $\not{k}$ to the right and all the $\not{k}'$ to the left, this yields

$$N_\mu = \bar{u}(k')\Big(- 2[i\not{k}'(1-x) - i\not{q}(1+y-x)]\gamma_\mu[i\not{k}(1-x) + i\not{q}(1-y)] -$$
$$2M\{\gamma_\mu[i\not{k}(1-x) - i\not{q}y] + [i\not{k}'(1-x) - i\not{q}(1+y-x)]\gamma_\mu\} -$$
$$2M\{\gamma_\mu[i\not{k}(1-x) + i\not{q}(1-y)] + [i\not{k}'(1-x) + i\not{q}(x-y)]\gamma_\mu\} -$$
$$2M^2\gamma_\mu - (2-n)^2 n^{-1} t^2\gamma_\mu \Big)u(k) \qquad (9.64)$$

The Dirac equation can now be used to the right and the left to replace $(i\not{k}, i\not{k}')$ by $-M$. This gives

$$N_\mu = \bar{u}(k')\Big(- 2[-M(1-x) - i\not{q}(1+y-x)]\gamma_\mu[-M(1-x) + i\not{q}(1-y)] -$$
$$2M\{\gamma_\mu[-M(1-x) - i\not{q}y] + [-M(1-x) - i\not{q}(1+y-x)]\gamma_\mu\} -$$
$$2M\{\gamma_\mu[-M(1-x) + i\not{q}(1-y)] + [-M(1-x) + i\not{q}(x-y)]\gamma_\mu\} -$$
$$2M^2\gamma_\mu - (2-n)^2 n^{-1} t^2\gamma_\mu \Big)u(k) \qquad (9.65)$$

---

[13]See Prob. 8.3.

Introduce

$$\gamma_\mu \gamma_\nu \equiv \frac{1}{2}\{\gamma_\mu, \gamma_\nu\} + \frac{1}{2}[\gamma_\mu, \gamma_\nu]$$
$$= \delta_{\mu\nu} + i\sigma_{\mu\nu}$$
$$\gamma_\nu \gamma_\mu = \delta_{\mu\nu} - i\sigma_{\mu\nu} \tag{9.66}$$

and use the fact that $\bar{u}(k')\not{q}u(k) = 0$. A collection of terms then gives

$$N_\mu = \gamma_\mu[-(2-n)^2 n^{-1} t^2 + 4M^2(1 - x - x^2/2)] + 2M\sigma_{\mu\nu}q_\nu x(1-x) +$$
$$2q^2\gamma_\mu(1 + y - x)(1 - y) - 2iMq_\mu(1+x)(x - 2y) \tag{9.67}$$

The integral over $d^n t$ of the remaining terms now follows from Eq. (G.24)

$$\bar{u}(k')\Lambda_\mu(k', k)u(k) = -\frac{\alpha}{2\pi^3}\int_0^1 dx \int_0^x dy \frac{n(\pi a^2)^{n/2}}{a^6}\bar{u}(k')\left\{\frac{\Gamma(3 - n/2)}{(n/2)2\Gamma(3)} \times\right.$$
$$[4M^2\gamma_\mu(1 - x - x^2/2) + 2M\sigma_{\mu\nu}q_\nu\, x(1-x) + 2q^2\gamma_\mu(1 + y - x)(1 - y) -$$
$$\left. 2iMq_\mu(1+x)(x - 2y)] - a^2\gamma_\mu\frac{(2-n)^2}{n}\frac{\Gamma(2 - n/2)}{2\Gamma(3)}\right\}u(k) \qquad ;$$
$$a^2 = q^2 y(x - y) + M^2 x^2 \tag{9.68}$$

9.4.4 *The Constant L*

The constant L follows from Eq. (9.49) as

$$\bar{u}(k)\Lambda_\mu(k, k)u(k) = \bar{u}(k)L\gamma_\mu u(k) \qquad ; q = 0 \tag{9.69}$$

Thus

$$L = -\frac{\alpha}{2\pi^3}\int_0^1 dx \int_0^x dy \frac{n(\pi a_0^2)^{n/2}}{a_0^6}\left\{\frac{\Gamma(3 - n/2)}{(n/2)2\Gamma(3)} \times\right.$$
$$\left. [4M^2\gamma_\mu(1 - x - x^2/2)] - a_0^2\frac{(2-n)^2}{n}\frac{\Gamma(2 - n/2)}{2\Gamma(3)}\right\} \qquad ; a_0^2 = M^2 x^2 \tag{9.70}$$

With dimensional regularization, L is a finite constant with an isolated singularity at $n = 4$. The singularity at $n = 4$ comes from the factor $\Gamma(2 - n/2)$ in the last term. With the definition $n = 4 - \epsilon$ and the use of $\Gamma(\epsilon/2) = (2/\epsilon)\Gamma(1 + \epsilon/2)$, that singularity is identified as

$$L = \frac{\alpha}{2\pi}\frac{2}{\epsilon}\int_0^1 dx \int_0^x dy = \frac{\alpha}{2\pi\epsilon} \qquad ; n = 4 - \epsilon \to 4 \tag{9.71}$$

In parallel with the previous discussion, simply cutting off the momentum integral at some high value Λ would give

$$L = \frac{\alpha}{2\pi} \ln \frac{\Lambda}{M} \qquad (9.72)$$

and we again note the weakness of this logarithmic divergence.

If $\bar{u}(k')L\gamma_\mu u(k)$ is now subtracted from Eq. (9.68), then by Eq. (9.49) the resulting expression is convergent as $n \to 4$, and that limit can then be taken. We have to be a little careful in treating the singularity at $n = 4$ in the difference, since

$$a^{-\epsilon}\Gamma(\epsilon/2) = \frac{2}{\epsilon}\Gamma(1 + \epsilon/2)e^{-\epsilon \ln a} = \frac{2}{\epsilon} + \Gamma'(1) - 2\ln a + O(\epsilon) \quad (9.73)$$

Hence

$$\bar{u}(k')\Lambda_{c\mu}(k', k)u(k) = -\frac{\alpha}{4\pi}\int_0^1 dx \int_0^x dy\, \bar{u}(k')\left\{\frac{1}{a^2}\left[\left(1 - \frac{a^2}{a_0^2}\right) \times\right.\right.$$
$$4M^2\gamma_\mu(1 - x - x^2/2) + 2M\sigma_{\mu\nu}q_\nu\, x(1 - x) + 2q^2\gamma_\mu(1 + y - x)(1 - y) -$$
$$\left.2iMq_\mu(1 + x)(x - 2y)\right] + 2\gamma_\mu \ln \frac{a^2}{a_0^2}\left.\right\}u(k) \qquad ;$$
$$a^2 = q^2y(x - y) + M^2x^2 \qquad (9.74)$$

Furthermore, the term in q_μ does not contribute, since a change of variables to $u = y(x - y)$ with $du = (x - 2y)dy$ gives

$$\int_0^x \frac{(x - 2y)dy}{q^2y(x - y) + M^2x^2} = \int_0^0 \frac{du}{q^2u + M^2x^2} \equiv 0 \qquad (9.75)$$

Thus the convergent part of the vertex, finite at $n = 4$, is given by

$$\bar{u}(k')\Lambda_{c\mu}(k', k)u(k) = \frac{-\alpha}{2\pi}\int_0^1 dx \int_0^x dy\, \bar{u}(k')\left\{\frac{1}{a^2}\left[q^2\gamma_\mu(1 + y - x)(1 - y)\right.\right.$$
$$\left.+M\sigma_{\mu\nu}q_\nu\, x(1 - x) + \left(1 - \frac{a^2}{a_0^2}\right)2M^2\gamma_\mu\left(1 - x - \frac{x^2}{2}\right)\right] + \gamma_\mu \ln \frac{a^2}{a_0^2}\left.\right\}u(k) \quad ;$$
$$a^2 = q^2y(x - y) + M^2x^2 \qquad ; \ a_0^2 = M^2x^2 \quad (9.76)$$

9.4.5 *The Infrared Problem*

There is a remaining difficulty that we have not yet addressed. A little study of the expression in Eq. (9.76) indicates that in some cases the final parametric integrals $\int_0^1 dx$ *diverge as* $x \to 0$. It is easy to see the source of this problem. If the photon is given a tiny mass M_γ, so that the photon propagator becomes $D_{\mu\nu}^F(l) = \delta_{\mu\nu}/(l^2 + M_\gamma^2)$, then from Eq. (9.56) the result is to replace $a^2 \to a^2 + M_\gamma^2(1 - x)$ in Eq. (9.60)

$$D_{\mu\nu}^F(l) \to \frac{\delta_{\mu\nu}}{l^2 + M_\gamma^2} \qquad ; \; M_\gamma \text{ is "photon mass"}$$

$$a^2 \to a^2 + M_\gamma^2(1 - x) \tag{9.77}$$

and similarly for a_0^2. All the denominators are then *protected* as $x \to 0$. The source of this *infrared problem* is our attempt to treat radiation at very long wavelength photon-by-photon. This can be done, as first shown by Schwinger [Schwinger (1958)]; one just has to make sure that the fictitious photon mass disappears from any physical result (see the discussion of radiative corrections in [Bjorken and Drell (1964); Bjorken and Drell (1965)]).[14]

The infrared problem is really a *technical* problem,[15] since at very long wavelength, the radiation is really that coming from a specified current distribution, and, in many cases, that problem can be solved exactly with a canonical transformation [Bloch and Nordsieck (1937)]. This solution is presented in the last chapter of this book, and we leave any further discussion of the infrared problem until we get there.

9.4.6 *Schwinger Moment*

The general form of the convergent part of the vertex in Eq. (9.76) is

$$\bar{u}(k')\Lambda_{c\mu}(k',k)u(k) = \bar{u}(k')\left[F_E^{(2)}(q^2)\gamma_\mu - \frac{1}{2M}\sigma_{\mu\nu}q_\nu\, F_M^{(2)}(q^2)\right]u(k) \tag{9.78}$$

where $F_{E,M}^{(2)}(q^2)$ are the contributions of $O(e^2)$ to the *electric and magnetic form factors*. The further development of $F_E^{(2)}(q^2)$ requires the evaluation of several well-defined integrals, and we will pursue that no further. The final expression for $F_E^{(2)}(q^2)$ can always be found in any good book on QED, for example [Bjorken and Drell (1964); Bjorken and Drell (1965)]. In contrast, the result for the *magnetic* form factor, which involves neither

---

[14]See also [Walecka (2001)]. The rules for converting from the metric used by Bjorken and Drell are given in appendix I.

[15]This does not imply that it is always easy!

ultraviolet nor infrared divergences, is given quite simply in Eq. (9.76) as

$$F_M^{(2)}(q^2) = \frac{\alpha}{2\pi} \left[2M^2 \int_0^1 dx \int_0^x dy \frac{x(1-x)}{M^2 x^2 + q^2 y(x-y)} \right]$$

$$F_M^{(2)}(0) = \frac{\alpha}{2\pi} \tag{9.79}$$

As we shall now demonstrate, the quantity $F_M^{(2)}(0)$ is the $O(\alpha)$ contribution to the *anomalous magnetic moment of the electron* in QED. This term was first calculated by Schwinger [Schwinger (1958)], and was discussed in Vol. I.

Suppose one adds to the Dirac lagrangian density an additional effective Lorentz-invariant, gauge invariant, anomalous *Pauli-moment* interaction of the electron with an external electromagnetic field

$$\mathcal{L}_P(x) = \left(\frac{\lambda e}{2M} \right) \bar{\psi}(x) \frac{1}{2} \sigma_{\mu\nu} F_{\mu\nu}^{\text{ext}} \psi(x) \qquad ; \text{Pauli moment} \quad (9.80)$$

The result is to add the following term to the quantized Dirac hamiltonian in the interaction picture

$$\hat{H}_P = -\frac{e\lambda}{2M} \int d^3x \, \hat{\bar{\psi}}(x) \frac{1}{2} \sigma_{\mu\nu} F_{\mu\nu}^{\text{ext}} \hat{\psi}(x)$$

$$\frac{1}{c} F_{\mu\nu}^{\text{ext}} = \frac{\partial A_\nu^{\text{ext}}}{\partial x_\mu} - \frac{\partial A_\mu^{\text{ext}}}{\partial x_\nu} \tag{9.81}$$

As justification, observe that this has the correct non-relativistic limit

$$\hat{H}_P = \int d^3x \, \hat{\psi}^\dagger(x) \left(-\frac{e\hbar\lambda}{2m_e} \boldsymbol{\sigma} \cdot \mathbf{B}^{\text{ext}} \right) \hat{\psi}(x) \qquad ; \text{NRL} \quad (9.82)$$

The magnetic moment of a particle is defined by doing a scattering experiment with $q^2 \to 0$, in a *weak* magnetic field, so that A_μ^{ext} can be treated in lowest order. The contribution of the interaction in Eq. (9.81) to the S-matrix is then

$$S_{fi}^P = \left(\frac{-i}{\hbar c} \right) \int d^4x \left(-\frac{e\lambda}{2M} \right) \langle f| \hat{\bar{\psi}}(x) \sigma_{\mu\nu} \hat{\psi}(x) |i\rangle \frac{c}{2} \left[\frac{\partial A_\nu^{\text{ext}}(x)}{\partial x_\mu} - \frac{\partial A_\mu^{\text{ext}}(x)}{\partial x_\nu} \right]$$

$$= \left(\frac{ie\lambda}{2M\hbar c} \right) \frac{1}{\Omega} \bar{u}(k') \sigma_{\mu\nu} u(k) \frac{1}{2\sqrt{\varepsilon_0}} \left[iq_\mu a_\nu(q) - iq_\nu a_\mu(q) \right] \tag{9.83}$$

where we have again used

$$c A_\mu^{\text{ext}}(x) = \int \frac{d^4q}{(2\pi)^4} \frac{a_\mu(q)}{\sqrt{\varepsilon_0}} e^{iq\cdot x} \qquad ; q = k' - k \quad (9.84)$$

Now use the antisymmetry of $\sigma_{\mu\nu}$ to arrive at

$$\left[\hat{S}_P^{\text{ext}}\right]_{fi} = \left(\frac{-e}{\hbar c\sqrt{\varepsilon_0}}\right)\frac{1}{\Omega}\bar{u}(k')\left(-\frac{\lambda}{2M}\sigma_{\mu\nu}q_\nu\right)u(k)a_\mu(q) \qquad (9.85)$$

This is now in precisely the same form as Eq. (9.3), and hence we can identify from Eq. (9.78)

$$\lambda = F_M^{(2)}(0) = \frac{\alpha}{2\pi} \qquad (9.86)$$

This is the *Schwinger moment.*

Higher-order contributions have been calculated in QED. Through $O(\alpha^3)$, the result is [Cvitanovik and Kinoshita (1974)]

$$\mu_{\text{th}} = \mu_B\left[1 + \frac{\alpha}{2\pi} - 0.328,48\left(\frac{\alpha}{\pi}\right)^2 + 1.195(26)\left(\frac{\alpha}{\pi}\right)^3\right] \qquad (9.87)$$

where $\mu_B = |e|\hbar/2m_e$ is the Bohr magneton. The comparison of theory with experiment is awe-inspiring [Cvitanovik and Kinoshita (1974); Van Dyck, Schwinberg, and Dehmelt (1977)]

$$a \equiv \frac{\mu - \mu_B}{\mu_B}$$

$$a_{\text{th}} = 1,159,651,700(2200) \times 10^{-12} \qquad \text{; theory}$$

$$a_{\text{exp}} = 1,159,652,410(200) \times 10^{-12} \qquad \text{; experiment} \qquad (9.88)$$

Some comments:

- Because of the virtual emission and absorption of photons, the electron has a little larger magnetic moment that just the Bohr magneton;
- This effect is calculable in QED; indeed, we have just derived the Schwinger term;
- The agreement between theory and experiment for the anomalous magnetic moment of the electron is the *greatest success of pure QED;*
- In order to obtain the theoretical number, one needs an accurate value for the fine-structure constant. The above result uses a value obtained from Josephson junction measurements

$$\alpha^{-1} = 137.036,08(26) \qquad \text{; Josephson junction} \qquad (9.89)$$

- One needs to be able to isolate the divergences in higher order, as we have done in second order, to obtain the theoretical result. Thus one needs *general renormalization theory* (see later);

- Through an *heroic effort*, the theoretical and experimental results continue to be improved (see [Gabrielse (2009)]). The current results can always be found in the latest version of [Particle Data Group (2009)].

9.5 External Lines and Wavefunction Renormalization

Consider the diagram in Fig. 9.1(e), together with the mass counter-term, whose contribution to the S-matrix is given in Eq. (9.7)

$$\left[e^3 \hat{S}_{13}^{\text{eat}}\right]_{fi} = \frac{1}{\Omega} \left(\frac{-e}{\hbar c \sqrt{\varepsilon_0}}\right) \bar{u}(\mathbf{k}'\lambda')\gamma_\mu \left\{\frac{1}{i\slashed{k} + M}[\delta M - \Sigma(k)]\right\} u(\mathbf{k}\lambda)a_\mu(q)$$

(9.90)

Insert the general form of the electron self-energy in Eq. (9.36), and use the fact that the last term does not contribute since $(i\slashed{k} + M)u(k) = 0$

$$\left[e^3 \hat{S}_{13}^{\text{ext}}\right]_{fi} = \frac{1}{\Omega} \left(\frac{-e}{\hbar c \sqrt{\varepsilon_0}}\right) \bar{u}(\mathbf{k}'\lambda')\gamma_\mu \left\{\frac{-B}{i\slashed{k} + M}[i\slashed{k} + M]\right\} u(\mathbf{k}\lambda)a_\mu(q) \quad (9.91)$$

Now look at what sits next to the initial Dirac spinor:

(1) If the $i\slashed{k} + M$ acts to the *right*, this term vanishes;
(2) If the $i\slashed{k} + M$ acts to the *left*, one simply has $\{-B\}$.

How can this be? The answer is that this expression is really $0/0$, and hence it is *undefined* for a real particle. The goal in this section is to define it properly.[16]

Where does this term come from? The vertex in Fig. 9.1(b), which has all the physics in it, represents the interaction of an electron with the external field while it is accompanied by a virtual photon. The first-order term in Fig. 9.1(a) represents the interaction when there is no virtual photon present. Since the theory is *unitary* it conserves probability, and thus *the amplitude that the electron interacts when there is no virtual photon around must be reduced.*[17] This is known as *wavefunction renormalization.*

If one starts with an electron with the correct mass but no photons around, then as the interaction is turned on and the electron surrounds itself with photons, the amplitude that it is just a bare electron with no photons around is *reduced*.

---

[16]As we shall see, the answer is $-B/2$, which is a nice number between 0 and 1!
[17]Note the sign; the term is proportional to $-B = -L$.

To determine the correct number to use in front of $-B$, start by looking at the *insertion* of the electron self-energy and mass counter-term in a higher-order diagram, as illustrated in Fig. 9.2.   With the aid of the

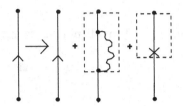

Fig. 9.2   Self-energy insertion together with mass counter-term.

Feynman rules, one observes that the effect of including this insertion is to replace

$$\frac{-1}{i\not p + M} \to \frac{-1}{i\not p + M} + \frac{-1}{i\not p + M}[\Sigma(p) - \delta M]\frac{-1}{i\not p + M} \qquad (9.92)$$

Define

$$S_F(p) \equiv -\frac{1}{i\not p + M} \qquad (9.93)$$

The effect of the self-energy insertion in Fig. 9.2 is then to replace

$$\begin{aligned}
S_F(p) \to S_F'(p) &= S_F(p) + S_F(p)[\Sigma(p) - \delta M]S_F(p) \\
&= S_F(p)(1 - B) + \Sigma_f(p) \qquad (9.94)
\end{aligned}$$

where the last line follows from the general form of $\Sigma(p) - \delta M$ in Eq. (9.36). Thus $(1 - B)$ is the *Green's function renormalization* (recall appendix F). Since the Green's function comes from a bilinear combination of fields, this says that each field should be renormalized by a factor of $(1 - B)^{1/2}$, and to this order $(1 - B)^{1/2} = 1 - B/2$. Thus the wavefunction renormalization from each leg should be[18]

$$-\frac{1}{2}B \qquad ; \text{ wave function renormalization from each leg} \qquad (9.95)$$

Although this argument actually suffices, we proceed to an analytic proof of this result within the framework of what we have done so far.[19] The difficulty is that the external particle is real so that its propagator

---

[18]An analysis exactly paralleling the above holds for the outgoing leg in Fig. 9.1(d).

[19]This proof is taken from [Schweber (1961)].

blows up. We can define things properly for the external particle by going back to our old friend, the adiabatic switch-off. Write

$$e^{-\epsilon|t|} \equiv g(t) = \int_{-\infty}^{\infty} G(\Gamma_0) e^{-i\Gamma_0 t} \, d\Gamma_0 = \int_{-\infty}^{\infty} G(\Gamma_0) e^{i\Gamma \cdot x} \, d\Gamma_0 \quad (9.96)$$

where the general properties of this Fourier transform are

$$g(0) = 1 \quad\quad \Rightarrow \quad\quad \int_{-\infty}^{\infty} G(\Gamma_0) d\Gamma_0 = 1$$

$$g(t) \approx 1 \quad\quad \Rightarrow \quad\quad G(\Gamma_0) \approx \delta(\Gamma_0)$$

$$\Gamma_\mu \equiv (0, i\Gamma_0) \quad\quad\quad\quad\quad\quad\quad (9.97)$$

This is actually all that we will need.[20] Since the adiabatic-damping factor in Eq. (9.96) multiplies the interaction $H_I(t)$, it can be looked at as bringing in an additional four-momentum Γ into each electromagnetic vertex.[21] Define $\Sigma(k) - \delta M \equiv \tilde{\Sigma}(k)$ in Eq. (9.36). Then, when the adiabatic-damping factor is included, one has the situation in Fig. 9.3 for the incoming leg, and the analytic expression for this diagram is

$$\left[e^3 \hat{S}_{13}^{\text{ext}}\right]_{fi} = \frac{1}{\Omega} \left(\frac{-e}{\hbar c \sqrt{\varepsilon_0}}\right) \bar{u}(\mathbf{k}'\lambda') \gamma_\mu \left\{ \int d\Gamma_0 \int d\Gamma_0' \, G(\Gamma_0) G(\Gamma_0') \times \right.$$

$$\left. \frac{1}{i(\slashed{k} + \slashed{\Gamma} + \slashed{\Gamma}') + M} [-\tilde{\Sigma}(k + \Gamma)] \right\} u(\mathbf{k}\lambda) a_\mu(q)$$

$$\tilde{\Sigma}(k) \equiv \Sigma(k) - \delta M = B(i\slashed{k} + M) + (i\slashed{k} + M)\Sigma_f(k)(i\slashed{k} + M) \quad (9.98)$$

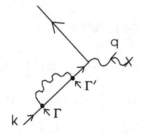

Fig. 9.3 Self-energy insertion on incoming leg with adiabatic damping.

---

[20] In fact, one can invert the Fourier transform in Eq. (9.96) to give (see Prob. 9.6) $G(\Gamma_0) = \epsilon/\pi(\Gamma_0^2 + \epsilon^2)$.

[21] This is not needed for the external field, since one can get away from the external field; however, an electron can never get away from its own virtual photon cloud.

As $\epsilon \to 0$, only small Γ_0 contributes [see Eqs. (9.97)]. We can then expand around $\Gamma_0 = 0$, and by keeping the first correction term in the numerator and denominator, we are no longer faced with an expression that is $0/0$. The expansion of $\tilde{\Sigma}(k + \Gamma)$ gives

$$\tilde{\Sigma}(k + \Gamma) = \tilde{\Sigma}(k) + \Gamma_\mu \frac{\partial \Sigma(k)}{\partial k_\mu} \qquad ; \Gamma_0 \to 0$$

$$= \tilde{\Sigma}(k) + i\Gamma_\mu \Lambda_\mu(k, k) \qquad (9.99)$$

where the second line follows from Ward's identity in Eq. (9.8). Rationalization of the denominator in Eq. (9.98) gives

$$\frac{1}{i(\not{k} + \not{\Gamma} + \not{\Gamma}') + M} = \frac{-[i(\not{k} + \not{\Gamma} + \not{\Gamma}') - M]}{k^2 + 2k \cdot (\Gamma + \Gamma') + (\Gamma + \Gamma')^2 + M^2}$$

$$= \frac{-(i\not{k} - M) - i(\not{\Gamma} + \not{\Gamma}')}{-2k_0(\Gamma_0 + \Gamma_0')} \qquad ; \Gamma_0 \to 0 \quad (9.100)$$

where we retain only the leading term in Γ in the denominator. We must then evaluate

$$[-(i\not{k} - M) - i(\not{\Gamma} + \not{\Gamma}')][\tilde{\Sigma}(k) + i\Gamma_\mu \Lambda_\mu(k, k)]u(\mathbf{k}\lambda) =$$
$$-(i\not{k} - M)[i\Gamma_\mu \Lambda_\mu(k, k)]u(\mathbf{k}\lambda) \qquad ; \Gamma_0 \to 0 \quad (9.101)$$

where we have used the fact that $\tilde{\Sigma}(k)u(\mathbf{k}\lambda) = 0$ and retained only the leading term in Γ.

Now $(i\not{k} - M)\Lambda_\mu(k, k)u(k)$ is just like $\bar{u}(k)\Lambda_\mu(k, k)u(k)$ since $(i\not{k} - M)(i\not{k} + M) = 0$, and therefore from Eq. (9.49), one can replace $\Lambda_\mu(k, k) \to L\gamma_\mu$. Hence, restoring Eq. (9.98) to its full form,

$$\left[e^3 \hat{S}_{13}^{\text{ext}}\right]_{fi} = \frac{1}{\Omega} \left(\frac{-e}{\hbar c \sqrt{\varepsilon_0}}\right) \bar{u}(\mathbf{k}'\lambda') \gamma_\mu \left\{ \int d\Gamma_0 \int d\Gamma_0' \, G(\Gamma_0) G(\Gamma_0') \times \right.$$

$$\left. \frac{1}{i(\not{k} + \not{\Gamma} + \not{\Gamma}') + M} [-iL\not{\Gamma}] \right\} u(\mathbf{k}\lambda) a_\mu(q) \qquad (9.102)$$

Now, using a change in integration variables, one can replace

$$i\not{\Gamma} \to \frac{i}{2}(\not{\Gamma} + \not{\Gamma}') = \frac{1}{2}[i(\not{k} + \not{\Gamma} + \not{\Gamma}') + M] \qquad (9.103)$$

where the last equality follows since $(i\not{k} + M)u(k) = 0$. At this point we can indeed write

$$\frac{1}{i(\not{k} + \not{\Gamma} + \not{\Gamma}') + M}[i(\not{k} + \not{\Gamma} + \not{\Gamma}') + M] = 1 \qquad (9.104)$$

since this expression is no longer $0/0$! The integrals over $\int d\Gamma_0 \int d\Gamma'_0$ can then be performed with the aid of Eqs. (9.97) to give

$$
\begin{aligned}
\left[e^3 \hat{S}_{13}^{\text{ext}}\right]_{fi} &= \frac{1}{\Omega}\left(\frac{-e}{\hbar c \sqrt{\varepsilon_0}}\right) \bar{u}(\mathbf{k}'\lambda')\gamma_\mu \left[-\frac{1}{2}L\right] u(\mathbf{k}\lambda)a_\mu(q) \\
&= \frac{1}{\Omega}\left(\frac{-e}{\hbar c \sqrt{\varepsilon_0}}\right) \bar{u}(\mathbf{k}'\lambda')\gamma_\mu \left[-\frac{1}{2}B\right] u(\mathbf{k}\lambda)a_\mu(q) \qquad (9.105)
\end{aligned}
$$

This is the previously stated result.

9.5.1 Cancellation of Divergences

Let us combine the contribution of diagrams (a,b,d,e) in Fig. 9.1 together with the mass counter-terms, which simply eliminate the constant A in $\Sigma(k)$. The above results then give, through $O(e^3)$,

$$
\left[e\hat{S}_{11}^{\text{ext}} + e^3 \hat{S}_{13}^{\text{ext}}\right]_{fi} = \frac{1}{\Omega}\left(\frac{-e}{\hbar c \sqrt{\varepsilon_0}}\right)\bar{u}(\mathbf{k}'\lambda') \times \qquad (9.106)
$$

$$
\left\{\gamma_\mu \left[1 - \frac{1}{2}L - \frac{1}{2}L + L\right] + \Lambda_{c\mu}(k',k)\right\} u(\mathbf{k}\lambda)a_\mu(q)
$$

$$
= \frac{1}{\Omega}\left(\frac{-e}{\hbar c \sqrt{\varepsilon_0}}\right)\bar{u}(\mathbf{k}'\lambda')\left[\gamma_\mu + \Lambda_{c\mu}(k',k)\right] u(\mathbf{k}\lambda)a_\mu(q)
$$

We observe that

> *Ward's identity implies that the wavefunction renormalization just cancels the divergent part of the vertex, and all that remains from these diagrams, through $O(e^3)$, is the convergent part of the vertex $\bar{u}(k')\gamma_\mu u(k) \rightarrow \bar{u}(k')[\gamma_\mu + \Lambda_{c\mu}(k',k)]u(k)$.*

The quantities (L, B), which with dimensional regularization are finite constants with an isolated singularity at $n = 4$, *disappear from the theory.* The constant A has been eliminated with mass renormalization. It remains to investigate the role of the vacuum polarization contribution in Fig. 9.1(c).

9.6 Vacuum Polarization

We turn now to the evaluation of the vacuum polarization contribution in Fig. 9.1(c) and Eq. (9.4). With dimensional regularization, the vacuum polarization tensor is given by

$$\Pi_{\mu\nu}(q) = \frac{e^2}{i\hbar c\varepsilon_0} \int \frac{d^n k}{(2\pi)^4} \text{Tr} \frac{1}{i(\not{k} - \not{q}/2) + M} \gamma_\mu \frac{1}{i(\not{k} + \not{q}/2) + M} \gamma_\nu \quad (9.107)$$

It is shown in appendix G and Prob. G.1 that this quantity satisfies current conservation

$$q_\mu \Pi_{\mu\nu}(q) = \Pi_{\mu\nu}(q) q_\nu = 0 \quad (9.108)$$

and hence has the general form

$$\Pi_{\mu\nu}(q) = (q_\mu q_\nu - q^2 \delta_{\mu\nu}) C(q^2) \quad (9.109)$$

It is also shown in appendix G that with dimensional regularization

$$\Pi_{\mu\mu}(0) = (1 - n)[q^2 C(q^2)]_{q^2=0} = 0 \quad (9.110)$$

which implies that the photon remains massless, and hence requires no mass renormalization (see Prob. G.2).

9.6.1   *Evaluation*

It is evident from Eqs. (9.109)– (9.110) that it is enough to compute $\Pi_{\mu\mu}(q)$ to determine $C(q^2)$, and thus $\Pi_{\mu\nu}(q)$. Since the steps are familiar by now, we can go through them rather quickly. Rationalization of the denominators in Eq. (9.107), and the gamma-matrix algebra, give

$$(1 - n)q^2 C(q^2) = -\frac{i\alpha}{4\pi^3} \int d^n k \, \text{Tr} \left[\frac{i(\not{k} - \not{q}/2) - M}{(k - q/2)^2 + M^2} \gamma_\mu \frac{i(\not{k} + \not{q}/2) - M}{(k + q/2)^2 + M^2} \gamma_\mu \right]$$

$$= -\frac{i\alpha}{4\pi^3} \tau \int d^n k \frac{[-(2 - n)(k^2 - q^2/4) + nM^2]}{[(k - q/2)^2 + M^2][(k + q/2)^2 + M^2]} \quad (9.111)$$

where we have employed

$$\gamma_\mu \not{a} \gamma_\mu = (2 - n)\not{a} \qquad ; \; \gamma_\mu \gamma_\mu = n$$
$$\text{Tr} \, \not{a} \not{b} = \tau(a \cdot b) \qquad ; \; \text{Tr} \, \not{a} = 0 \quad (9.112)$$

The Feynman parameterization of the denominators in Eq. (9.25) then combines them into

$$[(k + q/2)^2 + M^2]x + [(k - q/2)^2 + M^2](1 - x)$$
$$= k^2 + k \cdot q(2x - 1) + q^2/4 + M^2$$
$$= [k + (x - 1/2)\, q]^2 + M^2 + q^2 x(1 - x) \quad (9.113)$$

A change of variables in the integral from $k \to t = k + (x - 1/2)\,q$, and use of symmetric integration to discard any terms linear in t, then give

$$(1-n)q^2 C(q^2) = -\frac{i\alpha}{4\pi^3}\tau \int_0^1 dx \int d^n t \frac{\{(n-2)[t^2 - x(1-x)q^2] + nM^2\}}{(t^2 + a^2 - i\eta)^2} \ ;$$

$$a^2 = q^2 x(1-x) + M^2 \qquad (9.114)$$

Now make use of the basic relation in Eq. (G.24) to obtain

$$(1-n)q^2 C(q^2) = \frac{\alpha}{4\pi^3}\tau \int_0^1 dx \frac{n(\pi a^2)^{n/2}}{a^4} \frac{\Gamma(2-n/2)}{2\Gamma(2)} \times$$

$$\left\{ -2a^2 + \frac{2}{n}[nM^2 - (n-2)x(1-x)q^2] \right\} \qquad (9.115)$$

It is evident that if $q^2 = 0$, this expression vanishes identically for any n [see Eq. (9.110)]. A little algebra allows us to rewrite Eq. (9.115) as

$$C(q^2) = \frac{\alpha}{2\pi^3}\tau \int_0^1 x(1-x)dx \frac{(\pi a^2)^{n/2}}{a^4}\Gamma(2-n/2) \qquad (9.116)$$

9.6.2 *General Form*

The quantity $C(0)$ has a singularity at $n = 4$ arising from the factor $\Gamma(2 - n/2)$. Its behavior at the singularity can be determined by again defining $n = 4 - \epsilon$ and using $\Gamma(\epsilon/2) = (2/\epsilon)\Gamma(1 + \epsilon/2)$. Then as $\epsilon \to 0$

$$C(0) = \frac{4\alpha}{\pi\epsilon} \int_0^1 x(1-x)dx$$

$$C(0) = \frac{2\alpha}{3\pi\epsilon} \qquad \qquad ; \ n = 4 - \epsilon \to 4 \qquad (9.117)$$

where $\tau = \text{Tr}\,\underline{1} = 4$ (see appendix G). As before, if we had simply cut off the momentum integrals at a large momentum Λ, we would have obtained

$$C(0) = \frac{2\alpha}{3\pi} \ln \frac{\Lambda}{M} \qquad (9.118)$$

and we again note the very weak logarithmic dependence on the cutoff.[22]

---

[22]Simple power counting would indicate a *quadratic* divergence in the integral in Eq. (9.107) for $n = 4$; however, dimensional regularization, which preserves current conservation and gauge invariance, eliminates that divergence identically for all n. In addition, symmetric integration then eliminates any possible *linear* divergence, and when the dust clears, as in the previous constants (A, B, L), we are simply left with a real, weak *logarithmic* divergence in the vacuum polarization constant $C(0)$.

Define the *general form* of the vacuum polarization amplitude as

$$C(q^2) \equiv C(0) - q^2 \Pi_f(q^2) \qquad (9.119)$$

Take the difference of Eq. (9.116) with the same result at $q^2 = 0$

$$q^2 \Pi_f(q^2) = \frac{\alpha}{2\pi^3} \tau \int_0^1 x(1-x)dx \, \Gamma(2-n/2) \left[\frac{(\pi M^2)^{n/2}}{M^4} - \frac{(\pi a^2)^{n/2}}{a^4} \right] \qquad (9.120)$$

This expression is perfectly well behaved at $n = 4$ and we can proceed to now take that limit using

$$\left(\frac{a}{M} \right)^{-\epsilon} = 1 - \epsilon \ln \frac{a}{M} + O(\epsilon^2) \qquad (9.121)$$

Hence

$$q^2 \Pi_f(q^2) = \frac{2\alpha}{\pi} \int_0^1 dx \, x(1-x) \ln \left[1 + x(1-x) \frac{q^2}{M^2} \right] \qquad (9.122)$$

9.6.3   *Limiting Cases*

Equation (9.122) gives the general result for $\Pi_f(q^2)$ to $O(\alpha)$. The limits for $q^2 \ll M^2$ and $q^2 \gg M^2$ are relatively simple to evaluate, and are very useful

$$q^2 \Pi_f(q^2) = \frac{\alpha}{15\pi} \frac{q^2}{M^2} \qquad\qquad ; q^2 \ll M^2 \quad ; \text{Uehling term}$$

$$q^2 \Pi_f(q^2) = \frac{\alpha}{3\pi} \left[\ln \frac{q^2}{M^2} - \frac{5}{3} \right] \qquad ; q^2 \gg M^2 \qquad (9.123)$$

The expression for the vacuum polarization contribution at low q^2 is referred to as the "Uehling term".[23]

9.6.4   *Insertion*

The contribution of the vacuum polarization diagram in Fig. 9.1(c) involves an insertion in the lowest order result in Fig. 9.1(a). The contribution to the scattering of an electron in an external field from diagram (c) in Fig. 9.1 is given by Eq. (9.4) as

$$[e^3 \hat{S}_{13}^{\text{ext}}]_{fi} = \frac{1}{\Omega} \left(\frac{-e}{\hbar c \sqrt{\varepsilon_0}} \right) \bar{u}(k') \gamma_\mu \left[D_{\mu\nu}^F(q) \Pi_{\nu\lambda}(q) \right] u(x) a_\lambda(q) \qquad (9.124)$$

---

[23] Note that $\Pi_f(0)$ is a finite number [and, for later, recall the second of Eqs. (9.49)].

where $D_{\mu\nu}^F(q) = \delta_{\mu\nu}/q^2$. It follows from the form of $\Pi_{\nu\lambda}(q)$ in Eq. (9.109), and the fact that $\bar{u}(k')\not{q}u(k) = 0$, that this expression reduces to

$$[e^3 \hat{S}_{13}^{\text{ext}}]_{fi} = \frac{1}{\Omega} \left(\frac{-e}{\hbar c \sqrt{\varepsilon_0}} \right) \bar{u}(k')\gamma_\mu \left[-C(q^2) \right] u(k) a_\mu(q) \; ; \; q = k' - k \quad (9.125)$$

Hence the contribution from the sum of the diagrams in Fig. 9.1(a) and (c) takes the form

$$[e\hat{S}_{11}^{\text{ext}} + e^3 \hat{S}_{13}^{\text{ext}}]_{fi} = \frac{1}{\Omega} \left(\frac{-e}{\hbar c \sqrt{\varepsilon_0}} \right) \bar{u}(k')\gamma_\mu \left[1 - C(q^2) \right] u(k) a_\mu(q) \quad (9.126)$$

$$= \frac{1}{\Omega} \left(\frac{-e}{\hbar c \sqrt{\varepsilon_0}} \right) \bar{u}(k')\gamma_\mu \left[1 - C(0) + q^2 \Pi_f(q^2) \right] u(k) a_\mu(q)$$

where we have inserted the general form of $C(q^2)$ from Eq. (9.119).

The Feynman rules imply that insertion of the vacuum polarization loop into an internal photon line in a diagram involves the replacement (see Fig. 9.4)

$$D_{\mu\nu}^F(q) \to D_{\mu\nu}^F(q) + D_{\mu\lambda}^F(q)\Pi_{\lambda\sigma}(q)D_{\sigma\nu}^F(q) \quad (9.127)$$

where

$$D_{\mu\nu}^F(q) = \frac{\delta_{\mu\nu}}{q^2} \equiv D_F(q^2)\delta_{\mu\nu} \quad (9.128)$$

This defines the scalar function $D_F(q^2) = 1/q^2$.

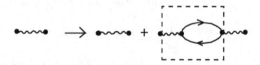

Fig. 9.4 Vacuum polarization insertion on an internal photon line.

The form of $\Pi_{\lambda\sigma}(q)$ in Eq. (9.109) can now be inserted in Eq. (9.127). Since terms proportional to q_μ or q_ν in the photon propagator $D_{\mu\nu}^F(q)$ do not contribute to the scattering amplitude,[24] one can use the *effective* polarization tensor

$$\Pi_{\lambda\sigma}(q) \doteq -q^2 C(q^2)\delta_{\lambda\sigma} \qquad ; \text{ effective vacuum polarization} \quad (9.129)$$

---

[24]Compare with Prob. 9.5.

Hence one can rewrite Eq. (9.127) as a relation involving scalar amplitudes

$$D_F(q^2) \to D'_F(q^2) = D_F(q^2) + D_F(q^2)[-q^2 C(q^2)] D_F(q^2)$$
$$= D_F(q^2)[1 - C(0)] + \Pi_f(q^2) \qquad (9.130)$$

where the last lime follows from insertion of the general from of the vacuum polarization amplitude in Eq. (9.119)

9.6.5 Charge Renormalization

We must now interpret these results. We know the vacuum is a dynamical quantity in a relativistic quantum field theory, in the sense that there is a rich variety of virtual processes that can take place in it. The vacuum-polarization insertion discussed above takes into account the virtual creation and destruction of $(e^- e^+)$ pairs. The vacuum is *polarizable*, and in QED it acts like a *dielectric medium* where the virtual pairs shield the original point charges.[25] Let us go back, then, and carefully distinguish the *"bare charge"* e_0 *that we use in our starting lagrangian.* We make this charge explicit and write $C(q^2) \equiv e_0^2 \bar{C}(q^2)$.[26] Since there will always be a charge buried in the external field [witness Eq. (8.124)], let us also make that charge explicit by writing $a_\mu(q) \equiv (-e_0/\sqrt{\varepsilon_0}) \bar{a}_\mu(q)$. Hence Eq. (9.126) should really be written as

$$[e_0 \hat{S}_{11}^{\text{ext}} + e_0^3 \hat{S}_{13}^{\text{ext}}]_{fi} = \frac{1}{\Omega} \left(\frac{1}{\hbar c \varepsilon_0} \right) \bar{u}(k') \gamma_\mu \times$$
$$\{ e_0^2 [1 - e_0^2 \bar{C}(0)] + e_0^4 q^2 \bar{\Pi}_f(q^2) \} u(k) \bar{a}_\mu(q)$$
$$a_\mu(q) \equiv \left(\frac{-e_0}{\sqrt{\varepsilon_0}} \right) \bar{a}_\mu(q) \qquad (9.131)$$

Now we have to ask ourselves, how do we actually determine the *observed charge* e? One way is to make use of the singularity in $\bar{a}_\mu(q)$ for an external Coulomb field where [see Eq. (8.124)]

$$\bar{a}_\mu(q) = i\delta_{\mu 4} \, \delta(q_0) \left(\frac{2\pi Z}{\mathbf{q}^2} \right) = i\delta_{\mu 4} \, \delta(q_0) \left(\frac{2\pi Z}{q^2} \right) \qquad (9.132)$$

and define the charge by what is observed in the limit as $q^2 \to 0$. This is just forward Rutherford scattering. Since $\bar{\Pi}_f(0)$ is a finite constant, one

---

[25] See the discussion in Vol. I.
[26] The quantities $(\bar{\Pi}_f, \bar{\Lambda}_{c\mu})$, and the constants $(\bar{A}, \bar{B}, \bar{L})$, are similarly defined.

has for $q^2 \to 0$

$$[\hat{S}_1^{\text{ext}}]_{fi} = \frac{1}{\Omega} \left(\frac{1}{\hbar c \varepsilon_0} \right) \bar{u}(k) \gamma_\mu \left\{ e_0^2 [1 - e_0^2 \bar{C}(0)] \right\} u(k) \bar{a}_\mu(q) \quad ; q^2 \to 0$$

$$\equiv \frac{1}{\Omega} \left(\frac{e^2}{\hbar c \varepsilon_0} \right) \bar{u}(k) \gamma_\mu u(k) \bar{a}_\mu(q) \quad ; \text{ through } O(e_0^4) \qquad (9.133)$$

where the last relation now *defines* the observed charge e^2. One then identifies[27]

$$e^2 \equiv e_0^2 [1 - e_0^2 \bar{C}(0)] \qquad ; \text{ observed charge through } O(e_0^4) \qquad (9.134)$$

Now one has to consistently replace the bare charge e_0^2 with the observed charge e^2 to the order in which one is working [here, with the charge extracted from $a_\mu(q)$, that would be to $O(e_0^4)$], but observe that *to this order, one can use the charge e^2 in all the corrections calculated in Fig. 9.1.* Thus, when written in terms of the *observed charge* through $O(e_0^4)$, Eq. (9.131) for the sum of the contributions of the diagrams in Fig. 9.1(a,c) becomes

$$[\hat{S}_1^{\text{ext}}]_{fi} = \frac{4\pi\alpha}{\Omega} \bar{u}(k') \gamma_\mu \left[1 + \frac{\alpha}{15\pi} \frac{q^2}{M^2} \right] u(k) \bar{a}_\mu(q) \quad ; \frac{q^2}{M^2} \ll 1$$

$$= \frac{4\pi\alpha}{\Omega} \bar{u}(k') \gamma_\mu \left[1 + \frac{\alpha}{3\pi} \ln \frac{q^2}{M^2} \right] u(k) \bar{a}_\mu(q) \quad ; \frac{q^2}{M^2} \gg 1 \quad (9.135)$$

where we have inserted the limiting results in Eq. (9.123). Here $\alpha = e^2/4\pi\hbar c\varepsilon_0$ is the *observed* fine-structure constant, and these expressions are *exact through* $O(e_0^4)$.

Some comments:

- The charge is redistributed by vacuum polarization, one no longer has a point charge, and Coulomb's law is modified at short distances.[28] This is an observable effect, for example in the level shifts in the low-lying levels in μ^--atoms, and it is *calculable in QED*;
- Note that charge renormalization (at least to the order through which we have calculated it) comes *entirely from vacuum polarization* in QED. The electron self-energy graphs and vertex insertion in Fig. 9.1 play no role in charge renormalization. Ward's identity eliminates any potential charge renormalization from the constants L and B;

---

[27] Alternatively, one can look at Møller scattering with the vacuum-polarization insertion in the internal photon line, where in the same limit $e_0^2 D'_F(q^2) \to e^2/q^2$ as $q^2 \to 0$. Equation (9.130) then leads to an identical result.

[28] The Fourier transform of $(4\pi/\mathbf{q}^2)[1 + \mathbf{q}^2 \Pi_f(\mathbf{q}^2)]$ is *not* just $1/|\mathbf{x}|$.

- Since there will be a vacuum-polarization contribution from *any* charged particle-antiparticle pair, including those that interact strongly, the complete calculation of vacuum polarization effects involves evaluating the contribution of these *strongly interacting pairs*.[29] Hence it is impossible *in principle* to consistently analyze QED by itself in an isolated sector of the theory.

9.6.6 *Charge Strength*

Let us rewrite the second of Eqs. (9.135) as

$$[\hat{S}_1^{\text{ext}}]_{fi} = \frac{4\pi\alpha(q^2)}{\Omega}\bar{u}(k')\gamma_\mu u(k)\bar{a}_\mu(q)$$

$$\alpha(q^2) \equiv \alpha\left[1 + \frac{\alpha}{3\pi}\ln\frac{q^2}{M^2}\right] \qquad ; \; \frac{q^2}{M^2} \gg 1 \qquad (9.136)$$

Since vacuum polarization is universal, no matter what process one computes, this same charge $\alpha(q^2)$ will always appear. For example, suppose one includes a muon in the theory, as we did before, and then computes the same set of QED diagrams in Fig. 9.1 for the much heavier μ^-. The calculation is exactly the same as that carried out above, only the mass M_μ^2 appears everywhere; *however, one then also has to include the vacuum polarization diagram in Fig. 9.1(c) for the lightest pair of leptons* (e^-e^+). The result of this calculation can be written as

$$[\hat{S}_1^{\text{ext}}]_{fi} = \frac{4\pi\alpha(q^2)}{\Omega}\bar{u}(k')\left\{\gamma_\mu\left[1 + \alpha(q^2)q^2\bar{\Pi}(q^2)\right] + \alpha(q^2)\bar{\Lambda}_{c\mu}(k',k)\right\}u(k)\bar{a}_\mu(q) \qquad ;$$

$$\mu^- \; ; \text{ through } O(\alpha^2)$$

$$\alpha(q^2) \equiv \alpha\left[1 + \frac{\alpha}{3\pi}\ln\frac{q^2}{M_e^2}\right] \qquad ; \; \frac{q^2}{M_e^2} \gg 1 \qquad (9.137)$$

where the first line is calculated entirely for the μ^-. This expression is *exact through order* α^2.[30] The coupling constant $\alpha(q^2)$ to be used in the muon calculation is larger than α, growing logarithmically with q^2. How do we understand this?

Recall that $q^2 = \mathbf{q}^2$ in the static Coulomb field. Equation (9.134) implies that the observed charge e^2, measured at $\mathbf{q}^2 = 0$, is smaller than

---

[29]The lightest mass such pair is $(\pi^+\pi^-)$. From Eq. (9.135) one can estimate these contributions to be of order $\alpha q^2/M_\pi^2$ at low q^2, and hence reduced from the vacuum-polarization effects of the much lighter electron pair by a factor of m_e^2/m_π^2.

[30]See Prob. 9.19; it really should have $\alpha(l^2)/l^2$ in the photon propagator in the vertex, but that makes no difference through this order.

the bare charge e_0^2, used in the starting lagrangian; it is shielded by vacuum polarization. At $\mathbf{q}^2 = 0$, with forward scattering, one measures the total integrated charge. It is clear from the discussion of electron scattering in Vol. I that as $\mathbf{q}^2$ increases, one probes structure at smaller and smaller distance scales. Here $\alpha(\mathbf{q}^2)/\alpha$ plays the role of an *intrinsic form factor*, and as $\mathbf{q}^2$ increases, one is *displaying the structure of the vacuum-polarization cloud and its bare point charge e_0 core*. The long-range part of the cloud is determined by the lightest pair of particles in the vacuum polarization loop, here the $(e^- e^+)$ with $M_e^2/M_\mu^2 \ll 1$.

Through order α^2, the result in Eq. (9.137) can be rewritten as

$$\alpha(q^2) \equiv \frac{\alpha}{1 - (\alpha/3\pi)\ln(q^2/M_e^2)} \quad ; \text{ through } O(\alpha^2) \; ; \; \frac{q^2}{M_e^2} \gg 1 \quad (9.138)$$

This is actually a correct result to all orders in α, obtained by summing leading logarithms in higher-order perturbation theory (a *tour-de-force* calculation), or through the use of the renormalization group (see later).

9.7 Renormalization Theory

We have analyzed the basic building blocks, encountered the basic divergences, and shown, at least through $O(\alpha^2)$, that when re-expressed in terms of the observed mass and charge (m, e), QED makes finite, accurate predictions for experimental observables. The question arises as to whether this renormalization program can be carried out consistently to higher order in α. It was shown in the classic papers by [Dyson (1949); Ward (1951)] that this is indeed possible, and we sketch that analysis.

9.7.1 *Proper Self-Energies*

A *proper* self-energy insertion is one that cannot be separated into two self-energy insertions by cutting a single line. We denote the entire set of proper self-energies for the electron and photon by $[\Sigma^\star(k), \Pi^\star(q)]$, where for the photon we use the *effective vacuum polarization* $\Pi^\star_{\mu\nu}(q) \doteq \Pi^\star(q)\delta_{\mu\nu}$ and *henceforth suppress the unit tensor*. The entire set of self-energy insertions is then obtained by converting Eqs. (9.94) and (9.130) to the following algebraic equations (see Fig. 9.5)

$$S_F'(k) = S_F(k) + S_F(k)[\Sigma^\star(k) - \delta M]S_F'(k) \quad ; \; S_F(k) = -[i\rlap{/}{k} + M]^{-1}$$

$$D_F'(q) = D_F(q) + D_F(q)\Pi^\star(q)D_F'(q) \quad ; \; D_F(q) = q^{-2} \quad (9.139)$$

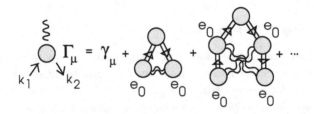

Fig. 9.5 Diagrammatic representation of Dyson's Eqs. (9.139) for the full electron and photon propagators (double lines). The circles here indicate the proper self-energies.

When these equations are iterated by continually substituting for the primed quantities on the r.h.s., one generates all the self-energy insertions, proper and improper. These are *Dyson's equations* for the propagators. Since they are simply algebraic equations in momentum space, Dyson's equations for the propagators are immediately solved to give

$$[S'_F(k)]^{-1} = [S_F(k)]^{-1} - [\Sigma^\star(k) - \delta M]$$
$$[D'_F(q)]^{-1} = [D_F(q)]^{-1} - \Pi^\star(q) \qquad (9.140)$$

9.7.2 Proper Vertex

A vertex part is one connected to the rest of the diagram by two electron and one photon lines. A *proper* vertex part has no self-energy insertions on its legs, and henceforth the term vertex will refer to proper vertex. The full vertex is written as

$$\Gamma_\mu(k_2, k_1) = \gamma_\mu + \Lambda_\mu(k_2, k_1) \qquad ; \text{ full vertex} \qquad (9.141)$$

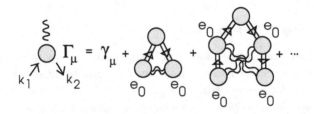

Fig. 9.6 Diagrammatic representation of Dyson's equation for the vertex $\Gamma_\mu(k_2, k_1)$ (circle), where (k_2, k_1) are the outgoing and incoming electron four-momenta. The double lines represent the full electron and photon propagators (see Fig. 9.5.) The charge e_0 is now explicit, and we show the equation up through $O(e_0^4)$. To convert to the present conventions with S_F, D_F, Γ_μ, the Feynman rules require a factor of $(i/\hbar c \varepsilon_0)$ for each pair of charged vertices in the graphs.

Dyson's equation for the vertex is an equation that, when iterated, yields all possible vertex parts. It cannot be written in closed form, but its general structure is evident. One first removes all self-energy and vertex insertions to produce an *irreducible vertex part*, and then reinserts the full propagators and vertices in the irreducible vertices to reproduce *all* graphs. The charge e_0 is now made explicit, and Fig. 9.6 exhibits Dyson's equation for the vertex up through $O(e_0^4)$. When this equation is iterated consistently up through $O(e_0^4)$, including the self-energy insertions on the electron and photon lines, one reproduces all proper vertex parts up through that order.[31]

9.7.3 Ward's Identity

We previously proved Ward's identity relating the electron self-energy $\Sigma(k)$ and the vertex part $i\Lambda(k,k)$ in order e_0^2. It is clear from Eq. (9.12) that differentiating with respect to the external momentum k_μ running along an electron line S_F is equivalent to inserting a zero-energy photon on that line with a vertex $i\gamma_\mu$. In fact, Ward's identity is a *general result* relating the electron proper self-energy and vertex. This can be ascertained by staring at enough graphs, and we are content to illustrate it diagramatically with one contribution of order e_0^4 in Fig. 9.7.[32]

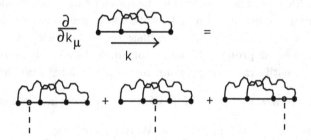

Fig. 9.7   Illustration of one contribution to Ward's identity in order $O(e_0^4)$. Here the dashed line indicates the insertion into S_F of a zero-energy photon with vertex $i\gamma_\mu$.

The analytic statement that Ward's identity holds to all orders is

$$\frac{\partial}{\partial k_\mu}\Sigma^\star(k) = i\Lambda_\mu(k,k) \qquad ; \text{ Ward's identity} \qquad (9.142)$$

We shall integrate this result to relate the electron proper self-energy to

---

[31] See Prob. 9.14.

[32] The demonstration of Ward's identity through $O(e_0^4)$ is assigned as Prob. 9.15.

the vertex. The *motivation* is that it will then be clear how to apportion the renormalization constants in the vertex when we come to the proof of renormalizability. Given a relation of the form $\mathbf{v} = \boldsymbol{\nabla}\phi$, then $\mathbf{v} \cdot d\mathbf{l}$ is a perfect differential, and integration along some path from points 1 to 2 gives

$$\mathbf{v} = \boldsymbol{\nabla}\phi$$
$$\int_1^2 \mathbf{v} \cdot d\mathbf{l} = \int_1^2 \boldsymbol{\nabla}\phi \cdot d\mathbf{l} = \phi(2) - \phi(1) \tag{9.143}$$

Similarly, integrate Ward's identity along a path from k_0 to k in four-dimensional Minkowski space

$$\Sigma^\star(k) - \Sigma^\star(k_0) = i \int_{k_0}^k \Lambda_\mu(k', k') dk'_\mu \tag{9.144}$$

Take the path to be a straight line where

$$k' = kx + k_0(1 - x) \equiv k(x) \qquad ; \text{ four-vectors}$$
$$dk'_\mu = (k - k_0)_\mu \, dx$$
$$\Sigma^\star(k) - \Sigma^\star(k_0) = i \int_0^1 dx \, (k - k_0)_\mu \Lambda_\mu[k(x), k(x)] \tag{9.145}$$

Now the l.h.s. of this relation, and hence the r.h.s. also, is the difference of two matrix functions. From Lorentz invariance, this relation takes the form $\Sigma^\star(i\not{k}, k^2) - \Sigma^\star(i\not{k}_0, k_0^2)$. The mass counter-term δM is to be chosen to just cancel the proper self-energy for an electron on the mass shell, and hence we identify the starting point in the above integration as

$$\Sigma^\star[i\not{k}_0 + M = 0, \ k_0^2 = -M^2] = \delta M \qquad ; \text{ mass renormalization} \tag{9.146}$$

With this choice, the integration of Ward's identity gives

$$\Sigma^\star(k) - \delta M = i \int_0^1 dx \, (k - k_0)_\mu \Lambda_\mu[k(x), k(x)] \Big|_{i\not{k}_0 + M = 0} \tag{9.147}$$

where the four-vector $k(x)$ is given by the first of Eqs. (9.145).

The derivative of the first of Eqs. (9.140) with respect to k_μ gives

$$\frac{\partial}{\partial k_\mu} [S'_F(k)]^{-1} = -i[\gamma_\mu + \Lambda_\mu(k, k)] = -i\Gamma_\mu(k, k) \tag{9.148}$$

where Γ_μ is the full vertex. Substitution of Eq. (9.147) into the first of Eqs. (9.140) provides an explicit expression for the full electron propagator itself

$$[S'_F(k)]^{-1} = [S_F(k)]^{-1} - i \int_0^1 dx \, (k - k_0)_\mu \Lambda_\mu[k(x), k(x)] \Big|_{i\not{k}_0+M=0} \quad (9.149)$$

Thus Ward's identity allows us to determine the electron propagator from the vertex!

9.7.4 *Ward's Vertex Construct*

The question now arises as to what to do with the photon propagator and vacuum polarization. In order to again deal with vertices, Ward argued by strict analogy to the above, and he *constructed a vertex* that gives the proper photon self-energy through a second identity

$$\frac{\partial \Pi^\star(q)}{\partial q_\mu} \equiv i\Delta_\mu(q, q) \quad ; \text{ Ward's second identity} \quad (9.150)$$

We give one example in Fig. (9.8).

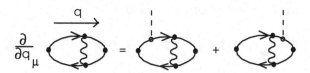

Fig. 9.8   Illustration of one contribution to Ward's second identity in order $O(e_0^4)$. Here the dashed line indicates the insertion into S_F of a zero-energy photon with vertex $i\gamma_\mu$.

We note the following:

- $\Delta_\mu(q, q)$ will be a *three-photon* vertex;
- Furry's theorem (Prob. 8.13) does not apply to the overall vertex here since only one direction of the electron loop is retained;
- Since photon lines do not run continuously through the diagrams, one can no longer use the previous device that the electron line simply carry the incoming four-momentum all the way through. Photon lines can end, and thus to generate more complicated proper photon self-energy parts, *it is necessary to specify just how the incoming four-momentum is carried through the proper self-energy and vertex;*[33]
- We are always dealing with the *effective* vacuum polarization tensor $\Pi_{\mu\nu}(q) \doteq \Pi(q^2)\delta_{\mu\nu}$.

---

[33]See Fig. 9.9 for an example.

Just as before, the integration of Eq. (9.150) gives

$$\Pi^\star(q) - \Pi^\star(0) = i \int_0^q \Delta_\mu(q',q') dq'_\mu$$

$$q' \equiv qy \qquad\qquad ; \text{ four-vectors}$$

$$\Pi^\star(q) - \Pi^\star(0) = i q_\mu \int_0^1 dy\, \Delta_\mu(qy, qy) \qquad (9.151)$$

Just as was true in second order, current conservation implies that $\Pi^\star(0)$ vanishes, so that no photon mass renormalization is required. Furthermore, from Lorentz covariance $\Delta_\mu(q,q) = i q_\mu T(q)$. Thus

$$\Pi^\star(0) = 0 \qquad\qquad ; \text{ current conservation}$$

$$\Delta_\mu(q,q) = i q_\mu T(q) \qquad\qquad ; \text{ Lorentz covariance} \qquad (9.152)$$

Hence

$$\Pi^\star(q) = -q^2 \int_0^1 y\, dy\, T(qy) \qquad (9.153)$$

It follows from the second of Eqs. (9.140) that

$$\frac{\partial}{\partial q_\mu} [D'_F(q)]^{-1} = 2q_\mu - i\Delta_\mu(q,q) \equiv W_\mu(q,q)$$

$$[D'_F(q)]^{-1} = [D_F(q)]^{-1} + q^2 \int_0^1 y\, dy\, T(qy) \qquad (9.154)$$

The first relation defines *Ward's full vertex construct* $W_\mu(q,q)$. The function $T(q)$ follows from that definition and the second of Eqs. (9.152)

$$W_\mu(q,q) = 2q_\mu + q_\mu T(q) \qquad (9.155)$$

It remains to write the analog of Dyson's equation for Ward's vertex construct, and this is done graphically in Fig. 9.9. (See also Prob. 9.16.)

9.7.5 *Finite Parts*

From simple power counting, one observes that increasing the *order* of a diagram will not increase the degree of divergence of the momentum-space integrals (see Probs. 9.12–9.13). There are only three so-called *primitive divergences* in the theory, and we have already met all of them in second-order in the electron and photon self-energies and the vertex. In that discussion, we identified the general form of these insertions, and developed a procedure for isolating the divergences through subtractions in the integrand.

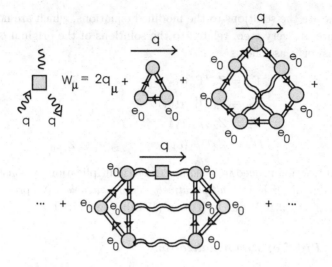

Fig. 9.9 Diagrammatic representation of Dyson's equation for Ward's full vertex construct $W_\mu(q,q)$ (square). The outgoing and incoming photon four-momenta are q. The double lines and circles represent the full propagators and vertex (see Figs. 9.5–9.6). The charge e_0 is now explicit, and we show the equation up through $O(e_0^6)$. One contribution of $O(e_0^8)$ is shown, where the vertex W_μ first again appears. To convert to the present conventions with $S_F, D_F, \Gamma_\mu, W_\mu$, the Feynman rules require a factor of $(i/\hbar c \varepsilon_0)$ for each pair of charged electron vertices in a graph, and a factor of $(-i)$ for the odd one.

Since the degree of divergence will not increase as more complicated graphs are included, we can use those same methods on the full self-energies and vertices.

Ward's definition of the new, finite functions is as follows

$$\Gamma_{\mu 1}(k_2, k_1) = \gamma_\mu + \left[\Lambda^{(1)}(k_2, k_1) - \Lambda^{(1)}(k_0, k_0) \right]\Big|_{i\slashed{k}_0 + M = 0} \qquad ; \text{ finite parts}$$

$$W_{\mu 1}(q, q) = 2q_\mu + q_\mu \left[T^{(1)}(q) - T^{(1)}(0) \right]$$

$$[S_{F1}(k)]^{-1} = [S_F(k)]^{-1} -$$
$$\left[i \int_0^1 dx \, (k - k_0)_\mu \left\{ \Lambda^{(1)}[k(x), k(x)] - \Lambda^{(1)}(k_0, k_0) \right\} \right]_{i\slashed{k}_0 + M = 0}$$

$$[D_{F1}(q)]^{-1} = [D_F(q)]^{-1} + q^2 \int_0^1 y \, dy \left[T^{(1)}(qy) - T^{(1)}(0) \right] \qquad (9.156)$$

where the (1) *functions are obtained by substituting these finite expressions into Dyson's equations, and using the renormalized charge e_1 at the vertices.*

How are the solutions to the modified equations, which are now explicitly finite at every step, related to the solutions of the original equations? The answer is as follows

$$\Gamma_\mu = Z_2^{-1}\Gamma_{\mu 1}(e_1)$$
$$W_\mu = Z_3^{-1}W_{\mu 1}(e_1)$$
$$S_F' = Z_2 S_{F1}(e_1)$$
$$D_F' = Z_3 D_{F1}(e_1) \qquad ; \, e_1^2 \equiv Z_3 e_0^2 \qquad (9.157)$$

The subtraction procedure is equivalent to multiplication by constants, and the theory of QED is *multiplicatively renormalizable*. We proceed to the proof of Eqs. (9.157) due to [Dyson (1949); Ward (1951)].

9.7.6 *Proof of Renormalization*

Substitute the ansatz in Eqs. (9.157) into Dyson's non-linear equations for the vertices illustrated in Fig. 9.6 and Fig. 9.9, and then

- Collect the renormalization factors at each vertex to give e_1;
- Note that all graphs in Dyson's equations for the vertices in Figs. 9.6 and 9.9 then *have the same overall structure*:

  (1) Graphs for Γ_μ lack one factor of Z_2 to convert e_0 to e_1 everywhere;
  (2) Graphs for W_μ lack one factor of Z_3 to convert e_0 to e_1 everywhere.

If those factors are supplied to produce the correct (1) functions, then Dyson's equations for the vertices in Figs. 9.6 and 9.9 become

$$\Gamma_\mu = \gamma_\mu + \Lambda_\mu^{(1)}(k_2, k_1)Z_2^{-1} = Z_2^{-1}\Gamma_{\mu 1}(e_1)$$
$$W_\mu = 2q_\mu + q_\mu T^{(1)}(q)Z_3^{-1} = Z_3^{-1}W_{\mu 1}(e_1) \qquad (9.158)$$

where the final relations must be satisfied. Substitution of the ansatz in Eqs. (9.157) into the equations for the propagators in Eqs. (9.149) and (9.154), and the same observations on the vertex graphs, leads to the conditions

$$Z_2 S_{F1}(e_1) = S_F + S_F \times$$
$$\left\{ i \int_0^1 dx\, (k - k_0)_\mu \Lambda_\mu^{(1)}[k(x), k(x)]Z_2^{-1} \right\}_{i\not{k}_0 + M = 0} Z_2 S_{F1}(e_1)$$
$$Z_3 D_{F1}(e_1) = D_F + D_F \left[-q^2 \int_0^1 y\, dy\, T^{(1)}(qy)Z_3^{-1} \right] Z_3 D_{F1}(e_1) \qquad (9.159)$$

Now make use of Ward's definition of the finite parts in Eqs. (9.156)

$$\Gamma_{\mu 1}(e_1) = \gamma_\mu + \Lambda_\mu^{(1)}(k_2, k_1) - \Lambda_\mu^{(1)}(k_0, k_0)\Big|_{i\not{k}_0 + M = 0}$$

$$W_{\mu 1}(e_1) = 2q_\mu + q_\mu \left[T^{(1)}(q) - T^{(1)}(0) \right]$$

$$S_{F1}(e_1) = S_F + S_F \times$$

$$\left[i \int_0^1 dx\, (k - k_0)_\mu \left\{ \Lambda_\mu^{(1)}[k(x), k(x)] - \Lambda_\mu^{(1)}(k_0, k_0) \right\} \right]_{i\not{k}_0 + M = 0} S_{F1}$$

$$D_{F1}(e_1) = D_F + D_F \left\{ -q^2 \int_0^1 ydy \left[T^{(1)}(qy) - T^{(1)}(0) \right] \right\} D_{F1} \qquad (9.160)$$

Compare Eqs. (9.160) with Eqs. (9.158)–(9.159):

(1) Multiplication of the first of Eqs. (9.158) by Z_2 shows the equations for the vertex $\Gamma_{\mu 1}(e_1)$ will coincide if

$$Z_2 \gamma_\mu = \gamma_\mu - \Lambda_\mu^{(1)}(k_0, k_0)\Big|_{i\not{k}_0 + M = 0} \qquad (9.161)$$

Just as in Eq. (9.48), the last term must be of the form

$$\Lambda_\mu^{(1)}(k_0, k_0)\Big|_{i\not{k}_0 + M = 0} = L(e_1)\gamma_\mu \qquad (9.162)$$

Hence the equations for the vertex $\Gamma_{\mu 1}(e_1)$ will be identical if the renormalization constant Z_2 is chosen as

$$Z_2 = 1 - L(e_1) \qquad (9.163)$$

(2) Take the difference of the first of Eqs. (9.159) and the third of Eqs. (9.160) to obtain the condition on the electron propagator

$$(Z_2 - 1)S_{F1} = S_F \left[i \int_0^1 dx\, (k - k_0)_\mu \Lambda_\mu^{(1)}(k_0, k_0) \right]_{i\not{k}_0 + M = 0} S_{F1} \qquad (9.164)$$

The integral is readily evaluated with the aid of Eq. (9.162)

$$\left[i \int_0^1 dx\, (k - k_0)_\mu \Lambda_\mu^{(1)}(k_0, k_0) \right]_{i\not{k}_0 + M = 0} = L(e_1)(i\not{k} + M)$$

$$= L(e_1)[-S_F]^{-1} \qquad (9.165)$$

Thus Eq. (9.164) reduces to

$$(Z_2 - 1)S_{F1} = -L(e_1)S_{F1}$$

$$\text{or ;} \qquad Z_2 = 1 - L(e_1) \qquad (9.166)$$

which is identical to the previous result in Eq. (9.163).

(3) Equate the second of Eqs. (9.160) with Z_3 times the second of Eqs. (9.158), which shows the equations for the vertex $W_{\mu 1}(e_1)$ will coincide if

$$2q_\mu Z_3 = 2q_\mu - q_\mu T^{(1)}(0)$$

$$\text{or ;} \quad Z_3 = 1 - \frac{1}{2}T^{(1)}(0) \tag{9.167}$$

This determines the renormalization constant Z_3.

(4) Take the difference of the second of Eqs. (9.159) and the last of Eqs. (9.160) to obtain the condition on the photon propagator

$$(Z_3 - 1)D_{F1} = (q^2 D_F)\left[-\frac{1}{2}T^{(1)}(0)\right]D_{F1}$$

$$\text{or ;} \quad Z_3 = 1 - \frac{1}{2}T^{(1)}(0) \tag{9.168}$$

which is identical to Eq. (9.167). This completes the proof.[34]

In *summary*, the ansatz in Eqs. (9.157) satisfies Dyson's equations, with finite (1) functions defined through Eqs. (9.156), provided the renormalization constants are chosen as in Eqs. (9.163) and (9.167). It is now straightforward to show that the S-matrix for any physical process is given by the finite functions and the renormalized charge e_1, and we leave that for Prob. 9.17.

9.7.7 *The Renormalization Group*

If the subtraction point used to define the finite functions is shifted, then the same analysis goes through, but with a different set of renormalization constants. Since the theory is multiplicatively renormalizable, this set of operations forms a group, the *renormalization group* of [Gell-Mann and Low (1954)]. These authors used the renormalization group to study the asymptotic behavior of QED; one of their results was Eq. (9.138). The use of renormalization-group techniques now forms one of the cornerstones of modern physics.[35] In particular, it played a key role in establishing the asymptotic freedom of QCD (see Vol. I).

---

[34] Confirmation that an arbitrary diagram, with all its nested loop integrals, is rendered finite through the subtraction procedure goes way beyond the scope of this book.

[35] See, for example, [Itzykson and Zuber(1980); Banks (2008)].

Chapter 10

Path Integrals

We next turn to the topic of functional methods and path integrals. There are many reasons for becoming familiar with these methods:

- This approach unites quantum mechanics, field theory, and statistical mechanics;
- It provides an alternative to doing quantum mechanics with canonical quantization;
- Exact expressions are obtained for quantum mechanical transition amplitudes;
- Everything is written in terms of *classical quantities*, in particular, the classical lagrangian and classical action;
- The price for this is that one has to consider *other dynamical paths* than the classical one given by Hamilton's principle of stationary action;
- With this approach, one can readily study the implications of various symmetries of the lagrangian, even with highly non-linear interactions with derivative couplings, where canonical quantization becomes prohibitively difficult.

The basic references here are [Feynman and Hibbs (1965); Abers and Lee (1973)].[1]

10.1 Non-Relativistic Quantum Mechanics with One Degree of Freedom

Let us go back to the beginning. Consider the non-relativistic quantum mechanics of a particle of mass m with coordinate q moving in a potential

---

[1]See also [Itzykson and Zuber(1980)].

$V(q)$ (Fig. 10.1).

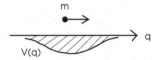

Fig. 10.1 Non-relativistic particle of mass m and coordinate q moving in a potential $V(q)$ in one dimension.

10.1.1 *General Relations*

The hamiltonian and canonical commutation relation are

$$\hat{H} = \frac{1}{2m}\hat{p}^2 + V(\hat{q}) \qquad ; \ [\hat{p}, \hat{q}] = \frac{\hbar}{i} \qquad (10.1)$$

Introduce the complete, orthonormal set of *eigenstates of position*[2]

$$\hat{q}|q\rangle = q|q\rangle$$
$$\langle q|q'\rangle = \delta(q - q')$$
$$\int dq\, |q\rangle\langle q| = \hat{1} \qquad (10.2)$$

Introduce also the corresponding complete, orthonormal set of *eigenstates of momentum in the continuum limit* [compare Eq. (2.46)]

$$\hat{p}|p\rangle = p|p\rangle$$
$$\langle p|p'\rangle = \delta(p - p')$$
$$\int dp\, |p\rangle\langle p| = \hat{1} \qquad (10.3)$$

As seen previously, the inner product of these states is just a plane wave

$$\langle q|p\rangle = \frac{1}{\sqrt{2\pi\hbar}}e^{\frac{i}{\hbar}pq} \qquad (10.4)$$

The Schrödinger wave function and its probability interpretation are[3]

$\langle q|\Psi(t)\rangle = \Psi(q, t)$; Schrödinger wave function

$|\Psi(q, t)|^2 = |\langle q|\Psi(t)\rangle|^2$; probability of observing q at time t   (10.5)

---

[2]See chapter 2; and for the momentum, recall $p = \hbar k$.
[3]The second is actually the probability *density*.

The formal solution to the time-dependent Schrödinger equation that reduces to the particle localized at position q_1 at time t_1 can be written as

$$|\Psi_{q_1}(t)\rangle = e^{-\frac{i}{\hbar}\hat{H}(t-t_1)}|q_1\rangle$$
$$|\Psi_{q_1}(t_1)\rangle = |q_1\rangle \tag{10.6}$$

From the general principles of quantum mechanics, the probability amplitude for finding the particle at position q_2 at a time t_2 if it started at q_1 at time t_1 is given by the inner product (see Fig. 10.2)

$$\langle q_2|\Psi_{q_1}(t_2)\rangle = \langle q_2|e^{-\frac{i}{\hbar}\hat{H}(t_2-t_1)}|q_1\rangle \quad ; \text{ transition amplitude} \tag{10.7}$$

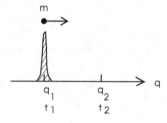

Fig. 10.2   Particle localized at q_1 at initial time t_1 propagates to the point q_2 at t_2.

The states $|q_1(t)\rangle$ in the *Heisenberg picture* are defined by

$$|q_1(t)\rangle = e^{\frac{i}{\hbar}\hat{H}t}|\Psi_{q_1}(t)\rangle \quad ; \text{ Heisenberg picture} \tag{10.8}$$

They have the following properties

$$i\hbar\frac{\partial}{\partial t}|q_1(t)\rangle = 0 \quad ; \text{ time-independent}$$
$$|q_1(t_1)\rangle \equiv |q_1 t_1\rangle = e^{\frac{i}{\hbar}\hat{H}t_1}|q_1\rangle \quad ; \text{ initial condition}$$
$$\int dq_1\,|q_1 t_1\rangle\langle q_1 t_1| = \hat{1} \quad ; \text{ complete}$$
$$\langle q_2 t_1|q_1 t_1\rangle = \langle q_2|q_1\rangle = \delta(q_1 - q_2) \quad ; \text{ orthonormal} \tag{10.9}$$

Note that the second line defines the Heisenberg state at the initial time $|q_1 t_1\rangle \equiv |q_1(t_1)\rangle = e^{\frac{i}{\hbar}\hat{H}t_1}|q_1\rangle$; it is a complicated quantity. The last two relations follow immediately from this definition and the properties of the eigenstates of position. The probability transition amplitude in Eq. (10.7) is just the inner product of two of these Heisenberg states

$$\langle q_2|\Psi_{q_1}(t_2)\rangle = \langle q_2 t_2|q_1 t_1\rangle \quad ; \text{ transition amplitude} \tag{10.10}$$

10.1.2 *Infinitesimals*

We will again build the general expression from infinitesimals. Write $t_2 = t_1 + \Delta t$ and let $\Delta t \to 0$. Equation (10.7) then becomes

$$\langle q_2 | \Psi_{q_1}(t_2) \rangle = \langle q_2 | \exp\left\{ -\frac{i}{\hbar} \Delta t \hat{H} \right\} | q_1 \rangle \approx \langle q_2 | 1 - \frac{i}{\hbar} \Delta t \hat{H} | q_1 \rangle \quad ; \; \Delta t \to 0$$

$$= \langle q_2 | 1 - \frac{i}{\hbar} \Delta t \left\{ \frac{1}{2m} \hat{p}^2 + V(\hat{q}) \right\} | q_1 \rangle$$

$$\approx \exp\left\{ -\frac{i}{\hbar} \Delta t V(q_1) \right\} \langle q_2 | \exp\left\{ -\frac{i}{\hbar} \Delta t \frac{\hat{p}^2}{2m} \right\} | q_1 \rangle \qquad (10.11)$$

Provided that $V(q)$ is appropriately slowly-varying, this expression is exact as $\Delta t \to 0$.

The final matrix element in Eq. (10.11) defines the *free-particle propagator*. It is evaluated by inserting a complete set of eigenstates of momentum (twice)

$$I \equiv \langle q_2 | \exp\left\{ -\frac{i}{\hbar} \Delta t \frac{\hat{p}^2}{2m} \right\} | q_1 \rangle$$

$$= \int dp \int dp' \langle q_2 | p \rangle \langle p | \exp\left\{ -\frac{i}{\hbar} \Delta t \frac{\hat{p}^2}{2m} \right\} | p' \rangle \langle p' | q_1 \rangle$$

$$= \int \frac{dp}{2\pi\hbar} \exp\left\{ \frac{i}{\hbar}(q_2 - q_1)p - \frac{i}{\hbar} \Delta t \frac{p^2}{2m} \right\} \qquad (10.12)$$

where $\hat{p}$ has been replaced by its eigenvalue, the orthonormality of the eigenstates of momentum invoked, and Eq. (10.4) used for the remaining inner products. To perform the final integral, introduce

$$p = \hbar k \qquad\qquad ; \; q_2 - q_1 = x$$
$$\frac{\hbar \Delta t}{2m} = \alpha \qquad\qquad ; \; k - \frac{x}{2\alpha} = t \qquad (10.13)$$

Then, by completing the square in the exponent,

$$I = \int_{-\infty}^{\infty} \frac{dk}{2\pi} e^{ikx} e^{-i\alpha k^2} = \int_{-\infty}^{\infty} \frac{dk}{2\pi} e^{-i\alpha(k - x/2\alpha)^2} e^{i\alpha x^2/4\alpha^2}$$

$$= e^{ix^2/4\alpha} \int_{-\infty}^{\infty} \frac{dt}{2\pi} e^{-i\alpha t^2} \qquad (10.14)$$

The following integral is readily evaluated for $\text{Re}\, z > 0$

$$\frac{1}{2\pi} \int_{-\infty}^{\infty} dt\, e^{-zt^2} = \frac{1}{2\pi\sqrt{z}} \int_{-\infty}^{\infty} dx\, e^{-x^2} = \frac{1}{\sqrt{4\pi z}} \quad ; \; \text{Re}\, z > 0 \quad (10.15)$$

Now analytically continue to the imaginary axis $z \to i\alpha = \alpha e^{i\pi/2}$ to obtain the free-particle propagator[4]

$$I = \frac{1}{\sqrt{4\pi\alpha e^{i\pi/2}}} e^{ix^2/4\alpha} \tag{10.16}$$

Hence, inserting the definitions of (α, x), the free-particle propagator in Eq. (10.12) is

$$\langle q_2| \exp\left\{-\frac{i}{\hbar}\Delta t\,\frac{\hat{p}^2}{2m}\right\}|q_1\rangle = \left(\frac{me^{-i\pi/2}}{2\pi\hbar\Delta t}\right)^{1/2} \exp\left\{\frac{i}{\hbar}\frac{m(q_2 - q_1)^2}{2\Delta t}\right\} \quad;$$
$$\Delta t = t_2 - t_1 \to 0 \tag{10.17}$$

The full infinitesimal propagator in Eq. (10.11) correspondingly becomes[5]

$$\langle q_2 t_2|q_1 t_1\rangle = \left(\frac{me^{-i\pi/2}}{2\pi\hbar\Delta t}\right)^{1/2} \exp\left\{\frac{i}{\hbar}\left[\frac{m(q_2 - q_1)^2}{2\Delta t} - \Delta t\, V(q_1)\right]\right\} \quad;$$
$$\Delta t = t_2 - t_1 \to 0 \tag{10.18}$$

These expressions are exact as $\Delta t = t_2 - t_1 \to 0$.

Let us rewrite this result. Define the quantity $\dot{q}$ for small Δt, by the following

$$\frac{q_2 - q_1}{\Delta t} \equiv \dot{q}_1 \qquad ; \text{ classical velocity} \tag{10.19}$$

If the particle is at q_2 at time t_2 when it started at q_1 at t_1, then this is just the *classical particle velocity*. Equation (10.18) then takes the form

$$\langle q_2 t_2|q_1 t_1\rangle = \left(\frac{me^{-i\pi/2}}{2\pi\hbar\Delta t}\right)^{1/2} \exp\left\{\frac{i}{\hbar}\Delta t\left[\frac{m}{2}\dot{q}_1^2 - V(q_1)\right]\right\}$$
$$= \left(\frac{me^{-i\pi/2}}{2\pi\hbar\Delta t}\right)^{1/2} \exp\left\{\frac{i}{\hbar}\int_{t_1}^{t_1+\Delta t} dt\, L(q_1, \dot{q}_1)\right\} \quad;$$
$$\Delta t = t_2 - t_1 \to 0 \tag{10.20}$$

where we now identify the *classical lagrangian and classical action*

$$L(q, \dot{q}) = \frac{m}{2}\dot{q}^2 - V(q) \qquad ; \text{ lagrangian}$$
$$S(t + \Delta t, t) = \int_t^{t+\Delta t} dt\, L(q, \dot{q}) \qquad ; \text{ action} \tag{10.21}$$

---

[4]See Prob. B.5.

[5]This result can be made to look a little more symmetric by replacing $V(q_1) \to [V(q_1) + V(q_2)]/2$; for our purposes, this nicety will not matter.

10.1.3 *Transition Amplitude and Path Integral*

The above result is exact as $\Delta t \to 0$. One can use *superposition* to build the amplitude $\langle q_f t_f | q_i t_i \rangle$ for finite times from the infinitesimals. Insert the completeness statement in Eqs. (10.9) for $n - 1$ ordered, equally-spaced, intermediate times to obtain

$$
\langle q_f t_f | q_i t_i \rangle = \int \cdots \int dq_1 \cdots dq_{n-1} \, \langle q_f t_f | q_{n-1} t_{n-1} \rangle \langle q_{n-1} t_{n-1} | q_{n-2} t_{n-2} \rangle \times
$$
$$
\cdots \langle q_2 t_2 | q_1 t_1 \rangle \langle q_1 t_1 | q_i t_i \rangle \qquad \text{; completeness} \qquad (10.22)
$$

In order to make use of the infinitesimal result for $\Delta t \to 0$, *one must take the limit $n \to \infty$ in this expression, so that $(t_f - t_i)/n = \Delta t \to 0$.* Hence completeness must actually be used an infinite number of times here

$$
\langle q_f t_f | q_i t_i \rangle = \mathrm{Lim}_{n \to \infty} \int \cdots \int \prod_{m=1}^{n-1} dq_m \prod_{p=0}^{n-1} \langle q_{p+1}, t_{p+1} | q_p t_p \rangle \qquad ;
$$
$$
|q_0 t_0\rangle \equiv |q_i t_i\rangle \qquad ; \; |q_n t_n\rangle \equiv |q_f t_f\rangle \qquad (10.23)
$$

There are n intervals with $(t_f - t_i)/n = \Delta t$, and $n - 1$ intermediate integrations in this expression.

Insertion of the result for infinitesimals from Eq. (10.18) now gives

$$
\langle q_f t_f | q_i t_i \rangle = \mathrm{Lim}_{n \to \infty} \left[\frac{nme^{-i\pi/2}}{2\pi\hbar(t_f - t_i)} \right]^{n/2} \int \cdots \int \prod_{m=1}^{n-1} dq_m \prod_{p=0}^{n-1} \times
$$
$$
\exp\left\{ \frac{i}{\hbar} \left[\frac{mn}{2} \frac{(q_{p+1} - q_p)^2}{(t_f - t_i)} - \frac{(t_f - t_i)}{n} V(q_p) \right] \right\} \quad ; \; \Delta t = \frac{(t_f - t_i)}{n} \quad (10.24)
$$

Since the exponents *add* in the product, one has

$$
\prod_{p=0}^{n-1} \exp\left\{ \frac{i}{\hbar} \left[\frac{mn}{2} \frac{(q_{p+1} - q_p)^2}{(t_f - t_i)} - \frac{(t_f - t_i)}{n} V(q_p) \right] \right\} =
$$
$$
\exp\left\{ \frac{i}{\hbar} \sum_{p=0}^{n-1} \Delta t \left[\frac{m}{2} \dot{q}_p^2 - V(q_p) \right] \right\} \qquad ; \; \Delta t = \frac{(t_f - t_i)}{n} \quad (10.25)
$$

Hence

$$\langle q_f t_f | q_i t_i \rangle = \mathrm{Lim}_{n \to \infty} \left[\frac{nme^{-i\pi/2}}{2\pi\hbar(t_f - t_i)} \right]^{n/2} \int \cdots \int \prod_{m=1}^{n-1} dq_m \times$$

$$\exp\left\{ \frac{i}{\hbar} \sum_{p=0}^{n-1} \Delta t \left[\frac{m}{2} \dot{q}_p^2 - V(q_p) \right] \right\} \tag{10.26}$$

This expression defines the *path integral*

$$\langle q_f t_f | q_i t_i \rangle \equiv \int \mathcal{D}(q) \exp\left\{ \frac{i}{\hbar} \int_{t_i}^{t_f} dt\, L(q, \dot{q}) \right\} \qquad ; \text{ path integral} \tag{10.27}$$

Let us discuss this result. The situation is illustrated in Fig. 10.3.

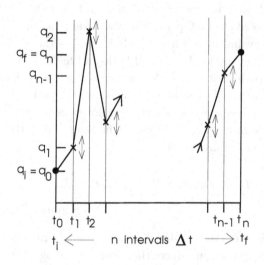

Fig. 10.3　Path integral (see text); here $q_i \equiv q_0$ and $q_f \equiv q_n$.

The rules for evaluating $\langle q_f t_f | q_i t_i \rangle$ as a path integral are as follows:

(1) Split up the total time interval into n equal subintervals with $\Delta t = (t_f - t_i)/n$;

(2) Write out the action $S(t_f, t_i) = \int_{t_i}^{t_f} dt\, L(q, \dot{q})$ as a finite sum in the time, where $q_i \equiv q_0$ and $q_f \equiv q_n$ [see Eq. (10.26)];

(3) Do the multiple integral $\int dq_1 \cdots \int dq_{n-1}$ *over the coordinates at each intermediate time*;

(4) Define the classical velocity between neighboring coordinates as $\dot{q}_p \equiv$ $(q_{p+1} - q_p)/\Delta t$;

(5) Use a weighting factor $[nme^{-i\pi/2}/2\pi\hbar(t_f - t_i)]^{1/2}$ for each *interval*;

(6) Take the limit $n \to \infty$ at the end.

Everything is now classical, all the integrals are over commuting c-numbers. The only quantum mechanics in the phase is the factor of $i/\hbar$ in front!

In the limit $n \to \infty$, this expression gives the exact quantum mechanical probability amplitude for a non-relativistic particle of mass m to propagate from a point q_i at time t_i to a point q_f at a time t_f while moving in a potential $V(q)$.

10.1.4 *Classical Limit*

The classical limit of this result is now obtained by letting $\hbar \to 0$. In this limit, the phase in Eq. (10.27) varies very rapidly. There a classical approach to evaluating such integrals known as the *method of stationary phase* (see, for example, [Fetter and Walecka (2003)]). With this technique, one finds a path that leaves the phase stationary, and the integral then receives all of its contributions from along this path. The procedure for finding this path is precisely the one employed in the application of Hamilton's principle in chapter 5. Thus, in the limit $\hbar = 0$, the integral is given by that path satisfying

$$\delta \int_{t_i}^{t_f} dt\, L(q, \dot{q}) = 0 \qquad ; \hbar \to 0$$

Hamilton's principle!   (10.28)

In the limit $\hbar \to 0$, the path-integral description of the quantum mechanical transition amplitude reproduces Hamilton's principle of classical mechanics!

10.1.5 *Superposition*

Suppose t is one of the intermediate times in Eq. (10.27) and Fig. 10.3. First compute the path integral for *fixed* $q(t)$, and then perform $\int dq(t)$. Since the phase is additive, and the number of intervals is unchanged, the path integral *factors* into

$$\langle q_f t_f | q_i t_i \rangle = \int \mathcal{D}(q) e^{\frac{i}{\hbar} S(f,i)} = \int dq(t) \int \mathcal{D}(q) e^{\frac{i}{\hbar} S(f,t)} \int \mathcal{D}(q) e^{\frac{i}{\hbar} S(t,i)} \qquad ;$$

superposition   (10.29)

This is the principle of *superposition*. It is the path-integral statement of the completeness relation

$$\langle q_f t_f | q_i t_i \rangle = \int dq \, \langle q_f t_f | qt \rangle \langle qt | q_i t_i \rangle \qquad \text{; completeness} \quad (10.30)$$

10.1.6 Matrix Elements

Consider two solutions to the Schrödinger equation, which at the appropriate times are the localized particle states

$$|\Psi_{q_1}(t)\rangle = e^{-\frac{i}{\hbar}\hat{H}(t-t_1)}|q_1\rangle \qquad \text{; solutions to S-eqn}$$

$$|\Psi_{q_2}(t)\rangle = e^{-\frac{i}{\hbar}\hat{H}(t-t_2)}|q_2\rangle$$

$$|\Psi_{q_1}(t_1)\rangle = |q_1\rangle \qquad \qquad ; |\Psi_{q_2}(t_2)\rangle = |q_2\rangle \quad (10.31)$$

The matrix element at the time t of the Schrödinger-picture operator $O(\hat{q})$ between these states is given by

$$\langle \Psi_{q_2}(t) | O(\hat{q}) | \Psi_{q_1}(t) \rangle = \langle q_2 | e^{-\frac{i}{\hbar}\hat{H}t_2} e^{\frac{i}{\hbar}\hat{H}t} O(\hat{q}) e^{-\frac{i}{\hbar}\hat{H}t} e^{\frac{i}{\hbar}\hat{H}t_1} | q_1 \rangle \quad (10.32)$$

In the Heisenberg picture the time-dependent operator $\hat{q}(t)$ is given by

$$\hat{q}(t) = e^{\frac{i}{\hbar}\hat{H}t} \hat{q} \, e^{-\frac{i}{\hbar}\hat{H}t} \qquad \text{; Heisenberg picture} \quad (10.33)$$

Hence the matrix element in Eq. (10.32) is rewritten in the Heisenberg picture as

$$\langle \Psi_{q_2}(t) | O(\hat{q}) | \Psi_{q_1}(t) \rangle = \langle q_2 t_2 | O[\hat{q}(t)] | q_1 t_1 \rangle \quad (10.34)$$

Now insert a complete set of eigenstates of position $|qt\rangle$ in this expression,[6] and use

$$\hat{q}(t)|qt\rangle = \left[e^{\frac{i}{\hbar}\hat{H}t} \hat{q} e^{-\frac{i}{\hbar}\hat{H}t} \right] e^{\frac{i}{\hbar}\hat{H}t} | q \rangle = q|qt\rangle \quad (10.35)$$

Hence the matrix element in Eq. (10.34) takes the form

$$\langle \Psi_{q_2}(t) | O(\hat{q}) | \Psi_{q_1}(t) \rangle = \langle q_2 t_2 | O[\hat{q}(t)] | q_1 t_1 \rangle$$

$$= \int dq \, O(q) \langle q_2 t_2 | qt \rangle \langle qt | q_1 t_1 \rangle$$

$$\equiv \int dq(t) \, O[q(t)] \langle q_2 t_2 | qt \rangle \langle qt | q_1 t_1 \rangle \quad (10.36)$$

---

[6] Recall $\int dq \, |qt\rangle\langle qt| = \hat{1}$.

where we write $q(t)$ in the last line to remind ourselves that the coordinate q is associated with the time t in the first. The integration in the last line is then just over the c-number coordinate $q(t)$. The superposition principle in Eq. (10.29) can now be used to rewrite this expression as a single path integral

$$\langle \Psi_{q_2}(t)|O(\hat{q})|\Psi_{q_1}(t)\rangle = \langle q_2 t_2|O[\hat{q}(t)]|q_1 t_1\rangle = \int \mathcal{D}(q)O[q(t)]e^{\frac{i}{\hbar}S(2,1)} \quad (10.37)$$

This result tells us to simply weight the coordinate $q(t)$ with the additional factor $O[q(t)]$ when performing the path integral.

This analysis can be extended and applied to the *time-ordered product of two coordinate operators*

$$\langle q_f t_f|T[\hat{q}(t_1)\hat{q}(t_2)]|q_i t_i\rangle = \int \mathcal{D}(q)q(t_1)q(t_2)e^{\frac{i}{\hbar}S(f,i)} \quad (10.38)$$

No time-ordering symbol is needed in the integrand on the right since the coordinates there $q(t_1)q(t_2)$ are c-numbers and commute; furthermore, by *definition*, these coordinates will be weighted in the path integral in a time-ordered fashion. This result is immediately generalized to the time-ordered product of any number of coordinate operators

$$\langle q_f t_f|T[\hat{q}(t_1)\cdots\hat{q}(t_p)]|q_i t_i\rangle = \int \mathcal{D}(q)q(t_1)\cdots q(t_p)e^{\frac{i}{\hbar}S(f,i)} \quad (10.39)$$

Now let $|\psi_0\rangle$ be the *exact ground state* of the system

$$\hat{H}|\psi_0\rangle = E_0|\psi_0\rangle \qquad ;\ \text{exact ground state} \qquad (10.40)$$

Then, using completeness twice and employing the above results, one has

$$\langle\psi_0|O[\hat{q}(t)]|\psi_0\rangle = \int\int dq_1 dq_2\,\langle\psi_0|q_2 t_2\rangle\langle q_2 t_2|O[\hat{q}(t)]|q_1 t_1\rangle\langle q_1 t_1|\psi_0\rangle$$

$$= \int\int dq_1 dq_2\,\psi_0^*(q_2,t_2)\int\mathcal{D}(q)O[q(t)]e^{\frac{i}{\hbar}S(2,1)}\psi_0(q_1,t_1)$$

$$(10.41)$$

where we have identified the ground-state wave function

$$\psi_0(q,t) = \langle qt|\psi_0\rangle = \langle q|e^{-\frac{i}{\hbar}\hat{H}t}|\psi_0\rangle = e^{-\frac{i}{\hbar}E_0 t}\langle q|\psi_0\rangle = e^{-\frac{i}{\hbar}E_0 t}\psi_0(q) \quad (10.42)$$

Equation (10.41) provides a path-integral expression for the *ground-state expectation value of the Heisenberg operator* $O[\hat{q}(t)]$. This result is again immediately generalized to the time-ordered product of several coordinate operators

$$\langle\psi_0|T[\hat{q}(t_1)\cdots\hat{q}(t_p)]|\psi_0\rangle =$$
$$\int\int dq dq'\,\psi_0^\star(q',t')\int\mathcal{D}(q)q(t_1)\cdots q(t_p)e^{\frac{i}{\hbar}S(t',t)}\,\psi_0(q,t)\quad(10.43)$$

10.1.7 *Crucial Theorem of Abers and Lee*

Suppose the system is in its ground state at the time T in the distant past. Let us calculate the amplitude for the system to remain in that state at a time T' in the distant future when an *arbitrary external source $J(t)q$ is added to the lagrangian between intermediate times t and t'* (Fig. 10.4).

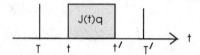

Fig. 10.4 External source $J(t)q$ added to the system between t and t'.

The path integral expression for the transition amplitude between the Heisenberg states $|QT\rangle$ and $|Q'T'\rangle$ in the presence of the source is

$$\langle Q'T'|QT\rangle^J = \int\mathcal{D}(q)\exp\left\{\frac{i}{\hbar}\int_T^{T'}dt\,[L(q,\dot{q}) + J(t)q]\right\}\quad(10.44)$$

Now use completeness twice at the times t and t' (see Fig. 10.4)

$$\langle Q'T'|QT\rangle^J = \int\int dq dq'\langle Q'T'|q't'\rangle\langle q't'|qt\rangle^J\langle qt|QT\rangle\quad(10.45)$$

The amplitudes on the ends are independent of J, and since the source is turned off there, they are given by

$$\langle qt|QT\rangle = \langle q|e^{-\frac{i}{\hbar}\hat{H}(t-T)}|Q\rangle\quad(10.46)$$

The insertion of a complete set of the *exact energy eigenstates* $|\psi_n\rangle$ of $\hat{H}$ with $\hat{H}|\psi_n\rangle = E_n|\psi_n\rangle$ then leads to

$$\langle qt|QT\rangle = \sum_n \psi_n(q)\psi_n^\star(Q)e^{-\frac{i}{\hbar}E_n(t-T)}\quad(10.47)$$

Now multiply by $e^{-\frac{i}{\hbar}E_0 T}$ and take the limit $T\to +i\infty$ (notice the i!). Since in this limit $e^{\frac{i}{\hbar}(E_n-E_0)T}\to 0$ for all $E_n > E_0$, only the ground state survives in the sum

$$\text{Lim}_{T \to i\infty} \, e^{-\frac{i}{\hbar}E_0 T} \langle qt | QT \rangle = \psi_0(q)\psi_0^\star(Q) e^{-\frac{i}{\hbar}E_0 t}$$

$$= \psi_0(q,t)\psi_0^\star(Q) \qquad (10.48)$$

Similarly

$$\text{Lim}_{T' \to -i\infty} \, e^{\frac{i}{\hbar}E_0 T'} \langle Q'T' | q't' \rangle = \psi_0^\star(q')\psi_0(Q') e^{\frac{i}{\hbar}E_0 t'}$$

$$= \psi_0^\star(q',t')\psi_0(Q') \qquad (10.49)$$

In this fashion one obtains the following relation

$$\text{Lim}_{T' \to -i\infty} \text{Lim}_{T \to i\infty} \frac{\langle Q'T' | QT \rangle^J}{e^{-\frac{i}{\hbar}E_0(T'-T)}\psi_0^\star(Q)\psi_0(Q')} =$$

$$\int dq \int dq' \, \psi_0^\star(q',t') \langle q't' | qt \rangle^J \psi_0(q,t) \qquad (10.50)$$

The insertion on both sides of the path-integral expression for the transition amplitude in the presence of the source [see Eq. (10.44)] then gives

$$W(J) \equiv \text{Lim}_{T' \to -i\infty} \text{Lim}_{T \to i\infty} \frac{1}{e^{-\frac{i}{\hbar}E_0(T'-T)}\psi_0^\star(Q)\psi_0(Q')} \times$$

$$\int \mathcal{D}(q) \exp\left\{ \frac{i}{\hbar} \int_T^{T'} dt \, [L(q,\dot{q}) + J(t)q] \right\}$$

$$= \int dq \int dq' \, \psi_0^\star(q',t') \left[\int \mathcal{D}(q) \exp\left\{ \frac{i}{\hbar} \int_t^{t'} dt'' \, [L(q,\dot{q}) + J(t'')q] \right\} \right] \psi_0(q,t)$$

$$(10.51)$$

where this expression defines $W(J)$. This is the crucial theorem of [Abers and Lee (1973)]. It allows us to extract the *ground-state expectation value on the r.h.s.* [compare Eq. (10.43)] from the path integral from QT to $Q'T'$ through this unusual limiting procedure. The utility of this result, and its interpretation, will become clearer when we actually apply the theorem.

10.1.8   *Functional Derivative*

The way we will extract physical results from these path-integral expressions is through the use of *functional* or *variational* derivatives. Let $W(f)$

be a functional of f,[7] for example

$$W(f) = \int dx\, K(x) f(x) \qquad (10.52)$$

where $K(x)$ is some specified kernel. Let $\eta(x)$ be an *arbitrary function* of x and λ be an *infinitesimal*. The functional, or variational, derivative of $W(f)$ is then defined through (see Fig. 10.5)

$$\mathrm{Lim}_{\lambda \to 0}\, \frac{W(f + \lambda\eta) - W(f)}{\lambda} \equiv \int \frac{\delta W(f)}{\delta f(x)} \eta(x)\, dx \qquad ;$$

variational derivative   (10.53)

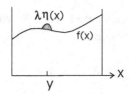

Fig. 10.5   Definition of the variational derivative.

Since $\eta(x)$ is arbitrary, take $\eta(x) = \delta(x - y)$ so that the variation of $f(x)$ is localized at the point y. One then obtains an explicit expression for the variational derivative

$$\mathrm{Lim}_{\lambda \to 0}\, \frac{W[f + \lambda\delta(x - y)] - W(f)}{\lambda} = \frac{\delta W(f)}{\delta f(y)} \qquad (10.54)$$

where x is the integration variable in the functional.

We give some examples:

(1) The simplest functional derivatives are

$$W(f) = \int dx\, f(x) \qquad ; \quad \frac{\delta W(f)}{\delta f(y)} = 1$$

$$W_z(f) = \int dx\, K(z, x) f(x) \qquad ; \quad \frac{\delta W_z(f)}{\delta f(y)} = K(z, y) \quad (10.55)$$

---

[7] *Functional* means a "function of a function".

(2) If $K_2(x_1, x_2) = K_2(x_2, x_1)$ in the following functional, then the second variational derivative is computed with the same rules

$$W(f) = \int \int dx_1 dx_2 \, K_2(x_1, x_2) f(x_1) f(x_2)$$

$$\frac{\delta^2 W(f)}{\delta f(x_1) \delta f(x_2)} = 2! \, K_2(x_1, x_2) \tag{10.56}$$

(3) To generalize this result, suppose $K_n(x_1, \cdots, x_n)$ is a totally symmetric function of its arguments. Then

$$W(f) = \int \cdots \int dx_1 \cdots dx_n \, K_n(x_1, \cdots, x_n) f(x_1) \cdots f(x_n)$$

$$\frac{\delta^n W(f)}{\delta f(x_1) \cdots \delta f(x_n)} = n! \, K_n(x_1, \cdots, x_n) \tag{10.57}$$

(4) A somewhat more complicated example, which forms a key element of the subsequent analysis, is

$$\frac{\hbar}{i} \frac{\delta}{\delta J(t_1)} \exp \left\{ \frac{i}{\hbar} \int dt \, J(t) q(t) \right\} = \tag{10.58}$$

$$\frac{\hbar}{i} \mathrm{Lim}_{\lambda \to 0} \frac{1}{\lambda} \left[e^{\frac{i}{\hbar} \int dt \, J(t) q(t)} e^{\frac{i}{\hbar} \lambda q(t_1)} - e^{\frac{i}{\hbar} \int dt \, J(t) q(t)} \right]$$

$$= q(t_1) \exp \left\{ \frac{i}{\hbar} \int dt \, J(t) q(t) \right\}$$

This variational derivative with respect to $J(t_1)$ simply brings down a factor of $q(t_1)$.

Repeated application of this last result to the functional $W(J)$ in Eq. (10.51), with the source J set equal to zero at the end, gives

$$\left(\frac{\hbar}{i} \right)^n \frac{\delta^n W(J)}{\delta J(t_1) \cdots \delta J(t_n)} \bigg|_{J=0} =$$

$$\int dq \int dq' \, \psi_0^\star(q', t') \int \mathcal{D}(q) q(t_1) \cdots q(t_n) e^{\frac{i}{\hbar} S(t', t)} \psi_0(q, t) \tag{10.59}$$

From Eq. (10.43) this is identified as the ground-state expectation value of the time-ordered product

$$\left(\frac{\hbar}{i} \right)^n \frac{\delta^n W(J)}{\delta J(t_1) \cdots \delta J(t_n)} \bigg|_{J=0} = \langle \psi_0 | T[\hat{q}(t_1) \cdots \hat{q}(t_n)] | \psi_0 \rangle \tag{10.60}$$

and these are just the Green's functions of the theory!

10.1.9   *Generating Functional*

Let us divide $W(J)$ by the ground-state to ground-state amplitude in the absence of J, since one always measures *relative* to this. This amplitude is just our previous result in Eqs. (10.50)–(10.51) with $J = 0$.[8]

Take the ratio of Eq. (10.51) to the corresponding expression with $J = 0$, and then proceed to the limit. This produces the generating functional $\tilde{W}(J)$

$$\tilde{W}(J) = \text{Lim}_{T' \to -i\infty} \text{Lim}_{T \to +i\infty} \frac{\int \mathcal{D}(q) \exp\left\{ \frac{i}{\hbar} \int_T^{T'} dt \, [L(q, \dot{q}) + J(t)q] \right\}}{\int \mathcal{D}(q) \exp\left\{ \frac{i}{\hbar} \int_T^{T'} dt \, [L(q, \dot{q})] \right\}} \quad ;$$

$$\text{generating functional} \qquad (10.61)$$

where we have cancelled common factors in the numerator and denominator of this ratio.[9]

When we get to field theory, the ground-state to ground-state amplitude with $J = 0$ here becomes the vacuum-vacuum amplitude, and from the analysis in chapter 7, division by this amplitude removes the disconnected diagrams. The variational derivatives in Eq. (10.60) applied to the generating functional $\tilde{W}(J)$ now produce the *connected* Green's functions

$$\left(\frac{\hbar}{i}\right)^n \frac{\delta^n \tilde{W}(J)}{\delta J(t_1) \cdots \delta J(t_n)} \bigg|_{J=0} = \langle \psi_0 | T[\hat{q}(t_1) \cdots \hat{q}(t_n)] | \psi_0 \rangle_C \quad (10.62)$$

10.2   Many Degrees of Freedom

So far the analysis has focused on a system with just one degree of freedom. Consider now a system with *many degrees of freedom* $(q_1, q_2, \cdots, q_N)$; for example, mass points on a massless string (Fig. 10.6).

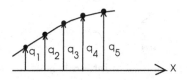

Fig. 10.6   Many degrees of freedom; for example, mass points on a massless string.

---

[8]To interpret the denominator in the r.h.s. of this ratio, just set $O = 1$ in Eq. (10.41).

[9]It is tacitly assumed here, and in the following, that the limit of the ratio is the ratio of the limits (where they exist).

We label the coordinates in the path integral by

$$q_{pr} \qquad ; \; p = 1, \cdots, N$$
$$r = 1, \cdots, n - 1 \qquad (10.63)$$

Here the first index p labels the coordinate, and the second index r refers to the particular time slice in Fig. 10.3. The measure in the path integral then simply becomes the product of the measures for each coordinate

$$\mathcal{D}(q) = \prod_{p=1}^{N} \mathcal{D}(q_p) \qquad ; \; \mathcal{D}(q_p) = \prod_{r=1}^{n-1} dq_{pr} \qquad (10.64)$$

The many-coordinate transition amplitude then takes the form

$$\langle q_{1f}, q_{2f}, \cdots, q_{Nf}; t' | q_{1i}, q_{2i}, \cdots, q_{Ni}; t \rangle =$$
$$\int \cdots \int \prod_{p=1}^{N} \mathcal{D}(q_p) \exp\left\{ \frac{i}{\hbar} \int_{t}^{t'} dt'' \, L(q_1, \cdots, q_N, \dot{q}_1, \cdots, \dot{q}_N) \right\} \quad (10.65)$$

There is now an additional (particle) label on the coordinates, and the path integral goes over *all* the coordinates at each time.

10.2.1 *Gaussian Integrals*

There is a result from classical analysis on "gaussian integrals" that proves indispensable in evaluating path integrals. Suppose that one has the following quadratic form in the coordinates $(q_1, \cdots, q_{n-1})$

$$\underline{q}^T \underline{M} \, \underline{q} = \sum_{i=1}^{n-1} \sum_{j=1}^{n-1} q_i M_{ij} q_j \qquad ; \; M_{ij} = M_{ji} \quad ; \; M_{ij} \text{ real} \quad (10.66)$$

where $M_{ij} = M_{ji}$ is a real, symmetric matrix. Suppose that there is also an additional "source term"

$$\underline{J}^T \underline{q} = \sum_{i=1}^{n-1} J_i q_i \qquad ; \; J_i \text{ real} \qquad (10.67)$$

Then the result for the gaussian integrals is as follows

$$\int dq_1 \cdots \int dq_{n-1} \exp\left\{ \frac{i}{\hbar} \left[a \, \underline{q}^T \underline{M} \, \underline{q} + b \, \underline{J}^T \underline{q} \right] \right\} = \qquad (10.68)$$

$$\frac{1}{(\det \underline{M})^{1/2}} \left(\frac{\pi \hbar e^{\pm i\pi/2}}{|a|} \right)^{(n-1)/2} \exp\left\{ -\frac{i}{\hbar} \frac{b^2}{4a} \underline{J}^T \underline{M}^{-1} \underline{J} \right\} \qquad ; \; a \gtrless 0$$

where (a, b) are real numbers, and the integrals run from $-\infty$ to ∞. We proceed to derive this expression.

(1) It is a result from algebra that a real symmetric matrix can be diagonalized with an orthogonal transformation satisfying $\underline{U}^T \underline{U} = 1$, which implies the determinant is given by $\det \underline{U} = 1$,

$$\underline{U}\, \underline{M}\, \underline{U}^T = \underline{M}_D \qquad ; \text{ diagonal} \qquad (10.69)$$

Here $\underline{M}_D$ is a diagonal matrix with elements $(M_1, \cdots, M_{n-1})$;[10]

(2) The jacobian for the corresponding coordinate transformation $\underline{\zeta} = \underline{U}\,\underline{q}$ or $\underline{q} = \underline{U}^T \underline{\zeta}$ is computed to be

$$\frac{\partial(\zeta_1, \cdots, \zeta_{n-1})}{\partial(q_1, \cdots, q_{n-1})} = \det \underline{U} = 1 \qquad (10.70)$$

(3) Thus the integral I on the l.h.s. of Eq. (10.68) is reduced to

$$I = \int d\zeta_1 \cdots \int d\zeta_{n-1} \exp\left\{ \frac{i}{\hbar} a \sum_{p=1}^{n-1} M_p \left[\zeta_p^2 + \frac{b}{aM_p} \sum_{l=1}^{n-1} J_l U_{lp}^T \zeta_p \right] \right\} (10.71)$$

(4) Now translate the coordinates, and complete the squares in the exponent,

$$\xi_p \equiv \zeta_p + \frac{b}{2aM_p} \sum_{l=1}^{n-1} J_l U_{lp}^T$$

$$I = \int d\xi_1 \cdots \int d\xi_{n-1} \times$$

$$\exp\left\{ \frac{i}{\hbar} a \sum_{p=1}^{n-1} \left(M_p \xi_p^2 - \frac{b^2}{4a^2 M_p} \sum_{l=1}^{n-1} \sum_{m=1}^{n-1} J_l U_{lp}^T J_m U_{mp}^T \right) \right\} (10.72)$$

(5) The multiple integral involving the first term in the exponent now factors into the product of the standard gaussian integrals in Eqs. (10.14)–(10.16)

$$\int_{-\infty}^{\infty} d\xi_p\, e^{\frac{i}{\hbar} a M_p \xi_p^2} = \left(\frac{\pi \hbar}{M_p |a| e^{\mp i\pi/2}} \right)^{1/2} \qquad ; a \gtrless 0 \qquad (10.73)$$

Furthermore

$$(M_1 M_2 \cdots M_{n-1})^{-1/2} = (\det \underline{M}_D)^{-1/2} = (\det \underline{M})^{-1/2} \quad (10.74)$$

---

[10]This result is proven within the framework of mechanics in [Fetter and Walecka (2003)].

The last equality follows since the determinant of $\underline{M}$ is unchanged under the orthogonal transformation

$$\det \underline{M} = \det \underline{U}^T \underline{M}_D \underline{U} = \det \underline{M}_D \qquad (10.75)$$

(6) The second term in the exponent in Eq. (10.72) gives rise to an expression that factors from the integrals. The sums in this second term can be written in matrix notation as

$$\sum_{p=1}^{n-1} \frac{1}{M_p} \sum_{l=1}^{n-1} \sum_{m=1}^{n-1} J_l U_{lp}^T J_m U_{mp}^T = \underline{J}^T \underline{U}^T \underline{M}_D^{-1} \underline{U}\, \underline{J} \qquad (10.76)$$

However $\underline{M}_D = \underline{U}\,\underline{M}\,\underline{U}^T$, which implies $\underline{M}_D^{-1} = \underline{U}\,\underline{M}^{-1}\underline{U}^T$, and hence

$$\underline{J}^T \underline{U}^T \underline{M}_D^{-1} \underline{U}\, \underline{J} = \underline{J}^T \underline{M}^{-1} \underline{J} \qquad (10.77)$$

We have now reproduced Eq. (10.68), and the proof is complete.

10.3   Field Theory

10.3.1   *Fields as Coordinates*

For concreteness, let us start with a neutral scalar field ϕ. A field is a system with many degrees of freedom. If we imagine space-time divided into small units, as we shall do in the evaluation of the path integral, and label each element with the index α, then the coordinates specifying the configuration of the system are ϕ_α where α runs over all the elements in space-time (see Fig. 10.7).

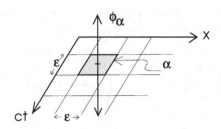

Fig. 10.7   Fields as coordinates. Here α labels a given element in space-time. Only one spatial dimension is shown.

10.3.2 Measure

The measure in the path integral correspondingly becomes

$$\prod_p \prod_r dq_{pr} \to \prod_\alpha d\phi_\alpha \equiv \mathcal{D}(\phi) \qquad ; \text{ field measure} \qquad (10.78)$$

10.3.3 Generating Functional

The lagrangian now becomes an integral over all space of the lagrangian density $L = \int d^3x\, \mathcal{L}(\phi, \boldsymbol{\nabla}\phi, \dot\phi)$; the action becomes an integral over all space-time $S = (1/c) \int d^4x\, \mathcal{L}(\phi, \boldsymbol{\nabla}\phi, \dot\phi)$; and the generating functional then takes the form [see Eq. (10.61)]

$$\tilde{W}(J) = \frac{\int \mathcal{D}(\phi) \exp\left\{\frac{i}{\hbar c} \int d^4x \left[\mathcal{L}(\phi, \boldsymbol{\nabla}\phi, \dot\phi) + J(x)\phi\right]\right\}}{\int \mathcal{D}(\phi) \exp\left\{\frac{i}{\hbar c} \int d^4x \left[\mathcal{L}(\phi, \boldsymbol{\nabla}\phi, \dot\phi)\right]\right\}} \qquad (10.79)$$

10.3.4 Convergence

Suppose one starts in *Minkowski space*, with a lagrangian density

$$\mathcal{L} = \mathcal{L}_0 + \mathcal{L}_1$$
$$\mathcal{L}_0 = -\frac{c^2}{2}\left[\left(\frac{\partial\phi}{\partial x_\mu}\right)^2 + m^2\phi^2\right]$$
$$\mathcal{L}_1 = \mathcal{L}_1(\phi) \qquad (10.80)$$

It soon becomes apparent that the quantity $\tilde{W}(J)$ is *ill-defined* due to infinite phases and oscillating integrands, for both large values of the time t, and for large values of the field ϕ.

10.3.4.1 Euclidicity Postulate

To circumvent this difficulty, one can appeal to the *euclidicity postulate*, which states that the Green's functions can be analytically continued in the time. One can start by computing the generating functional in Eq. (10.79) for *imaginary time*[11]

$$t \equiv -i\tau$$

$$i \int_{i\infty}^{-i\infty} dt \quad \rightarrow \quad \int_{-\infty}^{\infty} d\tau \qquad ; \text{ euclidicity postulate} \qquad (10.81)$$

Notice that this *builds in the conditions of the crucial theorem of Abers and Lee in Eq. (10.61)!*

$$T' \rightarrow -i\infty \qquad ; T \rightarrow i\infty \qquad ; \text{ crucial theorem} \qquad (10.82)$$

This procedure gives an integrand that is exponentially damped for large ϕ and large τ, and leads to a convergent generating functional, since the exponent is now[12]

$$\frac{i}{\hbar c} \int d^4x \left[\mathcal{L} + J\phi \right] \rightarrow \frac{1}{\hbar c} \int d^3x \, d\tau \, \left[\mathcal{L}\left(\phi, \boldsymbol{\nabla}\phi, i\frac{\partial\phi}{\partial\tau} \right) + J\phi \right] =$$

$$-\frac{1}{\hbar c} \int d^3x \, d\tau \, \left\{ \frac{1}{2} \left[\left(\frac{\partial\phi}{\partial\tau} \right)^2 + c^2(\boldsymbol{\nabla}\phi)^2 + m^2c^2\phi^2 \right] - \mathcal{L}_1(\phi) - J\phi \right\} (10.83)$$

One can then use the functional $\tilde{W}(J)$ in Eq. (10.79) to generate the connected Green's functions; however, the Green's functions are now computed in *euclidian space*. They must then be analytically continued (assumed possible) back to Minkowski space.

10.3.4.2 *Adiabatic Damping*

There is an equivalent alternative approach, and that is to stay in Minkowski space and add an *adiabatic damping factor* in the exponent

$$\frac{i}{\hbar c} \int d^4x \left[\mathcal{L} + J\phi \right] \rightarrow \frac{i}{\hbar c} \int d^4x \left[i\eta \frac{c^2\phi^2}{2} + \mathcal{L} + J\phi \right] \qquad ;$$

$$\text{adiabatic damping} \quad ; \eta \rightarrow 0^+ \qquad (10.84)$$

The path integral is again convergent for large ϕ and large t. The addition of this adiabatic damping term can be accomplished through the following replacement in the lagrangian density of Eq. (10.80)

$$m^2 \rightarrow m^2 - i\eta \qquad ; \text{ adiabatic damping} \qquad (10.85)$$

---

[11]Recall what we did with the momentum-space integrals in Fig. G.2. The rotation of the contour took $k_0 \rightarrow ik_0$, and then $k_\mu^2 = \mathbf{k}^2 - k_0^2 \rightarrow \mathbf{k}^2 + k_0^2$ was converted from Minkowski to euclidean form.

[12]We assume that $\mathcal{L}_1(\phi) < 0$, for example, $\mathcal{L}_1(\phi) = -\lambda\phi^4/4!$.

The great advantage of this second approach, which we shall actually employ, is that one can remain in Minkowski space throughout.[13]

10.4 Scalar Field

Let us start with the generating functional for the free scalar field with an additional source term $J(x)\phi$

$$\tilde{W}_0(J) = \frac{\int \mathcal{D}(\phi) \exp\left\{\frac{i}{2\hbar c}\int d^4x\left[\dot{\phi}^2 - c^2(\boldsymbol{\nabla}\phi)^2 - m^2c^2\phi^2 + i\eta c^2\phi^2 + 2J\phi\right]\right\}}{\int \mathcal{D}(\phi) \exp\left\{\frac{i}{2\hbar c}\int d^4x\left[\dot{\phi}^2 - c^2(\boldsymbol{\nabla}\phi)^2 - m^2c^2\phi^2 + i\eta c^2\phi^2\right]\right\}}$$

(10.86)

where we work in Minkowski space and include an adiabatic damping factor, as discussed above.

10.4.1 *Generating Functional for Free Scalar Field*

First, perform a partial integration in space-time and replace $-(\boldsymbol{\nabla}\phi)^2 \to \phi\nabla^2\phi$ and $\dot{\phi}^2 \to -\phi(\partial^2/\partial t^2)\phi$ in Eq. (10.86)

$$\tilde{W}_0(J) = \frac{\int \mathcal{D}(\phi) \exp\left\{\frac{i}{\hbar c}\int d^4x\left[\frac{c^2}{2}\phi(\Box - m^2 + i\eta)\phi + J\phi\right]\right\}}{\int \mathcal{D}(\phi) \exp\left\{\frac{i}{\hbar c}\int d^4x\left[\frac{c^2}{2}\phi(\Box - m^2 + i\eta)\phi\right]\right\}}$$

(10.87)

This is justified by our familiar continuum boundary conditions:

- It is assumed the disturbance is confined to a *finite region* of space-time;
- The boundary contributions can consequently be discarded.

Next, in line with our discussion of the path integral and fields, divide space into small cubes of volume ε^3, and time into small intervals with spacing $\Delta t = \varepsilon/c$, so that the space-time volume corresponding to d^4x is $\Delta V = \varepsilon^4$. Simultaneously, label the units in space-time with the index α (see Fig. 10.7)

$$\Delta V = \varepsilon^4 \qquad ; \text{ space-time volume at unit } \alpha \qquad (10.88)$$

---

[13]The role of this adiabatic damping factor is to put the correct Feynman boundary conditions in the propagators (see later).

Now compute the ratio, and at the very end, take the limit $\varepsilon \to 0$. In this manner, the generating functional in Eq. (10.87) becomes

$$\tilde{W}_0(J) = \mathrm{Lim}_{\varepsilon \to 0} \left(\int \cdots \int \prod_\alpha d\phi_\alpha \times \right.$$

$$\left. \exp\left\{ \frac{i}{\hbar c} \left[\sum_\beta \varepsilon^4 \sum_\gamma \varepsilon^4 \frac{c^2}{2} \phi_\beta K_{\beta\gamma} \phi_\gamma + \sum_\beta \varepsilon^4 J_\beta \phi_\beta \right] \right\} \right) \bigg/ (\cdots)_{J=0} \quad (10.89)$$

This relation *defines* $K_{\beta\gamma}$. In the limit $\varepsilon \to 0$ one has

$$\beta \to x, \qquad \gamma \to y \qquad\qquad ; \varepsilon \to 0 \quad (10.90)$$

Hence

$$\sum_\gamma \varepsilon^4 \frac{1}{2} \phi_\beta K_{\beta\gamma} \phi_\gamma \to \frac{1}{2} \int d^4y\, \phi(x) K(x,y) \phi(y) \qquad ; \varepsilon \to 0$$

$$= \frac{1}{2} \phi(x)(\Box_x - m^2 + i\eta)\phi(x) \quad (10.91)$$

Thus one can identify

$$\mathrm{Lim}_{\varepsilon \to 0}\, K_{\beta\gamma} = K(x,y) = \left(\Box_x - m^2 + i\eta\right)\delta^{(4)}(x-y) \quad (10.92)$$

This is readily verified by substitution into the first of Eqs. (10.91) and integration over d^4y.

We now observe the very important feature that *any common factors in the numerator and denominator cancel in the ratio in the generating functional.*

Our result for gaussian integrals in Eq. (10.68) then immediately allows us to evaluate $\tilde{W}_0(J)$ in Eq. (10.89) as

$$\tilde{W}_0(J) = \mathrm{Lim}_{\varepsilon \to 0} \left[\frac{1}{(\det \underline{K})^{1/2}} \prod_\alpha \left(\frac{2\pi\hbar c\, e^{i\pi/2}}{c^2 \varepsilon^8} \right)^{1/2} \times \right.$$

$$\left. \exp\left\{ -\frac{i}{\hbar} \frac{\varepsilon^8}{2c^3 \varepsilon^8} \sum_\beta \sum_\gamma J_\beta [\underline{K}^{-1}]_{\beta\gamma} J_\gamma \right\} \right] \bigg/ [\cdots]_{J=0} \quad (10.93)$$

It remains to identify the matrix $\underline{K}^{-1}$, which is the inverse of the matrix $\underline{K}$. By definition

$$\sum_\rho K_{\beta\rho} K_{\rho\gamma}^{-1} = \delta_{\beta\gamma} \quad (10.94)$$

Therefore

$$\sum_{\rho} \varepsilon^4 K_{\beta\rho} \frac{1}{\varepsilon^8} K_{\rho\gamma}^{-1} = \frac{1}{\varepsilon^4} \delta_{\beta\gamma} \qquad (10.95)$$

In the limit $\varepsilon \to 0$

$$\rho \to z \qquad ; \sum_{\rho} \varepsilon^4 \to \int d^4 z \qquad ; \frac{1}{\varepsilon^4} \delta_{\beta\gamma} \to \delta^{(4)}(x-y) \quad (10.96)$$

where the last relation follows from the definition of $\delta_{\beta\gamma}$. Now *define*

$$\mathrm{Lim}_{\varepsilon \to 0} \frac{1}{\varepsilon^8} K_{\rho\gamma}^{-1} \equiv -\Delta_F(z-y) \qquad (10.97)$$

Hence, in the limit, Eq. (10.95) becomes

$$\int d^4 z [(\Box_x - m^2 + i\eta)\delta^{(4)}(x-z)][-\Delta_F(z-y)] = \delta^{(4)}(x-y)$$

$$\Rightarrow \qquad (\Box_x - m^2 + i\eta)\, \Delta_F(x-y) = -\delta^{(4)}(x-y) \quad (10.98)$$

The solution to this last equation is just

$$\Delta_F(x-y) = \int \frac{d^4 k}{(2\pi)^4} \frac{e^{ik\cdot(x-y)}}{k^2 + m^2 - i\eta} \qquad ; \text{Feynman propagator} \quad (10.99)$$

This is the *Feynman propagator for the scalar field!*

With the cancellation of common factors in the numerator and denominator, the generating functional is given in the same limit by

$$\tilde{W}_0(J) = \exp\left\{ \frac{i}{2\hbar c^3} \int d^4 x \int d^4 y\, J(x)\Delta_F(x-y)J(y) \right\} \quad (10.100)$$

Notice that, with this procedure, we have carried out an exact evaluation of the ratio of field-theory path-integrals for the free scalar field.

10.4.1.1 *Applications*

The exponential in Eq. (10.100) can be expanded as

$$e^x = 1 + x + \frac{1}{2!}x^2 + \cdots \qquad (10.101)$$

(1) As a first application, we compute the second variational derivative of $\tilde{W}_0(J)$ with respect to the source $J(x)$, and then set the source equal to zero. In this case, it is only the term bilinear in J that makes

a non-vanishing contribution, and this arises from the second term in the expansion in Eq. (10.101). With the labeling $\tilde{W}^{(2)}(J)$, this term is

$$\tilde{W}_0^{(2)}(J) = \left(\frac{i}{\hbar c}\right)^2 \int d^4x_1 \int d^4x_2 \frac{1}{2!} K_2(x_1, x_2) J(x_1) J(x_2)$$

$$K_2(x_1, x_2) \equiv \frac{\hbar}{ic} \Delta_F(x_1 - x_2) \tag{10.102}$$

Note that $K_2(x_1, x_2)$ is a symmetric function of its arguments. Hence from Eq. (10.56)

$$\left(\frac{\hbar c}{i}\right)^2 \frac{\delta^2 \tilde{W}_0^{(2)}(J)}{\delta J(x_1) \delta J(x_2)} = K_2(x_1, x_2) \tag{10.103}$$

Now it is evident from Eqs. (10.58) and (10.87) that $(\hbar c/i)\delta/\delta J(x)$ brings down a factor of $\phi(x)$ in the generating functional. It is also clear from Eq. (10.62) that the second variational derivative, with J set equal to zero at the end, produces the ground-state expectation value of the time-ordered product of the fields. Thus the result in Eq. (10.103) implies

$$\left(\frac{\hbar c}{i}\right)^2 \frac{\delta^2 \tilde{W}_0(J)}{\delta J(x_1) \delta J(x_2)}\Big|_{J=0} = \frac{\hbar}{ic} \Delta_F(x_1 - x_2)$$

$$= \langle \psi_0 | T[\hat{\phi}(x_1)\hat{\phi}(x_2)] | \psi_0 \rangle \tag{10.104}$$

This is precisely Eq. (F.44) for the Feynman propagator of the scalar field!

(2) Let us extend the above to the corresponding fourth-order calculation, where the relevant contribution this time arises from the third term in the expansion in Eq. (10.101)

$$\tilde{W}_0^{(4)}(J) = \frac{1}{2!}\left(\frac{i}{2\hbar c^3}\right)^2 \int d^4x_1 \cdots \int d^4x_4 \, \Delta_F(x_1 - x_2)\Delta_F(x_3 - x_4) \times$$

$$J(x_1) \cdots J(x_4) \tag{10.105}$$

Utilizing a change of dummy integration variables, this can be written as a functional with a completely symmetric kernel

$$\tilde{W}_0^{(4)}(J) = \left(\frac{i}{\hbar c}\right)^4 \int d^4x_1 \cdots \int d^4x_4 \frac{1}{4!} K_4(x_1, \cdots, x_4) \times$$

$$J(x_1) \cdots J(x_4)$$

$$K_4(x_1, \cdots, x_4) = \left(\frac{\hbar}{ic}\right)^2 [\Delta_F(x_1 - x_2)\Delta_F(x_3 - x_4) +$$

$$\Delta_F(x_1 - x_3)\Delta_F(x_2 - x_4) + \Delta_F(x_1 - x_4)\Delta_F(x_2 - x_3)] \tag{10.106}$$

The use of Eqs. (10.62) and (10.57) then gives

$$\left(\frac{\hbar c}{i}\right)^4 \frac{\delta^4 \tilde{W}_0(J)}{\delta J(x_1)\delta J(x_2)\delta J(x_3)\delta J(x_4)}\bigg|_{J=0} = \langle\psi_0|T[\hat{\phi}(x_1)\hat{\phi}(x_2)\hat{\phi}(x_3)\hat{\phi}(x_4)]|\psi_0\rangle_C$$

$$= \left(\frac{\hbar}{ic}\right)^2 [\Delta_F(x_1 - x_2)\Delta_F(x_3 - x_4)$$

$$+\Delta_F(x_1 - x_3)\Delta_F(x_2 - x_4) + \Delta_F(x_1 - x_4)\Delta_F(x_2 - x_3)] \quad (10.107)$$

This is just the fully contracted term in Wick's theorem for the free scalar field!

It is evident that in this path-integral approach, rather than forming the starting point, the Feynman rules are obtained at the end.

10.4.2 *Interactions*

Let us extend the analysis to include interactions of the scalar field, for example, with a term of the form

$$\mathcal{L}_1(\phi) = -\frac{\lambda}{4!}\phi^4 \qquad (10.108)$$

This can be done through a series of steps:

(1) It is important to remember that ϕ is just a c-number field. The term arising from $\mathcal{L}_1(\phi)$ in the action therefore *factors* in the integrand of the path integral.

$$\exp\left\{\frac{i}{\hbar c}\int d^4x\,\mathcal{L}_1(\phi)\right\} \qquad ;\text{factors} \qquad (10.109)$$

(2) The field can now be replaced by a functional derivative, since

$$\phi(x)\exp\left\{\frac{i}{\hbar c}\int d^4y\,J(y)\phi(y)\right\} = \frac{\hbar c}{i}\frac{\delta}{\delta J(x)}\exp\left\{\frac{i}{\hbar c}\int d^4y\,J(y)\phi(y)\right\}$$

$$(10.110)$$

In this manner, the generating functional with interactions becomes

$$\tilde{W}(J) = \left(\int \mathcal{D}(\phi)\exp\left\{\frac{i}{\hbar c}\int d^4x\,\mathcal{L}_1\left[\frac{\hbar c}{i}\frac{\delta}{\delta J(x)}\right]\right\}\times\right.$$

$$\left.\exp\left\{\frac{i}{\hbar c}\int d^4y\,[\mathcal{L}_0 + J\phi]\right\}\right)\bigg/(\cdots)_{J=0} \qquad (10.111)$$

(3) The first term in the numerator is now *independent of ϕ and factors out of the path integral.* The remaining expression in the numerator is just $\tilde{W}_0(J) \times$ [factor independent of J] . Since any common factor in the numerator and denominator *cancels in the ratio,* one has

$$\tilde{W}(J) = \left(\exp\left\{ \frac{i}{\hbar c} \int d^4x\, \mathcal{L}_1 \left[\frac{\hbar c}{i} \frac{\delta}{\delta J(x)} \right] \right\} \tilde{W}_0(J) \right) \Big/ (\cdots)_{J=0} \quad (10.112)$$

where the denominator is to be evaluated at the end.

(4) We have thus done the major part of the work by computing $\tilde{W}_0(J)$, and it remains to just evaluate the appropriate functional derivatives in $\mathcal{L}_1$ to determine $\tilde{W}(J)$.[14]

(5) From $\tilde{W}(J)$, one generates all the Feynman rules for the connected diagrams of the fully interacting $\lambda\phi^4/4!$ field theory!

10.5 Fermions

There is an essential complication with fermions. The fields in the quantum field theory *anticommute*. One must have a corresponding mechanism in the path integrals, where the fields are *c*-numbers, to keep track of signs. We describe such a procedure.

10.5.1 *Grassmann Algebra*

Introduce a *Grassmann algebra of anticommuting c-numbers* c_i, with $i = 1, \cdots, n$, satisfying

$$\{c_i,\, c_j\} \equiv c_i c_j + c_j c_i = 0 \qquad ; (i,j) = 1, \cdots, n$$
$$c_i^2 = 0 \qquad \text{Grassmann algebra} \quad (10.113)$$

In the continuum limit, we identify

$$c_i \to c_x = c(x) \qquad ; \text{continuum limit} \quad (10.114)$$

---

[14]See Prob. 10.3.

10.5.2 *Functional Derivative*

Assume one has two distinct algebras $c(x)$ and $\bar{c}(x)$, with all elements mutually anticommuting, and a functional $P[\bar{c}, c]$ given by

$$P[\bar{c}, c] = \int dx_1 \cdots dx_n \int dy_1 \cdots dy_n \, \bar{c}(x_1) \cdots \bar{c}(x_n) \frac{1}{n!} \frac{1}{n!} \times$$
$$K_n(x_1, \cdots, x_n; y_1, \cdots, y_n) c(y_1) \cdots c(y_n) \qquad (10.115)$$

where the kernel K_n is completely antisymmetric both in $(x_1, \cdots, x_n)$ and in $(y_1, \cdots, y_n)$. The *left*- and *right*-variational derivatives are then defined in the following fashion

$$\frac{\delta^n}{\delta \bar{c}(x_n) \cdots \delta \bar{c}(x_1)} P[\bar{c}, c] \frac{\delta^n}{\delta c(y_n) \cdots \delta c(x_1)} \equiv K_n(x_1, \cdots, x_n; y_1, \cdots, y_n)$$
$$(10.116)$$

Note that it is crucial to keep track of the *order* in which the variational derivatives are taken; each side of this equation is antisymmetric both in $(x_1, \cdots, x_n)$ and in $(y_1, \cdots, y_n)$.

10.5.3 *Functional Integration*

Divide space-time up into units of size ε^4, and label each element by α exactly as in Fig. 10.7. Identify distinct Grassmann algebras with ψ and $\bar{\psi}$ through

$$\psi(x) \to \psi_\alpha$$
$$\bar{\psi}(x) \to \bar{\psi}_\alpha \qquad ; \text{ Grassmann algebras} \qquad (10.117)$$

The differentials of these quantities $(d\psi_\alpha, d\bar{\psi}_\alpha)$ are assumed to form similar Grassmann algebras, and everything anticommutes. Thus

$$\{\psi_\alpha, \psi_\beta\} = 0 \qquad ; \{d\psi_\alpha, d\psi_\beta\} = 0$$
$$\{\bar{\psi}_\alpha, \bar{\psi}_\beta\} = 0 \qquad ; \{d\bar{\psi}_\alpha, d\bar{\psi}_\beta\} = 0$$
$$\{\psi_\alpha, d\psi_\beta\} = 0 \qquad ; \{\bar{\psi}_\alpha, d\bar{\psi}_\beta\} = 0 \qquad ; \text{ etc.} \qquad (10.118)$$

Introduce the *measure* for the path integrals as follows

$$\mathcal{D}(\bar{\psi})\mathcal{D}(\psi) \equiv d\bar{\psi}_n \cdots d\bar{\psi}_1 \, d\psi_1 \cdots d\psi_n \qquad ; \text{ measure} \qquad (10.119)$$

Note that the order is again important since the infinitesimals anticommute.

10.5.4   *Integrals*

Integrals are then *defined* in the following fashion

$$\int d\bar{\psi}_\alpha = \int d\psi_\beta \equiv 0 \qquad ; \text{ definitions}$$

$$\int \bar{\psi}_\alpha d\bar{\psi}_\alpha = \int \psi_\beta d\psi_\beta \equiv 1 \qquad (10.120)$$

Several comments:

- There is no sum over repeated indices here;
- If we remember that $\psi_\alpha \psi_\alpha = 0$, then something different is clearly called for in the definition of integration;
- This is the non-intuitive part of the development;
- The definition of integration will be justified *a posteriori*, through the results to which it leads.

10.5.5   *Basic Results*

The basic result for fermions, which corresponds to the previous gaussian integrals, is

$$\int d\bar{c}_n \cdots d\bar{c}_1 \int dc_1 \cdots dc_n \, \exp\left\{ -\sum_{i=1}^n \sum_{j=1}^n \bar{c}_i N_{ij} c_j \right\} = \det \underline{N} \quad (10.121)$$

Some comments:

- Note the order of the differentials on the l.h.s., which anticommute;
- $\det \underline{N}$ sits upstairs on the r.h.s., in contrast to the previous $(\det \underline{M})^{-1/2}$;
- There are no factors in the path integral measure produced on the r.h.s., as there were before. In any event, additional factors would cancel in the *ratio* that will be of interest to us;
- We proceed to a proof of Eq. (10.121).

We prove the result for the simplest case of $n = 2$, assign the case $n = 3$ as Prob. 10.9, and leave the generalization as an exercise for the dedicated reader. Consider the following integral over the distinct sets of Grassmann variables $(\bar{c}, c)$

$$I \equiv \int d\bar{c}_2 d\bar{c}_1 \int dc_1 dc_2 \, \exp\left\{ -[\bar{c}_1 N_{11} c_1 + \bar{c}_1 N_{12} c_2 + \bar{c}_2 N_{21} c_1 + \bar{c}_2 N_{22} c_2] \right\}$$

$$(10.122)$$

Expand the exponential as in Eq. (10.101). One only needs to keep the terms up through x^2, since with any higher powers of x, the individual factors can be anticommuted until they sit next to each other, and then at least two of them will lead to a vanishing contribution since

$$c_1^2 = c_2^2 = \bar{c}_1^2 = \bar{c}_2^2 = 0 \tag{10.123}$$

The terms $(1, x)$ in the expansion give a vanishing result by the definition of integration in Eqs. (10.120). Hence the integral in Eq. (10.122) is given by

$$I = \int d\bar{c}_2 d\bar{c}_1 \int dc_1 dc_2 \, \frac{(-1)^2}{2!} \left[(\bar{c}_1 N_{11} c_1 + \bar{c}_1 N_{12} c_2 + \bar{c}_2 N_{21} c_1 + \bar{c}_2 N_{22} c_2) \times \right.$$
$$\left. (\bar{c}_1 N_{11} c_1 + \bar{c}_1 N_{12} c_2 + \bar{c}_2 N_{21} c_1 + \bar{c}_2 N_{22} c_2) \right] \tag{10.124}$$

Note[15]

$$(-1) \int \int d\bar{c}_1 dc_1 \, (\bar{c}_1 c_1) = 1 \tag{10.125}$$

Now use Eqs. (10.120) to obtain

$$I = \frac{2!}{2!} \, [N_{11} N_{22} - N_{21} N_{12}] = \det \underline{N} \tag{10.126}$$

This is Eq. (10.121) for $n = 2$.

The generalization of this result to the case where there are real external *sources* $(\bar{\xi}, \xi)$, which themselves form distinct sets of Grassmann variables, is

$$\int d\bar{c}_n \cdots d\bar{c}_1 \int dc_1 \cdots dc_n \exp \left\{ - \sum_{i=1}^{n} \sum_{j=1}^{n} \bar{c}_i N_{ij} c_j + \sum_{i=1}^{n} (\bar{c}_i \xi_i + \bar{\xi}_i c_i) \right\} =$$
$$(\det \underline{N}) \exp \left\{ \sum_{i=1}^{n} \sum_{j=1}^{n} \bar{\xi}_i \left[\underline{N}^{-1} \right]_{ij} \xi_j \right\} \tag{10.127}$$

To prove this result, rewrite the exponent on the l.h.s. in matrix notation as

$$-\bar{\underline{c}}^T \underline{N} \, \underline{c} + \bar{\underline{c}}^T \underline{\xi} + \bar{\underline{\xi}}^T \underline{c} = - \left(\bar{\underline{c}}^T - \bar{\underline{\xi}}^T \underline{N}^{-1} \right) \underline{N} \left(\underline{c} - \underline{N}^{-1} \underline{\xi} \right) + \bar{\underline{\xi}}^T \underline{N}^{-1} \underline{\xi} \tag{10.128}$$

---

[15] Recall $\{c_i, \bar{c}_j\} = \{dc_i, d\bar{c}_j\} = \{dc_i, \bar{c}_j\} = \{d\bar{c}_i, c_j\} = 0$.

This is an algebraic identity. Now change variables in the integral according to

$$\eta \equiv \underline{c} - \underline{N}^{-1}\underline{\xi}$$
$$\bar{\eta}^T \equiv \underline{\bar{c}}^T - \underline{\bar{\xi}}^T\underline{N}^{-1} \tag{10.129}$$

The final terms on the r.h.s. are just constants, and this is just an ordinary change of variables that preserves the definition of integration

$$d\eta_i = dc_i \qquad ; \qquad d\bar{\eta}_i = d\bar{c}_i$$

$$\int d\eta_i = 0 \qquad ; \qquad \int d\bar{\eta}_i = 0$$

$$\int \eta_i d\eta_i = \int (c_i - [\underline{N}^{-1}\underline{\xi}]_i)dc_i = 1$$

$$\int \bar{\eta}_i d\bar{\eta}_i = \int (\bar{c}_i - [\underline{\bar{\xi}}^T\underline{N}^{-1}]_i)d\bar{c}_i = 1 \tag{10.130}$$

Hence, with the aid of Eq. (10.128), the integral $\mathcal{I}$ on the l.h.s. of Eq. (10.127) takes the form

$$\mathcal{I} = \int d\bar{\eta}_n \cdots d\bar{\eta}_1 \int d\eta_1 \cdots d\eta_n \exp\left\{-\bar{\eta}^T\underline{N}\,\eta + \bar{\xi}^T\underline{N}^{-1}\xi\right\} \tag{10.131}$$

The last term in the exponent is bilinear in $(\bar{\xi}, \xi)$, and the corresponding exponential then factors out of the integral. The first term is bilinear in $(\bar{\eta}, \eta)$, and it gives rise to the integral done previously in Eq. (10.121). The result is just Eq. (10.127), which was to be proven.

10.5.6   *Generating Functional for Free Dirac Field*

The lagrangian density for the free Dirac field is

$$\mathcal{L}_0 = -\hbar c\bar{\psi}\left(\gamma_\mu\frac{\partial}{\partial x_\mu} + M\right)\psi \tag{10.132}$$

The generating functional is correspondingly defined by

$$\tilde{W}_0[\bar{\zeta}, \zeta] \equiv \left(\int\int \mathcal{D}(\bar{\psi})\mathcal{D}(\psi)\exp\left\{\frac{i}{\hbar c}\int d^4x\,[\mathcal{L}_0 + \bar{\zeta}(x)\psi + \bar{\psi}\,\zeta(x)]\right\}\right)$$
$$\Big/(\cdots)_{\bar{\zeta}=\zeta=0} \tag{10.133}$$

We again divide space-time into small cells of volume ε^4 and label each element with the index α just as in Fig. 10.7, so the Dirac field becomes

$\psi(x) \to \psi_\alpha$. The generating functional is computed as with the scalar field, only now using the properties of the Grassmann variables. The numerator $\mathcal{N}$ of the generating functional is given by

$$\mathcal{N} = \int d\bar{\psi}_n \cdots d\bar{\psi}_1 \int d\psi_1 \cdots d\psi_n \times \tag{10.134}$$

$$\exp\left\{-i\left[\sum_\alpha \sum_\beta \varepsilon^8 \bar{\psi}_\alpha K_{\alpha\beta} \psi_\beta - \frac{1}{\hbar c} \sum_\alpha \varepsilon^4 \left(\bar{\zeta}_\alpha \psi_\alpha + \bar{\psi}_\alpha \zeta_\alpha\right)\right]\right\}$$

Here the matrix $\underline{K}$ is defined so that in the continuum limit

$$\sum_\beta \varepsilon^4 \bar{\psi}_\alpha K_{\alpha\beta} \psi_\beta \to \int d^4 y \, \bar{\psi}(x) K(x,y) \psi(y)$$

$$= \bar{\psi}(x) \left[\gamma_\mu \frac{\partial}{\partial x_\mu} + M - i\eta\right] \psi(x) \tag{10.135}$$

where the adiabatic damping factor has again been included in the mass. This implies

$$K(x,y) = \left[\gamma_\mu \frac{\partial}{\partial x_\mu} + M - i\eta\right] \delta^{(4)}(x-y) \tag{10.136}$$

Now compare Eq. (10.134) with the basic result in Eq. (10.127), and identify

$$N_{\alpha\beta} = i\varepsilon^8 K_{\alpha\beta} \qquad ; [\underline{N}^{-1}]_{\alpha\beta} = \frac{1}{i\varepsilon^8}[\underline{K}^{-1}]_{\alpha\beta}$$

$$\xi_\alpha = \frac{i}{\hbar c}\varepsilon^4 \zeta_\alpha \qquad ; \bar{\xi}_\alpha = \frac{i}{\hbar c}\varepsilon^4 \bar{\zeta}_\alpha \tag{10.137}$$

Equation (10.127) then evaluates the generating functional as

$$\tilde{W}_0[\bar{\zeta}, \zeta] = \mathrm{Lim}_{\varepsilon \to 0}\left[(\det \underline{N}) \exp\left\{\sum_\alpha \sum_\beta \left(\frac{i\varepsilon^4}{\hbar c}\right)^2 \bar{\zeta}_\alpha \left(\frac{1}{i\varepsilon^8}[\underline{K}^{-1}]_{\alpha\beta}\right) \zeta_\beta\right\}\right]$$

$$\Big/ [\cdots]_{\bar{\zeta}=\zeta=0} \tag{10.138}$$

The inverse $\underline{K}^{-1}$ of the matrix $\underline{K}$ is again defined by

$$\sum_\gamma K_{\alpha\gamma} K_{\gamma\beta}^{-1} = \delta_{\alpha\beta} \tag{10.139}$$

Now define, in the limit $\varepsilon \to 0$,

$$\frac{1}{\varepsilon^8} K_{\alpha\beta}^{-1} \equiv -S_F(x-y) \qquad ; \alpha \to x \;\;, \beta \to y \tag{10.140}$$

Just as before

$$\int d^4z\, K(x-z)[-S_F(z-y)] = \delta^{(4)}(x-y)$$

$$\Rightarrow \quad \left[\gamma_\mu \frac{\partial}{\partial x_\mu} + M - i\eta\right] S_F(x-y) = -\delta^{(4)}(x-y) \quad (10.141)$$

The solution to this equation is just the *Feynman propagator for the Dirac field*

$$S_F(x-y) = -\int \frac{d^4k}{(2\pi)^4} \frac{e^{ik\cdot(x-y)}}{i\slashed{k} + M - i\eta} \quad ; \text{ Feynman propagator} \quad (10.142)$$

The generating functional for the free Dirac field then becomes

$$\tilde{W}_0[\bar{\zeta}, \zeta] = \exp\left\{-\frac{i}{(\hbar c)^2}\int d^4x \int d^4y\, \bar{\zeta}(x)S_F(x-y)\zeta(y)\right\} \quad ;$$

$$\text{generating functional} \quad (10.143)$$

We note that, once again, the ratio of path integrals has been evaluated exactly.

10.5.6.1 *Applications*

It is clear from Eq. (10.133) that a left-variational derivative $(\hbar c/i)\delta/\delta\bar{\zeta}(x)$ brings down a factor of $\psi(x)$ and a right-variational derivative $(\hbar c/i)\delta/\delta\zeta(y)$ brings down a factor of $\bar{\psi}(y)$ in the generating functional.[16] Furthermore, since the fields are Grassmann variables, the ground-state expectation value is that of the P-product of the fields (rather than just the T-product).

(1) *Two Sources.* The exponential in Eq. (10.143) can again be expanded as in Eq. (10.101). The contribution bilinear in the sources arises from the term x in this expansion.

$$\tilde{W}_0^{(2)}[\bar{\zeta}, \zeta] = -\frac{i}{(\hbar c)^2}\int d^4x \int d^4y\, \bar{\zeta}(x)S_F(x-y)\zeta(y) \quad (10.144)$$

Hence

$$\left(\frac{\hbar c}{i}\right)^2 \left[\frac{\delta}{\delta\bar{\zeta}(x)}\tilde{W}_0[\bar{\zeta}, \zeta]\frac{\delta}{\delta\zeta(y)}\right]_{\bar{\zeta}=\zeta=0} = \langle\psi_0|P[\hat{\psi}(x), \hat{\bar{\psi}}(y)]|\psi_0\rangle$$

$$= iS_F(x-y) \quad (10.145)$$

This is precisely the Feynman propagator of Eqs. (F.49)–(F.50)!

---

[16]Since the source terms are bilinear in Grassmann variables, they commute with everything else. (Compare with Prob. 10.2).

(2) *Four Sources.* The contribution quartic in the sources arises from the x^2 in Eq. (10.101) and takes the form

$$\tilde{W}_0^{(4)}[\bar{\zeta}, \zeta] = \left[\frac{-i}{(\hbar c)^2}\right]^2 \frac{1}{2!} \int d^4x_1 d^4x_2 d^4y_1 d^4y_2 \, (-1)\bar{\zeta}(x_1)\bar{\zeta}(x_2)\zeta(y_1)\zeta(y_2) \times$$
$$[S_F(x_1 - y_1)S_F(x_2 - y_2)] \tag{10.146}$$

Use a change in dummy variables to antisymmetrize in (x_1, x_2) and (y_1, y_2)

$$[S_F(x_1 - y_1)S_F(x_2 - y_2)] \rightarrow$$
$$\frac{1}{2!}[S_F(x_1 - y_1)S_F(x_2 - y_2) - S_F(x_2 - y_1)S_F(x_1 - y_2)] \tag{10.147}$$

Then use the functional derivative relation in Eq. (10.115)–(10.116) to obtain

$$\left(\frac{\hbar c}{i}\right)^4 \left[\frac{\delta}{\delta\bar{\zeta}(x_2)} \frac{\delta}{\delta\bar{\zeta}(x_1)} \tilde{W}_0[\bar{\zeta}, \zeta] \frac{\delta}{\delta\zeta(y_2)} \frac{\delta}{\delta\zeta(y_1)}\right]_{\bar{\zeta}=\zeta=0} = \tag{10.148}$$
$$= K_2(x_1, x_2; y_1, y_2)$$
$$= \langle\psi_0|P[\hat{\psi}(x_2)\hat{\psi}(x_1)\hat{\bar{\psi}}(y_2)\hat{\bar{\psi}}(y_1)]|\psi_0\rangle_C$$
$$= i^2[S_F(x_2 - y_1)S_F(x_1 - y_2) - S_F(x_1 - y_1)S_F(x_2 - y_2)]$$

This is just the fully contracted term in Wick's theorem for the Dirac fields![17]

In *summary*, the results in these two applications:

- Justify the definition of functional integration for fermions;
- Justify the definition of functional differentiation for fermions;
- Justify the use of Grassmann algebras for the sources;
- Give Wick's theorem for fermions, with the correct signs and factors.

10.5.7 *Interactions (Dirac-Scalar)*

Consider the Dirac-scalar theory with the interaction lagrangian density of Eq. (5.119)

$$\mathcal{L}_1 = g\bar{\psi}\psi\phi \tag{10.149}$$

The previous arguments on interactions in the scalar case can be repeated step-by-step, and the analog of Eq. (10.112) for the generating functional

---

[17]We leave the restoration of the Dirac indices in these expressions as an exercise (see Prob. 10.1).

in the interacting field theory in this case is

$$\tilde{W}[\bar{\zeta}, \zeta, J] = \left(\exp\left\{ \frac{ig}{\hbar c} \int d^4x \left[-\left(\frac{\hbar c}{i}\right)^3 \frac{\delta}{\delta\zeta(x)} \frac{\delta}{\delta\bar{\zeta}(x)} \frac{\delta}{\delta J(x)} \right] \right\} \times$$

$$\tilde{W}_0^{\text{Dirac}}[\bar{\zeta}, \zeta] \, \tilde{W}_0^{\text{scalar}}[J] \right) \Big/ (\cdots)_{\bar{\zeta}=\zeta=J=0} \qquad (10.150)$$

Here the variational derivatives of the Dirac sources are to be interpreted as left-derivatives.[18]

10.6 Electromagnetic Field

The derivation of the Feynman rules in a gauge-invariant, covariant, and unitary form for a non-abelian gauge theory is a highly non-trivial task.[19] This problem was first solved by [Faddeev and Popov (1967)] using a path-integral approach. The gauge invariance of the theory presents an essential complication since in performing a path integral over all components of the fields, one includes *gauge-equivalent contributions that do not correspond to distinct physical configurations*. Faddeev and Popov showed how to *factor* the gauge-equivalent contributions from the path integrals, leaving a covariant, gauge-invariant generating functional. Although derivation of the results for a non-abelian gauge theory goes well beyond the confines of this book,[20] one can illustrate the principles within the framework of QED, and the path-integral treatment of QED, while challenging in itself, is discussed in appendix H.

---

[18]This is the reason for the minus sign – see Prob. 10.2. The Feynman rules following from this generating functional are discussed in appendix C of [Serot and Walecka (1986)].

[19]The reference [Feynman (1963)] is a verbatim transcript of lectures given by Feynman as he struggled to solve this problem. Every student of physics should read this paper for the insight it provides as to how a truly great mind works.

[20]See, for example, [Abers and Lee (1973); Itzykson and Zuber(1980); Cheng and Li (1984)].

Chapter 11

Canonical Transformations for Quantum Systems

As a final topic, we return to the study of Bose and Fermi quantum fluids. Although one can selectively sum sets of graphs, Feynman diagrams and Feynman rules are based on a perturbation expansion in the coupling constant. There are many physical systems for which a perturbation expansion is inadequate, for example, the Bose fluid ^{4}He, superconductors, nuclei in the pairing regime, and QED systems at long-wavelength. In these cases, it is extremely valuable to have *model problems that can be solved exactly*. Such models both provide insight and serve as a starting point for a more rigorous analysis. Canonical transformations provide us with just such a tool.

The basic idea is that, as we have seen, the properties of the creation and destruction operators in the abstract Hilbert space follow *entirely from the (anti)commutation relations*. The introduction of combinations of the original operators which *preserve* these (anti)commutation relations, allows one to include interactions between the quanta of the many-body system on a consistent basis.

The basic properties of quantum fluids were summarized in chapter 11 of Vol. I. In the present chapter, we first examine the behavior of a Bose system with a short-range repulsive interaction between the constituents using the canonical transformation of [Bogoliubov (1947)]. We then use the canonical transformation of [Bogoliubov (1958); Valatin (1958)] to examine a Fermi system with bound Cooper pairs [Cooper (1956)], which serves as the underlying basis of the BCS theory of superconductivity [Bardeen, Cooper, and Schrieffer (1957)].[1]

The problem of the quantized E-M field interacting with a specified external current, which provides a model for the infrared limit of QED,

---

[1] The material in this chapter is taken from chapter 10 of [Fetter and Walecka (2003a)].

was first solved with a canonical transformation by [Bloch and Nordsieck (1937)]. The reader is taken through that solution in Prob. 11.2. Problem 11.1 takes the reader through the analysis of a quantized real massive scalar field interacting with a specified static source, which provides an exact derivation of the strong-coupling Yukawa potential of nuclear physics.

11.1 Interacting Bose System

11.1.1 *Pseudopotential*

Let us start with a simplified problem. One expects the low-density properties of the many-body system to be determined by the *s*-wave scattering amplitude. We introduce a *pseudopotential*, which is chosen to give the correct *s*-wave scattering. Take the many-body hamiltonian $\hat{H} = \hat{T} + \hat{V}$ for a collection of identical spin-zero bosons to be[2]

$$\hat{H} \approx \sum_{\mathbf{k}} \hbar\omega_k a_{\mathbf{k}}^{\dagger} a_{\mathbf{k}} + \frac{G}{2\Omega} \sum_{\mathbf{k}_1} \cdots \sum_{\mathbf{k}_4} \delta_{\mathbf{k}_1+\mathbf{k}_2, \mathbf{k}_3+\mathbf{k}_4}\, a_{\mathbf{k}_3}^{\dagger} a_{\mathbf{k}_4}^{\dagger} a_{\mathbf{k}_2} a_{\mathbf{k}_1} \quad (11.1)$$

To identify G, we compute the two-body scattering amplitude in the C-M system with this interaction. The Golden Rule gives the transition rate as

$$\omega_{fi} = \frac{2\pi}{\hbar} |\langle f|\hat{V}|i\rangle|^2 \frac{\Omega d^3 k'}{(2\pi)^3} \delta\left(\frac{\hbar^2}{m}[\mathbf{k}^2 - \mathbf{k}'^2]\right) \quad ; \; \mu_r = \frac{m}{2} \quad (11.2)$$

where the reduced mass in $\mu_r = m/2$. The initial and final states are (compare Fig. 8.5)

$$|i\rangle = |\mathbf{k}, -\mathbf{k}\rangle \qquad ; \; |f\rangle = |\mathbf{k}', -\mathbf{k}'\rangle$$
$$\Rightarrow \qquad \langle \mathbf{k}', -\mathbf{k}'|\hat{V}|\mathbf{k}, -\mathbf{k}\rangle = \frac{4G}{2\Omega} \quad (11.3)$$

The required matrix element of the interaction in the second line follows immediately (see Prob. 11.3). The incident flux in the C-M system is

$$I_{\text{inc}} = \frac{2}{\Omega}\frac{\hbar k}{m} \quad (11.4)$$

Hence the differential cross section is

$$\frac{d\sigma}{d\Omega} \equiv |f|^2 = \frac{2\pi}{\hbar}\frac{\Omega}{(2\pi)^3}\left(\frac{km}{2\hbar^2}\right)\left(\frac{m\Omega}{2\hbar k}\right)\left(\frac{2G}{\Omega}\right)^2 = \left|\frac{2Gm}{4\pi\hbar^2}\right|^2 \quad (11.5)$$

---

[2]We again use a big box of volume Ω with p.b.c., and the result in Prob. 11.6(a). The remaining matrix element of V in Eqs. (2.113)–(2.114) is then replaced by a constant.

For identical spin-zero bosons, since *either* of the final particles can get into the detector, the low-energy scattering amplitude is given by (see Prob. 11.4)

$$|f|^2 = |f(\theta) + f(\pi - \theta)|^2 = |2a|^2 \qquad ; \text{ scattering length} \qquad (11.6)$$

where a is the *scattering length*. Thus, choosing a sign appropriate to a repulsive potential, we have the following relation between the strength of the pseudopotential G and the s-wave scattering length a [3]

$$G = \frac{4\pi\hbar^2 a}{m} \qquad ; \text{ pseudopotential} \qquad (11.7)$$

11.1.2 Special Role of the Ground State

In the non-interacting ground state of the system with N bosons, all the particles are condensed into the mode with $\mathbf{k} = 0$

$$|\Phi_0(N)\rangle = |N, 0, 0, \cdots\rangle \qquad ; \text{ non-interacting ground state} \quad (11.8)$$

The creation and destruction operators for the $\mathbf{k} = 0$ mode acting on this state give

$$
\begin{aligned}
a_0 |\Phi_0(N)\rangle &= \sqrt{N}\, |\Phi_0(N-1)\rangle \\
a_0^\dagger |\Phi_0(N)\rangle &= \sqrt{N+1}\, |\Phi_0(N+1)\rangle
\end{aligned}
\qquad (11.9)
$$

We make some observations:

- Define new operators

$$\xi_0 \equiv \frac{a_0}{\sqrt{\Omega}} \qquad ; \xi_0^\dagger \equiv \frac{a_0^\dagger}{\sqrt{\Omega}} \qquad (11.10)$$

  Now consider the *thermodynamic limit* where $(N, \Omega) \to \infty$ at finite density $\rho = N/\Omega$. While the new operators have a finite effect in Eqs. (11.9), the *commutator* of these new operators vanishes

$$[\xi_0, \xi_0^\dagger] = \frac{1}{\Omega} \to 0 \qquad ; (N, \Omega) \to \infty \quad ; \rho = \frac{N}{\Omega} \text{ finite} \quad (11.11)$$

- This suggests that one can replace the new operators by *c-numbers*

$$(\xi_0, \xi_0^\dagger) \to \left(\frac{N}{\Omega}\right)^{1/2} = \sqrt{\rho} \qquad (11.12)$$

---

[3] Note that the relation to a is here only obtained to lowest order in G; the calculation can clearly be extended to higher order in G.

- Now turn on the interaction. The replacement in Eq. (11.11) will still be valid in the interacting theory as long as the interaction is weak enough so that a finite fraction of the particles N_0/N *remains* in the ground state $|\Phi_0\rangle$;
- In general, there will be a *depletion* of the state $|\Phi_0\rangle$ in the exact ground state $|\Psi_0\rangle$ of the interacting system so that

$$\langle\Psi_0|\xi_0^\dagger\xi_0|\Psi_0\rangle = \frac{N_0}{\Omega} = \rho_0 < \rho \qquad (11.13)$$

- Thus, in summary, one is justified in treating $(\xi_0, \xi_0^\dagger)$ as *c*-numbers in the thermodynamic limit provided a finite fraction of the particles continue to occupy the state $|\Phi_0\rangle$ in the interacting system;
- If this is true, one can then make the following replacement to simplify the hamiltonian in Eq. (11.1)

$$(a_0, a_0^\dagger) \to \sqrt{N_0} \qquad ; \text{ } c\text{-numbers} \qquad (11.14)$$

11.1.3   *Effective Hamiltonian*

We will now assume for the weakly interacting Bose gas that *most* of the particles remain in the condensate. We thus assume that $(N - N_0)/N \ll 1$ and neglect interactions between those particles above the condensate. With the retention of terms quartic and quadratic in $(a_0, a_0^\dagger)$, the interaction hamiltonian in Eq. (11.1) becomes

$$\hat{V} = \frac{G}{2\Omega}\left\{ a_0^\dagger a_0^\dagger a_0 a_0 + \sum_{k\neq 0}\left[2(a_k^\dagger a_k a_0^\dagger a_0 + a_{-k}^\dagger a_{-k}a_0^\dagger a_0)+\right.\right.$$

$$\left.\left. a_k^\dagger a_{-k}^\dagger a_0 a_0 + a_0^\dagger a_0^\dagger a_k a_{-k}\right]\right\} \qquad (11.15)$$

Substitution of Eq. (11.14) then leads to

$$\hat{V} = \frac{G}{2\Omega}\left[N_0^2 + 2N_0\sum_{k\neq 0}(a_k^\dagger a_k + a_{-k}^\dagger a_{-k}) + N_0\sum_{k\neq 0}(a_k^\dagger a_{-k}^\dagger + a_k a_{-k})\right] \qquad (11.16)$$

The number operator for the system is given by

$$\hat{N} = \sum_{\mathbf{k}} a_{\mathbf{k}}^{\dagger} a_{\mathbf{k}} \tag{11.17}$$

With the same set of assumptions as above, this becomes

$$\hat{N} = N_0 + \frac{1}{2} \sum_{\mathbf{k} \neq 0} \left(a_{\mathbf{k}}^{\dagger} a_{\mathbf{k}} + a_{-\mathbf{k}}^{\dagger} a_{-\mathbf{k}} \right) \tag{11.18}$$

This equation can be used to eliminate N_0 in terms of the number operator $\hat{N}$. There is a slight difficulty here since the effective interaction hamiltonian in Eq. (11.16) *no longer commutes with the number operator*, which is therefore no longer a constant of the motion.[4] In the thermodynamic limit, when the fluctuations are small, it should be a very good approximation to replace $\hat{N}$ by its expectation value N, and we shall assume this to be the case.

Thus, with the substitution of Eq. (11.18) for N_0, the replacement $\hat{N} \rightarrow N$, and the inclusion of the kinetic energy,[5] the effective hamiltonian for the weakly interacting Bose system becomes

$$\hat{H} = \frac{GN^2}{2\Omega} + \frac{1}{2} \left[\sum_{\mathbf{k} \neq 0} \left(\frac{\hbar^2 k^2}{2m} + \frac{NG}{\Omega} \right) \left(a_{\mathbf{k}}^{\dagger} a_{\mathbf{k}} + a_{-\mathbf{k}}^{\dagger} a_{-\mathbf{k}} \right) + \right.$$

$$\left. \frac{NG}{\Omega} \sum_{\mathbf{k} \neq 0} (a_{\mathbf{k}}^{\dagger} a_{-\mathbf{k}}^{\dagger} + a_{\mathbf{k}} a_{-\mathbf{k}}) \right] \quad ; \text{ effective hamiltonian} \tag{11.19}$$

We have again neglected terms of $O\left[\left(\sum_{\mathbf{k} \neq 0} a_{\mathbf{k}}^{\dagger} a_{\mathbf{k}} \right)^2 \right]$ on the assumption that $(N - N_0)/N \ll 1$.

11.1.4   *Bogoliubov Transformation*

The model problem governed by the effective hamiltonian in Eq. (11.19) *can be solved exactly with a canonical transformation* [Bogoliubov (1947)].

---

[4]See Prob. 2.4. Our effective hamiltonian for the condensed Bose system spontaneously breaks the phase symmetry that leads to number conservation.

[5]Note there is no condensate contribution to the kinetic energy since $\varepsilon_0(a_0^{\dagger} a_0) = 0$, where we use the notation $\hbar \omega_k \equiv \varepsilon_k = \hbar^2 k^2 / 2m$.

Define new operators $(\alpha_{\mathbf{k}}, \alpha_{\mathbf{k}}^{\dagger})$ through the relations

$$a_{\mathbf{k}} = u_k \alpha_{\mathbf{k}} - v_k \alpha_{-\mathbf{k}}^{\dagger} \qquad ; (u_k, v_k) \text{ real}$$
$$a_{\mathbf{k}}^{\dagger} = u_k \alpha_{\mathbf{k}}^{\dagger} - v_k \alpha_{-\mathbf{k}} \qquad u_k^2 - v_k^2 = 1 \qquad (11.20)$$

These relations are readily inverted to give

$$\alpha_{\mathbf{k}} = u_k a_{\mathbf{k}} + v_k a_{-\mathbf{k}}^{\dagger}$$
$$\alpha_{\mathbf{k}}^{\dagger} = u_k a_{\mathbf{k}}^{\dagger} + v_k a_{-\mathbf{k}} \qquad (11.21)$$

It follows that for any (u_k, v_k) satisfying the conditions in Eqs. (11.20), this transformation *preserves the canonical commutation relations*

$$[\alpha_{\mathbf{k}}, \alpha_{\mathbf{k}'}^{\dagger}] = \delta_{\mathbf{k},\mathbf{k}'} \qquad ; \text{ canonical commutation relations} \qquad (11.22)$$

Since we have shown that all the properties of the creation and destruction operators follow from these commutation relations,[6] *we know all the properties of the new operators* $(\alpha_{\mathbf{k}}, \alpha_{\mathbf{k}}^{\dagger})$!

Substitution of Eqs. (11.20) into Eq. (11.19) gives

$$\hat{H} = \frac{GN^2}{2\Omega} + \sum_{\mathbf{k} \neq 0} \left[\left(\varepsilon_k + \frac{NG}{\Omega} \right) v_k^2 - \frac{NG}{\Omega} u_k v_k \right] + \qquad (11.23)$$

$$\frac{1}{2} \sum_{\mathbf{k} \neq 0} \left[\left(\varepsilon_k + \frac{NG}{\Omega} \right) (u_k^2 + v_k^2) - \frac{NG}{\Omega} (2 u_k v_k) \right] (\alpha_{\mathbf{k}}^{\dagger} \alpha_{\mathbf{k}} + \alpha_{-\mathbf{k}}^{\dagger} \alpha_{-\mathbf{k}}) +$$

$$\frac{1}{2} \sum_{\mathbf{k} \neq 0} \left[\left(\varepsilon_k + \frac{NG}{\Omega} \right) (-2 u_k v_k) + \frac{NG}{\Omega} (u_k^2 + v_k^2) \right] (\alpha_{\mathbf{k}}^{\dagger} \alpha_{-\mathbf{k}}^{\dagger} + \alpha_{\mathbf{k}} \alpha_{-\mathbf{k}})$$

We now make use of the freedom in (u_k, v_k) to choose them so that *the last line vanishes*. The hamiltonian will then be *diagonal* in terms of the new number operator $\alpha_{\mathbf{k}}^{\dagger} \alpha_{\mathbf{k}}$ with spectrum $(0, 1, 2, \cdots)$, and we have solved the problem. The conditions that the last line in Eq. (11.23) vanish are

$$2 u_k v_k \left(\varepsilon_k + \frac{NG}{\Omega} \right) = (u_k^2 + v_k^2) \frac{NG}{\Omega}$$
$$u_k^2 - v_k^2 = 1 \qquad (11.24)$$

Define

$$v_k \equiv \sinh \phi_k \qquad ; u_k \equiv \cosh \phi_k \qquad (11.25)$$

---

[6]See Prob. 2.1; note $[\alpha_{\mathbf{k}}^{\dagger}, \alpha_{\mathbf{k}'}^{\dagger}] = [\alpha_{\mathbf{k}}, \alpha_{\mathbf{k}'}] = 0$.

Then the conditions in Eqs. (11.25) are parameterized by ϕ_k according to

$$\cosh^2 \phi_k - \sinh^2 \phi_k = 1 \tag{11.26}$$

$$\frac{\sinh 2\phi_k}{\cosh 2\phi_k} = \tanh 2\phi_k = \frac{NG/\Omega}{\varepsilon_k + NG/\Omega}$$

$$\cosh 2\phi_k = \cosh^2 \phi_k + \sinh^2 \phi_k = 1 + 2\sinh^2 \phi_k = (1 - \tanh^2 2\phi_k)^{-1/2}$$

Define the energy E_k by

$$E_k \equiv \left[\left(\varepsilon_k + \frac{NG}{\Omega} \right)^2 - \left(\frac{NG}{\Omega} \right)^2 \right]^{1/2} \tag{11.27}$$

Equations (11.25)–(11.26) then give

$$u_k^2 + v_k^2 = \cosh 2\phi_k = \frac{(\varepsilon_k + NG/\Omega)}{E_k}$$

$$2u_k v_k = \sinh 2\phi_k = \frac{NG/\Omega}{E_k}$$

$$v_k^2 = \sinh^2 \phi_k = \frac{1}{2} \left[\frac{(\varepsilon_k + NG/\Omega)}{E_k} - 1 \right] \tag{11.28}$$

In terms of these new quantities, the hamiltonian in Eq. (11.23) takes the form

$$\hat{H} = \frac{GN^2}{2\Omega} + \frac{1}{2} \sum_{k \neq 0} \left[E_k - \left(\varepsilon_k + \frac{NG}{\Omega} \right) \right] + \frac{1}{2} \sum_{k \neq 0} E_k (\alpha_{\mathbf{k}}^\dagger \alpha_{\mathbf{k}} + \alpha_{-\mathbf{k}}^\dagger \alpha_{-\mathbf{k}}) \tag{11.29}$$

We proceed to discuss these results.

11.1.5 *Discussion of Results*

11.1.5.1 *Excitation Spectrum*

It follows from the commutation relations that the new number operator $\alpha_{\mathbf{k}}^\dagger \alpha_{\mathbf{k}}$ has the eigenvalues $0, 1, 2, \cdots, \infty$

$$\alpha_{\mathbf{k}}^\dagger \alpha_{\mathbf{k}} |\underline{n}_k\rangle = n_k |\underline{n}_k\rangle \qquad ; \ n_k = 0, 1, 2, \cdots, \infty \tag{11.30}$$

The new quanta ("quasiparticles") therefore have the energy

$$E_k = \left[\left(\varepsilon_k + \frac{NG}{\Omega} \right)^2 - \left(\frac{NG}{\Omega} \right)^2 \right]^{1/2} \tag{11.31}$$

This expression has the following limits

$$E_k \rightarrow \frac{\hbar^2 \mathbf{k}^2}{2m} \qquad\qquad\qquad ; k \rightarrow \infty$$

$$\rightarrow \left(\frac{NG}{m\Omega}\right)^{1/2} \hbar k = \left(\frac{4\pi\hbar^2 a\rho}{m^2}\right)^{1/2} \hbar k \qquad ; k \rightarrow 0 \qquad (11.32)$$

Here Eq. (11.7) has been used in the second line and the density defined by

$$\rho \equiv \frac{N}{\Omega} \qquad\qquad ; \text{density} \qquad\qquad (11.33)$$

- As $k \rightarrow 0$, one obtains a *phonon spectrum*, with a sound velocity c_s given by[7]

$$\omega_k = c_s k$$

$$c_s = \left(\frac{4\pi\hbar^2 a\rho}{m^2}\right)^{1/2} \qquad ; \text{sound velocity} \qquad (11.34)$$

- This result only makes sense if the scattering length $a > 0$, that is, the interaction between the particles is *repulsive*. If $a < 0$, the excitation energies are imaginary and the ground-state is unstable;
- The sound velocity can be re-written as

$$c_s = 2\sqrt{\pi} \frac{\hbar}{ma} \sqrt{\rho a^3} \qquad\qquad (11.35)$$

This allows us to identify the *dimensionless parameter* $\sqrt{\rho a^3}$ in the analysis;
- The assumption of "low-density" clearly implies that

$$\sqrt{\rho a^3} \ll 1 \qquad ; \text{low-density} \qquad (11.36)$$

- Note that the canononical transformation has led us to results that are non-analytic in the combination ρa^3.[8]

11.1.5.2 *Depletion*

The ground state of the interacting Bose system is evidently defined by

$$a_{\mathbf{k}}^\dagger a_{\mathbf{k}} |\underline{0}\rangle = 0 \qquad ; \text{ground state} \qquad (11.37)$$

---

[7]This result was used in Probl. 11.2; note $E_k = \hbar\omega_k$ here.
[8]This implies they could not be obtained by a *perturbation expansion* in ρa^3.

The *depletion* of the ground state is then given by

$$N - N_0 = \sum_{\mathbf{k} \neq 0} \langle \underline{0} | a_\mathbf{k}^\dagger a_\mathbf{k} | \underline{0} \rangle$$

$$= \sum_{\mathbf{k} \neq 0} \langle \underline{0} | (u_k \alpha_\mathbf{k}^\dagger - v_k \alpha_{-\mathbf{k}})(u_k \alpha_\mathbf{k} - v_k \alpha_{-\mathbf{k}}^\dagger) | \underline{0} \rangle$$

$$= \sum_{\mathbf{k} \neq 0} v_k^2 \tag{11.38}$$

where the last line follows from the familiar properties of the creation and destruction operators, as derived from the commutation relations. With the conversion of the sum to an integral,[9] the last of Eqs. (11.28) then gives

$$N - N_0 = \frac{\Omega}{(2\pi)^3} \int d^3k \, \frac{1}{2} \left[\frac{(\varepsilon_k + NG/\Omega)}{E_k} - 1 \right] \tag{11.39}$$

Make use of Eqs. (11.31) and (11.33) to write this out in detail

$$N - N_0 = \frac{\Omega}{8\pi^3} 4\pi \int_0^\infty k^2 dk \, \frac{1}{2} \left[\frac{(\hbar^2 k^2/2m + G\rho)}{[(\hbar^2 k^2/2m + G\rho)^2 - (G\rho)^2]^{1/2}} - 1 \right] \tag{11.40}$$

Introduce

$$x \equiv \frac{\hbar k}{\sqrt{m\rho G}} \tag{11.41}$$

Equation (11.40) then becomes

$$\frac{N - N_0}{N} = \frac{1}{2\pi^2 \rho} \left(\frac{m\rho G}{\hbar^2} \right)^{3/2} \int_0^\infty x^2 dx \, \frac{1}{2} \left[\frac{(x^2/2 + 1)}{[(x^2/2 + 1)^2 - 1]^{1/2}} - 1 \right] \tag{11.42}$$

The integral is just $2/3$ (see Prob. 11.5), and thus the fraction of particles above the condensate is determined to be

$$\frac{N - N_0}{N} = \frac{1}{3\pi^2 \rho} \left(\frac{m\rho G}{\hbar^2} \right)^{3/2} \tag{11.43}$$

Now use the relation between the strength of the pseudopotential and the scattering length in Eq. (11.7) to obtain

$$\frac{N - N_0}{N} = \frac{8}{3\sqrt{\pi}} \sqrt{\rho a^3} \tag{11.44}$$

Two comments:

[9]Note that the deletion of *one state* with $\mathbf{k} = 0$ does not affect this conversion.

- The dimensionless parameter $\sqrt{\rho a^3}$ again appears;
- The analysis assumes $(N - N_0)/N$ is a small number; hence Eq. (11.36) must be satisfied.

11.1.5.3   *Ground-State Energy*

The ground-state energy is given by

$$
\begin{aligned}
E_0 &= \langle 0|\hat{H}|0\rangle \\
&= \frac{GN^2}{2\Omega} + \frac{1}{2}\sum_{k\neq 0}\left[E_k - \left(\varepsilon_k + \frac{NG}{\Omega}\right)\right]
\end{aligned}
\tag{11.45}
$$

To leading order in G only the first term contributes, and this gives an energy per particle of

$$
\frac{E_0}{N} = \frac{G\rho}{2} = \frac{2\pi\hbar^2 a\rho}{m}
\tag{11.46}
$$

The contribution of the second term in Eq. (11.45) is evaluated in [Fetter and Walecka (2003a)], and when written consistently in terms of the scattering length, one obtains

$$
\frac{E_0}{N} = \frac{2\pi\hbar^2 a\rho}{m}\left[1 + \frac{128}{15\sqrt{\pi}}(\rho a^3)^{1/2}\right]
\tag{11.47}
$$

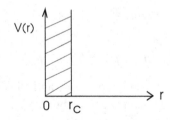

Fig. 11.1   An infinite barrier in the interparticle potential $V(r)$ out to a relative distance $r = r_c$.

A *hard sphere* represents an infinite barrier in the interparticle potential $V(r)$ out to a relative distance $r = r_c$ (Fig. 11.1). The s-wave solution to the Schrödinger equation outside the barrier is $u_0(r) = \sin k(r - r_c) \equiv \sin(kr + \delta_0)$ where δ_0 is the s-wave phaseshift; here $\delta_0 = -kr_c$. The scattering length is defined in terms of δ_0 according to $\delta_0 \to -ka$ as $k \to 0$.[10]

---

[10] See [Schiff (1968)].

Hence one has $a = r_c$ for a hard sphere. Equation (11.47) then represents the first two terms in the classic problem of the ground-state energy of the hard-sphere Bose gas. This energy is calculated to be

$$\frac{E_0}{N} = \frac{2\pi\hbar^2 a\rho}{m}\left[1 + \frac{128}{15\sqrt{\pi}}(\rho a^3)^{1/2} + 8\left(\frac{4\pi}{3} - \sqrt{3}\right)(\rho a^3)\ln(\rho a^3) + O(\rho a^3)\right]$$

(11.48)

Notice again the non-analyticity in the dimensionless parameter ρa^3. The first two terms in this series were originally obtained by [Lee and Yang (1957); Brueckner and Sawada (1957)], and the third term by [Wu (1959)].

11.1.6 *Superfluid ^{4}He*

The relation of the calculated spectrum in this model problem to that observed in superfluid ^{4}He, the relation of that spectrum to the property of superfluidity, the extension of the analysis to non-uniform systems, and the quantization of circulation in such systems are all discussed in chapter 11 of Vol. I.

11.2 Superconductors

11.2.1 *Cooper Pairs*

As first pointed out by [Cooper (1956)], a weak attractive two-body potential gives rise to unusual bound-states in the Fermi gas. In Vol. I, the existence of these bound states was demonstrated by solving the Bethe-Goldstone equation [Bethe and Goldstone (1957)] for an interacting pair with $(\mathbf{k}\uparrow, -\mathbf{k}\downarrow)$ at the Fermi surface, in the presence of the filled Fermi sea. The identification of these bound states is subtle, and an analytic solution was obtained for them in Vol. I by assuming a separable potential.

It was pointed out in Vol. I that these bound pairs yield a Fermi system that in many respects resembles a condensed Bose system, at least close to the Fermi surface, and this analogy was used there to derive some interesting properties of superconductors.

The distortion of the underlying lattice, as mediated by phonon exchange, gives an attractive interaction between electrons that can serve as a mechanism for forming these Cooper pairs in superconducting metals, and Cooper pairs form the basis for the extremely successful BCS theory of superconductivity [Bardeen, Cooper, and Schrieffer (1957)].

It was also pointed out in Vol. I that the consistent incorporation of these bound-pairs in the many-body system presents a formidable theoretical challenge, which BCS originally solved through the use of a variational wavefunction. A canonical transformation incorporating pairing presents an alternative approach to the interacting many-body problem, which has the added advantage of providing a great deal of physical insight. We proceed to a discussion of the Bogoliubov-Valatin canonical transformation for a Fermi system with pairing [Bogoliubov (1958); Valatin (1958)].

11.2.2 *Bogoliubov-Valatin Transformation*

11.2.2.1 *Pairing*

We start from the observation that *pairing* between fermions with $(\mathbf{k}\uparrow, -\mathbf{k}\downarrow)$ is important and can make the Fermi sea unstable if the interaction is attractive. The following canonical transformation is then performed on the hamiltonian for a system of interacting fermions in Eqs. (2.120)–(2.121),

$$\alpha_{\mathbf{k}} \equiv u_k a_{\mathbf{k}\uparrow} - v_k a^{\dagger}_{-\mathbf{k}\downarrow} \qquad ; \ (u_k, v_k) \ \text{real}$$
$$\beta_{-\mathbf{k}} \equiv u_k a_{-\mathbf{k}\downarrow} + v_k a^{\dagger}_{\mathbf{k}\uparrow} \qquad u_k^2 + v_k^2 = 1 \qquad (11.49)$$

These equations are readily inverted to give

$$a_{\mathbf{k}\uparrow} = u_k \alpha_{\mathbf{k}} + v_k \beta^{\dagger}_{-\mathbf{k}}$$
$$a_{-\mathbf{k}\downarrow} = u_k \beta_{-\mathbf{k}} - v_k \alpha^{\dagger}_{\mathbf{k}} \qquad (11.50)$$

or, equivalently,

$$a^{\dagger}_{\mathbf{k}\uparrow} = u_k \alpha^{\dagger}_{\mathbf{k}} + v_k \beta_{-\mathbf{k}}$$
$$a^{\dagger}_{-\mathbf{k}\downarrow} = u_k \beta^{\dagger}_{-\mathbf{k}} - v_k \alpha_{\mathbf{k}} \qquad (11.51)$$

The canonical anticommutation relations for the fermion creation and destruction operators, $\{a_{\mathbf{k}\lambda}, a^{\dagger}_{\mathbf{k}'\lambda'}\} = \delta_{\mathbf{k},\mathbf{k}'}\delta_{\lambda,\lambda'}$, are preserved under this transformation

$$\{\alpha_{\mathbf{k}}, \alpha^{\dagger}_{\mathbf{k}'}\} = \delta_{\mathbf{k},\mathbf{k}'} \qquad ; \ \{\beta_{\mathbf{k}}, \beta^{\dagger}_{\mathbf{k}'}\} = \delta_{\mathbf{k},\mathbf{k}'} \qquad ;$$
$$\text{all other anticommutators vanish} \qquad (11.52)$$

Since the behavior of the new operators follows entirely from these anticommutation relations, *we know all the properties of the new fermion creation and destruction operators.*

As *motivation* for the transformation in Eq. (11.49), we note that the bilinear combinations of quasiparticle operators $\alpha_{\mathbf{k}}^{\dagger}\alpha_{\mathbf{k}}$ and $\beta_{\mathbf{k}}^{\dagger}\beta_{\mathbf{k}}$ contain terms such as $a_{\mathbf{k}\uparrow}^{\dagger}a_{-\mathbf{k}\downarrow}^{\dagger}$ and $a_{\mathbf{k}\uparrow}a_{-\mathbf{k}\downarrow}$, which allows for the possibility of the *condensation* of these boson-like objects (Cooper pairs).

11.2.2.2 *Thermodynamic Potential*

As in the Bose case, our starting approximate (model) hamiltonian will not conserve the number of particles. In order to have a more consistent theoretical framework, we treat the system is an *open* one in the thermodynamic sense of appendix D in Vol. I. We then introduce the *thermodynamic potential*

$$\Omega(\mu, \mathcal{V}, T) \equiv E - TS - \mu N \qquad \text{; thermodynamic potential}$$

$$d\Omega = -SdT - Pd\mathcal{V} - Nd\mu \tag{11.53}$$

where μ is the *chemical potential*, and the second line follows immediately from the thermodynamic relations in that appendix. The number of particles can now be determined at the *end* of the calculation from the relation

$$\left[\frac{\partial \Omega(\mu, \mathcal{V}, T)}{\partial \mu}\right]_{\mathcal{V}, T} = -N \tag{11.54}$$

Here the thermodynamic volume is denoted by $\mathcal{V}$. Furthermore, we define the quantity K to be the thermodynamic potential at zero temperature

$$K(\mu, \mathcal{V}) \equiv \Omega(\mu, \mathcal{V}, 0) = E - \mu N \tag{11.55}$$

The fermion thermodynamic potential *operator* at $T = 0$ then follows from Eqs. (2.120)–(2.121) as

$$\hat{K} = \hat{H} - \mu\hat{N}$$

$$= \sum_{\mathbf{k}\lambda} a_{\mathbf{k}\lambda}^{\dagger}a_{\mathbf{k}\lambda}(\varepsilon_k^0 - \mu) - \frac{1}{2}\sum_{\mathbf{k}_1\lambda_1}\cdots\sum_{\mathbf{k}_4\lambda_4}\langle\mathbf{k}_1\lambda_1\mathbf{k}_2\lambda_2|V|\mathbf{k}_3\lambda_3\mathbf{k}_4\lambda_4\rangle \times$$

$$a_{\mathbf{k}_1\lambda_1}^{\dagger}a_{\mathbf{k}_2\lambda_2}^{\dagger}a_{\mathbf{k}_4\lambda_4}a_{\mathbf{k}_3\lambda_3} \tag{11.56}$$

Several comments:

- It is assumed that the equilibrium properties of a metallic superconductor are established through the Coulomb interaction between the electrons and the lattice. It is also assumed that at equilibrium the electrons, at least those near the Fermi surface, behave as a non-interacting

Fermi gas with a Hartree-Fock energy spectrum ε_k^0. Equation (11.56) then describes the *additional pairing interaction*;

- The sign of the potential term in Eq. (11.56) reflects an attractive pairing interaction with $V(\mathbf{x}, \mathbf{y}) > 0$;
- We assume a spin-independent interaction so that the matrix element of the potential gives M.E. $\propto \delta_{\lambda_1 \lambda_3} \delta_{\lambda_2 \lambda_4}$;
- The remaining spatial part of the matrix element of the potential satisfies (here the thermodynamic volume is the quantization volume Ω)

$$\langle \mathbf{k}_1 \mathbf{k}_2 | V | \mathbf{k}_3 \mathbf{k}_4 \rangle = \frac{1}{\Omega^2} \int d^3x \int d^3y \, e^{-i\mathbf{k}_1 \cdot \mathbf{x}} e^{-i\mathbf{k}_2 \cdot \mathbf{y}} V(\mathbf{x}, \mathbf{y}) e^{i\mathbf{k}_3 \cdot \mathbf{x}} e^{i\mathbf{k}_4 \cdot \mathbf{y}}$$
$$= \langle \mathbf{k}_2 \mathbf{k}_1 | V | \mathbf{k}_4 \mathbf{k}_3 \rangle = \langle -\mathbf{k}_3, -\mathbf{k}_4 | V | -\mathbf{k}_1, -\mathbf{k}_2 \rangle \quad (11.57)$$

- We again assume translational invariance with $V(\mathbf{x}, \mathbf{y}) = V(|\mathbf{x} - \mathbf{y}|)$, which implies $\langle \mathbf{k}_1 \mathbf{k}_2 | V | \mathbf{k}_3 \mathbf{k}_4 \rangle \propto \delta_{\mathbf{k}_1 + \mathbf{k}_2, \mathbf{k}_3 + \mathbf{k}_4}$ [see Prob. 11.6(a)];
- The task is to now substitute the canonical transformation in Eqs. (11.50)–(11.51) and normal-order the result with respect to the new operators (α, β). Wick's theorem simplifies that analysis.

11.2.2.3 *Wick's Theorem*

Wick's theorem states that[11]

$$a^\dagger a^\dagger a a = : \left(\sum \text{all possible pairs of contractions} \right) : \quad (11.58)$$

The contractions are again defined according to

$$a_i^\dagger a_j = : a_i^\dagger a_j : + a_i^\dagger \cdot a_j \cdot \quad (11.59)$$

Define the *new vacuum* by

$$\alpha |\underline{0}\rangle = \beta |\underline{0}\rangle = 0 \quad (11.60)$$

Then

$$a_i^\dagger \cdot a_j \cdot = \langle \underline{0} | a_i^\dagger a_j | \underline{0} \rangle \quad (11.61)$$

It follows that the non-zero contractions are given by

$$a_{\mathbf{k}\uparrow}^\dagger \cdot a_{\mathbf{k}'\uparrow} \cdot = \delta_{\mathbf{k}\mathbf{k}'} v_k^2 = a_{-\mathbf{k}\downarrow}^\dagger \cdot a_{-\mathbf{k}'\downarrow} \cdot$$
$$a_{\mathbf{k}\uparrow}^\dagger \cdot a_{-\mathbf{k}'\downarrow}^\dagger \cdot = \delta_{\mathbf{k}\mathbf{k}'} u_k v_k = a_{-\mathbf{k}\downarrow} \cdot a_{\mathbf{k}'\uparrow} \cdot \quad (11.62)$$

---

[11]To make the formal connection with Wick's theorem complete, one can associate a time with the operators on the l.h.s., making each one on the left infinitesimally later; however, a little thought will convince the reader that this artifice is unnecessary.

Note that all these non-zero contractions contain a factor $\delta_{\mathbf{kk'}}$.

One-Body Term. With the application of Wick's theorem, the one-body term in Eq. (11.56) then takes the form

$$\sum_{\mathbf{k}}(\varepsilon_k^0 - \mu)(a_{\mathbf{k}\uparrow}^\dagger a_{\mathbf{k}\uparrow} + a_{\mathbf{k}\downarrow}^\dagger a_{\mathbf{k}\downarrow}) =$$

$$= \sum_{\mathbf{k}}(\varepsilon_k^0 - \mu)\left[2v_k^2 + :a_{\mathbf{k}\uparrow}^\dagger a_{\mathbf{k}\uparrow}: + :a_{-\mathbf{k}\downarrow}^\dagger a_{-\mathbf{k}\downarrow}:\right]$$

$$= \sum_{\mathbf{k}}(\varepsilon_k^0 - \mu)\left[2v_k^2 + (u_k^2 - v_k^2)(\alpha_{\mathbf{k}}^\dagger \alpha_{\mathbf{k}} + \beta_{-\mathbf{k}}^\dagger \beta_{-\mathbf{k}}) + \right.$$

$$\left. 2u_k v_k(\beta_{-\mathbf{k}}\alpha_{\mathbf{k}} + \alpha_{\mathbf{k}}^\dagger \beta_{-\mathbf{k}}^\dagger)\right] \qquad (11.63)$$

Two-Body Term. The two-body term in Eq. (11.56) takes the form

$$-\frac{1}{2}\sum_{\mathbf{k}_1}\cdots\sum_{\mathbf{k}_4}\langle\mathbf{k}_1\mathbf{k}_2|V|\mathbf{k}_3\mathbf{k}_4\rangle\left[a_{\mathbf{k}_1\uparrow}^\dagger a_{\mathbf{k}_2\uparrow}^\dagger a_{\mathbf{k}_4\uparrow}a_{\mathbf{k}_3\uparrow} + \right. \qquad (11.64)$$

$$\left. a_{\mathbf{k}_1\downarrow}^\dagger a_{\mathbf{k}_2\downarrow}^\dagger a_{\mathbf{k}_4\downarrow}a_{\mathbf{k}_3\downarrow} + a_{\mathbf{k}_1\uparrow}^\dagger a_{\mathbf{k}_2\downarrow}^\dagger a_{\mathbf{k}_4\downarrow}a_{\mathbf{k}_3\uparrow} + a_{\mathbf{k}_1\downarrow}^\dagger a_{\mathbf{k}_2\uparrow}^\dagger a_{\mathbf{k}_4\uparrow}a_{\mathbf{k}_3\downarrow}\right] \equiv \hat{V}_a + \hat{V}_b$$

where $\hat{V}_a$ comes from the first two contributions, and $\hat{V}_b$ from the last two. Application of Wick's theorem to $\hat{V}_a$ gives (note Prob. 11.6)

$$\hat{V}_a = :\hat{V}_a: \ -\frac{1}{2}\sum_{\mathbf{k}}\sum_{\mathbf{k'}}\{\langle\mathbf{kk'}|V|\mathbf{kk'}\rangle - \langle\mathbf{kk'}|V|\mathbf{k'k}\rangle\} \times$$

$$\left\{\left[v_k^2 v_{k'}^2 + 2v_{k'}^2 :a_{\mathbf{k}\uparrow}^\dagger a_{\mathbf{k}\uparrow}:\right] + [\uparrow\rightleftharpoons\downarrow]\right\} \ (11.65)$$

A change in dummy summation variables and the use of the symmetry properties in Eqs. (11.57) shows that the two terms in $\hat{V}_b$ are identical. With the retention of just the first, with a factor of 2, one has

$$\hat{V}_b = :\hat{V}_b: \ -\sum_{\mathbf{k}}\sum_{\mathbf{k'}}\langle\mathbf{k},-\mathbf{k'}|V|\mathbf{k},-\mathbf{k'}\rangle v_{k'}^2\left[v_k^2 + :a_{\mathbf{k}\uparrow}^\dagger a_{\mathbf{k}\uparrow}: + :a_{\mathbf{k}\downarrow}^\dagger a_{\mathbf{k}\downarrow}:\right] -$$

$$\sum_{\mathbf{k}}\sum_{\mathbf{k'}}\langle\mathbf{k},-\mathbf{k}|V|\mathbf{k'},-\mathbf{k'}\rangle u_{k'}v_{k'}\left[u_k v_k + :a_{\mathbf{k}\uparrow}^\dagger a_{-\mathbf{k}\downarrow}^\dagger: + :a_{-\mathbf{k}\downarrow}a_{\mathbf{k}\uparrow}:\right] \ (11.66)$$

Now note:

(1) The expression $:a_{\mathbf{k}\uparrow}^\dagger a_{\mathbf{k}\uparrow}: + :a_{-\mathbf{k}\downarrow}^\dagger a_{-\mathbf{k}\downarrow}:$ is worked out in Eqs. (11.63);
(2) A similar evaluation gives

$$:a_{\mathbf{k}\uparrow}^\dagger a_{-\mathbf{k}\downarrow}^\dagger: + :a_{-\mathbf{k}\downarrow}a_{\mathbf{k}\uparrow}: = -2u_k v_k(\alpha_{\mathbf{k}}^\dagger \alpha_{\mathbf{k}} + \beta_{-\mathbf{k}}^\dagger \beta_{-\mathbf{k}}) +$$

$$(u_k^2 - v_k^2)(\alpha_{\mathbf{k}}^\dagger \beta_{-\mathbf{k}}^\dagger + \beta_{-\mathbf{k}}\alpha_{\mathbf{k}}) \ (11.67)$$

(3) The symmetry properties of the potential in Eqs. (11.57) imply

$$\langle \mathbf{k}, -\mathbf{k}'|V|\mathbf{k}, -\mathbf{k}'\rangle = \langle \mathbf{k}', -\mathbf{k}|V|\mathbf{k}', -\mathbf{k}\rangle$$

$$\langle \mathbf{k}, -\mathbf{k}|V|\mathbf{k}', -\mathbf{k}'\rangle = \langle \mathbf{k}', -\mathbf{k}'|V|\mathbf{k}, -\mathbf{k}\rangle \tag{11.68}$$

A combination of these results leads to the canonically transformed thermodynamic potential at zero temperature

$$\hat{K} = U + \hat{H}_1 + \hat{H}_2 + :\hat{V}:$$

$$U = 2\sum_{\mathbf{k}} \left(\varepsilon_k^0 - \mu\right) v_k^2$$

$$- \sum_{\mathbf{k}}\sum_{\mathbf{k}'} \left[\langle \mathbf{k}\mathbf{k}'|\bar{V}|\mathbf{k}\mathbf{k}'\rangle v_k^2 v_{k'}^2 + \langle \mathbf{k}, -\mathbf{k}|V|\mathbf{k}', -\mathbf{k}'\rangle u_k v_k u_{k'} u_{k'}\right]$$

$$\hat{H}_1 = \sum_{\mathbf{k}} \left(\alpha_{\mathbf{k}}^\dagger \alpha_{\mathbf{k}} + \beta_{-\mathbf{k}}^\dagger \beta_{-\mathbf{k}}\right) \left[\left(\varepsilon_k^0 - \mu - \sum_{\mathbf{k}'}\langle \mathbf{k}, \mathbf{k}'|\bar{V}|\mathbf{k}, \mathbf{k}'\rangle v_{k'}^2\right)(u_k^2 - v_k^2)\right.$$

$$\left. + \left(\sum_{\mathbf{k}'}\langle \mathbf{k}, -\mathbf{k}|V|\mathbf{k}', -\mathbf{k}'\rangle u_{k'} v_{k'}\right) 2u_k v_k\right]$$

$$\hat{H}_2 = \sum_{\mathbf{k}} \left(\alpha_{\mathbf{k}}^\dagger \beta_{-\mathbf{k}}^\dagger + \beta_{-\mathbf{k}}\alpha_{\mathbf{k}}\right) \left[\left(\varepsilon_k^0 - \mu - \sum_{\mathbf{k}'}\langle \mathbf{k}, \mathbf{k}'|\bar{V}|\mathbf{k}, \mathbf{k}'\rangle v_{k'}^2\right) 2u_k v_k\right.$$

$$\left. - \left(\sum_{\mathbf{k}'}\langle \mathbf{k}, -\mathbf{k}|V|\mathbf{k}', -\mathbf{k}'\rangle u_{k'} v_{k'}\right)\left(u_k^2 - v_k^2\right)\right] \tag{11.69}$$

Here we define

$$\langle \mathbf{k}, \mathbf{k}'|\bar{V}|\mathbf{k}, \mathbf{k}'\rangle \equiv \langle \mathbf{k}, \mathbf{k}'|V|\mathbf{k}, \mathbf{k}'\rangle - \langle \mathbf{k}, \mathbf{k}'|V|\mathbf{k}', \mathbf{k}\rangle + \langle \mathbf{k}, -\mathbf{k}'|V|\mathbf{k}, -\mathbf{k}'\rangle \tag{11.70}$$

We also define the expression in Eq. (11.69) as

$$\hat{K} \equiv \hat{K}_0 + :\hat{V}:$$

$$\hat{K}_0 = U + \hat{H}_1 + \hat{H}_2 \tag{11.71}$$

where $:V:$ is the normal-ordered quasiparticle interaction (see later).

11.2.2.4　*Diagonalization of* $\hat{K}_0$

Now choose the parameters (u_k, v_k) to diagonalize the operator $\hat{K}_0$ and hence solve that part of the problem exactly. Define the new Hartree-Fock fermion energy spectrum by

$$\varepsilon_k \equiv \varepsilon_k^0 - \sum_{\mathbf{k}'}\langle \mathbf{k}, \mathbf{k}'|\bar{V}|\mathbf{k}, \mathbf{k}'\rangle v_{k'}^2 \qquad ; \text{Hartree-Fock energy} \tag{11.72}$$

Further define the fermion energy with respect to the chemical potential and the gap function by

$$\xi_k \equiv \varepsilon_k - \mu \qquad\qquad ; \text{ fermion energy}$$

$$\Delta_k \equiv \sum_{k'} \langle \mathbf{k}, -\mathbf{k}|V|\mathbf{k}', -\mathbf{k}'\rangle u_{k'} v_{k'} \qquad ; \text{ gap function} \qquad (11.73)$$

The conditions that $\hat{H}_2$ *vanish* in Eq. (11.69) are then

$$2u_k v_k \,\xi_k = (u_k^2 - v_k^2)\Delta_k$$

$$u_k^2 + v_k^2 = 1 \qquad\qquad (11.74)$$

These conditions can be parameterized in terms of an angle χ_k according to

$$u_k = \cos \chi_k \qquad ; \; v_k = \sin \chi_k \qquad\qquad (11.75)$$

The first of Eqs. (11.74) then becomes (see Fig. 11.2)

$$\tan 2\chi_k = \frac{\Delta_k}{\xi_k}$$

$$\sin 2\chi_k = 2u_k v_k = \frac{\Delta_k}{E_k}$$

$$\cos 2\chi_k = u_k^2 - v_k^2 = \frac{\xi_k}{E_k} \qquad\qquad (11.76)$$

where we have defined

$$E_k \equiv (\Delta_k^2 + \xi_k^2)^{1/2}$$

$$= \xi_k \cos 2\chi_k + \Delta_k \sin 2\chi_k \qquad\qquad (11.77)$$

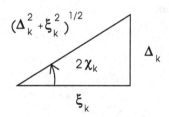

Fig. 11.2   Conditions for diagonalization of $\hat{K}_0$. We use $E_k \equiv (\Delta_k^2 + \xi_k^2)^{1/2}$.

The zero-temperature thermodynamic potential in Eq. (11.69) now takes the form[12]

$$\hat{K} = \hat{K}_0 + :\hat{V}:$$
$$\hat{K}_0 = U + \sum_{\mathbf{k}} E_k \left(\alpha_{\mathbf{k}}^\dagger \alpha_{\mathbf{k}} + \beta_{\mathbf{k}}^\dagger \beta_{\mathbf{k}} \right) \tag{11.78}$$

since $\hat{K}_0$ is diagonal in the new number operators, *we have solved this part of the problem exactly.*

11.2.2.5 *Gap Equation*

Substitution of the second of Eqs. (11.76) into the equation for the gap function Δ_k in Eqs. (11.73) leads to the *gap equation*

$$\Delta_k = \frac{1}{2} \sum_{\mathbf{k}'} \langle \mathbf{k}, -\mathbf{k} | V | \mathbf{k}', -\mathbf{k}' \rangle \frac{\Delta_{k'}}{(\xi_{k'}^2 + \Delta_{k'}^2)^{1/2}} \qquad ; \text{gap equation} \tag{11.79}$$

This is a non-linear relation that must be solved for the gap function; it depends on the pairing potential and the fermion spectrum. One solution is evidently $\Delta = 0$, corresponding to a normal phase. We anticipate a second solution with $\Delta \neq 0$, corresponding to the superconducting phase, and associate Δ in that case with the binding energy of a Cooper pair. Note that for small, constant Δ, the energy of the new quanta ("quasiparticles") in the diagonalized problem now exhibits a gap similar to the relativistic dispersion relation for a *massive* particle

$$E_k = (\Delta^2 + \xi_k^2)^{1/2} \qquad ; \text{quasiparticle energy} \tag{11.80}$$

11.2.3 *Discussion of Results*

Let us proceed to discuss some of these results.

11.2.3.1 *Particle Number*

The expectation value N of the particle number operator $\hat{N}$ in the ground state of $\hat{K}_0$ is given by

$$N \equiv \langle \underline{0} | \hat{N} | \underline{0} \rangle = \sum_{k\lambda} \langle \underline{0} | a_{\mathbf{k}\lambda}^\dagger a_{\mathbf{k}\lambda} | \underline{0} \rangle$$
$$= 2 \sum_{k} v_k^2 \tag{11.81}$$

---

[12]Note the second of Eqs. (11.77).

where the last expression follows from the two contractions in the first of Eqs. (11.62). Hence

$$N = 2 \sum_{k} \sin^2 \chi_k = \sum_{k} (1 - \cos 2\chi_k)$$

$$= 2 \sum_{k} \frac{1}{2} \left[1 - \frac{\xi_k}{(\xi_k^2 + \Delta_k^2)^{1/2}} \right]$$

$$\equiv 2 \sum_{k} n_k \tag{11.82}$$

The implied particle occupation number distribution n_k is sketched for small, constant Δ in Fig. 11.3.

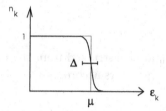

Fig. 11.3 Sketch of the occupation number distribution n_k in Eq. (11.82) for small constant Δ; here $\xi_k = \varepsilon_k - \mu$.

Three comments:

- The relation $\varepsilon_k = \mu$ evidently defines the Fermi surface;
- The pairing interaction smears this distribution out over a band of width Δ at the Fermi surface;
- Given the Hartree-Fock spectrum ε_k, the second of Eqs. (11.82) can be used to adjust the chemical potential μ to yield a specified number of particles N.

11.2.3.2 *Ground-State Thermodynamic Potential*

The expectation of $\hat{K}$ in the ground state of $\hat{K}_0$ yields an expression for the zero-temperature thermodynamic potential in Eq. (11.55)

$$\Omega(\mu, \mathcal{V}, 0) = \langle \underline{0} | \hat{K} | \underline{0} \rangle$$

$$= \langle \underline{0} | \hat{K}_0 | \underline{0} \rangle = U(\mu, \mathcal{V}) \tag{11.83}$$

The second line follows since $\langle \underline{0} | : \hat{V} : | \underline{0} \rangle = 0$.

11.2.3.3 *Ground-State Energy*

The expectation value of $\hat{H}$ in the ground state of $\hat{K}_0$ yields an expression for the ground-state energy of the system

$$E(N, \mathcal{V}) = \langle \underline{0} | \hat{H} | \underline{0} \rangle = \langle \underline{0} | (\hat{K} + \mu \hat{N}) | \underline{0} \rangle$$
$$= U(\mu, \mathcal{V}) + \mu N \tag{11.84}$$

where $\mathcal{V}$ is the thermodynamic volume (here identical to the quantization volume), and the chemical potential $\mu(N, \mathcal{V})$ is determined as described above. Since this is the expectation value of $\hat{H}$ in the normalized state $|\underline{0}\rangle$, this expression provides a *variational estimate* for the ground-state energy.[13]

11.2.3.4 *Excitation Spectrum*

The eigenstates of $\hat{K}_0$ are constructed from the properties of the new creation and destruction operators according to

$$|\underline{n}_1 \, \underline{n}_2 \cdots ; \, \underline{m}_1 \, \underline{m}_2 \cdots \rangle = (\alpha_1^\dagger)^{n_1} (\alpha_2^\dagger)^{n_2} \cdots (\beta_1^\dagger)^{m_1} (\beta_2^\dagger)^{m_2} \cdots |\underline{0}\rangle \tag{11.85}$$

To get the excitation spectrum *at fixed N*, consider a collection of systems at slightly different μ, chosen so that $\langle \hat{N} \rangle = N$ always. Then

$$\Delta E^\star(N) = \langle \underline{n}_1 \cdots ; \, \underline{m}_1 \cdots | \hat{H} | \underline{n}_1 \cdots ; \, \underline{m}_1 \cdots \rangle - \langle \underline{0} | \hat{H} | \underline{0} \rangle \tag{11.86}$$
$$= \sum_k E_k (n_k + m_k) + U(\mu_{\text{exc}}, \mathcal{V}) - U(\mu_0, \mathcal{V}) + N(\mu_{\text{exc}} - \mu_0)$$

Now use the thermodynamic relations in Eqs. (11.54), (11.55), and (11.83), which imply

$$\frac{\partial \Omega(\mu, \mathcal{V}, 0)}{\partial \mu} = \frac{\partial U(\mu, \mathcal{V})}{\partial \mu} = -N \tag{11.87}$$

The additional terms in the second of Eqs. (11.86) therefore *cancel* for small changes in μ, and hence the expectation spectrum at fixed N is given by

$$\Delta E^\star(N) = \sum_k E_k (n_k + m_k) \qquad ; \text{ excitation spectrum} \tag{11.88}$$

---

[13]See [Schiff (1968)] for the variational principle in quantum mechanics.

11.2.3.5 *Momentum Operator*

The momentum operator for the system is given

$$\hat{P} = \sum_{k\lambda} \hbar k \, a_{k\lambda}^{\dagger} a_{k\lambda}$$

$$= \sum_{k} \hbar k \left(: a_{k\uparrow}^{\dagger} a_{k\uparrow} : - : a_{-k\downarrow}^{\dagger} a_{-k\downarrow} : \right) \tag{11.89}$$

The second line follows from Eqs. (11.62) and the fact that $\sum_k \hbar k v_k^2 = 0$. Substitution of Eqs. (11.50)–(11.51) and normal ordering give

$$\hat{P} = \sum_{k} \hbar k \left(\alpha_{k}^{\dagger} \alpha_{k} + \beta_{k}^{\dagger} \beta_{k} \right) \tag{11.90}$$

We conclude from this that the new operators $(\alpha_{k}^{\dagger}, \beta_{k}^{\dagger})$ *create eigenstates of momentum* with $p = \hbar k$. We recall that ε_k is the Hartee-Fock energy of the single-particle state with that same momentum $\hbar k$.

11.2.3.6 *Quasiparticle Spectrum*

In general, the Hartree-Fock energy ε_k is an increasing function of $k = |k|$, and the quasiparticle spectrum for small, constant Δ in Eq. (11.80) is sketched in Fig. 11.4.

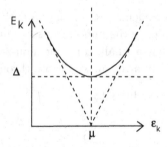

Fig. 11.4   Quasiparticle spectrum in Eq. (11.80) for small constant Δ; here $\xi_k = \varepsilon_k - \mu$.

We note that:

- The quasiparticle energy increases on both sides of μ (the Fermi energy), and the theory contains within it both particle and hole excitations (see Prob. 11.9);

- There is a gap in the spectrum, which reproduces the observed behavior of the low-temperature specific heat of superconductors $C_v \propto e^{-\Delta/k_B T}$;
- The quantity $E_k/\hbar k$ has a positive minimum $(E_k/\hbar k)_{min} = v_{critical} > 0$. As we saw in Vol. I, this provides an explanation for the superfluid flow, or vanishing resistance, in superconductors.

11.2.3.7 *Calculation of the Energy Gap* Δ

Let us try to solve the gap Eq. (11.79).

Normal Solution. There is one obvious solution, the *normal solution*, with

$$\Delta = 0 \qquad\qquad ; \text{ normal solution} \qquad (11.91)$$

This solution implies

$$v_k = \theta(\mu - \varepsilon_k) \qquad ; u_k = \theta(\varepsilon_k - \mu)$$
$$u_k v_k = 0 \qquad\qquad\qquad\qquad (11.92)$$

Equations (11.49) then simply represent a *canonical transformation to particles and holes* (see [Fetter and Walecka (2003a)]).

Superconducting Solution. It is possible to have a second solution to the gap Eq. (11.79) with $\Delta \neq 0$. Here we make a very simple model for the gap function in a superconducting metal:[14]

- Assume the matrix element G of the potential is a constant near the Fermi surface, which is the important wavenumber region for pairing;
- Assume the gap has a small, constant, non-zero value $\Delta_k \approx \Delta$ in the vicinity of $\varepsilon_k = \mu$, in a region we denote through $\sum'_{\mathbf{k}}$.

The gap equation then reduces to

$$\Delta = \frac{G}{2\Omega}\sum_{\mathbf{k}}{}' \frac{\Delta}{(\Delta^2 + \xi_k^2)^{1/2}}$$
$$\text{or;} \qquad 1 = \frac{G}{2\Omega}\sum_{\mathbf{k}}{}' \frac{1}{(\Delta^2 + \xi_k^2)^{1/2}} \qquad (11.93)$$

where the non-zero factor of Δ has been cancelled in the second line.

It is the phonon-exchange force that is responsible for the attractive pairing interaction between the electrons in the metallic superconductor. Since that force is no longer attractive when the frequency of the exchanged

---

[14]The energy gap Δ in nuclei is discussed in [Bohr, Mottelson, and Pines (1958)].

phonon is greater than the Debye frequency ω_D,[15] we shall only include an energy region near the Fermi surface satisfying $|\xi_k| < \hbar\omega_D$ in the sum over wavenumbers in Eq. (11.93). That equation then becomes

$$1 = \frac{G}{2\Omega}\frac{\Omega}{(2\pi)^3}\int \theta(\hbar\omega_D - |\xi_k|)\frac{k^2 dk d\Omega_k}{[\Delta^2 + (\varepsilon_k - \mu)^2]^{1/2}} \qquad (11.94)$$

With the realization that this is a very thin slice in momentum space, a change of integration variable to $x = \varepsilon_k - \mu$ gives

$$1 = \frac{G}{2}\frac{1}{2\pi^2}\left(k^2\frac{dk}{d\varepsilon_k}\right)_{\varepsilon_k = \mu} 2\int_0^{\hbar\omega_D}\frac{dx}{(x^2 + \Delta^2)^{1/2}} \qquad (11.95)$$

The integral is evaluated as

$$\int_0^{\hbar\omega_D}\frac{dx}{(x^2 + \Delta^2)^{1/2}} = \ln\left(t + \sqrt{t^2 + 1}\,\right)\Big|_0^{\hbar\omega_D/\Delta}$$

$$\approx \ln\frac{2\hbar\omega_D}{\Delta} \qquad ; \Delta \ll \hbar\omega_D \qquad (11.96)$$

where the last relation holds if $\Delta \ll \hbar\omega_D$, which is the case in a typical superconductor. The number of states per unit energy, and per unit volume, for one spin projection at the Fermi surface is given by

$$\rho(\mu) = \frac{1}{(2\pi)^3}4\pi\left(k^2\frac{dk}{d\varepsilon_k}\right)_{\varepsilon_k = \mu} \qquad ; \text{density of states} \quad (11.97)$$

Equation (11.95) thus becomes

$$1 = G\rho(\mu)\ln\frac{2\hbar\omega_D}{\Delta}$$

$$\Rightarrow \quad \Delta = 2\hbar\omega_D\, e^{-1/G\rho(\mu)} \qquad ; \Delta \ll \hbar\omega_D \qquad (11.98)$$

Two comments:

- The energy gap Δ is related to the temperature T_c at which the transition from the normal to the superconducting state occurs by

$$\Delta \approx k_B T_c \qquad ; \text{transition temperature} \quad (11.99)$$

- Since $\omega_D \propto M_{\rm ion}^{-1/2}$, where $M_{\rm ion}$ is the mass of an ion in the underlying lattice in the metal, Eqs. (11.98)–(11.99) reproduce the isotope effect in superconductors

$$T_c \propto M_{\rm ion}^{-1/2} \qquad ; \text{isotope effect} \quad (11.100)$$

[15] See Vol. I and [Fetter and Walecka (2003a)].

It was the isotope effect that provided an important clue to understanding the mechanism for the attractive pairing interaction in metallic superconductors, which arises from the distortion of the underlying lattice as mediated by phonon exchange;

- An estimate for the energy difference between the superconducting and normal states can be obtained by associating Δ with the binding energy of a Cooper pair, as analyzed in Vol. I, and then taking $E_s - E_n =$ (binding-energy/pair) × (number of Cooper pairs). The number of Cooper pairs can be taken as one-half the number of electrons in a shell of thickness Δ at the Fermi surface

$$E_s - E_n \approx -\Delta \left[\frac{\Delta}{2} \Omega \rho(\mu) \right] \qquad (11.101)$$

The paired, superconducting state has a lower energy, and the variational principle implies that it therefore provides a better description of the ground-state.

11.2.3.8 *Quasiparticle Interactions*

So far we have explored the model problem obtained by diagonalizing the operator $\hat{K}_0$ in Eqs. (11.69) and (11.71). The full pairing problem also contains the term $:\hat{V}:$. When expressed in terms of the new quasiparticle creation and destruction operators, this term contains the processes illustrated in Fig. 11.5. The number of quasiparticles is not conserved by this interaction.

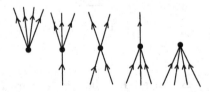

Fig. 11.5 Quasiparticle processes described by $:\hat{V}:$. The bold dot stands for the appropriate matrix element of the pairing potential in Eq. (11.56).

One can analyze the effects of the quasiparticle interactions arising from $:\hat{V}:$ by using the solutions to the problem with $\hat{K}_0$ as the starting point. It is commonly believed that while a realistic inclusion of these interactions may shift the spectrum, it will not destroy the energy gap, which forms the basis for the superconducting state.

Chapter 12

Problems

2.1 Repeat ProbsI. 4.17–4.18 in abstract Hilbert space.

2.2 Take matrix element of the completeness relation in Eq. (2.65) between the states $|\xi'\rangle$ and $|\xi''\rangle$, and show that one obtains the correct expression for $\langle\xi'|\xi''\rangle$.

2.3 (a) Let $\hat{O}$ be some time-independent linear hermitian operator in abstract Hilbert space. Prove Ehrenfest's theorem[1]

$$\frac{d}{dt}\langle\Psi(t)|\hat{O}|\Psi(t)\rangle = \frac{i}{\hbar}\langle\Psi(t)|[\hat{H},\hat{O}]|\Psi(t)\rangle$$

(b) Hence conclude that if $[\hat{H},\hat{O}] = 0$, then the observable O represented by $\hat{O}$ will be a constant the motion.

2.4 The number operator that is companion to the hamiltonian $\hat{H}$ in Eq. (2.113) is

$$\hat{N} = \int d^3x\, \hat{\psi}^\dagger(\mathbf{x})\hat{\psi}(\mathbf{x}) = \sum_{i=1}^{\infty} \hat{N}_i \qquad ;\text{ number operator}$$

(a) Show $[\hat{N},\hat{H}] = 0$ in this case;

(b) Use the results in Prob. 2.3 to conclude that if the relation in part (a) is satified, then $N = \sum_{i=1}^{\infty} n_i$ is a constant of the motion.

2.5 One of the most useful operator relations in quantum mechanics is the following

$$e^{i\hat{A}}\hat{B}e^{-i\hat{A}} = \hat{B} + i[\hat{A},\hat{B}] + \frac{i^2}{2!}[\hat{A},[\hat{A},\hat{B}]] + \frac{i^3}{3!}[\hat{A},[\hat{A},[\hat{A},\hat{B}]]] + \cdots$$

---

[1]Compare with ProbI. 4.9.

This expresses the l.h.s. as a series of repeated commutators that can be evaluated using the canonical commutation relations.

Expand the exponentials on the l.h.s., and prove this relation through the indicated order.

2.6 The translation operator is defined as $\hat{U}(a) \equiv \exp\{-i\hat{p}a/\hbar\}$ where $\hat{p}$ is the momentum operator with $[\hat{p}, \hat{x}] = \hbar/i$, and a is a real number.

(a) Show that $\hat{U}(a)$ is a unitary operator $\hat{U}(a)^{\dagger} = \hat{U}(a)^{-1}$;

(b) Use the result in Prob. (2.5) to show $\hat{U}(a)\hat{x}\,\hat{U}(a)^{-1} = \hat{x} - a$ and $\hat{U}(a)^{-1}\hat{x}\,\hat{U}(a) = \hat{x} + a$;

(c) Apply the last relation in part (b) to an eigenstate of position $|\xi\rangle$ and show $\hat{U}(a)|\xi\rangle = |\xi + a\rangle$.

2.7 Consider the matrix element $\langle k|\hat{U}(a)|\xi\rangle$.

(a) Use the results from Prob. 2.6(c) and the analysis in the text to show that

$$\langle k|\hat{U}(a)|\xi\rangle = \langle k|\xi + a\rangle = e^{-ika}\langle k|\xi\rangle$$

(b) Now let $\xi \to 0$, and then relable $a \to \xi$. Hence prove the last of Eqs. (2.51) entirely with operator methods.

2.8 Suppose one looks for an approximate solution to the time-independent Schrödinger equation $Hu(x) = Eu(x)$ as a finite sum of a complete orthonormal set of wave functions $\phi_n(x)$ which satisfy boundary conditions appropriate to the problem at hand

$$u(x) = \sum_{n=1}^{N} c_n \phi_n(x)$$

The completeness statement in Eq. (2.14) indicates that in the limit $N \to \infty$, the true solution can, in fact, be reproduced this way.[2]

(a) Substitute this expression in the Schrödinger equation, use the orthonormality of the eigenfunctions, and derive the following set of matrix equations

$$\sum_{n=1}^{N} [\langle m|H|n\rangle - E\delta_{mn}]c_n = 0 \qquad ; m = 1, 2, \cdots, N$$

(b) Linear algebra teaches one that this set of linear homogeneous algebraic equations for the set of amplitudes $\{c_n\}$ will only have a non-trivial solution if the determinant of their coefficients vanishes. Show this leads to

---

[2] Here n labels an ordered set of states.

a polynomial expression for the eigenvalues E with precisely N roots $E^{(s)}$, where $s = 1, 2, \cdots, N$;

(c) Show that if H is hermitian with $H^\star_{mn} = H_{nm}$, then the eigenvalues $E^{(s)}$ will be real;

(d) Prove that the solutions $\{c_n^{(s)}\}$ for the amplitudes corresponding to different values of $E^{(s)}$ are orthogonal. Since degenerate solutions can always be orthogonalized, and one is free to choose the normalization of the $\{c_n\}$, show that

$$\sum_{n=1}^{N} c_n^{(s)\star} c_n^{(t)} = \delta_{st} \tag{12.1}$$

This analysis provides a powerful and widely-used approximation method for any problem in quantum mechanics.

2.9 This problem concerns the projection of the abstract time-independent Schrödinger Eq. (2.66) onto eigenstates of momentum; it makes liberal use of the completeness of both the eigenstates of momentum [Eq. (2.63)] and of position [Eq. (2.65)].

(a) Show that the projection of the state vector $|\psi\rangle$ on the state $|k\rangle$ is[3]

$$\langle k|\psi\rangle = \frac{1}{\sqrt{L}} \int d\xi e^{-ik\xi} \psi(\xi) \equiv A(k)$$

(b) Define $\tilde{V}(q) \equiv (2\pi)^{-1/2} \int d\xi\, e^{-iq\xi} V(\xi)$ and show

$$\langle k|\hat{H}|k'\rangle = \frac{\hbar^2 k^2}{2m} \delta_{kk'} + \frac{\sqrt{2\pi}}{L} \tilde{V}(k - k')$$

(c) Hence show that $\langle k|\hat{H}|\psi\rangle = E\langle k|\psi\rangle$ implies

$$\left(E - \frac{\hbar^2 k^2}{2m}\right) A(k) = \frac{\sqrt{2\pi}}{L} \sum_{k'} \tilde{V}(k - k') A(k')$$

(d) Define $\tilde{\psi}(k) \equiv (L/2\pi)^{1/2} A(k)$, and take the limit $L \to \infty$. Show the Schrödinger equation in momentum space is an *integral equation*

$$\left(E - \frac{\hbar^2 k^2}{2m}\right) \tilde{\psi}(k) = \frac{1}{\sqrt{2\pi}} \int dk'\, \tilde{V}(k - k') \tilde{\psi}(k')$$

---

[3] We first use the periodic boundary conditions to convert the coordinate interval to $[-L/2, L/2]$.

where $\tilde{\psi}(k)$ is the Fourier transform of the coordinate space wave function, and $\tilde{V}(q)$ is the Fourier transform of the potential;

(e) Re-derive this result by taking the Fourier transform of the coordinate-space Schrödinger equation.

2.10 Consider the rotations of a particle in the x-y plane. The abstract eigenstates of the z-component of angular momentum satisfy

$$\hat{L}_z|m\rangle = m|m\rangle \qquad ; \langle m|m'\rangle = \delta_{mm'} \qquad ; \sum_m |m\rangle\langle m| = \hat{1}$$

The eigenfunctions satisfying p.b.c. in this case were found in Vol. I

$$\langle\phi|m\rangle = \frac{1}{\sqrt{2\pi}}e^{im\phi} \qquad ; m = 0, \pm 1, \pm 2, \cdots$$

where ϕ is the polar angle. As in the text, use these wavefunctions as a basis for arriving at each of the above abstractions.

3.1 Operator commutation relations play a central role here:
(a) Verify Eqs. (3.19) for the vectors $\hat{\mathbf{v}} = (\hat{\mathbf{x}}, \hat{\mathbf{p}})$;
(b) Verify the re-writing in Eqs. (3.25);
(c) Verify Eqs. (3.30) for the raising and lowering operators.

3.2 Finite transformations follow from the commutation relations:
(a) Verify the 3-D translation operator in Eq. (3.10);
(b) Verify the finite form of the rotation in Eq. (3.20);
(c) Verify the finite rotation in Eqs. (3.22).

3.3 Use $\hat{J}_1 = (\hat{J}_+ + \hat{J}_-)/2$, $\hat{J}_2 = (\hat{J}_+ - \hat{J}_-)/2i$, and Eqs. (3.59):
(a) Compute the matrix $\underline{\mathbf{s}} \equiv \boldsymbol{\sigma}/2$ where

$$[\underline{\mathbf{s}}]_{m'm} \equiv \left\langle \frac{1}{2}m' \left| \hat{\mathbf{J}} \right| \frac{1}{2}m \right\rangle \qquad ; \underline{\mathbf{s}} \equiv \frac{1}{2}\boldsymbol{\sigma}$$

Show that $(\sigma_x, \sigma_y, \sigma_z)$ are just the Pauli matrices[4]

$$\sigma_x \equiv \begin{pmatrix} 0 & 1 \\ 1 & 0 \end{pmatrix} \qquad \sigma_y \equiv \begin{pmatrix} 0 & -i \\ i & 0 \end{pmatrix} \qquad \sigma_z \equiv \begin{pmatrix} 1 & 0 \\ 0 & -1 \end{pmatrix} \quad ; \text{Pauli matrices}$$

(b) Show that as *matrices*, the $\underline{s}_i$ obey the commutation relations of angular momentum

$$[\underline{s}_i, \underline{s}_j] = i\epsilon_{ijk}\underline{s}_k \qquad ; \text{spin}$$

---

[4]Here $(x, y, z) \equiv (1, 2, 3)$; as with the Dirac matrices, the underlining of the Pauli matrices is suppressed.

(c) Hence conclude that for the quantum mechanics of a single non-relativistic particle one can append a direct product two-component spin wave function and extend the set of vectors to $(\mathbf{p}, \mathbf{x}, \mathbf{s})$ (see Vol. I).

3.4 Repeat Prob. 3.3 for spin $s = 1$, and derive the 3×3 matrix representation of the commutation relations.

3.5 Derive the orthonormality and completeness relations for the C-G coefficients in Eqs. (3.75) on the basis of the general assumptions concerning operators and state vectors.

3.6 (a) Verify the orthonormality and completeness relations in Eqs. (3.97);
(b) Verify that Eqs. (3.107) and (3.110) define ITO's of rank one;
(c) Verify the finite rotation in Eq. (3.123).

3.7 (a) Use the relation between the C-G coefficients and 3-j symbols given in Eq. (3.94), and the symmetry properties of the 3-j symbols stated there, to verify the following properties of the C-G coefficients

$$\langle j_1 m_1 j_2 m_2 | j_1 j_2 j m \rangle = (-1)^{j_1 + j_2 - j} \langle j_2 m_2 j_1 m_1 | j_2 j_1 j m \rangle$$

$$\langle j_1 m_1 j_2 m_2 | j_1 j_2 j m \rangle = (-1)^{j_1 + j_2 - j} \langle j_1, -m_1, j_2, -m_2 | j_1 j_2 j, -m \rangle$$

(b) Extend the results in part (a) to include the relations

$$\langle j_1 m_1 j_2 m_2 | j_1 j_2 j_3 m_3 \rangle = (-1)^{j_2 + m_2} \left(\frac{2 j_3 + 1}{2 j_1 + 1} \right)^{1/2} \langle j_2, -m_2, j_3 m_3 | j_2 j_3 j_1 m_1 \rangle$$

$$\langle j_1 m_1 j_2 m_2 | j_1 j_2 j_3 m_3 \rangle = (-1)^{j_1 - m_1} \left(\frac{2 j_3 + 1}{2 j_2 + 1} \right)^{1/2} \langle j_3 m_3 j_1, -m_1 | j_3 j_1 j_2 m_2 \rangle$$

3.8 Use the definition of the 3-j symbol in Eq. (3.94) to express the 6-j symbol in Eqs. (3.100) as a sum over four 3-j symbols. Verify with [Edmonds (1974)].

3.9 Use the construction of the states in Eqs. (3.96) and (3.98), and the definition of the 6-j symbol in Eq. (3.99), to prove the following recoupling relation

$$\sum_{m_{12}} \langle j_1 m_1 j_2 m_2 | j_1 j_2 j_{12} m_{12} \rangle \langle j_{12} m_{12} j_3 m_3 | j_{12} j_3 j m \rangle =$$

$$\sum_{j_{23}, m_{23}} (-1)^{j_1 + j_2 + j_3 + j} \sqrt{(2 j_{12} + 1)(2 j_{23} + 1)} \begin{Bmatrix} j_1 & j_2 & j_{12} \\ j_3 & j & j_{23} \end{Bmatrix} \times$$

$$\langle j_2 m_2 j_3 m_3 | j_2 j_3 j_{23} m_{23} \rangle \langle j_1 m_1 j_{23} m_{23} | j_1 j_{23} j m \rangle$$

3.10 Start from the result in Prob. 3.9, and prove the additional recoupling relation[5]

$$\sum_{m_2,m_3,m_{12}} \langle j_1 m_1 j_2 m_2 | j_1 j_2 j_{12} m_{12}\rangle \langle j_{12} m_{12} j_3 m_3 | j_{12} j_3 j m\rangle \langle j_2 m_2 j_3 m_3 | j_2 j_3 j_{23} m_{23}\rangle$$

$$= (-1)^{j_1+j_2+j_3+j} \sqrt{(2j_{12}+1)(2j_{23}+1)} \left\{ \begin{matrix} j_1 & j_2 & j_{12} \\ j_3 & j & j_{23} \end{matrix} \right\} \langle j_1 m_1 j_{23} m_{23} | j_1 j_{23} j m\rangle$$

3.11 Write out the linear algebraic Eqs. (3.87) that determine the C-G coefficients for $\mathbf{j} = \mathbf{l} + \mathbf{s}$ in the case $s = 1$.

(a)Show the analogue of the eigenvalue Eq. (3.92) for α_j is now

$$(\alpha_j - 2l)(\alpha_j + 2)(\alpha_j + 2l + 2) = 0$$

This equation has the roots $\alpha_j = (2l, -2, -2l - 2)$. Show this implies $j = (l + 1, l, l - 1)$;

(b) The solutions to the linear equations in each case determine the C-G coefficients in Table 12.1.[6] Pick any row in this table, and verify the result.

Table 12.1   Clebsch-Gordan coefficients $\langle l, m - m_s, 1, m_s | l, 1, j, m\rangle$ for $s = 1$.

	$m_s = 1$	$m_s = 0$	$m_s = -1$
$j = l+1$	$\left[\frac{(l+m)(l+m+1)}{(2l+1)(2l+2)}\right]^{1/2}$	$\left[\frac{(l+m+1)(l-m+1)}{(2l+1)(l+1)}\right]^{1/2}$	$\left[\frac{(l-m)(l-m+1)}{(2l+1)(2l+2)}\right]^{1/2}$
$j = l$	$-\left[\frac{(l+m)(l-m+1)}{2l(l+1)}\right]^{1/2}$	$\frac{m}{\sqrt{l(l+1)}}$	$\left[\frac{(l-m)(l+m+1)}{2l(l+1)}\right]^{1/2}$
$j = l-1$	$\left[\frac{(l-m)(l-m+1)}{2l(2l+1)}\right]^{1/2}$	$-\left[\frac{(l-m)(l+m)}{l(2l+1)}\right]^{1/2}$	$\left[\frac{(l+m)(l+m+1)}{2l(2l+1)}\right]^{1/2}$

3.12 The rotation matrix $[\underline{d}^j(\beta)]$ is defined through its matrix elements $d^j_{m'm}(\beta) = \langle jm' | e^{i\beta \hat{J}_2} | jm\rangle$. Write $\hat{J}_2 = (\hat{J}_+ - \hat{J}_-)/2i$, use Eqs. (3.59), and show that for $j = 1/2$ the rotation matrix takes the form

$$\underline{d}^{1/2}(\beta) = \left[\begin{matrix} \cos \beta/2 & \sin \beta/2 \\ -\sin \beta/2 & \cos \beta/2 \end{matrix} \right]$$

What is this expression if $\beta = 2\pi$? If $\beta = 4\pi$? Discuss.

3.13 Repeat Prob. 3.12 for $j = 1$. Show through the first two

---

[5]Note that there is no easy way to get these results just starting from Eqs. (3.100).
[6]See [Edmonds (1974)].

contributing powers in β in each term that

$$
\underline{d}^1(\beta) = \begin{bmatrix} (1+\cos\beta)/2 & (\sin\beta)/\sqrt{2} & (1-\cos\beta)/2 \\ -(\sin\beta)/\sqrt{2} & \cos\beta & (\sin\beta)/\sqrt{2} \\ (1-\cos\beta)/2 & -(\sin\beta)/\sqrt{2} & (1+\cos\beta)/2 \end{bmatrix}
$$

3.14 If $\hat{T}(\kappa, q)$ is an ITO working on the first part of a coupled scheme $|\gamma j_1 j_2 jm\rangle$, and $\hat{U}(\kappa, q)$ is an ITO working on the second part, then [Edmonds (1974)]

$$
\langle \gamma' j_1' j_2 j' || T(\kappa) || \gamma j_1 j_2 j \rangle = (-1)^{j_1' + j_2 + j + \kappa} \sqrt{(2j+1)(2j'+1)} \begin{Bmatrix} j_1' & j' & j_2 \\ j & j_1 & \kappa \end{Bmatrix} \times
$$

$$
\langle \gamma' j_1' || T(\kappa) || \gamma j_1 \rangle
$$

$$
\langle \gamma' j_1 j_2' j' || U(\kappa) || \gamma j_1 j_2 j \rangle = (-1)^{j_1 + j_2 + j' + \kappa} \sqrt{(2j+1)(2j'+1)} \begin{Bmatrix} j_2' & j' & j_1 \\ j & j_2 & \kappa \end{Bmatrix} \times
$$

$$
\langle \gamma' j_2' || U(\kappa) || \gamma j_2 \rangle
$$

These relations are proven by writing out the coupling of the direct-product states using C-G coefficients, using the W-E theorem on the one-body matrix element, then using a recoupling relation from Probs. 3.7–3.10 on the sum over the three C-G coefficients, and then identifying the reduced matrix element. Prove either one of these relations.[7]

3.15 The general result for the addition of two angular momenta can be obtained from the *weight diagram*. Suppose one wants to add the angular momenta of the states $|j_1 m_1\rangle$ and $|j_2 m_2\rangle$, where, for concreteness, we assume that both are half-integral and $j_1 \geq j_2$. Make a plot where the ordinate is j and the abcissa is m and place a cross on it for each state. From the general theory of angular momentum one knows that $m = m_1 + m_2$ for the state $|j_1 j_2 jm\rangle$. One also knows that the allowed values of m for a given j are $-j \leq m \leq j$ in integer steps, with each value occuring once.

(a) The maximum value of m is $j_1 + j_2$. Place a cross at this point, and then at all of the other requisite values of m at the appropriate value of j on the weight diagram.[8] What is the next highest possible value of m? Show that it occurs twice in the direct-product basis, and observe that it has already been used once in the above. Place a cross on this m, and all the other requisite values of m at the appropriate value of j;

(b) Repeat the process in (a);

---

[7]This problem requires somewhat more algebra, but the results are extremely useful.

[8]Note that these states can actually be constructed by applying $\hat{J}_- = (\hat{j}_1 + \hat{j}_2)_-$ repeatedly to the state with $m = m_{\text{max}}$.

(c) Show that the value $m = 0$ occurs $2j_2+1$ times in the direct-product basis. Since one of these is used once in each step, show that the process in (b) must terminate after $2j_2 + 1$ steps. Hence conclude that the correct addition law for angular momenta in this case is

$$j_1 + j_2 \leq j \leq |j_1 - j_2| \qquad ; \text{ integer steps}$$

$$\text{each value occurs once}$$

3.16 (a) Show that the definitions of an ITO in Eqs. (3.106) and (3.112) are equivalent;

(b) Show that the tensor product in Eq. (3.147) satisfies the defining relation for an ITO in Eq. (3.148).

4.1 (a) Write out the $n = 2$ term in Eq. (4.15). There will be two contributions, one for $t_1 > t_2$ and one for $t_1 < t_2$. Change dummy integration variables in the second term, and show that it is identical to the first. Hence show that the $n = 2$ term in Eq. (4.15) explicitly reproduces the $n = 2$ contribution in Eq. (4.14);

(b) Demonstrate the relation in Eq. (4.22) for $n = 2$.

4.2 Write out the proof of the group property in Eq. (4.26) in detail for $n = 3$.

4.3 Carry out an argument parallel to that in the text for $|\psi_i^{(+)}\rangle$, and derive Eq. (4.69) for the incoming scattering state $|\psi_f^{(-)}\rangle$.

4.4 Consider the integral for the Green's function $G_0(\mathbf{x} - \mathbf{y})$ in Eq. (4.104). Show that the contribution around the large semicircle in the complex t-plane in Fig. 4.1 vanishes as $R \to \infty$, and hence conclude that in that limit, the integral is identical to the integral around the contour C.

4.5 (a) Prove the following operator relations

$$\frac{1}{\hat{A} - \hat{B}} = \frac{1}{\hat{A}} + \frac{1}{\hat{A} - \hat{B}} \hat{B} \frac{1}{\hat{A}} = \frac{1}{\hat{A}} + \frac{1}{\hat{A}} \hat{B} \frac{1}{\hat{A} - \hat{B}}$$

(b) Now iterate to obtain a power series in $\hat{B}$

$$\frac{1}{\hat{A} - \hat{B}} = \frac{1}{\hat{A}} + \frac{1}{\hat{A}} \hat{B} \frac{1}{\hat{A}} + \cdots = \sum_{n=0}^{\infty} \left(\frac{1}{\hat{A}} \hat{B} \right)^n \frac{1}{\hat{A}} = \sum_{n=0}^{\infty} \frac{1}{\hat{A}} \left(\hat{B} \frac{1}{\hat{A}} \right)^n$$

This produces, for example, a power series in the coupling constant, and when the l.h.s. can be iterated, it must be equal to the r.h.s.

4.6 (a) Use the results in Prob. 4.5 to prove the following relation

$$\frac{1}{E_0 - \hat{H}_0 + i\varepsilon - \hat{H}_1} = \sum_{n=0}^{\infty} \left(\frac{1}{E_0 - \hat{H}_0 + i\varepsilon} \hat{H}_1 \right)^n \frac{1}{E_0 - \hat{H}_0 + i\varepsilon}$$

(b) Hence conclude that

$$\frac{1}{E_0 - \hat{H} + i\varepsilon} \hat{H}_1 = \sum_{n=1}^{\infty} \left(\frac{1}{E_0 - \hat{H}_0 + i\varepsilon} \hat{H}_1 \right)^n$$

where $\hat{H} = \hat{H}_0 + \hat{H}_1$ is the full hamiltonian, and the sum starts with $n = 1$.

4.7 Use the results in Prob. 4.6 to obtain *explicit* expressions for the scattering states $|\psi_i^{(+)}\rangle$ and $|\psi_f^{(-)}\rangle$ in terms of the full hamiltonian $\hat{H}$

$$|\psi_i^{(+)}\rangle = \left\{ 1 + \frac{1}{E_0 - \hat{H} + i\varepsilon} \hat{H}_1 \right\} |\psi_i\rangle$$

$$|\psi_f^{(-)}\rangle = \left\{ 1 + \frac{1}{E_f - \hat{H} - i\varepsilon} \hat{H}_1 \right\} |\psi_f\rangle$$

4.8 Derive the following relations from the results in Prob. 4.7.

(a) Show the incoming and outgoing scattering states $|\psi_f^{(-)}\rangle$ and $|\psi_f^{(+)}\rangle$ are explicity related by

$$|\psi_f^{(-)}\rangle = |\psi_f^{(+)}\rangle + \left(\frac{1}{E_f - \hat{H} - i\varepsilon} - \frac{1}{E_f - \hat{H} + i\varepsilon} \right) \hat{H}_1 |\psi_f\rangle$$

(b) Show the S-matrix, as expressed in Eq. (4.74), can be written as

$$\langle \psi_f | \hat{S} | \psi_i \rangle = \langle \psi_f^{(-)} | \psi_i^{(+)} \rangle$$

$$= \langle \psi_f^{(+)} | \psi_i^{(+)} \rangle + \langle \psi_f | \hat{H}_1 \left(\frac{1}{E_f - \hat{H} + i\varepsilon} - \frac{1}{E_f - \hat{H} - i\varepsilon} \right) | \psi_i^{(+)} \rangle$$

(c) Make use of the limiting relation in Eq. (4.58), and then prove the following [compare Eq. (4.90)]

$$\text{Lim}_{\varepsilon \to 0} \left(\frac{1}{E_f - E_0 + i\varepsilon} - \frac{1}{E_f - E_0 - i\varepsilon} \right) = \text{Lim}_{\varepsilon \to 0} \frac{-2i\varepsilon}{(E_f - E_0)^2 + \varepsilon^2}$$

$$= -2\pi i \delta(E_f - E_0)$$

(d) Now use the results in part (c) to reproduce the general expression for the S-matrix

$$\langle \psi_f | \hat{S} | \psi_i \rangle = \langle \psi_f | \psi_i \rangle - 2\pi i \delta(E_f - E_0) \langle \psi_f | \hat{H}_1 | \psi_i^{(+)} \rangle$$

4.9 Consider the *incoming* scattering wave funtion $\psi_f^{(-)}(\mathbf{x})$ with $E_f = E_0$ in potential scattering. For this, one needs the Green's function $G_0^{(-)}(\mathbf{x} - \mathbf{y})$ given by

$$G_0^{(-)}(\mathbf{x} - \mathbf{y}) \equiv \langle \mathbf{x} | \frac{1}{\hat{H}_0 - E_0 + i\varepsilon} | \mathbf{y} \rangle$$

(a) Follow the arguments in the text for $G_0(\mathbf{x} - \mathbf{y}) \equiv G_0^{(+)}(\mathbf{x} - \mathbf{y})$. Convert to a contour integral, locate the singularities as in Fig. 4.1, and use the method of residues to show

$$G_0^{(-)}(\mathbf{x} - \mathbf{y}) = \frac{2m}{\hbar^2} \frac{e^{-ikr}}{4\pi r} \qquad ; \mathbf{r} \equiv \mathbf{x} - \mathbf{y}$$

(b) Show the wave function $\psi_f^{(-)}(\mathbf{x})$ satisfies the integral equation

$$\psi_f^{(-)}(\mathbf{x}) = e^{i\mathbf{k}_f \cdot \mathbf{x}} - \frac{2m}{\hbar^2} \int d^3y \, \frac{e^{-ik|\mathbf{x} - \mathbf{y}|}}{|\mathbf{x} - \mathbf{y}|} V(y) \, \psi_f^{(-)}(\mathbf{y})$$

(c) Show the T-matrix can be written

$$T_{fi} = \frac{1}{\Omega} \int d^3y \, \psi_f^{(-)}(\mathbf{y})^* V(y) \, e^{i\mathbf{k} \cdot \mathbf{y}}$$

4.10 (a) Consider the asymptotic form of the Green's function $G_0(\mathbf{x} - \mathbf{y})$ in Eq. (4.107), where $x = |\mathbf{x}| \to \infty$ while $\mathbf{y}$ is confined to the region of the potential. Show

$$G_0(\mathbf{x} - \mathbf{y}) \to \frac{2m}{\hbar^2} \frac{e^{ikx}}{4\pi x} e^{-i\mathbf{k}_f \cdot \mathbf{y}} \qquad ; x = |\mathbf{x}| \to \infty$$

$$\mathbf{y} \text{ in potential}$$

Here $\mathbf{k}_f = k(\mathbf{x}/|\mathbf{x}|)$ is a vector of length k that points in the direction of observation.

(b) Show that the corresponding asymptotic form of the scattering wave function in Eq. (4.118) is

$$\psi_i^{(+)}(\mathbf{x}) \to e^{i\mathbf{k} \cdot \mathbf{x}} + f(k, \theta) \frac{e^{ikx}}{x}$$

where the scattering amplitude is defined in Eq. (4.117).

4.11 Verify Eq. (4.115).

5.1 Consider lagrangian particle mechanics with a set of generalized coordinates $(q^1, \cdots, q^n)$. Assume the kinetic energy is a quadratic form in the generalized velocities $\dot{q}^i \equiv dq^i(t)/dt$ so that

$$L(q, \dot{q}) = T - V = \sum_i \sum_j t_{ij}(q)\dot{q}^i\dot{q}^j - V(q)$$

(a) Show that one can assume that $t_{ij}(q)$ is symmetric in (i, j);
(b) Prove that the hamiltonian is now the energy $H = T + V = E$.

5.2 Work in the coordinate representation. Assume a generalized coordinate q and its conjugate momentum p satisfying the canonical commutation relations $[p, q] = \hbar/i$.
(a) Prove $[p, F(q)] = -i\hbar\, \partial F(q)/\partial q$;
(b) Prove $[q, G(p)] = i\hbar\, \partial G(p)/\partial p$;
For parts (a,b) you may assume that the functions (F, G) have power series expansions in their respective arguments [*Hint*: try a few terms].

(c) Extend these results to the case where the series contain additional inverse powers of the arguments.

5.3 Ehrenfest's theorem relates the time derivative of the expectation value of an operator $\hat{O}$ to the expectation value of the commutator of the operator with the hamiltonian. Show that in the abstract Hilbert space it reads (compare Probl. 4.9)[9]

$$\frac{d}{dt}\langle\Psi(t)|\hat{O}|\Psi(t)\rangle = \langle\Psi(t)|\frac{i}{\hbar}[\hat{H}, \hat{O}]|\Psi(t)\rangle \qquad ; \text{Ehrenfest's theorem}$$

(a) Now work in the coordinate representation. Assume a hermitian hamiltonian $H(p, q)$ with a specified, ordered power-series expansion in (p, q), and write out Ehrenfest's theorem for both (p, q). Use the canonical commutation relations $[p, q] = \hbar/i$, and the results in Prob. 5.2, to show that in each case the operator on the r.h.s. is the appropriate derivative of $H(p, q)$ appearing in Hamilton's equations;

(b) Explain under what circumstances one now satisfies the *correspondence principle*.

5.4 Consider the differential operator p_r^2 in polar coordinates (r, θ) defined in Eq. (5.43).

---

[9]This assumes that $\hat{O}$ has no *explicit* time dependence.

(a) Show it is hermitian with respect to the volume element in Eq. (5.41). State carefully any assumptions about the boundary contributions;

(b) Show it satisfies the commutation relation in the second line of Eq. (5.42).

5.5 (a) Use the expression for p_r^2 in Eq. (5.43), and show that the operator p_r determined from the first of Eqs. (5.42) is

$$p_r = \frac{\hbar}{i} \left[\frac{\partial}{\partial r} + \frac{\rho'(r)}{2\rho(r)} \right]$$

(b) Show this p_r satisfies the commutation relation $[p_r, r] = \hbar/i$;

(c) Show this p_r is hermitian with respect to the volume element in Eq. (5.41).

5.6 Given the free-field hamiltonian $\hat{H}_0$ in Eq. (5.50), and the commutation relations for the creation and destruction operators in Eq. (5.48),

(a) Show

$$[\hat{H}_0, a_k] = -\hbar\omega_k a_k \qquad ; \quad [\hat{H}_0, a_k^\dagger] = \hbar\omega_k a_k^\dagger$$

(b) Use the operator relation in Eq. (3.9) to then prove Eqs. (5.51);

(c) Take the time derivative of the l.h.s. of Eqs. (5.51) and show that

$$\frac{da_k(t)}{dt} = \frac{i}{\hbar}[\hat{H}_0, a_k(t)] = -i\omega_k a_k(t) \qquad ; \quad a_k(0) = a_k$$

$$\frac{da_k^\dagger(t)}{dt} = \frac{i}{\hbar}[\hat{H}_0, a_k^\dagger(t)] = i\omega_k a_k^\dagger(t) \qquad ; \quad a_k^\dagger(0) = a_k^\dagger$$

(d) Now integrate the results in part (c) with respect to time to rederive the result in part (b).

5.7 (a) Explicitly convert the Schrödinger-picture hamiltonian in Eq. (5.49) to the interaction picture, and derive the expression in Eq. (5.54);

(b) Explain why the expressions in Eqs. (5.50) and (5.55) turn out to be identical.

5.8 Start from the lagrangian for the massive scalar field in Eq. (5.63).

(a) Show that Lagrange's equation produces the Klein-Gordon equation $(\Box - m_s^2)\phi(\mathbf{x}, t) = 0$;

(b) Show the angular frequency appearing in the quantum fields and hamiltonian is $\omega_k = c\sqrt{\mathbf{k}^2 + m_s^2}$;

(c) Interpret the result in (b) in terms of the relativistic energy-momentum relation for the quanta (particles).

5.9 Despite the fact that the creation and destruction operators for the Dirac (fermion) field obey the *anticommutation* relations of Eq. (5.103), all of the results in Prob. 5.6 continue to hold for them when the hamiltonian $\hat{H}_0$ of Eq. (5.105) is employed. Prove this statement.

5.10 Start from the quantum field $\hat{\psi}(\mathbf{x})$ in the Schrödinger picture, and use the results in Prob. 5.9 to derive the field $\hat{\psi}(\mathbf{x}, t)$ in the interaction picture given in Eq. (5.102).

5.11 Prove that the second of Eqs. (5.94) is the proper adjoint of the Dirac equation in the first of Eqs. (5.94).

5.12 (a) Derive the second of Eqs. (5.79) for the momentum $\hat{\mathbf{P}}$ in the massive scalar field;
   (b) Derive Eq. (5.106) for the momentum $\hat{\mathbf{P}}$ in the Dirac field.

5.13 Show that if there are n generalized coordinates $q^i(\mathbf{x}, t)$ with $i = 1, 2, \cdots, n$, then the stress tensor in Eq. (5.69) gets generalized to

$$T_{\mu\nu} \equiv \mathcal{L}\,\delta_{\mu\nu} - \sum_{i=1}^{n} \frac{\partial \mathcal{L}}{\partial(\partial q^i / \partial x_\mu)} \frac{\partial q^i}{\partial x_\nu}$$

5.14 (a) Construct the stress tensor for the string. Use the two-vector (x, ict) where c is the wave velocity in the string;
   (b) Construct the two-vector $P_\mu = (P, iH/c)$;
   (c) Insert the interaction-picture expansion of the string field in Eq. (5.52) and rederive the previous expression for the hamiltonian $\hat{H}_0$ in Eq. (5.55);
   (d) Repeat part (c), and determine the momentum $\hat{P}$.

5.15 Problems 5.15–5.17 concern the complex, or charged, scalar field (one can only get a current with a complex field). Assume equal mass scalar fields $(\phi, \phi^\star)$. These are independent fields, since both the real and imaginary parts must be specified to specify both of them [compare part (a) below]. Assume a lagrangian density of the form

$$\mathcal{L}\left(\phi, \frac{\partial \phi}{\partial x_\mu}, \phi^\star, \frac{\partial \phi^\star}{\partial x_\mu}\right) = -c^2 \left[\frac{\partial \phi^\star}{\partial x_\mu} \frac{\partial \phi}{\partial x_\mu} + m_s^2 \phi^\star \phi\right]$$

a) Write out Hamilton's principle. Show that if $A\delta\phi + B\delta\phi^\star = 0$ then $A = B = 0$;[10]

---

[10] *Hint:* First take $\delta\phi$ to be a real variation, and then an imaginary one.

b) Show that Lagrange's equations are just the Klein-Gordon equations

$$(\Box - m_s^2)\phi = 0 \qquad ; \; \phi^* \text{ eqn.}$$
$$(\Box - m_s^2)\phi^* = 0 \qquad ; \; \phi \text{ eqn.}$$

c) Show the stress tensor is [note Prob. 5.13]

$$T_{\mu\nu} = \mathcal{L}\,\delta_{\mu\nu} + c^2 \frac{\partial \phi^*}{\partial x_\mu}\frac{\partial \phi}{\partial x_\nu} + c^2 \frac{\partial \phi^*}{\partial x_\nu}\frac{\partial \phi}{\partial x_\mu}$$

d) Show the momentum densities are[11]

$$\pi_\phi = \frac{\partial \phi^*}{\partial t} \qquad ; \; \pi_{\phi^*} = \frac{\partial \phi}{\partial t}$$

e) Show that the four-momentum gives

$$H = \int_\Omega d^3x \, \left(\pi_{\phi^*}\pi_\phi + c^2 \boldsymbol{\nabla}\phi^* \cdot \boldsymbol{\nabla}\phi + m_s^2 c^2 \phi^*\phi \right)$$
$$\mathbf{P} = -\int_\Omega d^3x \, \left(\pi_{\phi^*}\boldsymbol{\nabla}\phi^* + \pi_\phi \boldsymbol{\nabla}\phi \right)$$

5.16 Now quantize the continuum problem formulated in Prob. 5.15.[12]

(a) Write out the canonical equal-time commutation relations in the interaction picture for (π_{ϕ^*}, ϕ^*) and (π_ϕ, ϕ);

(b) Show the following expansions of the fields in the interaction picture satisfy these commutation relations

$$\hat{\phi}(\mathbf{x}, t) = \sum_\mathbf{k} \left(\frac{\hbar}{2\omega_k \Omega} \right)^{1/2} \left[a_\mathbf{k} e^{i(\mathbf{k}\cdot\mathbf{x} - \omega_k t)} + b_\mathbf{k}^\dagger e^{-i(\mathbf{k}\cdot\mathbf{x} - \omega_k t)} \right]$$
$$\hat{\pi}_\phi(\mathbf{x}, t) = i \sum_\mathbf{k} \left(\frac{\hbar\omega_k}{2\Omega} \right)^{1/2} \left[a_\mathbf{k}^\dagger e^{-i(\mathbf{k}\cdot\mathbf{x} - \omega_k t)} - b_\mathbf{k} e^{i(\mathbf{k}\cdot\mathbf{x} - \omega_k t)} \right]$$

Here p.b.c. are assumed, and $(a^\dagger, b^\dagger, a, b)$ are the usual boson creation and destruction operators. The operators $(\hat{\phi}^*, \hat{\pi}_{\phi^*})$ are just the adjoints of these expressions;

---

[11]Note carefully where the ϕ and ϕ^* appear in these equations.

[12]The interpretation of the Klein-Gordon equation as the field equation in quantum field theory is due to [Pauli and Weisskopf (1934)].

(c) Subtitute these expressions in part (e) of Prob. 5.15, and show the hamiltonian and momentum become

$$\hat{H}_0 = \sum_{\mathbf{k}} \hbar\omega_k \left(a_{\mathbf{k}}^\dagger a_{\mathbf{k}} + b_{\mathbf{k}}^\dagger b_{\mathbf{k}} + 1 \right)$$

$$\hat{\mathbf{P}} = \sum_{\mathbf{k}} \hbar\mathbf{k} \left(a_{\mathbf{k}}^\dagger a_{\mathbf{k}} + b_{\mathbf{k}}^\dagger b_{\mathbf{k}} \right)$$

5.17 (a) Show that the lagrangian density in Prob. 5.15 has an invariance under global phase transformations;

(b) Use Noether's theorem to construct the corresponding conserved current

$$j_\mu(x) = \frac{c}{i\hbar} \left[\phi^\star \frac{\partial\phi}{\partial x_\mu} - \frac{\partial\phi^\star}{\partial x_\mu}\phi \right]$$

(c) Show that the integral over all space of the fourth component of this current is a constant of the motion, and show that when quantized, this charge is given by[13]

$$\hat{Q} = \sum_{\mathbf{k}} \left(a_{\mathbf{k}}^\dagger a_{\mathbf{k}} - b_{\mathbf{k}}^\dagger b_{\mathbf{k}} \right)$$

(d) Hence conclude that this quantum field theory contains both *particles and antiparticles* with the opposite sign of this charge.

5.18 One would like to have a *symmetric* stess tensor $T_{\mu\nu} = T_{\nu\mu}$ in order to construct a conserved angular momentum density

$$M_{\mu\nu\nu'} = T_{\mu\nu}x_{\nu'} - T_{\mu\nu'}x_\nu \qquad ; \text{ angular momentum density}$$

(a) Show that $\partial M_{\mu\nu\nu'}/\partial x_\mu = 0$ if $T_{\mu\nu}$ is symmetric;

It is always possible to symmetrize $T_{\mu\nu}$, while not changing the integrated four-momentum in Eq. (5.74), and the general procedure for doing this is in an appendix in [Wentzel (1949)]. Here we simply note that the scalar stress tensors in Eq. (5.76) and Prob. 5.15(c) are already symmetric, and we develop a procedure for symmetrizing the Dirac stress tensor.

(b) The Dirac $\mathcal{L}$ vanishes along the dynamical trajectory, and $T_{\mu\nu}$ in Eq. (5.96) can equally well be written as[14]

$$\tilde{T}_{\mu\nu} = \frac{\hbar c}{2} \left[\bar{\psi}\gamma_\mu \frac{\partial\psi}{\partial x_\nu} - \frac{\partial\bar{\psi}}{\partial x_\nu}\gamma_\mu\psi \right]$$

---

[13] Here we write $j_\mu = (\mathbf{j}/c, i\rho)$, where ρ is the charge density.

[14] This comes from the equivalent lagrangian density $\tilde{\mathcal{L}} = \mathcal{L}_{\text{Dirac}} + (\hbar c/2)\partial(\bar{\psi}\gamma_\mu\psi)/\partial x_\mu$.

Show this $\tilde{T}_{\mu\nu}$ has the following properties:

$$\frac{\partial \tilde{T}_{\mu\nu}}{\partial x_\mu} = \frac{\partial \tilde{T}_{\mu\nu}}{\partial x_\nu} = 0$$

$$\int_\Omega d^3x\, \tilde{T}_{4\nu} = \int_\Omega d^3x\, T_{4\nu}$$

(c) Now take $\theta_{\mu\nu} \equiv \tilde{T}_{\mu\nu} + (\tilde{T}_{\nu\mu} - \tilde{T}_{\mu\nu})/2$. Show[15]

$$\frac{\partial \theta_{\mu\nu}}{\partial x_\mu} = 0 \qquad\qquad ; \ \theta_{\mu\nu} = \theta_{\nu\mu}$$

$$\int_\Omega d^3x\, \theta_{4\nu} = \int_\Omega d^3x\, T_{4\nu}$$

Hence conclude that $\theta_{\mu\nu}$ has all the desired properties for the stress tensor.

5.19 Assume that under a Lorentz transformation the Dirac field transforms as $\psi(x) \to S\psi(x')$ where S is a 4×4 matrix with the properties in Prob. E.4.

(a) Show the lagrangian density is Eq. (5.92) is then a Lorentz scalar;

(b) Show the resulting Dirac equation is then Lorentz invariant.

5.20 Show that the term that is subtracted off in the normal-ordered current in Eq. (5.116) simply counts the number of negative energy-states.

5.21 Show that $g^2/4\pi\hbar c^3$ is *dimensionless* in Eq. (5.119).

5.22 Insert the field expansions, and discuss the processes described by the interaction hamiltonian $\hat{H}_I(t)$ for the Dirac-scalar theory in Eq. (5.121).

5.23 Show that the canonical equal-time commutation relations in Eq. (5.53) are picture-independent.

5.24 Consider the interacting Dirac-scalar field theory of Eqs. (5.117) and (5.118).

(a) Write Lagrange's equations for each of the fields;

(b) In the Schrödinger picture, operators are independent of time, and their time derivative is defined through Ehrenfest's theorem[16]

$$\frac{d\hat{O}}{dt} \equiv \frac{i}{\hbar}[\hat{H}, \hat{O}]$$

---

[15] *Hint:* Show that the extra term $(\tilde{T}_{\nu\mu} - \tilde{T}_{\mu\nu})/2$ does not contribute in the integral.

[16] See Prob. 2.3. It is assumed here that $\hat{O}$ has no additional explicit time dependence.

The expectation value of this operator taken with the time-dependent Schrödinger state vector gives the time derivative of the expectation value of the operator. Take one time derivative of the scalar field, make use of the canonical commutation relations in the Schrödinger picture [the analogs of Eqs. (5.61)], and derive the operator form of Eq. (5.77);

(c) Compute the second time derivative, and derive the operator form of the scalar field equation in part (a);

(d) Discuss under what conditions the correspondence principle applies, that is, when can the quantum field be replaced by the corresponding classical field. (Compare Prob. 5.3.)

5.25 There is an alternate formulation of classical mechanics in terms of *Poisson brackets*, which bears a close analogy to quantum mechanics. Given a function $F(q^1, \cdots, q^n; p^1, \cdots, p^n; t)$, and a similar function G, their Poisson bracket is defined by

$$[F, G]_{\text{PB}} \equiv \sum_{\sigma=1}^{n} \left(\frac{\partial F}{\partial q_\sigma} \frac{\partial G}{\partial p_\sigma} - \frac{\partial F}{\partial p_\sigma} \frac{\partial G}{\partial q_\sigma} \right) \quad ; \text{ Poisson bracket}$$

where the partial derivatives imply that all the other variables in the arguments of $F(q^1, \cdots, q^n; p^1, \cdots p^n; t)$ and G are to be held fixed.

(a) Show that Hamilton's equations take the form

$$\dot{q}_\sigma = -[H, q_\sigma]_{\text{PB}} = \frac{\partial H}{\partial p_\sigma} \qquad\qquad ; \sigma = 1, \cdots, n$$

$$\dot{p}_\sigma = -[H, p_\sigma]_{\text{PB}} = -\frac{\partial H}{\partial q_\sigma}$$

(b) Show that the total time derivative of F is given by

$$\frac{dF}{dt} = -[H, F]_{\text{PB}} + \frac{\partial F}{\partial t}$$

(c) Show that

$$[p_\sigma, q_\rho]_{\text{PB}} = -\delta_{\sigma\rho}$$
$$[p_\sigma, p_\rho]_{\text{PB}} = [q_\sigma, q_\rho]_{\text{PB}} = 0$$

(d) Compare with the expressions in Eqs. (5.37) and Prob. 5.24(b). Hence make the argument that the formal transition from classical to quantum mechanics is made through the replacement of a Poisson bracket by a commutator as follows[17]

$$[A, B]_{\text{PB}} \rightarrow \frac{1}{i\hbar}[\hat{A}, \hat{B}]$$

6.1 Write out the commutator in the first line of Eqs. (6.19) in detail. Use the canonical commutation relations to reorder the fields to get two cancelling contributions. Show that one is left with the contribution in the second line of Eqs. (6.19).

6.2 Show by direct matrix multiplication that the 3×3 matrices defined in Eq. (6.13), and presented explicitly in Eqs. (6.14), satisfy the commutation relations for the generators of SU(2) in Eq. (6.21).

6.3 Repeat Prob. 6.1 starting with the charges in Eq. (6.37). Use the canonical *anticommutation* relations in Eq. (6.36). Show that one again obtains the relation in Eq. (6.38).

6.4 Show that Eq. (6.48) holds even when the fields obey canonical *anticommutation* relations.

6.5 Suppose $\hat{\phi}(\mathbf{x})$ is an n-component, column-vector, real scalar field; $\hat{\underline{\Pi}}(\mathbf{x})$ is the conjugate momentum density (a row matrix); and $\underline{M}^a$ with $a = (1, \cdots, p)$ is a set of $n \times n$ matrices. Define the charge $\hat{Q}^a$ by

$$\hat{Q}^a = \frac{1}{i\hbar} \int_\Omega d^3x \, \hat{\underline{\Pi}} \, \underline{M}^a \, \hat{\underline{\phi}}$$

Use the canonical commutation relations to show that

$$[\hat{Q}^a, \hat{Q}^b] = \frac{1}{i\hbar} \int_\Omega d^3x \, \hat{\underline{\Pi}} \, [\underline{M}^a, \underline{M}^b] \, \hat{\underline{\phi}} \qquad ; (a,b) = 1, \cdots, p$$

6.6 (a) Extend the interaction-picture representation of the massive scalar fields in Eqs. (5.60) to the isovector case by adding a subscript for the internal space. Insert these field expansions in the definition of the isospin operator $\hat{\mathbf{T}} = (\hat{T}_1, \hat{T}_2, \hat{T}_3)$ in Eq. (6.16), and express this operator in terms of the creation and destruction operators;[18]

(b) Repeat for the Dirac fields in Eqs. (5.102) and (6.28) and the isospin operator in Eq. (6.37).

6.7 Make a transformation to a new set of real (hermitian) fields in Probs. 5.15–5.16, with $\phi^* \equiv (\phi_1 + i\phi_2)/\sqrt{2}$ and $\phi \equiv (\phi_1 - i\phi_2)/\sqrt{2}$.

(a) Compute the new lagrangian density and new conserved currents;

---

[17]See [Fetter and Walecka (2003)].
[18]Recall Eqs. (5.65) and (6.23).

(b) Find the new commutation relations and interaction-picture fields;

(c) Add a neutral scalar field ϕ_3 of the same mass, and recover the configuration in Prob. 6.6(a).

6.8 (a) Use the repeated commutator relation in Prob. 2.5 to establish the following matrix identity

$$\exp\left\{-\frac{i}{2}\boldsymbol{\omega}\cdot\boldsymbol{\tau}\right\}\tau_i\exp\left\{\frac{i}{2}\boldsymbol{\omega}\cdot\boldsymbol{\tau}\right\} = \exp\left\{i\boldsymbol{\omega}\cdot\mathbf{t}\right\}_{ij}\tau_j$$

(b) Show $\exp\left\{i\boldsymbol{\omega}\cdot\mathbf{t}\right\}_{ij} = a_{ij}(\boldsymbol{\omega})$ is a real, orthogonal, 3×3 rotation matrix [compare Eq.(3.22)];

(c) Hence prove that the combination $\bar{\psi}\boldsymbol{\tau}\psi$ transforms as an isovector under isospin rotations.

6.9 Consider a field theory composed of an isovector scalar field with the lagrangian density of Eq. (6.7), an isospinor Dirac field with the lagrangian density of Eq. (6.29), and an *interaction* of the form

$$\mathcal{L}_1 = g\,\bar{\psi}\,\frac{1}{2}\boldsymbol{\tau}\,\psi\cdot\boldsymbol{\phi}$$

(a) Use the result in Prob. 6.8 to show this interaction is invariant under isospin rotations;

(b) Construct the total isospin operator $\hat{\mathbf{T}}$ for the combined system. What are the eigenvalues of $\hat{\mathbf{T}}^2$ in the interacting system? What is the degeneracy of each isospin multiplet $|TM_T\rangle$?

(c) A scattering state starts as one free isovector scalar meson and one free isospinor Dirac particle. What values of T are accessed in the scattering process?

(d) Show the scattering operator $\hat{S}$ commutes with all components of $\hat{\mathbf{T}}$.[19] Use the Wigner-Eckart theorem to deduce the consequences for the matrix elements $\langle T'M_T'|\hat{S}|TM_T\rangle$.

6.10 Show that the Dirac mass term $\mathcal{L}_{\text{mass}} = -\hbar c\bar{\psi}M\psi$ is *not* invariant under the chiral transformation in Eq. (6.138).

6.11 This problem concerns the σ-model. While each part is a significant problem on its own, it was decided to group them together because they are clearly related, and because the required algebra is, in fact, carried out in detail in [Walecka (2004)] (albeit with units where $\hbar = c = 1$).

(a) Set $\delta V_{\text{csb}} = 0$ and use Noether's therem to derive the conserved currents in Eqs. (6.157);

---

[19] *Hint*: start from Eq. (4.17).

(b) Now include $\delta V_{\text{csb}} = \epsilon \sigma$, and derive the field equations. Start from the currents in Eqs. (6.157) and derive the CVC and PCAC relations in Eqs. (6.167);

(d) Verify the form of the lagrangian density in Eq. (6.166) obtained by using the expansions about the value $\sigma = \sigma_0$ in Eqs. (6.164).

6.12 This problem explicitly exhibits the $SU(2)_L \otimes SU(2)_R$ symmetry of the σ-model. Define left- and right-handed Dirac fields, and a meson matrix, by [recall Eqs. (6.145)]

$$\hat{\psi}_L \equiv P_L \hat{\psi} \qquad\qquad ; \hat{\psi}_R \equiv P_R \hat{\psi}$$

$$\underline{\chi} \equiv \frac{1}{\sqrt{2}}\left(\underline{1}\,\sigma + i\underline{\tau}\cdot\boldsymbol{\pi}\right)$$

(a) Show the σ-model lagrangian can be written as

$$\mathcal{L} = -\hbar c\left[\hat{\bar{\psi}}_L \gamma_\mu \frac{\partial}{\partial x_\mu}\hat{\psi}_L + \hat{\bar{\psi}}_R \gamma_\mu \frac{\partial}{\partial x_\mu}\hat{\psi}_R - \sqrt{2}\,g\left(\hat{\bar{\psi}}_R\,\underline{\chi}\,\hat{\psi}_L + \hat{\bar{\psi}}_L\,\underline{\chi}^\dagger\,\hat{\psi}_R\right)\right]$$
$$-\frac{c^2}{2}\operatorname{tr}\left[\left(\frac{\partial\underline{\chi}}{\partial x_\mu}\right)^*\left(\frac{\partial\underline{\chi}}{\partial x_\mu}\right)\right] - V\left[\operatorname{tr}\left(\underline{\chi}^\dagger\underline{\chi}\right)\right]$$

Here "tr" is the trace of the isospin matrix, and $\underline{v}_\mu^* \equiv (\underline{v}^\dagger, +i\underline{v}_0^\dagger)$;[20]

(b) Let $\underline{R} = \exp\{\tfrac{i}{2}\boldsymbol{\omega}\cdot\underline{\tau}\}$ be a global SU(2) matrix. Show that $\mathcal{L}$ is invariant under the SU(2) transformation

$$\hat{\psi}_R \to \underline{R}\,\hat{\psi}_R \qquad ; \underline{\chi} \to \underline{R}\,\underline{\chi} \qquad ; \hat{\psi}_L \to \hat{\psi}_L$$

(c) Show that $\mathcal{L}$ is also invariant under the *independent* global SU(2) transformation

$$\hat{\psi}_L \to \underline{L}\,\hat{\psi}_L \qquad ; \underline{\chi} \to \underline{\chi}\,\underline{L}^\dagger \qquad ; \hat{\psi}_R \to \hat{\psi}_R$$

6.13 The generator operators $\hat{G}^a$ for the SU(3) symmetry of the Sakata model are obtained from the fundamental matrices in Eqs. (6.74) through Eq. (6.79).

(a) The *rank* of a group is the number of mutually commuting generators. Show from the matrices involved that SU(3) is a rank 2 group;

(b) The baryon number $\hat{B}$ in the Sakata model arises from the unit 3×3 matrix $\underline{1}$. Conclude that the strangeness operator $\hat{S}$ is obtained from

---

[20]We set $\delta V_{\text{csb}} \equiv 0$, and the metric is not complex conjugated in $\underline{v}_\mu^*$. Note that the trace is invariant under cyclic permutations of its arguments, $\operatorname{tr}\underline{\tau} = 0$, and $\operatorname{tr}\underline{1} = 2$.

$(\sqrt{3}\underline{\lambda}^8 - \underline{1})/3$, and the hypercharge operator $\hat{Y} = (\hat{B} + \hat{S})$ from $(2\underline{1} + \sqrt{3}\underline{\lambda}^8)/3$;

(c) The third component of isospin $\hat{T}_3$ arises from $\underline{\lambda}^3/2$. Hence prove from the matrices involved that in the Sakata model the charge operator $\hat{Q}$ is given by

$$\hat{Q} = \hat{T}_3 + \frac{1}{2}\hat{Y}$$

This is the Gell-Mann–Nishijima relation.

6.14 Consider the proof of the fundamental SU(3) matrix relations in Eqs. (6.76), which is here carried out to first order in ε^a where $\omega^a \equiv \varepsilon^a \to 0$. Write

$$\underline{r}(\varepsilon) = 1 + \frac{i}{2}\varepsilon^a \underline{\lambda}^a$$

(a) Show to $O(\varepsilon)$ that

$$\underline{r}(\varepsilon)^\dagger = \underline{r}(\varepsilon)^{-1} \qquad ; \ \underline{r}(\varepsilon)^{-1} = \underline{r}(-\varepsilon) \qquad ; \ \underline{r}(0) = 1$$
$$\underline{r}(\varepsilon)\underline{r}(\varepsilon') = \underline{r}(\varepsilon'') \qquad ; \ \varepsilon'' = \varepsilon' + \varepsilon$$

(b) Use the definition of the determinant in Eq. (D.16) to show to $O(\varepsilon)$

$$\det \underline{r}(\varepsilon) = 1 + \frac{i}{2}\varepsilon^a \lambda^a_{i_p i_p} = 1 + \frac{i}{2}\varepsilon^a \, \mathrm{tr}\,[\underline{\lambda}^a] = 1$$

7.1 Write out the fields in Eqs. (7.2)–(7.3), and verify Eqs. (7.10).

7.2 (a) Evaluate the matrix elements of the scalar field in Eqs. (7.12), and show that one obtains the appropriate sums over modes for the Feynman propagator as given in Eqs. (F.43) and (F.39);

(b) Use the distributive law to then verify Eq. (7.13);

(c) Repeat parts (a) and (b) for the Dirac field in Eqs. (7.15)–(7.18), where the appropriate sums over modes are given in Eqs. (F.47)–(F.48).

7.3 Show that if the massive particles (N, ϕ) are stable, then the first-order scattering operator $\hat{S}^{(1)}$ vanishes in the Dirac-scalar theory.

7.4 (a) Find the matrix elements S_{fi} to second order in g in the Dirac-scalar theory for the processes $N + \bar{N} \to \phi + \phi$ and $\phi + \phi \to N + \bar{N}$;

(b) Use these results to extend the Feynman rules in momentum space to take into account incoming and outgoing antifermion lines in this theory:

• Include a factor of $(1/\Omega)^{1/2}\bar{v}(-\mathbf{k}\lambda)$ for each incoming $\bar{N}$-line with incoming four-momentum k (opposite to the line's direction);

- Include a factor of $(1/\Omega)^{1/2} v(-\mathbf{k}\lambda)$ for each outgoing $\bar{N}$-line with outgoing four-momentum k (opposite to the line's direction). See Fig. 8.7.

7.5 (a) Draw all connected Feynman diagrams through order g^4 in the Dirac-scalar theory for the processes $\phi + N \to \phi + N$, $N + N \to N + N$, and the self-energies of (N, ϕ);

(b) Use the Feynman rules in momentum space to identify all the elements in the diagrams for $N(k_1) + N(k_2) \to N(k_3) + N(k_4)$.

7.6 (a) Consider the Dirac-(scalar)^2 theory containing two independent scalar fields (ϕ_1, ϕ_2), with coupling constants and masses (g_1, m_1) and (g_2, m_2) respectively. Assume that $m_1 \neq m_2$ and that all the particles are stable. The interaction hamiltonian density is now

$$\mathcal{H}_1 = \mathcal{H}_1^{(1)} + \mathcal{H}_1^{(2)}$$
$$\mathcal{H}_1^{(i)} = -g_i \bar{\psi}\psi\phi_i \qquad ; i = 1, 2$$

Show that the scattering operator takes the form

$$\hat{S} = \sum_{l=0}^{\infty} \sum_{m=0}^{\infty} \left(\frac{ig_1}{\hbar c}\right)^l \left(\frac{ig_2}{\hbar c}\right)^m \frac{1}{l!} \frac{1}{m!} \int d^4x_1 \cdots \int d^4x_l \int d^4y_1 \cdots \int d^4y_m \times$$
$$P[:\hat{\bar{\psi}}(x_1)\hat{\psi}(x_1): \cdots :\hat{\bar{\psi}}(x_l)\hat{\psi}(x_l)::\hat{\bar{\psi}}(y_1)\hat{\psi}(y_1): \cdots :\hat{\bar{\psi}}(y_m)\hat{\psi}(y_m):] \times$$
$$P[\hat{\phi}_1(x_1) \cdots \hat{\phi}_1(x_l)] P[\hat{\phi}_2(y_1) \cdots \hat{\phi}_2(y_m)]$$

The convention here is that the term with $l = 0$ is the scattering operator with $\mathcal{H}_1^{(1)} = 0$, and the term with $m = 0$ is the scattering operator with $\mathcal{H}_1^{(2)} = 0$.[21]

(b) Apply Wick's theorem up through second order in the coupling constants (that is, up through order $l + m = 2$).

7.7 (a) Use the results in Prob. 7.6(b) to calculate the amplitudes for the processes $\phi_1 + N \to \phi_1 + N$, $\phi_1 + N \to \phi_2 + N$, $N + N \to N + N$, and the self-energies of the (ϕ_1, ϕ_2, N);

(b) As in the text, use these results to deduce the Feynman rules for this theory.

7.8 An effective field theory that has proven useful in nuclear physics is composed of a Dirac nucleon described by the field in Eq. (6.28), interacting both with the neutral scalar field of Eqs. (5.117)–(5.118), and the

---

[21] *Hint*: Show all the terms with a given partition $(\hat{\phi}_1)^l (\hat{\phi}_2)^m$ are identical by a change of dummy variables, and then appeal to the binomial theorem.

neutral vector meson field of Probs. C.6–C.7.[22] The lagrangian density for this composite system is taken to be

$$\mathcal{L} = -\hbar c \bar{\underline{\psi}} \left[\gamma_\mu \left(\frac{\partial}{\partial x_\mu} - \frac{ig_v}{\hbar c} V_\mu \right) + \left(M - \frac{g_s}{\hbar c} \phi \right) \right] \underline{\psi} + \mathcal{L}_0^S + \mathcal{L}_0^V$$

(a) Show the interaction hamiltonian density is

$$\mathcal{H}_1 = -g_s \bar{\underline{\psi}} \underline{\psi} \phi - i g_v \bar{\underline{\psi}} \gamma_\mu \underline{\psi} V_\mu$$

(b) Show the Feynman propagator for the nucleon field is

$$\langle 0 | P[\hat{\psi}_i(x) \hat{\bar{\psi}}_j(y)] | 0 \rangle = i \delta_{ij} S_F(x-y) \qquad ; (i,j) = 1, 2$$

where these indices refer to isospin;

(c) Write out the scattering operator in the interaction picture to second order in the coupling constants (g_s, g_v);

(d) Apply Wick's theorem to the result in (c).

7.9 Make use of the results in Prob. 7.8, the vector fields in Probs. C.6–C.7, and the vector propagator in Prob. F.5(b).

(a) Take a matrix element of the second-order scattering operator, and extend the construction of the coordinate-space amplitude for the process $N + N \to N + N$;

(b) Construct the coordinate-space amplitudes for the processes $V + N \to V + N$, and $\phi + N \to V + N$;

(c) Insert Fourier transforms, and construct the corresponding amplitudes in momentum space.

7.10 (a) Work in analogy to the analysis in the text, and interpret the results in in Prob. 7.9 in terms of Feynman diagrams;

(b) Use these results to deduce a set of Feynman rules for the Dirac-scalar-vector theory.

7.11 (a) Relax the assumption that the scalar density is normal-ordered in the Dirac-scalar example in Eq. (7.32). Show that the contraction *within* the scalar density is to be interpreted as follows

$$\hat{\bar{\psi}}(x)^{\cdot} \hat{\psi}(x)^{\cdot} = (-1) i \operatorname{Tr} S_F(0^-)$$

$$= \langle 0 | \hat{\bar{\psi}}(x) \hat{\psi}(x) | 0 \rangle \equiv \rho_S$$

Here $S_F(0^-)$ is $S_F(x - x')$ with $\mathbf{x}' \to \mathbf{x}$ and $t' \to t^+$. The scalar density ρ_S is a constant independent of x;

----

[22]See Probl. 9.5.

(b) Apply Wick's theorem, and show the result is to add the "tadpole" diagrams shown in Fig. 12.1. Assign all factors associated with the tadpole contribution to the self-energy in Fig. 12.1(a);

(c) Insert Fourier transforms, and show the only consequence of the tadpole diagram in Fig. 12.1(a) is to add the following constant to the self-mass of the N

$$\delta M_{\text{tad}} = -\frac{g^2}{\hbar c^3} \frac{\rho_S}{m^2}$$

(d) Show the additional term in part (c) will be completely cancelled by the counter-term used in mass renormalization;

(e) Argue that the additional vacuum-vacuum contribution also disappears from physical results.

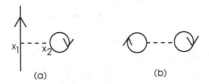

(a)   (b)

Fig. 12.1   Feynman diagrams for the tadpole contribution to: (a) the self-energy of N; (b) the vacuum-vacuum amplitude.

7.12 Discuss the mass renormalization of the ϕ in the Dirac-scalar theory.

7.13 Suppose the masses in the Dirac-scalar case are such that the ϕ is unstable, and the decay $\phi \to N + \bar{N}$ can actually take place.

(a) Show the amplitude S_{fi} for $\phi(p) \to N(k_1) + \bar{N}(k_2)$ is given to lowest order by

$$S_{fi} = (2\pi)^4 \delta^{(4)}(p - k_1 - k_2)\frac{1}{\Omega^{3/2}} \left(\frac{\hbar}{2\omega_p}\right)^{1/2} \left(\frac{ig}{\hbar c}\right) \bar{u}(k_1\lambda_1)v(-k_2\lambda_2)$$

(b) Show that before going to the continuum limit, in a big box of volume Ω with p.b.c., this expression is actually[23]

$$S_{fi} = 2\pi\delta(W_f - W_i)[\Omega\delta_{\mathbf{K}_f,\mathbf{K}_i}]\frac{1}{\Omega^{3/2}} \left(\frac{\hbar}{2\omega_p}\right)^{1/2} ig\bar{u}(k_1\lambda_1)v(-k_2\lambda_2)$$

---

[23]Compare with Probl. 7.2.

where (W_i, W_f) are the total initial and final energies, and $\hbar(\mathbf{K}_i, \mathbf{K}_f)$ are the total initial and final momenta.

7.14 (a) Use the arguments in chapter 7 of Vol. I to show that the differential decay rate for $\phi \to N + \bar{N}$ that follows from Prob. 7.13 is

$$ d\omega_{fi} = \frac{2\pi}{\hbar} \delta(W_f - W_i) \frac{1}{\Omega} |T_{fi}|^2 \frac{\Omega d^3 k_1}{(2\pi)^3} \qquad ; \; \mathbf{p} = \mathbf{k}_1 + \mathbf{k}_2 $$

Here we factor out $\Omega^{-(n_f + n_i)/2}$ in the definition of T_{fi} [see Eq. (7.38)].
(b) Show that for a ϕ at rest this gives

$$ d\omega_{fi} = \frac{1}{64\pi^2} \left(\frac{g^2}{\hbar c^3} \right) \left(\frac{m_\phi c^2}{\hbar} \right) \left(1 - \frac{4m_N^2}{m_\phi^2} \right)^{1/2} |\bar{u}(\mathbf{k}_1 \lambda_1) v(-\mathbf{k}_2 \lambda_2)|^2 d\Omega_{k_1} $$

7.15 (a) Verify the $N + N$ scattering amplitude in Eq. (7.44);
(b) Verify the expression for the ϕ self-energy in Eq. (7.53).

7.16 Use the notation for the creation and destruction operators in Eqs. (8.10). Evaluate the following matrix elements:

(a) $\langle 0 | a_{\mathbf{k}_4 \lambda_4} c_{\mathbf{k}_3 s_3} a_{\mathbf{k}_8 \lambda_8}^\dagger a_{\mathbf{k}_7 \lambda_7} c_{\mathbf{k}_6 s_6}^\dagger c_{\mathbf{k}_5 s_5} c_{\mathbf{k}_1 s_1}^\dagger a_{\mathbf{k}_2 \lambda_2}^\dagger | 0 \rangle$

(b) $\langle 0 | a_{\mathbf{k}_4 \lambda_4} a_{\mathbf{k}_3 \lambda_3} a_{\mathbf{k}_8 \lambda_8}^\dagger a_{\mathbf{k}_7 \lambda_7}^\dagger a_{\mathbf{k}_6 \lambda_6} a_{\mathbf{k}_5 \lambda_5} a_{\mathbf{k}_1 \lambda_1}^\dagger a_{\mathbf{k}_2 \lambda_2}^\dagger | 0 \rangle$

(c) $\langle 0 | a_{\mathbf{k}_4 \lambda_4} b_{\mathbf{k}_3 \lambda_3} b_{\mathbf{k}_8 \lambda_8}^\dagger a_{\mathbf{k}_7 \lambda_7}^\dagger b_{\mathbf{k}_6 \lambda_6} a_{\mathbf{k}_5 \lambda_5} b_{\mathbf{k}_1 \lambda_1}^\dagger a_{\mathbf{k}_2 \lambda_2}^\dagger | 0 \rangle$

8.1 (a) Verify that the interaction-picture Dirac current is conserved in Eqs. (8.33);
(b) Show that the equal-time interaction-picture commutator of the charge and current densities vanishes in Eq. (8.38).

8.2 Carry out the four-dimensional Fourier transform in Minkowski space, and show that the Coulomb interaction can be written as in Eq. (8.18).[24]

8.3 Prove the following relations involving the Dirac matrices[25]

$$ \slashed{a}\slashed{b} + \slashed{b}\slashed{a} = 2(a \cdot b) $$
$$ \gamma_\mu \slashed{a} \gamma_\mu = -2\slashed{a} $$
$$ \gamma_\mu \slashed{a}\slashed{b} \gamma_\mu = 2(\slashed{a}\slashed{b} + \slashed{b}\slashed{a}) = 4(a \cdot b) $$
$$ \gamma_\mu \slashed{a}\slashed{b}\slashed{c} \gamma_\mu = -2\slashed{c}\slashed{b}\slashed{a} $$

---

[24] *Hint:* see Vol. I.
[25] The Feynman notation is $\slashed{a} = a_\mu \gamma_\mu$, where the four-vector $a_\mu = (\mathbf{a}, ia_0)$.

8.4 Consider the cross section for Compton scattering in the *laboratory* frame obtained from the amplitude $S_{fi}^{(2)}$ in Eq. (8.44). Denote the initial and final three-momenta of the photons in this frame by $(\mathbf{l}_1, \mathbf{l}_2)$. An essential component of this calculation is the calculation of

$$\left(\frac{\partial W_f}{\partial l_2}\right)_\theta = (\hbar c)^2 \frac{M l_1}{E_2 l_2}$$

Here the partial derivative is a reminder that the change in variables from l_2 to the total final energy W_f in Eq. (8.58) is to be carried out at constant angle, and the result is evaluated at energy conservation $W_i = W_f$. Here E_2 is the final electron energy. Prove this relation.

8.5 Consider the sum over the transverse polarizations of an external photon, as required, for example, in calculating the Compton cross section. Write the polarization four-vector as [see Eqs. (8.45)] $\varepsilon_\mu(\mathbf{k}s) = (\mathbf{e}_{\mathbf{k}s}, 0)$. Assume that in calculating the cross section one has to evaluate an expression of the form $\sum_{s=1}^{2} \varepsilon_\mu(\mathbf{k}s) M_{\mu\nu} \varepsilon_\nu(\mathbf{k}s)$ where the response tensor satisfies current conservation in momentum space[26]

$$k_\mu M_{\mu\nu} = M_{\mu\nu} k_\nu = 0 \qquad \text{; current conservation}$$

Derive the *covariant polarization sum*

$$\sum_{s=1}^{2} \varepsilon_\mu(\mathbf{k}s) M_{\mu\nu} \varepsilon_\nu(\mathbf{k}s) = M_{\mu\mu}$$

8.6 The cross section for Compton scattering in the laboratory frame is given by the Klein-Nishina formula

$$\frac{d\sigma}{d\Omega} = \frac{1}{2} r_0^2 \left(\frac{l_2}{l_1}\right)^2 \left[\left(\frac{l_2}{l_1}\right) + \left(\frac{l_1}{l_2}\right) - \sin^2\theta\right] \qquad ; r_0 \equiv \frac{e^2}{4\pi\hbar c\varepsilon_0} \frac{1}{M}$$

Use the results in Probs. 8.3–8.5 to calculate the contribution to this result coming from the square of the first Feynman diagram in Fig. 8.1. Make sure you understand where all the factors come from.

8.7 Take the limit of the result in Prob. 8.6 for a heavy target, and derive the Thomson cross section for photon scattering

$$\frac{d\sigma}{d\Omega} = r_0^2 \frac{1}{2}(1 + \cos^2\theta) \qquad \text{; Thomson cross section}$$

---

[26]This can, and should, be checked in any application; this relation generally holds only for the full amplitude.

8.8 (a) Construct $S_{fi}^{(2)}$ for the Bhabha scattering process $e^-(k_1) + e^+(k_2) \rightarrow e^-(k_3) + e^+(k_4)$;

(b) Draw the two momentum-space Feynman diagrams.

8.9 Make use of the Dirac equation on the external lepton legs, and show explicitly that the terms proportional to q_μ or q_ν in the photon propagator $\tilde{D}_{\mu\nu}^F(q)$ in Eq. (8.28) do not contribute to the scattering amplitudes in Eqs. (8.52) and (8.96).

8.10 Show that Eq. (8.93) is the Rutherford cross section (see Vol. I).

8.11 (a) Start from the scattering operator in Eq. (8.120), and derive the expression for the bremmstrahlung amplitude in Eq. (8.126);

(b) Repeat part (a) for the pair-production amplitude in Eq. (8.128).

8.12 It is a theorem that there is only one inequivalent irreducible representation of dimension-4 of the Clifford algebra of the Dirac gamma matrices $\gamma_\mu \gamma_\nu + \gamma_\nu \gamma_\mu = 2\delta_{\mu\nu}$. This implies that any other representation must be related to the standard representation by a similarity transformation $s\gamma_\mu s^{-1}$ where s is a non-singular 4×4 matrix.

(a) Show $-\gamma_\mu^T$ satisfies the same Clifford algebra;

(b) Show $s_c = \gamma_2 \gamma_4$ gives $s_c \gamma_\mu s_c^{-1} = -\gamma_\mu^T$.

8.13 *Furry's theorem* states that a closed fermion loop with an odd number of electromagnetic vertices makes no contribution to the scattering operator in QED.

(a) Consider a loop with three vertices (the proof is readily extended). Convince yourself from Wick's theorem that there will be two Feynman diagrams with the fermion line running in opposite directions around the loop. Show that the sum of these contributions appears as

$$S = -\mathrm{Tr}\, \gamma_\mu iS_F(x_1 - x_2)\gamma_\nu iS_F(x_2 - x_3)\gamma_\lambda iS_F(x_3 - x_1)$$
$$-\mathrm{Tr}\, \gamma_\nu iS_F(x_2 - x_1)\gamma_\mu iS_F(x_1 - x_3)\gamma_\lambda iS_F(x_3 - x_2)$$

where $S_F(x_i - x_j) = -\int d^4 q (2\pi)^{-4} [i\not{q} + M]^{-1} e^{iq \cdot (x_i - x_j)}$;

(b) Insert $s_c^{-1} s_c$ everywhere, where s_c is the "charge conjugation" matrix from Prob. 8.12, and use invariance of the trace under cyclic permutations;

(c) Show $s_c S_F(x_i - x_j) s_c^{-1} = S_F^T(x_j - x_i)$;

(d) Show $\mathrm{Tr}\,[a^T b^T \cdots y^T z^T] = \mathrm{Tr}\,[zy \cdots ba]$;

(e) Hence show $S = (-1)^3 S$ and conclude $S = 0$. This is Furry's theorem.

8.14 Assume a big box with p.b.c., and work in the Coulomb gauge.

(a) Suppose that for some reason the interaction-picture current were to be augmented by a term $\hat{\mathbf{j}}(x) \to \hat{\mathbf{j}}(x) - \boldsymbol{\nabla}\hat{\vartheta}(x)$. Show the scattering operator $\hat{S}$ is unchanged;

(b) How would current conservation then be maintained? (*Hint*: Recall $\hat{\rho}(x) = (i/\hbar)[\hat{H}_0, \hat{\rho}(x)]$.)

8.15 Work in a big box with p.b.c., where the spatial integrals are of the form $\int_\Omega d^3x \int_\Omega d^3y$. Assume that for some reason the equal-time commutator in Eq. (8.37) did not vanish, but had a non-vanishing remainder of the form[27]

$$\delta(t_x - t_y)[\hat{\rho}(x), \hat{\mathbf{j}}(y)]_{t_x=t_y} \doteq i\delta(t_x - t_y)\delta^{(3)}(\mathbf{x} - \mathbf{y})\,\boldsymbol{\nabla}_y\,\hat{\varphi}(y)$$

Show that $\delta\hat{S}^{(2)}$ in Eq. (8.31) still vanishes.

9.1 Assume the mass counter-term δM has a power-series expansion in e^2, and show that the additional interaction $\hat{H}_I^{\text{mass}}(t)$ in Eq. (9.6) gives rise to the additional terms in $[e^3 \hat{S}_{13}^{\text{ext}}]_{fi}$ in Eq. (9.7).

9.2 (a) Evaluate the integral, and prove the Feynman parameterization relation in Eq. (9.25) for the product of two factors in the form $1/ab$;

(b) Repeat part (a) for the product of three factors $1/abc$ in Eq. (9.55);

(c) Use these results to generalize the Feynman parameterization for the product of n factors.

9.3 Write $n = 4 - \epsilon$, make a consistent expansion in ϵ as $\epsilon \to 0$, and display the additional finite terms present when passing from Eq. (9.32) to its singular part in Eq. (9.33) [compare Eq. (9.73)].

9.4 Use the Dirac equation on both sides $(i\not{k} + M)u(k) = \bar{u}(k)(i\not{k} + M) = 0$, together with the relations $2k_\mu = \not{k}\gamma_\mu + \gamma_\mu\not{k}$ and $\not{k}\not{k} = k^2 = -M^2$, to prove Eq. (9.48).

9.5 This problem concerns an explicit proof that the terms proportional to l_μ or l_ν in the photon propagator in the Coulomb gauge $\tilde{D}_{\mu\nu}^F(l)$ in Eq. (8.133) do not contribute to the S-matrix element $[e^3 \hat{S}_{13}^{\text{ext}}]_{fi}$ for the scattering of an electron in an external field.

(a) Use the fact that the photon propagator is symmetric in $\mu\nu$ and an even function of l to rewrite the integrand $M_{\mu\nu}(k, q; l)$ into which the photon propagator $\tilde{D}_{\mu\nu}^F(l)$ is contracted in the sum of the vertex and electron self-energy diagrams as (see Fig. 12.2)

---

[27]Such a contribution is known as a "Schwinger term".

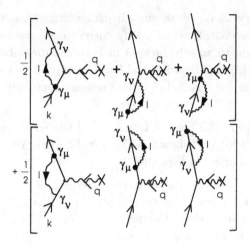

Fig. 12.2 Rewriting of the integrand $M_{\mu\nu}(k, q; l)$ in $[e^3 \hat{S}_{13}^{\text{ext}}]_{fi}$ for the diagrams in Fig. 9.1 (b,d,e).

$$M_{\mu\nu} = M_{\mu\nu}^{(i)} + M_{\mu\nu}^{(f)}$$

$$M_{\mu\nu}^{(i)} = \frac{1}{2}\left[\gamma_\nu \frac{1}{i(\not{k} - \not{l} + \not{q}) + M} \not{d} \frac{1}{i(\not{k} - \not{l}) + M}\gamma_\mu + \right.$$

$$\left. \not{d} \frac{1}{i\not{k} + M}\gamma_\nu \frac{1}{i(\not{k} - \not{l}) + M}\gamma_\mu + \not{d}\frac{1}{i\not{k} + M}\gamma_\mu \frac{1}{i(\not{k} + \not{l}) + M}\gamma_\nu\right]$$

$$M_{\mu\nu}^{(f)} = \frac{1}{2}\left[\gamma_\mu \frac{1}{i(\not{k} + \not{l} + \not{q}) + M}\not{d} \frac{1}{i(\not{k} + \not{l}) + M}\gamma_\nu + \right.$$

$$\gamma_\nu \frac{1}{i(\not{k} + \not{q} - \not{l}) + M}\gamma_\mu \frac{1}{i(\not{k} + \not{q}) + M}\not{d} + $$

$$\left. \gamma_\mu \frac{1}{i(\not{k} + \not{q} + \not{l}) + M}\gamma_\nu \frac{1}{i(\not{k} + \not{q}) + M}\not{d}\right]$$

In this fashion, the amplitude is separated into contributions where the virtual photon originating at the vertex γ_μ ends up at the vertex γ_ν located *in all possible positions along the electron line.*

(b) Now multiply by il_ν, and use the Dirac equation on the initial and final electron legs $\bar{u}(k + q)[i(\not{k} + \not{q}) + M] = (i\not{k} + M)u(k) = 0$. Show

$$M_{\mu\nu}^{(i)} l_\nu = M_{\mu\nu}^{(f)} l_\nu = 0$$

The term in l_μ is handled in a similar fashion.

This is very useful result since it demonstrates just what class of diagrams must be considered to satisfy current conservation. The extension of the argument to an arbitrary set of Feynman diagrams can be found in [Feynman (1949)] and [Bjorken and Drell (1964); Bjorken and Drell (1965)]. The key is that the charged electron lines run continuously through a diagram.

9.6 Invert the Fourier transform of the adiabatic damping factor $e^{-\epsilon|t|}$ in Eq. (9.96), and show $G(\Gamma_0) = \epsilon/\pi(\Gamma_0^2 + \epsilon^2)$. Verify that this $G(\Gamma_0)$ has all the properties in Eqs. (9.97).

9.7 (a) Consider the scattering operator $\hat{S}_0$ in Eq. (8.135) in order e^4. Use Wick's theorem, and compute the S-matrix element $[\hat{S}_{04}]_{fi}$ for photon-photon scattering $\gamma(l_1) + \gamma(l_2) \rightarrow \gamma(l_3) + \gamma(l_4)$;
(b) Draw the Feynman diagrams;
(c) Use the Feynman rules to reproduce the result in part (a);
(d) Use the analysis in Prob. 8.13 to relate the contributions where the electron runs in opposite directions around the loop.

Note that while simple power counting would imply that the amplitude for this process is logarithmically divergent, it is actually convergent.

9.8 (a) Consider a theory where a Dirac field $\hat{\psi}(x)$ is coupled to the massive vector field $\hat{V}_\mu(x)$ of Prob. C.6 with an interaction hamiltonian density $\hat{\mathcal{H}}_I = -g_v \hat{D}_\mu(x)\hat{V}_\mu(x)$ where $\hat{D}_\mu(x) = i:\hat{\bar{\psi}}(x)\gamma_\mu\hat{\psi}(x):$ is the conserved Dirac current. Construct the scattering operator $\hat{S}$ in order g_v^2;
(b) The Fourier transform of the vector propagator was calculated in Prob. F.5 to be $\tilde{\Delta}^F_{\mu\nu}(k) = [k^2 + m_v^2]^{-1}[\delta_{\mu\nu} + k_\mu k_\nu/m_v^2]$. Show that one can use an *effective* vector propagator $\tilde{\Delta}^F_{\mu\nu}(k) \doteq [k^2 + m_v^2]^{-1}\delta_{\mu\nu}$ in this theory.

9.9 (a) Verify the n-dimensional gamma-matrix relations used in Eq. (9.112);
(b) Verify that Eq. (9.122) follows from Eq. (9.120) in the limit $n \rightarrow 4$.

9.10 It is evident from Eq. (9.126) that the charge renormalization is given to $O(\alpha)$ by $e^2 = e_0^2[1 - C(0)]$. The sign here might be a cause for concern since bad things happen if $e^2 < 0$.[28] Use Eq. (9.118) to estimate the cutoff Λ for which $C(0) = 1$. Discuss.

9.11 One might be concerned that the same renormalized charge e should appear at a vertex involving a real photon.
(a) Argue in analogy to Eq. (9.95) that the correct wave function renor-

---

[28] In addition to having $e = i|e|$, there are the difficulties discussed in [Dyson (1952)].

malization for the external photon is to use a factor $-C(0)/2$;

(b) Hence show the external photon vertex gets modified to $e_0[1 - e_0^2\bar{C}(0)/2] = \sqrt{e^2}$ through $O(e_0^3)$. Discuss.

9.12 This problem concerns power-counting in the momentum integrals for an arbitrary Feynman diagram in QED. Assume the contours have all been rotated so that everything is in the euclidean metric. Let

$$\kappa = \# \text{ of extra powers of momenta in denominator}$$
$$F_e = \# \text{ of internal electron lines}$$
$$F_p = \# \text{ of internal photon lines}$$
$$E_e = \# \text{ of external electron lines}$$
$$E_p = \# \text{ of external photon lines}$$
$$n = \# \text{ of vertices (the order)}$$
$$n_s = \# \text{ of vertices without photons (mass counter-terms)}$$

(a) Since there are $F_e + F_p$ virtual momenta, with $n - 1$ $\delta^{(4)}$-functions at the vertices (one is overall), and $d^4q = q^3dq$ counts as 4 powers in the numerator, show

$$\kappa = F_e + 2F_p - 4[(F_e + F_p) - (n - 1)]$$

(b) Since every vertex has an electron line in and out, and an internal electron line counts as two vertices, show

$$2F_e + E_e = 2n$$

(c) Since every vertex (except for n_s) has a photon line, show

$$2F_p + E_p = n - n_s$$

(d) Hence show

$$\kappa = E_p + \frac{3}{2}E_e + n_s - 4$$

This is an important result. Note that κ is *independent of the order n!*

9.13 Apply the result in Prob. 9.12. A necessary condition for the convergence of the momentum integrals in an arbitrary Feynman diagram is $\kappa \geq 1$.[29] If $\kappa \leq 0$, the graph is *primitively divergent*. In QED there are four primitively divergent sets of graphs. What are they? Discuss.[30]

---

[29] The integral $\int q^p dq/q^{p+2}$ is convergent.
[30] Recall Probs. 8.13 and 9.7.

9.14 (a) Iterate Dyson's equation for the vertex $\Gamma_\mu(k_2, k_1)$ in Fig. 9.6 through $O(e_0^2)$, and reproduce the analytic expression for $\Lambda_\mu(k_2, k_1)$ in Eq. (9.3);

(b) Show that an iteration of Dyson's equation for the vertex $\Gamma_\mu(k_2, k_1)$ in Fig. 9.6 through $O(e_0^4)$, produces all the appropriate diagrams.

9.15 Verify Ward's identity in Eq. (9.142) through $O(e_0^4)$.

9.16 (a) Start from the following expression for the vacuum polarization tensor in $O(e_0^2)$ [compare Eq. (9.4)]

$$\Pi_{\lambda\sigma}(q) = \frac{e_0^2}{i\hbar c \varepsilon_0} \int \frac{d^n k}{(2\pi)^4} \text{Tr} \, \frac{1}{i\not{k} + M} \gamma_\lambda \frac{1}{i(\not{k} + \not{q}) + M} \gamma_\sigma$$

Use Eq. (9.12) to compute $(\partial/\partial q_\mu)\Pi_{\lambda\sigma}(q)$;

(b) Iterate Dyson's equation for the vertex $W_\mu(q, q)$ in Fig. 9.9 through $O(e_0^2)$ for fermion vertices (λ, σ). Show the $O(e_0^2)$ term $[W_\mu^{(2)}(q, q)]_{\lambda\sigma}$ produces the expression $-(\partial/\partial q_\mu)\Pi_{\lambda\sigma}(q)$;

(c) Use the form of $\Pi_{\lambda\sigma}(q)$ in Eq. (9.109) to solve for $\Pi(q)$, and then use Eq. (9.150) to show

$$(n - 1)\Pi(q) = \Pi_{\lambda\lambda}(q)$$
$$i\Delta_\mu(q, q) = \frac{\partial \Pi(q)}{\partial q_\mu} = \frac{1}{(n-1)} \frac{\partial}{\partial q_\mu} \Pi_{\lambda\lambda}(q)$$

(d) Hence conclude from Eqs. (9.154) that $W_\mu^{(2)}(q, q) \equiv (n - 1)^{-1}$ $[W_\mu^{(2)}(q, q)]_{\lambda\lambda}$ in part (b) produces the correct $i\Delta_\mu(q, q)$ through $O(e_0^2)$.[31]

9.17 The *skeleton* graphs for any physical process consist of the set of distinct diagrams obtained by removing all self-energy and vertex insertions from the Feynman diagrams for that process. All diagrams are then re-obtained by using the full propagators (S_F', D_F') and full vertex $e_0\Gamma_\mu$ in the skeletons. Show it then follows from Eqs. (9.157) that the S-matrix for any physical process is given by the use of the *finite functions* $(S_{F1}, D_{F1}, \Gamma_{\mu 1})$ and the *renormalized charge* e_1. You may assume the same wave function renormalizations $(Z_2^{1/2}, Z_3^{1/2})$ for the external electrons and photons that were established in the text to second order.

9.18 Suppose the additional term of Prob. 8.15 is actually present in the equal-time commutator, while the external field has the form of Eq. (8.121).

---

[31] Alternatively, one can just select the coefficient of $\delta_{\lambda\sigma}$ in $[W_\mu^{(2)}(q, q)]_{\lambda\sigma}$. Note that from Eq. (9.129) $\Pi_{\lambda\sigma} \doteq \Pi(q)\delta_{\lambda\sigma}$, and in $O(e_0^2)$, $\Pi(q) \equiv \Pi^\star(q)$.

(a) Write out the 3! possible time-orderings of the current operators in Eq. (8.132), and differentiate with respect to t_{x_1} as in Eqs. (8.34)–(8.37);

(b) Verify that the effective photon propagator is still given by Eq. (8.134).[32]

9.19 Verify Eqs. (9.137); remember to renormalize the charge.

10.1 Restore the Dirac indices and make them explicit in Eqs. (10.143)–(10.148). Explain the Dirac matrix structure of Eqs. (10.148).

10.2 Verify the following left-variational derivatives

$$\frac{1}{i}\frac{\delta}{\delta\bar{\zeta}(x)}\exp\left\{i\int d^4y\,[\bar{\zeta}(y)\psi]\right\} = \psi(x)\exp\left\{i\int d^4y\,[\bar{\zeta}(y)\psi]\right\}$$

$$\frac{1}{i}\frac{\delta}{\delta\zeta(x)}\exp\left\{i\int d^4y\,[\bar{\psi}\zeta(y)]\right\} = -\bar{\psi}(x)\exp\left\{i\int d^4y\,[\bar{\psi}\zeta(y)]\right\}$$

10.3 Use the scalar interaction in Eq. (10.108). Expand the free functional $W_0(J)$ in Eq. (10.100) and retain $W_0^{(6)}(J)$. Then evaluate the interacting functional $W(J)$ in Eq. (10.112) to $O(\lambda)$.

Problems 10.4–10.8 show how the partition function of statistical mechanics (see appendix E of Vol. I) can be written as a path integral.

10.4 Consider a single non-relativistic particle moving in a potential in one dimension (Fig. 10.1) with the hamiltonian of Eq. (10.1). The partition function in the microcanonical ensemble is defined by

$$Z \equiv \text{Trace}\left[e^{-\beta\hat{H}}\right] \qquad ;\beta \equiv 1/k_{\rm B}T$$

where "Trace" indicates the sum of the diagonal elements for a complete set of states. For example, if one uses the eigenstates of $\hat{H}$ with $\hat{H}|E_n\rangle = E_n|E_n\rangle$, then $Z = \sum_n e^{-\beta E_n}$. Use completeness to show that Z can also be computed with a complete set of *eigenstates of position*

$$Z = \int dq\,\langle q|e^{-\beta\hat{H}}|q\rangle$$

10.5 Use the arguments in the text to show that as $\varepsilon \to 0$

$$\langle q_l|e^{-(\varepsilon/\hbar)\hat{H}}|q_{l+1}\rangle = \left(\frac{m}{2\pi\varepsilon\hbar}\right)^{1/2}\exp\left\{-\frac{\varepsilon}{\hbar}\left[\frac{m}{2\varepsilon^2}(q_l - q_{l+1})^2 + V(q_l)\right]\right\}$$

10.6 Let the variable τ run over the interval $[0, \hbar\beta]$. Divide this interval into n-subintervals of length ε, and label the corresponding vari-

---

[32]Compare with Prob. 9.5.

ables by $(\tau_0, \tau_1, \cdots, \tau_n)$. Introduce coordinates $(q_0, q_1, \cdots, q_n)$ associated with these values, where $q_n \equiv q_0$. Write $e^{-\beta\hat{H}}$ as a product of n factors $e^{-(\varepsilon/\hbar)\hat{H}}$, and insert $n-1$ complete sets of eigenstates of position between the factors. Use the result in Prob. 10.5 to show

$$Z = \mathrm{Lim}_{\varepsilon \to 0} \left(\frac{m}{2\pi\varepsilon\hbar}\right)^{n/2} \int \prod_{l=0}^{n-1} dq_l \exp\left\{-\frac{\varepsilon}{\hbar} \sum_{p=0}^{n-1}\left[\frac{m}{2\varepsilon^2}(q_p - q_{p+1})^2 + V(q_p)\right]\right\}$$

10.7 The result in Prob. 10.6 is a path integral

$$Z = \int \mathcal{D}(q)\exp\left\{-\frac{1}{\hbar}\bar{S}(\beta\hbar, 0)\right\}$$

$$\bar{S}(\beta\hbar, 0) = \int_0^{\beta\hbar} d\tau\left[\frac{m}{2}\left(\frac{dq}{d\tau}\right)^2 + V(q)\right]$$

Here $i\bar{S}$ is the classical action computed for *imaginary time*, $t = -i\tau$. Construct the analogue of Fig. 10.3 and use the result in Prob. 10.6 to give the corresponding rules for evaluating this path integral. Note, in particular, the role here of the cyclic boundary condition $q_n = q_0$.

10.8 The *thermal average* of an operator $O(\hat{q})$ is defined by

$$\langle\langle O(\hat{q})\rangle\rangle \equiv \frac{\mathrm{Trace}\,[O(\hat{q})e^{-\beta\hat{H}}]}{\mathrm{Trace}\,[e^{-\beta\hat{H}}]}$$

(a) Show this is the ratio of two path integrals of the form in Prob. 10.7, where the path integral in the numerator has an additional factor of $O(q_0)$;

(b) Discuss the relation to the path-integral expression for the Green's functions in the euclidean metric [compare Eq. (10.81)]. What are the similarities? What are the differences?[33]

10.9 Derive Eq. (10.121) in the case $n = 3$.

10.10 Many physical systems can be described with *effective field theories* where one builds all the symmetries of the problem into an effective lagrangian density and then expands physical quantities in some meaningful small dimensionless ratio. The appropriate lagrangian density can involve nonlinear, derivative couplings, and it may be non-renormalizable; nevertheless, path integral techniques provide a means for quantizing the theory.

---

[33]Note that the reduction of any problem to convergent many-component multiple integrals, as in Probs. 10.7–10.8, lends itself to numerical approximation methods (see, for example, lattice gauge theory for QCD as described in [Walecka (2004)]).

(a) As an example of a model, non-linear, effective field theory consider the following lagrangian density involving nucleon, pion, and scalar fields $(\underline{\psi}, \boldsymbol{\pi}, \varphi)$

$$\mathcal{L}^{\text{eff}} = -\hbar c \left[\bar{\underline{\psi}}_L \gamma_\lambda \frac{\partial}{\partial x_\lambda} \underline{\psi}_L + \bar{\underline{\psi}}_R \gamma_\lambda \frac{\partial}{\partial x_\lambda} \underline{\psi}_R - g\sigma_0 \left(1 + \frac{\varphi}{\sigma_0} \right) \left(\bar{\underline{\psi}}_R \underline{U} \underline{\psi}_L + \bar{\underline{\psi}}_L \underline{U}^\dagger \underline{\psi}_R \right) \right]$$
$$- \frac{c^2}{2} \left(\frac{\partial \varphi}{\partial x_\lambda} \right)^2 - \frac{c^2 \sigma_0^2}{4} \text{tr} \left(\frac{\partial \underline{U}^\dagger}{\partial x_\lambda} \frac{\partial \underline{U}}{\partial x_\lambda} \right) - \mathcal{V} \left(\underline{U}, \frac{\partial \underline{U}}{\partial x_\lambda}; \varphi \right)$$

$$\underline{U} \equiv \exp \left\{ \frac{i}{\sigma_0} \boldsymbol{\tau} \cdot \boldsymbol{\pi} \right\}$$

Show that the lagrangian density, without $\mathcal{V}$, is invariant under the following transformation

$$\begin{aligned} \underline{\psi}_L &\to \underline{L}\,\underline{\psi}_L & &; \underline{U} \to \underline{R}\,\underline{U}\,\underline{L}^\dagger \\ \underline{\psi}_R &\to \underline{R}\,\underline{\psi}_R & &; \varphi \to \varphi \end{aligned}$$

where $(\underline{L}, \underline{R})$ are independent, global SU(2) matrices. Hence conclude that as long as the potential $\mathcal{V}$ is constructed to be invariant, this lagrangian density has an exact $SU(2)_L \bigotimes SU(2)_R$ symmetry (compare Prob. 6.12);

(b) Take the limit $|\sigma_0| \to \infty$, assume that in this limit $\mathcal{V} \to (m_s^2 c^2/2)\varphi^2 + O(1/|\sigma_0|)$, and reproduce the chiral-symmetric σ-model lagrangian density of Eq. (6.166) with $\sigma_0 = -M/g$;

(c) Show that the following additional contribution

$$\mathcal{L}_\pi^{\text{mass}} = \frac{m_\pi^2 c^2 \sigma_0^2}{4} \text{tr} \left(\underline{U} + \underline{U}^\dagger - 2 \right)$$

reduces to the usual pion mass term in the limit $|\sigma_0| \to \infty$ and is *not* chiral invariant.[34]

11.1 (a) Consider a real, massive, scalar field ϕ interacting with a time-independent, localized, c-number source $\lambda s(\mathbf{x})$; the equation of motion is

$$(\Box - m^2)\phi(x) = \lambda s(\mathbf{x})$$

---

[34]The conventional notation in Prob. 10.10 is $-\sigma_0 = M/g \equiv f_\pi$. The most general form of $\mathcal{L}^{\text{eff}}$ here is the point of departure for a subsequent expansion of scattering amplitudes in powers of $(q_\lambda, m_\pi)/f_\pi$ where q_λ is an external four-momentum; this is the basic idea behind chiral perturbation theory (see [Donoghue, Golowich, and Holstein (1993); Walecka (2004)]).

Construct the lagrangian and hamiltonian for this system.

(b) Quantize this system in the Schrödinger picture (see Chap. 5). Work in a big box with p.b.c. Introduce the normal-mode expansion for $\hat{\phi}(\mathbf{x})$, and show

$$\hat{H} = \hat{H}_0 + \hat{H}_1$$
$$\hat{H}_0 = \sum_{\mathbf{k}} \hbar\omega_k \left(c_{\mathbf{k}}^\dagger c_{\mathbf{k}} + 1/2 \right)$$
$$\hat{H}_1 = \lambda c^2 \sum_{\mathbf{k}} \left(\frac{\hbar}{2\omega_k} \right)^{1/2} \left[c_{\mathbf{k}} \tilde{s}(\mathbf{k}) + c_{\mathbf{k}}^\dagger \tilde{s}^\dagger(\mathbf{k}) \right]$$

where $s(\mathbf{x})$ has the Fourier expansion $s(\mathbf{x}) = (1/\sqrt{\Omega}) \sum_{\mathbf{k}} \tilde{s}(\mathbf{k}) e^{-i\mathbf{k}\cdot\mathbf{x}}$.

(c) Show the following canonical transformation diagonalizes the hamiltonian

$$C_{\mathbf{k}} = c_{\mathbf{k}} + \lambda c^2 \left(\frac{1}{2\hbar\omega_k^3} \right)^{1/2} \tilde{s}^\dagger(\mathbf{k})$$
$$C_{\mathbf{k}}^\dagger = c_{\mathbf{k}}^\dagger + \lambda c^2 \left(\frac{1}{2\hbar\omega_k^3} \right)^{1/2} \tilde{s}(\mathbf{k})$$

(d) Assume two point sources with $s(\mathbf{x}) = \delta^{(3)}(\mathbf{x} - \mathbf{x}_1) + \delta^{(3)}(\mathbf{x} - \mathbf{x}_2)$. Identify the interaction energy, and show that the result is the *Yukawa potential*

$$E_{\text{int}} = -\frac{\lambda^2 c^2}{4\pi} \frac{e^{-m|\mathbf{x}_1 - \mathbf{x}_2|}}{|\mathbf{x}_1 - \mathbf{x}_2|}$$

to all orders in λ. Note the sign.

(e) Compute the overlap of the exact vacuum $|\underline{0}\rangle$ with the non-interacting occupation-number eigenstates of $\hat{H}_0$, and show that

$$\langle n_1 n_2 \cdots n_\infty | \underline{0} \rangle = \frac{(-1)^{n_1+n_2+\cdots}}{\sqrt{n_1! n_2! \cdots}} \left(\frac{\lambda c^2}{\sqrt{2\hbar\omega_1^3}} \tilde{s}^\dagger(\mathbf{k}_1) \right)^{n_1} \left(\frac{\lambda c^2}{\sqrt{2\hbar\omega_2^3}} \tilde{s}^\dagger(\mathbf{k}_2) \right)^{n_2} \times$$
$$\cdots \langle 0 | \underline{0} \rangle$$

(f) Use completeness to show that the overlap of the exact vacuum and the non-interaction vacuum is given by

$$|\langle 0 | \underline{0} \rangle|^2 = \exp\left\{ -\sum_{\mathbf{k}} \frac{\lambda^2 c^4}{2\hbar\omega_k^3} |\tilde{s}(\mathbf{k})|^2 \right\}$$

11.2 Consider the Bloch-Nordsieck problem of the quantized radiation field interacting with a time-independent, classical external current source $j_\mu^{\text{ext}}(\mathbf{x}) = [\mathbf{j}^{\text{ext}}(\mathbf{x})/c, \, i\rho^{\text{ext}}(\mathbf{x})]$.[35]

(a) The hamiltonian for this system in the interaction picture was derived in Eq. (C.73). Go back to the Schrödinger picture (see Chap. 5). Introduce the expansion $\mathbf{j}^{\text{ext}}(\mathbf{x}) = (c/\sqrt{\Omega}) \sum_\mathbf{q} \tilde{\mathbf{j}}(\mathbf{q}) e^{-i\mathbf{q}\cdot\mathbf{x}}$ and show that

$$\hat{H} = \sum_{\mathbf{k}\lambda} \hbar\omega_k \left(a_{\mathbf{k}\lambda}^\dagger a_{\mathbf{k}\lambda} + 1/2 \right) - e \sum_{\mathbf{k}\lambda} \left(\frac{\hbar c^2}{2\omega_k\varepsilon_0} \right)^{1/2} \left(j(\mathbf{k}\lambda) a_{\mathbf{k}\lambda} + j^\dagger(\mathbf{k}\lambda) a_{\mathbf{k}\lambda}^\dagger \right) +$$
$$\frac{e^2}{8\pi\varepsilon_0} \int d^3x \int d^3x' \frac{\rho^{\text{ext}}(\mathbf{x})\rho^{\text{ext}}(\mathbf{x}')}{|\mathbf{x} - \mathbf{x}'|}$$

where $j(\mathbf{k}\lambda) \equiv \mathbf{e}_{\mathbf{k}\lambda} \cdot \tilde{\mathbf{j}}(\mathbf{k})$. Here λ is the photon helicity;

(b) Show the following canonical transformation *diagonalizes* the hamiltonian

$$A_{\mathbf{k}\lambda} = a_{\mathbf{k}\lambda} - \left(\frac{c^2}{2\hbar\omega_k^3\varepsilon_0} \right)^{1/2} e j^\dagger(\mathbf{k}\lambda)$$

$$A_{\mathbf{k}\lambda}^\dagger = a_{\mathbf{k}\lambda}^\dagger - \left(\frac{c^2}{2\hbar\omega_k^3\varepsilon_0} \right)^{1/2} e j(\mathbf{k}\lambda)$$

(c) Expand the exact ground state $|\underline{0}\rangle$ of the diagonalized hamiltonian in terms of the eigenstates of the free photon field, and show that the expansion coefficients are given by

$$\langle n_1 n_2 \cdots n_\infty | \underline{0} \rangle = \frac{1}{\sqrt{n_1! n_2! \cdots}} \times$$
$$\left(\sqrt{\frac{c^2}{2\hbar\omega_1^3\varepsilon_0}} \, e j^\dagger(\mathbf{k}_1\lambda_1) \right)^{n_1} \left(\sqrt{\frac{c^2}{2\hbar\omega_2^3\varepsilon_0}} \, e j^\dagger(\mathbf{k}_2\lambda_2) \right)^{n_2} \cdots \langle 0 | \underline{0} \rangle$$

(d) Use completeness to demonstrate that

$$|\langle 0 | \underline{0} \rangle|^2 = \exp\left\{ -\frac{e^2 c^2}{\varepsilon_0} \sum_{\mathbf{k}\lambda} \frac{|j(\mathbf{k}\lambda)|^2}{2\hbar\omega_k^3} \right\}$$

(e) Show the number of photons with $(\mathbf{k}\lambda)$ in the true ground state is

$$\langle \underline{0} | a_{\mathbf{k}\lambda}^\dagger a_{\mathbf{k}\lambda} | \underline{0} \rangle = \frac{e^2 c^2}{\varepsilon_0} \frac{|j(\mathbf{k}\lambda)|^2}{2\hbar\omega_k^3}$$

---

[35]The reference is [Bloch and Nordsieck (1937)].

Hence conclude that the result in part (d) can be written

$$|\langle 0|\underline{0}\rangle|^2 = e^{-N}$$

(f) Convert $\sum_{\mathbf{k}} \to \Omega(2\pi)^{-3} \int d^3k$, and show that if $|j(0\lambda)|^2 \neq 0$, then the resulting integral in part (d) *diverges* at long wavelength. Hence conclude that there are an *infinite number* of long-wavelength photons in the true ground state. Show that the *energy* carried by these photons is finite.[36]

11.3 Verify the boson matrix element in Eq. (11.3).

11.4 Consider the scattering problem of two identical spin-zero bosons in the C-M system. The wave function must be symmetric under particle interchange, or under $\mathbf{r} \to -\mathbf{r}$ where $\mathbf{r}$ is the relative coordinate.

(a) Assume the prepared initial state has the form[37] $\psi_{\mathbf{k}}(\mathbf{r}) = e^{i\mathbf{k}\cdot\mathbf{r}} + e^{-i\mathbf{k}\cdot\mathbf{r}}$. Show this wave function has vanishing flux. Interpret the first term as representing an incident beam moving with relative momentum $\hbar\mathbf{k}$ and the second term as representing a beam moving with $-\hbar\mathbf{k}$. Argue that the appropriate incident flux can be calculated from the first term just as if the particles were distinguishable;

(b) In the scattering region, one must employ the full wavefunction $\psi_{\mathbf{k}}(\mathbf{r})$. Argue from Prob. 4.10 that the first term will give rise to a scattered wave $f(k,\theta)e^{ikr}/r$ while the second term gives rise to a similar scattered wave, but with $\mathbf{r} \to -\mathbf{r}$, or $f(k,\pi-\theta)e^{ikr}/r$;

(c) Hence conclude that the scattering amplitude for this problem is $f = f(k,\theta) + f(k,\pi-\theta)$, where $f(k,\theta)$ is calculated as if the particles were distinguishable.

11.5 Show the required integral in the calculation of the depletion of the ground state for the weakly interacting Bose system is

$$\int_0^\infty x^2 dx \, \frac{1}{2}\left[\frac{(x^2/2+1)}{[(x^2/2+1)^2-1]^{1/2}} - 1\right] = \frac{2}{3}$$

11.6 (a) Show that if the two-body potential is of the form $V(|\mathbf{x}-\mathbf{y}|)$, then $\langle \mathbf{k}_1\mathbf{k}_2|V|\mathbf{k}_3\mathbf{k}_4\rangle \propto \delta_{\mathbf{k}_1+\mathbf{k}_2,\mathbf{k}_3+\mathbf{k}_4}$ in Eqs. (11.1) and (11.57);

(b) Use a change of dummy summation variables and the symmetry

---

[36] As discussed in the text, this *infrared divergence* is usually handled by giving the photon a tiny mass M_γ, so that $\omega_k/c = \sqrt{\mathbf{k}^2 + M_\gamma^2}$, and then taking the limit $M_\gamma \to 0$ at the end of the calculation of any physical quantity (the energy, for example).

[37] Since the scattering cross section is the ratio of fluxes, the overall normalization is irrelevant.

properties of the potential to show that in Eq. (11.65)

$$\sum_{\mathbf{k}}\sum_{\mathbf{k'}}[\langle\mathbf{kk'}|V|\mathbf{kk'}\rangle - \langle\mathbf{kk'}|V|\mathbf{k'k}\rangle]\, v_{k'}^2 : a_{\mathbf{k}\downarrow}^\dagger a_{\mathbf{k}\downarrow} : =$$

$$\sum_{\mathbf{k}}\sum_{\mathbf{k'}}[\langle\mathbf{kk'}|V|\mathbf{kk'}\rangle - \langle\mathbf{kk'}|V|\mathbf{k'k}\rangle]\, v_{k'}^2 : a_{-\mathbf{k}\downarrow}^\dagger a_{-\mathbf{k}\downarrow} :$$

11.7 Verify Eqs. (11.69).

11.8 Obtain Eq. (11.1) from Eqs. (2.113)–(2.114). What assumptions have you made?

11.9 (a) Given the state $(1/\sqrt{2})(a_{\mathbf{k}}^\dagger + a_{-\mathbf{k}}^\dagger)|\underline{0}\rangle$ in a Fermi system, compute the expectation values of $\hat{K}_0$ and $\hat{P}$;
(b) Interpret $a_{\mathbf{k}}^\dagger|\underline{0}\rangle$ in terms of particles and holes.

A.1 The spherical harmonics satisfy $Y_{lm}^*(\theta,\phi) = (-1)^m Y_{l,-m}(\theta,\phi)$, and the spherical basis vector satisfy $\mathbf{e}_{\mathbf{k}\lambda}^\dagger = (-1)^\lambda \mathbf{e}_{\mathbf{k},-\lambda}$. Use these relations to prove Eq. (A.25).

A.2 Start from the transition amplitude for the photon coming off in an arbitrary direction with respect to the target quantization axis in Eq. (A.36), and reproduce Eq. (A.33) for the total transition rate.

A.3 (a) Verify the commutation relations in Eq. (A.21) for the creation and destruction operators for circularly polarized photons;
(b) Verify Eqs. (A.22).

B.1 (a) Use Cauchy's theorem to demonstrate that a contour integral of an analytic function $f(z)$ around a closed curve C can be deformed in any fashion through a region of analyticity;
(b) Consider the contour integral between two points of an analytic function $f(z)$ in a simply-connected region of analyticity. Demonstrate from Cauchy's theorem that the integral is independent of the path.

B.2 Consider the integral

$$I \equiv \int_{-\infty}^{\infty} \frac{dx}{1+x^2} = 2\int_0^{\infty} \frac{dx}{1+x^2}$$

(a) Convert this to the integral $\oint_C dz/(1+z^2)$ around a closed contour C by adding the contribution of a large semi-circle in either half-plane, and showing that this additional contribution is negligible in the limit that the radius of the semi-circle $R \to \infty$;

(b) Locate the singularity of the integrand within that contour, and make a Laurent expansion about that singularity;

(c) Use the theory of residues to show that $I = \pi$.

B.3 Use the theory of residues to verify the derivation of Eq. (4.107) starting from Eq. (4.100).

B.4 Consider the following integral along the contour C illustrated in Fig. 12.3. Assume that $f(z)$ is well-behaved at x_0.

$$\int_C dz \frac{f(z)}{z - x_0}$$

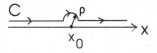

Fig. 12.3   Contour C in Prob. B.4. Here ρ is the radius of the small semi-circle in the upper-half plane centered on x_0.

(a) Break the integral up into three contributions: one up to $x_0 - \rho$; one from $x_0 + \rho$; and one around the small semicircle in the upper-half plane with radius ρ centered on x_0 . Define the Cauchy principal value by

$$\mathcal{P} \int dx \frac{f(x)}{x - x_0} \equiv \text{Lim}_{\rho \to 0} \left(\int^{x_0-\rho} + \int_{x_0+\rho} \right) dx \frac{f(x)}{x - x_0} \qquad ;$$
$$\text{Cauchy principal value}$$

(b) Explicitly evaluate the integral on the small semi-circle in polar coordinates and show that as $\rho \to 0$

$$\int_C dz \frac{f(z)}{z - x_0} = \mathcal{P} \int dx \frac{f(x)}{x - x_0} - i\pi f(x_0)$$

(c) Hence justify the symbolic replacement

$$\frac{1}{x - x_0 + i\eta} = \mathcal{P} \frac{1}{x - x_0} - i\pi \delta(x - x_0)$$

B.5 Consider the function $f(x) = 1/\sqrt{x}$ for $x > 0$. Show the analytic continuation of this function to the cut-plane, where the cut runs along the negative real axis, is $f(z) = 1/\sqrt{z}$. Evaluate this function along the positive imaginary axis and show $f(z) = 1/\sqrt{|z|e^{i\pi/2}}$.

B.6 Determine the analytic properties of $F_M^{(2)}(q^2)$ in Eq. (9.79).

C.1 Show that the action computed from the lagrangian density in Eq. (C.8) is gauge invariant. (*Hint*: make use of partial integration. Note that the addition of a constant to the action changes nothing.)

C.2 Show the energy-flux Poynting vector inplied by Eq. (C.23) reproduces the result quoted in Vol. I.

$$\mathbf{S} = \frac{1}{\mu_0}\mathbf{E} \times \mathbf{B}$$

C.3 Insert the interaction-picture representation of $\hat{A}(\mathbf{x}, t)$ and derive the expression for the total momentum contained in the free electromagnetic field in Eq. (C.36).

C.4 Insert the interaction-picture representation of $\hat{A}(\mathbf{x}, t)$ and discuss the processes described by the interaction $\hat{H}_I(t)$ in Eq. (C.71).

C.5 It is a theorem that any vector field can be uniquely separated into a part with zero divergence and a part with zero curl. That is, $\mathbf{v}(\mathbf{x}) = \mathbf{v}^T(\mathbf{x}) + \mathbf{v}^L(\mathbf{x})$ where $\nabla \cdot \mathbf{v}^T(\mathbf{x}) = 0$ and $\nabla \times \mathbf{v}^L(\mathbf{x}) = 0$.
(a) Provide a derivation of Eq. (C.46);
(b) Use that result to prove the theorem;
(c) Show $\int_\Omega d^3x\, \mathbf{v}^T(\mathbf{x}) \cdot \mathbf{v}^L(\mathbf{x}) = 0$.

C.6 Consider a neutral massive vector meson field V_μ with inverse Compton wavelength $m = m_0 c/\hbar$ and lagrangian density[38]

$$\mathcal{L} = -\frac{1}{4}V_{\mu\nu}V_{\mu\nu} - \frac{1}{2}m^2 V_\mu V_\mu \qquad ; \ V_{\mu\nu} = \frac{\partial V_\nu}{\partial x_\mu} - \frac{\partial V_\mu}{\partial x_\nu}$$

(a) Show the field equations are

$$\frac{\partial}{\partial x_\nu}V_{\mu\nu} + m^2 V_\mu = 0 \qquad\qquad ; \ (\mu,\nu) = 1,2,3,4$$

(b) Note that with $m^2 \neq 0$, the field equation for $\mu = 4$ can be solved for V_4, and the resulting expression used to eliminate V_4 as an independent dynamical variable.[39] Show that the momentum density and V_4 are then given by

$$\Pi_j = \frac{1}{ic}V_{j4} \qquad\qquad ; \ V_4 = \frac{ic}{m^2}\nabla \cdot \mathbf{\Pi}$$

---

[38] The analogy is to QED in H-L units.
[39] This is important in Prob. C.7, since here $\Pi_4 \equiv 0$. Although Probs. C.6–C.7 are algebraically challenging, they are very interesting.

(c) use the canonical procedure and show the hamiltonian is given by

$$H = \frac{1}{2} \int_{\Omega} d^3x \left[c^2 \mathbf{\Pi}^2 + (\mathbf{\nabla} \times \mathbf{V})^2 + m^2 \mathbf{V}^2 + \frac{c^2}{m^2} (\mathbf{\nabla} \cdot \mathbf{\Pi})^2 \right]$$

C.7 (a) Make use of the result in Prob. C.5(c), and show that the hamiltonian in Prob. C.6(c) separates into transverse and longitudinal contributions;

(b) Show that the following expansions of the field operators put the problem into normal modes

$$\mathbf{V}_T(\mathbf{x}, t) = \sum_{\mathbf{k}} \sum_{s=1}^{2} \left(\frac{\hbar c^2}{2\omega_k \Omega} \right)^{1/2} \left[c_{\mathbf{k}s} \mathbf{e}_{\mathbf{k}s} e^{i(\mathbf{k}\cdot\mathbf{x}-\omega_k t)} + c_{\mathbf{k}s}^{\star} \mathbf{e}_{\mathbf{k}s} e^{-i(\mathbf{k}\cdot\mathbf{x}-\omega_k t)} \right]$$

$$\mathbf{\Pi}_T(\mathbf{x}, t) = \frac{1}{i} \sum_{\mathbf{k}} \sum_{s=1}^{2} \left(\frac{\hbar \omega_k}{2c^2 \Omega} \right)^{1/2} \left[c_{\mathbf{k}s} \mathbf{e}_{\mathbf{k}s} e^{i(\mathbf{k}\cdot\mathbf{x}-\omega_k t)} - c_{\mathbf{k}s}^{\star} \mathbf{e}_{\mathbf{k}s} e^{-i(\mathbf{k}\cdot\mathbf{x}-\omega_k t)} \right]$$

$$\mathbf{V}_L(\mathbf{x}, t) = \sum_{\mathbf{k}} \left(\frac{\hbar \omega_k}{2m^2 \Omega} \right)^{1/2} \left[c_{\mathbf{k}3} \mathbf{e}_{\mathbf{k}3} e^{i(\mathbf{k}\cdot\mathbf{x}-\omega_k t)} + c_{\mathbf{k}3}^{\star} \mathbf{e}_{\mathbf{k}3} e^{-i(\mathbf{k}\cdot\mathbf{x}-\omega_k t)} \right]$$

$$\mathbf{\Pi}_L(\mathbf{x}, t) = \frac{1}{i} \sum_{\mathbf{k}} \left(\frac{\hbar m^2}{2\omega_k \Omega} \right)^{1/2} \left[c_{\mathbf{k}3} \mathbf{e}_{\mathbf{k}3} e^{i(\mathbf{k}\cdot\mathbf{x}-\omega_k t)} - c_{\mathbf{k}3}^{\star} \mathbf{e}_{\mathbf{k}3} e^{-i(\mathbf{k}\cdot\mathbf{x}-\omega_k t)} \right]$$

(c) Now identify the normal-mode amplitudes $(c_{\mathbf{k}s}^{\star}, c_{\mathbf{k}'s'})$ with the creation and destruction operators $(c_{\mathbf{k}s}^{\dagger}, c_{\mathbf{k}'s'})$. Show that imposing $[c_{\mathbf{k}s}, c_{\mathbf{k}'s'}^{\dagger}] = \delta_{\mathbf{k}'\mathbf{k}} \delta_{s's}$ produces the proper interaction-picture canonical commutation relations for the field operators $\hat{\mathbf{V}}(\mathbf{x}, t)$ and $\hat{\mathbf{\Pi}}(\mathbf{x}, t)$.

D.1 Show that Eq. (D.34) produces the singlet state [1] for any SU(n).

D.2 Consider an octet of scalar mesons interacting with a triplet of Dirac particles in the Sakata model. To find the irreducible representations of SU(3) available to this system, one must evaluate $[8] \otimes [3]$. This is done with the aid of the Young tableaux rules as shown in Fig. D.2. Compute the dimension of each of the resulting representations, and show that $[8] \otimes [3] = [3] \oplus [\bar{6}] \oplus [15]$.

D.3 Consider two nucleons outside of a light closed-shell nucleus and assume SU(4) symmetry with an inert core.

(a) Show that $[4] \otimes [4] = [6] \oplus [10]$, and hence determine the irreducible representations available to this system;

(b) Argue from the symmetry of the representations that the spin and isospin content $[(2S+1) \otimes (2T+1)]$ of these representations is $[6] = [(3) \otimes (1)] \oplus [(1) \otimes (3)]$ and $[10] = [(3) \otimes (3)] \oplus [(1) \otimes (1)]$;

(c) Given that the overall wave function of the pair must be antisymmetric, and that the attractive nuclear force favors a symmetric spatial state, which supermultiplet would you expect to lie lower in energy?

(d) The nuclei $({}^{6}_{4}\text{Be}, {}^{6}_{3}\text{Li}, {}^{6}_{2}\text{He})$ effectively consist of two nucleons, each in the $1p$-state and coupled to total orbital angular momentum $L = 0$, moving about an inert ${}^{4}_{2}\text{He}$ core. Compare your answer in (c) with the experimental spectra of these nuclei.[40]

D.4 Consider the Fermi-Yang model of mesons as $(\bar{B}B)$ bound states in the Sakata model.

(a) Use the Young tableaux to show that $[\bar{3}] \otimes [3] = [1] \oplus [8]$ (note Prob. D.1). Make a weight diagram for the octet where Y is plotted as the ordinate and T_3 as abcissa;

(b) Compare with the observed low-lying 0^- and 1^- mesons;[41]

(c) Explain why this is the same set of multiplets, with the same quantum numbers, that one obtains in the quark model.

E.1 (a) Use the expressions for $(P_\mu, M_{\mu\nu})$ in Eqs. (E.44), and derive the equations for the generators of the inhomogeneous Lorentz group in Eqs. (E.45);

(b) Show that replacing $M_{\mu\nu} \to M_{\mu\nu} + \hbar\sigma_{\mu\nu}/2$ leaves the commutation relations in part (a) unaltered;

(c) Show that γ_μ now satisfies the same commutation relation with $M_{\mu\nu}$ as does P_μ.

E.2 Start from Eqs. (E.55), go to infinitesimals as in the text, and derive Eqs. (E.54).

E.3 A two-component, spin-1/2, fermion field transforms in the following fashion under a real spatial rotation [compare Eq. (6.49)]

$$e^{-i\boldsymbol{\omega}\cdot\hat{\mathbf{J}}}\,\hat{\psi}(\mathbf{x})\,e^{i\boldsymbol{\omega}\cdot\hat{\mathbf{J}}} = \exp\left\{\frac{i}{2}\boldsymbol{\omega}\cdot\boldsymbol{\sigma}\right\}\hat{\psi}(\mathbf{x}') \qquad ; \; x_i' = a_{ij}(-\omega)\,x_j$$

where $\boldsymbol{\sigma}$ are the Pauli matrices, and the underlining of the column vectors and matrices is suppressed. Specialize the Lorentz transformation of the Dirac field in Eq. (E.55) to a real rotation, and show that both the upper

---

[40] Recall the discussion of "pairing" in Vol. I; for the data, see [National Nuclear Data Center (2009)].

[41] See [Particle Data Group (2009)]; note Fig. 7.2 and Table 7.4 in Vol. I.

and lower components individually transform in this fashion.[42]

E.4 The Dirac spinor transformation matrix in Eq. (E.55) is

$$\mathcal{S}(\Omega) \equiv \exp\left\{\frac{i}{4}\Omega\alpha_{\mu\nu}\sigma_{\mu\nu}\right\} \quad ; \ \sigma_{\mu\nu} = \frac{1}{2i}[\gamma_\mu, \gamma_\nu] \quad ;$$

$$\alpha_{\mu\nu} = (n_1)_\mu(n_2)_\nu - (n_2)_\mu(n_1)_\nu$$

(a) Show $\gamma_4\mathcal{S}(\Omega)^\dagger\gamma_4 = \mathcal{S}(\Omega)^{-1}$;

(b) It is a general result that

$$\mathcal{S}(\Omega)^{-1}\gamma_\mu\mathcal{S}(\Omega) = a_{\mu\nu}(-v)\gamma_\nu \quad ; \ \tan i\Omega = -iv/c$$

Establish this relation for a Lorentz transformation in the z-direction where the unit vectors characterizing the plane of rotation are $n_1 = (0,0,1,0)$ and $n_2 = (0,0,0,i)$. Use the expansion in repeated commutators, and the commutation relations in Prob. E.1(c), to show

$$\mathcal{S}(\Omega)^{-1}\gamma_3\mathcal{S}(\Omega) = \gamma_3 \cos i\Omega + \gamma_4 \sin i\Omega$$

$$\mathcal{S}(\Omega)^{-1}\gamma_4\mathcal{S}(\Omega) = -\gamma_3 \sin i\Omega + \gamma_4 \cos i\Omega$$

Hence reproduce the Lorentz transformation matrix of Eq. (E.2).

E.5 Use the results in Prob. E.4 to show that:

(a) The bilinear combination $\hat{\bar{\psi}}(x)\hat{\psi}(x)$ transforms as a scalar under homogeneous Lorentz transformations;

(b) The Dirac current $i\hat{\bar{\psi}}(x)\gamma_\mu\hat{\psi}(x)$ transforms as a four-vector [see Eq. (E.56)].

E.6 Given a set of generators $\hat{G}_\alpha$ in the abstract Hilbert space satisfying $[\hat{G}_\alpha, \hat{\phi}(x)] = -G_\alpha\hat{\phi}(x)$ where the G_α form a Lie algebra:

(a) Prove the relation $[[\hat{G}_\alpha, \hat{G}_\beta], \hat{\phi}(x)] = -[G_\alpha, G_\beta]\hat{\phi}(x)$;

(b) Hence verify the statement made before Eqs. (E.44).

F.1 Show that the commutator of the interaction-picture scalar field at two distinct space-time points $[\hat{\phi}(x^{(1)}), \hat{\phi}(x^{(2)})]$ depends only on the relative space-time coordinate $x = x^{(1)} - x^{(2)}$.

F.2 The completeness relation for the free Dirac spinors was derived in Probl. 12.7; it can be written as

$$\sum_\lambda \left[u_\alpha(\mathbf{k}\lambda)\, u_\beta^\dagger(\mathbf{k}\lambda) + v_\alpha(\mathbf{k}\lambda)\, v_\beta^\dagger(\mathbf{k}\lambda)\right] = \delta_{\alpha\beta}$$

---

[42]Recall that $\Sigma = \begin{pmatrix} \sigma & 0 \\ 0 & \sigma \end{pmatrix}$.

The positive and negative-energy projection matrices are constructed as

$$P_{\pm} = \left[\frac{E_k \pm H_D}{2E_k}\right] \qquad ; \; H_D = \hbar c(\boldsymbol{\alpha} \cdot \mathbf{k} + \beta M)$$

Multiply the completeness relation by these matrices on the left, and then by $\beta = \gamma_4$ on the right, to show that

$$\sum_{\lambda} u_{\alpha}(\mathbf{k}\lambda)\bar{u}_{\beta}(\mathbf{k}\lambda) = \left[\frac{M - i\gamma_{\mu}k_{\mu}}{2\omega_k/c}\right]_{\alpha\beta} \qquad ; \; k_{\mu} = (\mathbf{k}, i\omega_k/c)$$

$$\sum_{\lambda} v_{\alpha}(-\mathbf{k}\lambda)\bar{v}_{\beta}(-\mathbf{k}\lambda) = \left[\frac{-M - i\gamma_{\mu}k_{\mu}}{2\omega_k/c}\right]_{\alpha\beta} \qquad \omega_k = c\sqrt{\mathbf{k}^2 + M^2}$$

Here $\boldsymbol{\gamma} = i\boldsymbol{\alpha}\beta$, and we have taken $\mathbf{k} \to -\mathbf{k}$ in the second result.

F.3 Show directly that the Dirac Green's function constructed in Eq. (F.34) satisfies the defining Eq. (F.29).

F.4 (a) Fill in the steps in the derivation of the transverse photon propagator in Eq. (F.45);

(b) Show that one recovers the canonical commutation relations for the vector potential in the Coulomb gauge from this expression.

F.5 This problem evaluates the Feynman propagator for a massive vector field; it is more difficult, but the result is valuable.

(a) Start from the interaction-picture field expansions in Prob. C.7 and show that the vacuum expectation value of the time-ordered product of fields can be written as

$$\langle 0|T[\hat{V}_{\mu}(x)\,\hat{V}_{\nu}(0)]|0\rangle = \frac{1}{\Omega}\sum_{\mathbf{k}}\frac{\hbar c^2}{2\omega_k}\Delta^F_{\mu\nu}(\mathbf{k})e^{i\mathbf{k}\cdot\mathbf{x}}e^{-i\omega_k t} \qquad ; \; t > 0$$

$$= \frac{1}{\Omega}\sum_{\mathbf{k}}\frac{\hbar c^2}{2\omega_k}\Delta^F_{\mu\nu}(\mathbf{k})e^{-i\mathbf{k}\cdot\mathbf{x}}e^{i\omega_k t} \qquad ; \; t < 0$$

where $\hat{V}_4(x) \equiv (ic/m^2)\boldsymbol{\nabla}\cdot\hat{\boldsymbol{\Pi}}(x)$, and

$$\Delta^F_{ij}(\mathbf{k}) = \delta_{ij} + \frac{k_i k_j}{m^2} \qquad ; \; (i,j) = 1,2,3$$

$$\Delta^F_{j4}(\mathbf{k}) = \Delta^F_{4j}(\mathbf{k}) = \frac{ik_j\omega_k}{m^2 c} \qquad ; \; \Delta^F_{44}(\mathbf{k}) = -\frac{\mathbf{k}^2}{m^2}$$

(b) Show that in the limit $\Omega \to \infty$, this is the same result one gets by evaluating the following four-dimensional Fourier transform in Minkowski space using the techniques of contour integration, where the k_0-integral is defined in all cases by closing with a large semi-circle in the appropriate half-plane

$$\langle 0|T[\hat{V}_\mu(x)\,\hat{V}_\nu(0)]|0\rangle = \frac{\hbar c}{i}\Delta^F_{\mu\nu}(x;\,m^2)$$

$$\Delta^F_{\mu\nu}(x;\,m^2) = \frac{1}{(2\pi)^4}\int d^4k\,\frac{e^{ik\cdot x}}{k^2+m^2}\left(\delta_{\mu\nu}+\frac{k_\mu k_\nu}{m^2}\right) \qquad ;\, m\to m-i\eta$$

G.1 (a) Define the vacuum polarization tensor $\Pi_{\mu\nu}(l)$ in Eq. (G.29) through dimensional regularization, which allows one to manipulate it algebraically. Prove that it satisfies the current conservation relations $l_\mu\Pi_{\mu\nu}(l) = \Pi_{\mu\nu}(l)l_\nu = 0;$[43]

(c) Hence show that it has the general form in Eq. (G.31).

G.2 It was shown in Eq. (8.48) that the self-energy insertion for a photon in Fig. 8.3(b) leads to the following S-matrix element[44]

$$[e^2\hat{S}_{02}]_{fi} = (2\pi)^4\delta^{(4)}(l'-l)\frac{ic}{\Omega}\frac{1}{\sqrt{4\omega_1\omega_2}}\varepsilon^f_\mu\Pi_{\mu\nu}(l)\varepsilon^i_\nu$$

where the photon polarization vectors are of the form $\varepsilon = (\mathbf{e}_{ls},0)$ and satisfy $l\cdot\varepsilon_f = l\cdot\varepsilon_i = 0$. Insert the general form of the polarization tensor in Eq. (G.31), use the result in Eqs. (G.30)–(G.33), and show this matrix element *vanishes* for a real photon. Hence conclude that there is no mass renormalization for the photon to this order.

G.3 Start from the Clifford algebra in n-dimensions in Eq. (G.25), and show that Tr $\gamma_\mu = 0$. (*Hint*: make use of $\Gamma_2 \equiv \gamma_\mu\gamma_\lambda$ with $\lambda\neq\mu$.)

H.1 Show that the generating functional for the electromagnetic field is given by

$$\tilde{W}_0(J) = \exp\left\{\frac{i}{2\hbar c}\int d^4x\int d^4y\,J_\mu(x)D_{\mu\nu}(x-y)J_\nu(y)\right\} \qquad ;\,\text{E-M field}$$

Remember appendix H is in H-L units.

H.2 Re-express the result in Prob. H.1 in S-I units.

H.3 Show that the photon propagator in Eq. (H.26) satisfies the defining Eqs. (H.24)–(H.25).

H.4 Verify Eq. (H.21).

---

[43]*Hint*: First show that $[i\not{\partial}+M]^{-1}[i(\not{b}-\not{a})][i\not{b}+M]^{-1} = [i\not{a}+M]^{-1}-[i\not{b}+M]^{-1}$.
[44]Here $k\to l$ and $\mu\leftrightarrow\nu$.

Appendix A

Multipole Analysis of the Radiation Field

The general expression for the differential rate with which an isolated quantum mechanical system, with an electromagnetic current operator $\hat{\mathbf{J}}(\mathbf{x}) \equiv \hat{\mathbf{j}}(\mathbf{x})/c$, makes a transition from a state $|i\rangle$ to a state $|f\rangle$, and emits a photon with wave number and transverse polarization $(\mathbf{k}, s)$, is given in ProbsI. 12.3–12.4

$$d\omega_{fi} = \frac{\alpha\omega_k}{2\pi} \left| \langle f| \int d^3x \, e^{-i\mathbf{k}\cdot\mathbf{x}} \, \mathbf{e}_{\mathbf{k}s} \cdot \hat{\mathbf{J}}(\mathbf{x}) |i\rangle \right|^2 d\Omega_k \quad ; \text{photoemission}$$

$$\hat{\mathbf{J}}(\mathbf{x}) \equiv \frac{1}{c}\hat{\mathbf{j}}(\mathbf{x}) \quad \text{(A.1)}$$

One must therefore deal with matrix elements of the operator $\int d^3x \, e^{-i\mathbf{k}\cdot\mathbf{x}} \, \mathbf{e}_{\mathbf{k}s} \cdot \hat{\mathbf{J}}(\mathbf{x})$.[45] The goal of this appendix is to decompose this operator into ITO, so that one can use the W-E theorem on the matrix elements to both extract the explicit dependence on target orientation, and to obtain all of the accompanying angular momentum selection rules for a radiative transition (Fig. A.1).

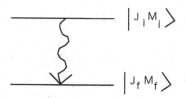

Fig. A.1   Radiative transition for an isolated system in quantum mechanics.

---

[45]The matrix elements of this operator are now dimensionless.

A.1 Vector Spherical Harmonics

Let us first examine a simpler problem. Consider $\int d^3x\, e^{-i\mathbf{k}\cdot\mathbf{x}}\hat{\rho}(\mathbf{x})$ where $e\hat{\rho}(\mathbf{x})$ is the charge density operator, whose matrix elements are required for inelastic scattering through the Coulomb interaction (see Probl. I.1). We start with the following expansion of a plane wave in spherical coordinates

$$e^{-i\mathbf{k}\cdot\mathbf{x}} = \sum_{L=0}^{\infty} \sqrt{4\pi(2L+1)}\,(-i)^L j_L(kr)\, Y_{L0}(\Omega_x) \qquad (A.2)$$

This is a basic result in classical physics; it has nothing to do with quantum mechanics.[46] Here $j_L(kr)$ is a spherical Bessel function, with

$$j_0(z) = \frac{\sin z}{z} \qquad\qquad ; \; j_1(z) = \frac{\sin z}{z^2} - \frac{\cos z}{z}$$

$$j_L(z) = (-z)^L \left(\frac{1}{z}\frac{d}{dz}\right)^L \frac{\sin z}{z} \qquad (A.3)$$

The spherical harmonic $Y_{L0}(\theta,\phi)$ is, within a factor, the Legendre polynomial $P_L(\cos\theta)$

$$Y_{L0}(\theta,\phi) = \left(\frac{2L+1}{4\pi}\right)^{1/2} P_L(\cos\theta)$$

$$P_0(x) = 1 \quad ; \; P_1(x) = x \quad ; \; P_2(x) = \frac{1}{2}(3x^2-1) \qquad (A.4)$$

Insertion of the expansion in Eq. (A.2) gives

$$\int d^3x\, e^{-i\mathbf{k}\cdot\mathbf{x}}\hat{\rho}(\mathbf{x}) = \sum_{L=0}^{\infty} \sqrt{4\pi(2L+1)}\,(-i)^L \hat{M}_{L0}(k)$$

$$\hat{M}_{LM}(k) \equiv \int d^3x\, j_L(kr) Y_{LM}(\Omega_x)\hat{\rho}(\mathbf{x}) \qquad (A.5)$$

We claim that $\hat{M}_{LM}(k)$ is an ITO of rank L. The general proof involves the analysis of this quantity under rotations [see Eq. (3.136)]; we shall be content here to demonstrate it in a special case. Suppose one is dealing with the non-relativistic quantum mechanics of a charged particle moving in a potential, as illustrated in Fig. A.2. The density operator in this case

---

[46] This expansion is derived, for example, in [Fetter and Walecka (2003)], appendix D; see also [Schiff (1968)].

is[47]

$$\hat{\rho}(\mathbf{x}) = \delta^{(3)}(\mathbf{x} - \hat{\mathbf{x}}_e) \tag{A.6}$$

Insertion of this expression in the second of Eqs. (A.5) gives, in the coordinate representation,

$$M_{LM}(k) = j_L(kr_e)Y_{LM}(\Omega_{x_e}) \qquad ; \text{ ITO } \qquad ;$$
$$\Pi = (-1)^L \tag{A.7}$$

and, indeed, this is just our prototype ITO [see Eqs. (3.109)]. Since the spherical harmonics simply pick up the phase $(-1)^L$ under spatial reflections (Vol. I), the *parity* of these multipoles is $\Pi = (-1)^L$.

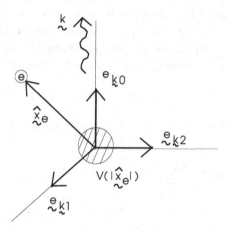

Fig. A.2 Example of a system with electromagnetic interactions in quantum mechanics. A particle of charge e, with position coordinate $\hat{x}_e$, moves in a potential $V(|\hat{\mathbf{x}}_e|)$. We here let the photon wave number $\mathbf{k}$ define the z-direction. The quantities $(\mathbf{e}_{\mathbf{k}0} = \mathbf{k}/|\mathbf{k}|, \mathbf{e}_{\mathbf{k}1}, \mathbf{e}_{\mathbf{k}2})$ then form an orthonormal set of unit vectors.

What happens with a *vector density* operator $\hat{\mathbf{v}}(\mathbf{x})$, as one has with the current in Eq. (A.1)? To deal with this situation, we define *vector spherical harmonics*

$$\boldsymbol{\mathcal{Y}}_{LJ}^M(\Omega_x) \equiv \sum_{M',q'} \langle LM'1q'|L1JM\rangle Y_{LM'}(\Omega_x)\mathbf{e}_{1q'} \qquad ;$$

$$\text{vector spherical harmonics} \tag{A.8}$$

---

[47]The expectation value of this quantity, in the coordinate representation, is then $\langle i|\hat{\rho}(\mathbf{x})|i\rangle = \int d^3x_e\, \psi_i^\star(\mathbf{x}_e)\delta^{(3)}(\mathbf{x}-\mathbf{x}_e)\psi_i(\mathbf{x}_e) = |\psi_i(\mathbf{x})|^2$, which is the proper expression.

Here $\langle LM'1q'|L1JM\rangle$ is a C-G coefficient, and the $\mathbf{e}_{1q'}$ are spherical basis vectors defined by (see Fig. A.2)

$$\mathbf{e}_{1,\pm 1} \equiv \mp \frac{1}{\sqrt{2}}(\mathbf{e}_{\mathbf{k}1} \pm i\mathbf{e}_{\mathbf{k}2}) \quad ; \text{ spherical basis vectors}$$

$$\mathbf{e}_{1,0} \equiv \mathbf{e}_{\mathbf{k}0} \tag{A.9}$$

In the present notation, $\mathbf{e}_{1\lambda} \equiv \mathbf{e}_{\mathbf{k}\lambda}$, with $\lambda = (0, \pm 1)$.

Consider the quantity

$$\hat{T}^M_{LJ}(k) \equiv \int d^3x \, j_L(kr)\boldsymbol{\mathcal{Y}}^M_{LJ}(\Omega_x) \cdot \hat{\mathbf{v}}(\mathbf{x}) \tag{A.10}$$

We claim that this is an ITO of rank J. The proof again involves an analysis of this quantity under rotations; however, we are content to again demonstrate this with an example. Suppose one has the configuration in Fig. A.2, and $\hat{\mathbf{v}}_e$ is one of the available particle *vector operators* $(\hat{\mathbf{x}}_e, \hat{\mathbf{p}}_e, \hat{\mathbf{s}}_e)$. Extend Eq. (A.6) to make it a *vector density* operator[48]

$$\hat{\mathbf{v}}(\mathbf{x}) = \delta^{(3)}(\mathbf{x} - \hat{\mathbf{x}}_e)\hat{\mathbf{v}}_e \tag{A.11}$$

Now substitute Eq. (A.11) into (A.10), and work in the coordinate representation,

$$T^J_{LM}(k) = j_L(kr_e)\sum_{M'}\sum_{q'}\langle LM'1q'|L1JM\rangle Y_{LM'}(\Omega_{x_e})(v_e)_{1q'} \quad ; \text{ ITO} \quad ;$$

$$\Pi = (-1)^{L+1} ; \; \hat{\mathbf{v}}_e \text{ a vector} \quad ;$$

$$\Pi = (-1)^L \quad ; \; \hat{\mathbf{v}}_e \text{ an axial vector} \tag{A.12}$$

Here the $(v_e)_{1q'}$ are the spherical components of the vector $\mathbf{v}_e$, which forms an ITO of rank 1 [see Eqs. (3.26) and (3.107)]. The above expression is now just the *tensor product* of $j_L(kr_e)Y_{LM}(\Omega_{x_e})$ and $(v_e)_{1q}$, and thus by Eq. (3.147), this is again an ITO of rank J.

If $\hat{\mathbf{v}}_e$ is a true vector whose components change sign under spatial reflection (*i.e.* $\hat{\mathbf{x}}_e, \hat{\mathbf{p}}_e$), then the parity of this multipole is $\Pi = (-1)^{L+1}$; if $\hat{\mathbf{v}}_e$ is an *axial vector* whose components *do not* change sign under spatial reflection (*i.e.* $\hat{\mathbf{s}}_e$), then the parity is $(-1)^L$.

A derivation of the properties of the vector spherical harmonics only involves (rather tedious) algebra, and is again simply classical analysis. We

---

[48]With $\hat{\mathbf{p}}_e$, to make the density hermitian, one must symmetrize and take $\hat{\mathbf{v}}(\mathbf{x}) \equiv [\hat{\mathbf{p}}_e\delta^{(3)}(\mathbf{x} - \hat{\mathbf{x}}_e) + \delta^{(3)}(\mathbf{x} - \hat{\mathbf{x}}_e)\hat{\mathbf{p}}_e]/2$.

are content here to just quote the following two results from [Edmonds (1974)],[49] which play an important role in what follows;

$$\boldsymbol{\nabla} \cdot \left[j_L(kr) \boldsymbol{\mathcal{Y}}_{LL}^M(\Omega_x) \right] = 0 \tag{A.13}$$

$$\frac{1}{k} \boldsymbol{\nabla} \times \left[j_L(kr) \boldsymbol{\mathcal{Y}}_{LL}^M(\Omega_x) \right] = i \left[\left(\frac{L+1}{2L+1} \right)^{1/2} j_{L-1}(kr) \boldsymbol{\mathcal{Y}}_{L-1,L}^M(\Omega_x) - \right.$$
$$\left. \left(\frac{L}{2L+1} \right)^{1/2} j_{L+1}(kr) \boldsymbol{\mathcal{Y}}_{L+1,L}^M(\Omega_x) \right]$$

Note that it is $j_L(kr)\boldsymbol{\mathcal{Y}}_{LL}^M(\Omega_x)$, with $J = L$, that appears on the l.h.s. of these expressions, and M is unchanged on the r.h.s.

In summary:

- Integration over a spherical harmonic projects an ITO from a scalar density operator;
- Integration over the dot-product with a *vector* spherical harmonic projects an ITO from a *vector* density operator.

A.2 Plane-Wave Expansion

Return to Fig. A.2, and consider the expression $e^{i\mathbf{k}\cdot\mathbf{x}}\,\mathbf{e}_{\mathbf{k}\lambda}$ for $\lambda = \pm 1$. We claim this can be written in the following form

$$e^{i\mathbf{k}\cdot\mathbf{x}}\,\mathbf{e}_{\mathbf{k}\lambda} = -\sum_{J=1}^{\infty} \sqrt{2\pi(2J+1)}\,i^J \left\{ \lambda j_J(kr) \boldsymbol{\mathcal{Y}}_{JJ}^\lambda(\Omega_x) + \right.$$
$$\left. \frac{1}{k} \boldsymbol{\nabla} \times \left[j_J(kr) \boldsymbol{\mathcal{Y}}_{JJ}^\lambda(\Omega_x) \right] \right\} \qquad ; \lambda = \pm 1 \tag{A.14}$$

Several comments before we prove this result:

- This is just an algebraic identity, but it has the invaluable merit of allowing us to project ITO from the current density operator;
- The relation holds for $\lambda = \pm 1$, which, as we will see, is what is needed for radiative decay;
- The sum starts with $J = 1$;
- The expression on the r.h.s. must have vanishing divergence since

$$\boldsymbol{\nabla} \cdot \left(e^{i\mathbf{k}\cdot\mathbf{x}}\,\mathbf{e}_{\mathbf{k}\lambda} \right) = 0 \qquad ; \lambda = \pm 1 \tag{A.15}$$

---

[49] Note that $[d/d\rho - L/\rho]j_L(\rho) = -j_{L+1}(\rho)$ and $[d/d\rho + (L+1)/\rho]j_L(\rho) = j_{L-1}(\rho)$.

The first of Eqs. (A.13), and the fact that $\nabla \cdot (\nabla \times \mathbf{v}) \equiv 0$ for any $\mathbf{v}$, guarantee that this is true;

- Note that there will only be *two* independent ITO, or two multipoles, for each J on the r.h.s., since the second of Eqs. (A.13) implies that these two terms always enter in the same combination.

Let us proceed to a proof of Eq. (A.14). From Eqs. (A.2) and (A.9), one has[50]

$$e^{i\mathbf{k}\cdot\mathbf{x}}\,\mathbf{e}_{k\lambda} = \sum_{L=0}^{\infty} \sqrt{4\pi(2L+1)}\, i^L j_L(kr) Y_{L0}(\Omega_x)\,\mathbf{e}_{1\lambda} \qquad (A.16)$$

The orthonormality of the C-G coefficients in Eq. (A.8) allows us to write

$$Y_{L0}(\Omega_x)\,\mathbf{e}_{1\lambda} = \sum_{JM}\langle L01\lambda|l1JM\rangle\,\boldsymbol{\mathcal{Y}}^M_{LJ}(\Omega_x)$$
$$= \sum_{J}\langle L01\lambda|L1J\lambda\rangle\,\boldsymbol{\mathcal{Y}}^\lambda_{LJ}(\Omega_x) \qquad (A.17)$$

Now substitute this into Eq. (A.16) and carry out the sum on L *first*, for a given J. The C-G coefficients imply $L = J-1,\ J,\ J+1$, and $J \geq 1$ for $\lambda = \pm 1$. Thus Eq. (A.16) becomes

$$e^{i\mathbf{k}\cdot\mathbf{x}}\,\mathbf{e}_{k\lambda} = \sum_{J\geq 1} \sqrt{4\pi}\ \times$$
$$\left\{ \sqrt{2J-1}\, i^{J-1} j_{J-1}(kr)\boldsymbol{\mathcal{Y}}^\lambda_{J-1,J}(\Omega_x)\langle J-1,01\lambda|J-1,1J\lambda\rangle + \right.$$
$$\sqrt{2J+1}\, i^J j_J(kr)\boldsymbol{\mathcal{Y}}^\lambda_{J,J}(\Omega_x)\langle J01\lambda|J1J\lambda\rangle +$$
$$\left. \sqrt{2J+3}\, i^{J+1} j_{J+1}(kr)\boldsymbol{\mathcal{Y}}^\lambda_{J+1,J}(\Omega_x)\langle J+1,01\lambda|J+1,1J\lambda\rangle \right\} \quad (A.18)$$

The requisite C-G coefficients are given in Table 12.1, and hence

$$e^{i\mathbf{k}\cdot\mathbf{x}}\,\mathbf{e}_{k\lambda} = \sum_{J\geq 1} \sqrt{4\pi}\ \times$$
$$\left\{ \sqrt{2J-1}\, i^{J-1} j_{J-1}(kr)\boldsymbol{\mathcal{Y}}^\lambda_{J-1,J}(\Omega_x)\left[\frac{(J+1)}{2(2J-1)}\right]^{1/2} + \right.$$
$$\sqrt{2J+1}\, i^J j_J(kr)\boldsymbol{\mathcal{Y}}^\lambda_{J,J}(\Omega_x)\left[\frac{-\lambda}{\sqrt{2}}\right] +$$
$$\left. \sqrt{2J+3}\, i^{J+1} j_{J+1}(kr)\boldsymbol{\mathcal{Y}}^\lambda_{J+1,J}(\Omega_x)\left[\frac{J}{2(2J+3)}\right]^{1/2} \right\} \quad (A.19)$$

---

[50] Recall $\mathbf{e}_{k\lambda} \equiv \mathbf{e}_{1\lambda}$.

An appeal to the second of Eqs. (A.13) now establishes the result in Eq. (A.14).

A.3 Transition Rate

We proceed to compute the general expression for the transition rate for photon emission. First, we want to rewrite Eq. (A.1) in terms of the circular polarization of the photon, rather than plane polarization. To do this, we make a simple *canonical transformation*. Redefine the photon creation and destruction operators in chapter 12 of Vol. I as

$$a^\dagger_{\mathbf{k},\pm 1} \equiv \mp \frac{1}{\sqrt{2}}(a^\dagger_{\mathbf{k}1} \pm ia^\dagger_{\mathbf{k}2}) \qquad ; \ \mathbf{e}_{\mathbf{k},\pm 1} = \mp \frac{1}{\sqrt{2}}(\mathbf{e}_{\mathbf{k}1} \pm i\mathbf{e}_{\mathbf{k}2})$$

$$a_{\mathbf{k},\pm 1} \equiv \mp \frac{1}{\sqrt{2}}(a_{\mathbf{k}1} \mp ia_{\mathbf{k}2}) \qquad ; \ \mathbf{e}^\dagger_{\mathbf{k},\pm 1} = \mp \frac{1}{\sqrt{2}}(\mathbf{e}_{\mathbf{k}1} \mp i\mathbf{e}_{\mathbf{k}2}) \ (A.20)$$

One immediately verifies that[51]

$$[a_{\mathbf{k}\lambda}, a^\dagger_{\mathbf{k}'\lambda'}] = \delta_{\mathbf{k}\mathbf{k}'}\delta_{\lambda\lambda'} \qquad ; \ (\lambda,\lambda') = \pm 1 \tag{A.21}$$

Since all our results on the creation and destruction operators followed from the commutation relations, one might just have well have used the states $|n_{\mathbf{k}\lambda}\rangle$ rather than the states $|n_{\mathbf{k}s}\rangle$ to describe the photon field. Furthermore, it is readily established that

$$\sum_{s=1}^{2} a_{\mathbf{k}s}\mathbf{e}_{\mathbf{k}s} = \sum_{\lambda=\pm 1} a_{\mathbf{k}\lambda}\mathbf{e}_{\mathbf{k}\lambda}$$

$$\sum_{s=1}^{2} a^\dagger_{\mathbf{k}s}\mathbf{e}_{\mathbf{k}s} = \sum_{\lambda=\pm 1} a^\dagger_{\mathbf{k}\lambda}\mathbf{e}^\dagger_{\mathbf{k}\lambda} \tag{A.22}$$

Hence one might also just as well have expanded the quantized vector potential in terms of the polarizations λ rather than s. As we shall see, λ will turn out to be the *helicity* of the emitted photon.

We assume the intial and final states are eigenstates of angular momentum (Fig. A.1). With these modifications, Eq. (A.1) becomes

$$d\omega_{fi} = \frac{\alpha\omega_k}{2\pi} \left| \langle J_f M_f| \int d^3x \, e^{-i\mathbf{k}\cdot\mathbf{x}} \, \mathbf{e}^\dagger_{\mathbf{k}\lambda} \cdot \hat{J}(\mathbf{x}) |J_i M_i\rangle \right|^2 d\Omega_k \qquad ;$$

$$\text{photon polarization } \lambda = \pm 1 \tag{A.23}$$

---

[51]See Prob. A.3; note $[a_{\mathbf{k}\lambda}, a_{\mathbf{k}'\lambda'}] = [a^\dagger_{\mathbf{k}\lambda}, a^\dagger_{\mathbf{k}'\lambda'}] = 0$.

We will calculate the transition rate for an initially unpolarized target, where the target and photon polarization are unobserved in the final state. In this case, we must *average over* initial orientations of the target and sum over everything that gets into our detector, that is, *sum over* final target orientations and photon polarizations. Equation (A.23) then becomes

$$d\omega_{fi} = \frac{\alpha\omega_k}{2\pi}\frac{1}{2J_i+1}\sum_{\lambda=\pm1}\sum_{M_i}\sum_{M_f}$$

$$\left|\langle J_fM_f|\int d^3x\,e^{-i\mathbf{k}\cdot\mathbf{x}}\,\mathbf{e}^{\dagger}_{\mathbf{k}\lambda}\cdot\hat{J}(\mathbf{x})\,|J_iM_i\rangle\right|^2 d\Omega_k \qquad (A.24)$$

Since we are here interested in only the total transition rate, it simplifies the calculation to choose the direction $\mathbf{e}_{\mathbf{k}0}$ as the z-axis for the quantization of the angular momentum of the target (Fig. A.3).

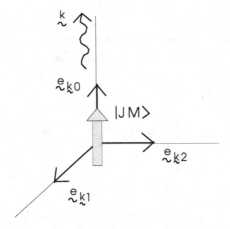

Fig. A.3   Quantize the target along the $\mathbf{e}_{\mathbf{k}0}$ direction in Fig. A.2.

For photon emission, one needs the adjoint of the expression in Eq. (A.14). This is obtained using the following property of the vector spherical harmonics (see Prob. A.1)

$$\mathbf{\mathcal{Y}}^M_{JJ}(\Omega_x)^{\dagger} = (-1)^{1+M}\mathbf{\mathcal{Y}}^{-M}_{JJ}(\Omega_x) \qquad (A.25)$$

Thus

$$e^{-i\mathbf{k}\cdot\mathbf{x}}\,\mathbf{e}_{\mathbf{k}\lambda}^{\dagger} = -\sum_{J=1}^{\infty} \sqrt{2\pi(2J+1)}\,(-i)^J \left\{ \lambda j_J(kr)\mathbf{\mathcal{Y}}_{JJ}^{-\lambda}(\Omega_x) + \right.$$

$$\left. \frac{1}{k}\mathbf{\nabla}\times\left[j_J(kr)\mathbf{\mathcal{Y}}_{JJ}^{-\lambda}(\Omega_x)\right]\right\} \qquad ; \lambda = \pm 1 \qquad \text{(A.26)}$$

The transition operator for radiative decay then takes the form

$$\int d^3x\, e^{-i\mathbf{k}\cdot\mathbf{x}}\,\mathbf{e}_{\mathbf{k}\lambda}^{\dagger}\cdot\hat{\mathbf{J}}(\mathbf{x}) =$$

$$-\sum_{J\geq 1}\sqrt{2\pi(2J+1)}\,(-i)^J\left[\lambda\hat{T}_{J,-\lambda}^{\mathrm{mag}}(k) + \hat{T}_{J,-\lambda}^{\mathrm{el}}(k)\right] \qquad \text{(A.27)}$$

Here the *transverse electric and magnetic multipoles* of the current are defined by

$$\hat{T}_{JM}^{\mathrm{el}}(k) \equiv \frac{1}{k}\int d^3x\,\left\{\mathbf{\nabla}\times\left[j_J(kr)\mathbf{\mathcal{Y}}_{JJ}^{M}(\Omega_x)\right]\right\}\cdot\hat{\mathbf{J}}(\mathbf{x})$$

$$\hat{T}_{JM}^{\mathrm{mag}}(k) \equiv \int d^3x\,\left[j_J(kr)\mathbf{\mathcal{Y}}_{JJ}^{M}(\Omega_x)\right]\cdot\hat{\mathbf{J}}(\mathbf{x}) \qquad \text{(A.28)}$$

Finally, since we have worked hard to reduce the transition amplitude to a sum of ITO, we are now in a position to invoke the *W-E theorem* in Eq. (3.113)

$$\langle J_f M_f|\hat{T}_{J,-\lambda}(k)|J_i M_i\rangle = (-1)^{J_f - M_f}\begin{pmatrix} J_f & J & J_i \\ -M_f & -\lambda & M_i \end{pmatrix}\langle J_f||T_J(k)||J_i\rangle$$

$$\text{(A.29)}$$

This does three things for us:

(1) It tells us that the matrix element *vanishes* unless the addition rule for angular momentum is satisfied

$$|J_f - J_i| \leq J \leq J_f + J_i \qquad \text{(A.30)}$$

(2) It allows us to perform the sums over target orientations in Eq. (A.24);
(3) It allows us to conclude that the angular momentum the photon carries off from the target along its direction of motion is $M_i - M_f = \lambda$; hence, λ is indeed the photon helicity.

The orthonormality statement for 3-j symbols follows immediately from that for the C-G coefficients (see [Edmonds (1974)])

$$\sum_{M_i}\sum_{M_f}\begin{pmatrix} J_f & J & J_i \\ -M_f & \lambda & M_i \end{pmatrix}\begin{pmatrix} J_f & J' & J_i \\ -M_f & \lambda' & M_i \end{pmatrix} = \frac{1}{2J+1}\delta_{\lambda\lambda'}\delta_{JJ'} \quad (A.31)$$

In this way, Eq. (A.24) is reduced to a simple sum of squares

$$d\omega_{fi} = \alpha\omega_k\frac{1}{2J_i+1}\sum_{\lambda=\pm1}\sum_{J\geq1}\left|\langle J_f||\lambda T_J^{\mathrm{mag}}(k) + T_J^{\mathrm{el}}(k)||J_i\rangle\right|^2 d\Omega_k \quad (A.32)$$

Only the terms satisfying the angular momentum selection rule contribute to the sum.

For the total rate, the final two steps are immediately carried out:

(1) The sum over circular photon polarizations gives $\sum_{\lambda=\pm1}|a+\lambda b|^2 = 2(|a|^2+|b|^2)$
(2) The resulting expression is independent of the overall orientation in space, and therefore $\int d\Omega_k = 4\pi$.

The final result for the total rate at which a target makes a transition from a state with (E_i, J_i) to a state with (E_f, J_f) while emitting a photon with energy $\hbar\omega = E_i - E_f$ is given by

$$\mathcal{R}_{fi} = 8\pi\alpha\,\omega\frac{1}{2J_i+1}\sum_{J\geq1}\left(|\langle J_f||T_J^{\mathrm{mag}}(k)||J_i\rangle|^2 + |\langle J_f||T_J^{\mathrm{el}}(k)||J_i\rangle|^2\right) \quad ;$$

$$\text{radiative decay rate} \quad (A.33)$$

Only the terms allowed by the angular momentum selection rules contribute to the sum.

The current density is a vector, while the charge density is a scalar. From our previous discussion, the parity of the multipoles in Eqs. (A.28) and (A.5) is therefore as given Table A.1. This gives the required parity change between the initial and final target states, since parity is conserved in the strong and electromagnetic interactions.

Table A.1 Parity of the multipole operators.

Operator	Parity
$\hat{M}_{JM}(k)$	$(-1)^J$
$\hat{T}_{JM}^{\mathrm{el}}(k)$	$(-1)^J$
$\hat{T}_{JM}^{\mathrm{mag}}(k)$	$(-1)^{J+1}$

While only a finite number of multipoles contribute to the sum in Eq. (A.33), the situation is generally even simpler than this. If the dimensionless quantities R/λ, λ_C/λ are small, where λ is the wavelength of the emitted photon, R is a typical target size, and λ_C is the Compton wavelength of a target particle, then one can make a long-wavelength expansion of the multipoles, and they will typically make a rapidly decreasing contribution with increasing J.[52] To see this in the simplest case, consider the long-wavelength expansion of the charge multipole in Eq. (A.5), and use[53]

$$j_L(kr) \to \frac{(kr)^L}{(2L+1)!!} \qquad ; \; kr \to 0 \qquad (A.34)$$

where $(2L+1)!! \equiv (2L+1)(2L-1)\cdots(3)(1)$. Then as $kR \to 0$,

$$\langle f|\hat{M}_{LM}(k)|i\rangle \to \frac{k^L}{(2L+1)!!} \int d^3x \, r^L Y_{LM}(\Omega_x)\langle f|\hat{\rho}(\mathbf{x})|i\rangle \qquad (A.35)$$

The long-wavelength characterization of all of the multipoles in a nuclear gamma decay with a localized source, originally presented in [Blatt and Weisskopf (1952)], is detailed in [Walecka (2004)].

A.4 Arbitrary Photon Direction

So far, everything has been done with $\mathbf{k}/|\mathbf{k}|$ as the z-axis — this is all we need if the total decay rate is the only quantity of interest. More generally, one is interested in the photon angular distribution with respect to a target that is quantized along an arbitrary direction in space. The general expression for that transition amplitude, which follows from our previous analysis of finite rotations, is

$$\langle \Psi_{J_f M_f}| \int d^3x \, e^{-i\mathbf{k}\cdot\mathbf{x}} \, \mathbf{e}_{\mathbf{k}\lambda}^\dagger \cdot \hat{\mathbf{J}}(\mathbf{x}) |\Psi_{J_i M_i}\rangle = -\sum_{J \geq 1} \sum_M \sqrt{2\pi(2J+1)} \, (-i)^J$$

$$\times \langle J_f M_f|\lambda \hat{T}_{JM}^{\mathrm{mag}}(k) + \hat{T}_{JM}^{\mathrm{el}}(k)|J_i M_i\rangle \mathcal{D}_{M,-\lambda}^J(-\phi_k, -\theta_k, \phi_k) \qquad (A.36)$$

Let us discuss this result before we derive it:

- The target states $|\Psi_{J_i M_i}\rangle$ and $|\Psi_{J_f M_f}\rangle$ are eigenstates of angular momentum, and are quantized along the z-direction in a cartesian (x, y, z) coordinate system (see Fig. A.4);

---

[52] A term in λ_C/λ will arise from the magnetic moment interaction.
[53] See [Schiff (1968)], or [Fetter and Walecka (2003)].

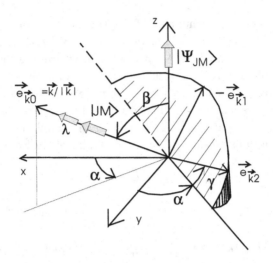

Fig. A.4   Configuration for the photon coming off in arbitrary direction with repect to the target quantization direction. The states $|J_i M_i\rangle, |J_f M_f\rangle$ are quantized along the $\mathbf{e_{k0}} = \mathbf{k}/|\mathbf{k}|$ direction, and the states $|\Psi_{J_i M_i}\rangle, |\Psi_{J_f M_f}\rangle$ are those rotated from the $(\mathbf{e_{k0}}, \mathbf{e_{k1}}, \mathbf{e_{k2}})$ to the new (x, y, z) frame. λ is the photon helicity. (α, β, γ) are the indicated Euler angles for this configuration, and we eventually identify $(\beta, \alpha) = (\theta_k, \phi_k)$.

- The photon comes off with wave vector $\mathbf{k}$, which is described by the usual polar and azimuthal angles (θ_k, ϕ_k) in this coordinate system;
- The photon has helicity $\lambda = \pm 1$ with respect to $\mathbf{k}$. This is the angular momentum carried away by the photon along its direction of motion;
- The state $|\Psi_{JM}\rangle$ is a rotated version of the target state $|JM\rangle$ quantized along $\mathbf{k}$ (see below);
- The quantum numbers (M_i, M_f, M) in the states $|J_i M_i\rangle, |J_f M_f\rangle$ and the ITO $\hat{T}_{JM}$ now refer to a common set of axes, and the dependence on these quantum numbers is consequently given by the W-E theorem, which implies

$$|J_i - J_f| \leq J \leq J_i + J_f \qquad ; \text{ W-E theorem}$$
$$M = M_f - M_i \tag{A.37}$$

The photon evidently carries away angular momentum $(J, -M)$ from the system;

- The factor $\mathcal{D}^J_{M,-\lambda}(-\phi_k, -\theta_k, \phi_k)$ is a rotation matrix; it provides the angular distribution of the photon corresponding to each term in the sum, and serves as a "photon wave function" for the transitions.[54]

We proceed to derive Eq. (A.36). We do this in a series of steps, while referring to Figure A.4:

- Expand the transition amplitude $\int d^3x\, e^{-i\mathbf{k}\cdot\mathbf{x}}\, \mathbf{e}_{\mathbf{k}\lambda}^{\dagger} \cdot \hat{J}(\mathbf{x})$ using $\mathbf{e}_{\mathbf{k}0}$ as the z-axis, as in Eq. (A.26);
- Rotate the physical state vectors $|JM\rangle$ to the new (x, y, z) coordinate system where it becomes $|\Psi_{JM}\rangle$, with the angular momentum quantized along the new z-axis — we know how to do this. The only subtlety here is that the rotations are to be carried out *with respect to the fixed-space, laboratory* $(\mathbf{e}_{\mathbf{k}0}, \mathbf{e}_{\mathbf{k}1}, \mathbf{e}_{\mathbf{k}2})$ *axes*;
- Significant concentration on Fig. A.4 will convince the reader that the following rotations, carried out in the indicated order, will produce the new (x, y, z) configuration:

(1) Rotate the state by $-\alpha$ about the $\mathbf{e}_{\mathbf{k}0} = \mathbf{k}/|\mathbf{k}|$ axis;
(2) Rotate the state by $-\beta$ about the $\mathbf{e}_{\mathbf{k}2}$ axis;
(3) Rotate the state by $-\gamma$ about the $\mathbf{e}_{\mathbf{k}0} = \mathbf{k}/|\mathbf{k}|$ axis.

From our discussion of finite rotations, the operator that produces this rotation for us is

$$\hat{\mathcal{R}}_{\gamma\beta\alpha} = e^{i\gamma \hat{J}_3} e^{i\beta \hat{J}_2} e^{i\alpha \hat{J}_3} \tag{A.38}$$

Thus

$$|\Psi_{J_i M_i}\rangle = \hat{\mathcal{R}}_{\gamma\beta\alpha} |J_i M_i\rangle$$
$$|\Psi_{J_f M_f}\rangle = \hat{\mathcal{R}}_{\gamma\beta\alpha} |J_f M_f\rangle \tag{A.39}$$

Note the the effect of this rotation is just to produce a linear combination of the original states

$$\hat{\mathcal{R}}_{\gamma\beta\alpha} |J_i M_i\rangle = \sum_{M_i'} \mathcal{D}^{J_i}_{M_i' M_i}(\gamma\beta\alpha) |J_i M_i'\rangle$$
$$\hat{\mathcal{R}}_{\gamma\beta\alpha} |J_f M_f\rangle = \sum_{M_f'} \mathcal{D}^{J_f}_{M_f' M_f}(\gamma\beta\alpha) |J_f M_f'\rangle \tag{A.40}$$

We have already worked out the transition amplitude for the operator in Eq. (A.27) between each of these original states. The calculation we need

---

[54]The helicity analysis of a general scattering, or reaction, S-matrix element is carried out in [Jacob and Wick (1959)]. This is an extremely useful paper; it is worth studying.

to do thus takes the form

$$\langle \Psi_{J_f M_f} | \hat{T}_{J,-\lambda} | \Psi_{J_i M_i} \rangle = \langle J_f M_f | \hat{\mathcal{R}}_{\gamma\beta\alpha}^{-1} \hat{T}_{J,-\lambda} \hat{\mathcal{R}}_{\gamma\beta\alpha} | J_i M_i \rangle \quad \text{(A.41)}$$

But now we can use a nice trick. Instead of rotating the *states*, we can just rotate the *ITO's* through Eq. (3.136). It is just necessary to get the rotation operator into the right form. To accomplish this, use

$$\hat{\mathcal{R}}_{\gamma\beta\alpha} = \hat{\mathcal{R}}_{-\alpha,-\beta,-\gamma}^{-1} \quad \text{(A.42)}$$

Then

$$\hat{\mathcal{R}}_{-\alpha,-\beta,-\gamma} \, \hat{T}_{J,-\lambda} \, \hat{\mathcal{R}}_{-\alpha,-\beta,-\gamma}^{-1} = \sum_M \mathcal{D}_{M,-\lambda}^J(-\alpha,-\beta,-\gamma) \, \hat{T}_{JM} \quad \text{(A.43)}$$

Finally, identify the usual polar and azimuthal angles (θ_k, ϕ_k) for the photon in the (x, y, z) coordinate system from Fig. A.4

$$\alpha \equiv \phi_k$$
$$\beta \equiv \theta_k$$
$$\gamma \equiv -\phi_k \qquad ; \text{(arbitrary)} \quad \text{(A.44)}$$

The last expression is a phase convention, which produces an *overall* phase $e^{-i\lambda\phi_k}$ in the amplitude.

The result of these steps is now Eq. (A.36).

Appendix B

Functions of a Complex Variable

A very brief introduction to the theory of functions of a complex variable is given in appendix A of Vol. I. Some additional features are required in the present volume, and the goal of this appendix is to review those topics. The theory of functions is one of the most beautiful parts of mathematics, and we certainly cannot do justice to it here. Fortunately, there are excellent, accessible books on the subject available to students (see, for example, [Titchmarch (1976)]).[55] An extensive introduction to this topic is contained in appendix A of [Fetter and Walecka (2003)], and that appendix forms the basis of the present discussion.

B.1 Convergence

The modulus of a complex number was defined in appendix A of Vol. I

$$|z| = |x + iy| = \sqrt{x^2 + y^2} \qquad ; \text{ modulus} \qquad (B.1)$$

This provides a measure of *distance* in the complex plane. With the use of readily established inequalities, such as $|a + b| \le |a| + |b|$, this concept of distance allows one to apply the standard notions of convergence.

B.2 Analytic Functions

A function $f(z)$ is *analytic* if it has a derivative

$$\frac{df(z)}{dz} = \text{Lim}_{\delta \to 0} \frac{f(z + \delta) - f(z)}{\delta} \qquad (B.2)$$

---

[55] See also [Morse and Feshbach (1953)].

383

that is *independent of the direction* in which the limit $\delta \to 0$ is taken in the complex z-plane. Examples of analytic functions are

$$f(z) = \sin z \qquad ; f'(z) = \cos z$$
$$f(z) = e^z \qquad ; f'(z) = e^z$$
$$f(z) = z^n \qquad ; f'(z) = nz^{n-1} \qquad ; n \text{ non-negative integer (B.3)}$$

The function $f(z) = 1/z$ is not analytic at $z = 0$ since it has no derivative there.

We also require that the analytic function be *single-valued*. The function $f(z) = \sqrt{z}$ changes sign as z circles the origin.[56] To eliminate this possibility, one takes a scissors and cuts the z-plane, say along the negative z-axis from $z = -\infty$ to $z = 0$. The function $\sqrt{z}$ is then analytic in the *cut-plane*.

B.3   Integration

Integration is defined through the multiplication of complex numbers and the addition of small elements. The integral will depend on just what curve C one moves along in the complex plane. If one moves along C in the opposite direction, the integral *changes sign*.

$$I = \int_{C \hookrightarrow} f(z)dz = -\int_{C \hookleftarrow} f(z)dz \qquad (B.4)$$

As an exercise, let us evaluate the integral of the function z^n in the positive direction of increasing ϕ around a circle centered on the origin. Here n is an integer (either sign). Then (see FigI. A.1)

$$z = \rho e^{i\phi}$$
$$dz = i\rho d\phi \, e^{i\phi}$$
$$\oint z^n \, dz = i\rho^{(n+1)} \int_0^{2\pi} d\phi \, e^{i(n+1)\phi} = 2\pi i \, \delta_{n,-1} \qquad ; \text{integer } n \quad (B.5)$$

We obtain the remarkable, and extremely useful, result that the integral *vanishes unless* $n = -1$, in which case it gives $2\pi i$.

---

[56]The origin in this case is said to form a *branch point*.

B.4 Cauchy's Theorem

The key to the theory of functions is Cauchy's theorem, which states that the integral of an analytic function around a closed curve in a simply-connected region *vanishes* (see Fig. B.1)[57]

$$\oint_C f(z)\,dz = 0 \qquad ;\ \text{Cauchy's theorem} \qquad (B.6)$$

We have already seen one application of the theorem for the monomials with $n = 0, 1, 2, \cdots, \infty$ in Eq. (B.5).

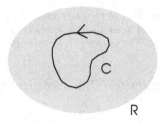

Fig. B.1   Conditions for Cauchy's theorem. $f(z)$ is analytic in the simply-connected region R, and C is a closed curve in R.

B.5 Cauchy's Integral

Suppose $f(\zeta)$ is an analytic function of ζ inside the region R in Fig. B.2, and $f(\zeta)/(\zeta - z)$ is then an analytic function of ζ inside the closed contour C in that figure, which excludes the point z. From Cauchy's theorem

$$\oint_C d\zeta \frac{f(\zeta)}{\zeta - z} = 0 \qquad ;\ \text{Cauchy's theorem} \qquad (B.7)$$

Consider the integral around the small circle C_2 where $\zeta = z + \rho e^{i\phi}$. Since $f(\zeta)$ is analytic at the point z, in a small region around z it can be expressed as

$$f(\zeta) = f(z) + (\zeta - z)f'(z) \qquad ;\ \rho \to 0 \qquad (B.8)$$

---

[57]Appendix A of [Fetter and Walecka (2003)] contains a proof of Cauchy's theorem, as well as of the subsequent results.

where this relation is exact as $\rho \to 0$.   Equations (B.5) imply that the

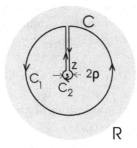

Fig. B.2   Contour for derivation of Cauchy's integral. $f(\zeta)$ is an analytic function of ζ in R, and then $f(\zeta)/(\zeta - z)$ is an analytic function of ζ inside the closed contour C, which excludes the point z. The contour C_1 denotes just the outer circle, and C_2 the inner circle.

integral around the small circle C_2 is then given by

$$\oint_{C_2} d\zeta \, \frac{f(\zeta)}{\zeta - z} = -2\pi i f(z) \tag{B.9}$$

where the minus sign appears because the integral around C_2 in Fig. B.2 runs in the negative direction. The integrals along the straight-line segments in Fig. B.2 cancel by Eq. (B.4) when these segments coincide. Thus one has

$$\oint_{C} d\zeta \frac{f(\zeta)}{\zeta - z} = \oint_{C_1} d\zeta \frac{f(\zeta)}{\zeta - z} - 2\pi i f(z) \tag{B.10}$$

where C_1 is the outer circle in Fig. B.2, and this relation is exact as $\rho \to 0$. Equations (B.7) and (B.10) imply that

$$f(z) = \frac{1}{2\pi i} \oint_{C_1} d\zeta \frac{f(\zeta)}{\zeta - z} \qquad ; \text{ Cauchy's integral} \tag{B.11}$$

This is *Cauchy's integral*. It is a remarkable relation that allows one to explicitly exhibit the z-dependence of $f(z)$ in terms of the values of $f(z)$ on a surrounding contour, which can be very far away, and have an arbitrary shape.[58]

As one application of Cauchy's integral, note that Eq. (B.11) can be differentiated any number of times with respect to z, establishing the fact

---

[58]The contour C_1 can be deformed into one of arbitrary shape through the use of Cauchy's theorem, provided one remains within the region of analyticity (see Prob. B.1).

that all derivatives of an analytic function *are themselves analytic.* Cauchy's integral is the starting point for the proof of many of the results quoted below.

B.6 Taylor's Theorem

Taylor's theorem states that if $f(z)$ is analytic in a circle about z_0, and z is any point inside the circle, then $f(z)$ has the following power-series expansion

$$f(z) = f(z_0) + (z - z_0)f'(z_0) + \cdots + \frac{(z - z_0)^n}{n!} \left[\frac{d^n f(z)}{dz^n} \right]_{z_0} + \cdots \quad (B.12)$$

This Taylor series *converges uniformly in a circle that extends out to the first singularity of $f(z)$.*[59] The series can be integrated and differentiated term-by-term within its circle of convergence.

B.7 Laurent Series

If a function $f(z)$ is analytic except for an isolated singularity at z_0, then the following *Laurent series* expansion holds in the vicinity of z_0

$$f(z) = \sum_{n=0}^{\infty} a_n (z - z_0)^n + \frac{b_1}{z - z_0} + \frac{b_2}{(z - z_0)^2} + \cdots \quad (B.13)$$

- If the series of inverse powers terminates at some finite N with a term $b_N/(z - z_0)^N$, then $f(z)$ is said to have a *pole of order N* at z_0;
- If the series of inverse powers does not terminate, then $f(z)$ has an *essential singularity* at z_0.[60]

B.8 Theory of Residues

Consider the integral around a closed contour of a function $f(z)$ that is analytic in a region R *except for a finite number of poles located at* $z = z_1, z_2, \cdots, z_p$. Let the contour C be that illustrated in Fig. B.3, which excludes the poles, and for which the function $f(z)$ is *analytic inside C.*

---

[59]For example, the Taylor series $1/(1+z^2) = 1-z^2+z^4+\cdots$ has a radius of convergence $R_c = 1$ since $1/(1 + z^2)$ has singularities at $z = \pm i$.

[60]Functions can manifest bizarre behavior in the neighborhood of an essential singularity.

Let $C_1, C_2, \cdots, C_p$ be small circles of radius ρ centered on each point.

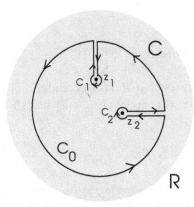

Fig. B.3   Contour for derivation of theory of residues. The function $f(z)$ is analytic inside the closed contour C that excludes the poles at $z = z_1, z_2, \cdots, z_p$ (here illustrated for $p = 2$). The curves $C_1, C_2, \cdots, C_p$ are small circles of radius ρ centered on the position of each pole. The curve C_0 is the outer circle.

Now make a Laurent series expansion in the vicinity of each pole and use the results in Eqs. (B.5) to explicitly perform the integral around each small circle in the contour C. Only the term $b_1/(z - z_0)$ will give the non-zero result $-2\pi i b_1$, the sign again coming from the direction. The contributions along the straight sections of C in Fig. B.3 again cancel in pairs.

Hence

$$\oint_C dz\, f(z) = 0 \qquad\qquad ; \text{ Cauchy's theorem}$$

$$= \oint_{C_0} dz\, f(z) - 2\pi i \sum_{n=1}^{p} b_{1n} \qquad\qquad (B.14)$$

Here the curve C_0 is the outer circle in Fig. B.3, and the quantity b_{1n}, the so-called *residue of $f(z)$ at the pole at z_n*, is the coefficient b_1 in the Laurent expansion of $f(z)$ about that pole. This expression is again exact as $\rho \to 0$. But now we have the lovely, powerful[61] result that

$$\oint_{C_0} dz\, f(z) = 2\pi i \left(\sum_{n=1}^{p} b_{1n} \right) \qquad\qquad ; \text{ theory of residues} \qquad (B.15)$$

---

[61] And quite unexpected!

To evaluate the integral around a closed contour C_0 of a function $f(z)$ that is analytic inside C_0 except for isolated poles at $z_1, z_2, \cdots, z_p$, one makes a Laurent series expansion of the function about each pole and identifies the residue b_1 at that pole. The integral is then simply given by $2\pi i \times$ (sum of the residues).

The contour C_0 can again be deformed in any fashion through a region of analyticity using Cauchy's theorem.

B.9 Zeros of an Analytic Function

The zeros of an analytic function are *isolated points.* It is relatively easy to establish the converse. *If a function is analytic in a circle R and vanishes on a continuous curve C passing through the origin, then it must vanish everywhere in R* (see Fig. B.4). This is shown as follows: The analytic

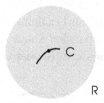

Fig. B.4   Function analytic in a circle R and vanishing along a continuous curve C through the origin; it must then vanish everywhere in R.

function can be expanded about the origin in a Taylor series. The coefficients in this expansion are the derivatives of the function evaluated at the origin. Since the function and its derivatives are analytic, all the derivatives can be evaluated *in any direction.* Evaluate them along C, where they all vanish. Hence the function is identically zero within the radius of convergence of the Taylor series. This result is extended to regions R of arbitrary shape using the approach described in the next section.

B.10 Analytic Continuation

One of the most unusual, and useful, characteristics of analytic functions is that of *analytic continuation.* As a familiar example, consider the geometric series

$$f(z) = 1 + z + z^2 + z^3 + \cdots \qquad ; \ |z| < 1 \qquad (B.16)$$

The radius of convergence of this series is $R_c = 1$. Introduce the function

$$f(z) = \frac{1}{1 - z} \qquad ; \text{ analytic continuation} \qquad (B.17)$$

This function agrees with the series in Eq. (B.16) within the circle of convergence, and provides the analytic continuation of $f(z)$ to the entire z-plane, where it has a simple pole at $z = 1$.

B.10.1   *Standard Method*

The standard method of analytic continuation is to pick a point in a region in which $f(z)$ is analytic and construct a Taylor series centered on that point. This series converges in a circle that extends out to the first singularity of $f(z)$. Now move to another point within that circle of convergence and make a Taylor series expansion about that new point. The new series has its own radius of convergence. Repeat the process (see Fig. B.5).   This

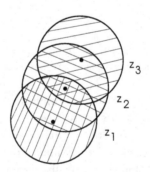

Fig. B.5   Standard method of analytic continuation. The function $f(z)$ is first expanded in a Taylor series about the point z_1, then in a Taylor series about a point z_2 that lies within the circle of convergence of the first series, and then this process is repeated.

new series converges out to the closest singularity in z. In this fashion, one covers the complex plane.[62]

---

[62]There are other procedures for analytic continuation (see appendix G).

B.10.2 *Uniqueness*

The key feature of analytic continuation is the *uniqueness* of the extended function. Consider three regions Γ_1, Γ_2, and Γ_3 with a part R in common (see Fig. B.6). Suppose that $[f_1(z), f_2(z), f_3(z)]$ are analytic in the regions

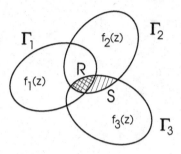

Fig. B.6 Uniqueness of analytic continuation. $f_2(z)$ provides an analytic continuation of $f_1(z)$ into Γ_2 and $f_3(z)$ provides an analytic continuation of $f_1(z)$ into Γ_3, where there is a common region R. Then $f_2(z) = f_3(z)$ in any further simply-connected common region S.

$(\Gamma_1, \Gamma_2, \Gamma_3)$ respectively , and that $f_2(z)$ provides an analytic continuation of $f_1(z)$ into Γ_2 and $f_3(z)$ provides an analytic continuation of $f_1(z)$ into Γ_3. Then $f_2(z) = f_3(z)$ in any further simply-connected common region S.

The proof of uniqueness follows immediately from the above results on the zeros of an analytic function. The function $f_2(z) - f_3(z)$ is analytic in the region $R \oplus S$. Furthermore $f_2(z) - f_3(z) = 0$ in the region R, since by assumption both are equal to $f_1(z)$ in that region. Thus by the previous arguments, the function $f_2(z) - f_3(z)$ is identically zero throughout $R \oplus S$.

Appendix C

Electromagnetic Field

The covariant formulation of Maxwell's equations was presented in Probls. 8.8–8.9. The electromagnetic field tensor $F_{\mu\nu}$ is defined by[63]

$$\underline{F} \equiv \begin{bmatrix} 0 & cB_3 & -cB_2 & -iE_1 \\ -cB_3 & 0 & cB_1 & -iE_2 \\ cB_2 & -cB_1 & 0 & -iE_3 \\ iE_1 & iE_2 & iE_3 & 0 \end{bmatrix} \qquad ;\text{ field tensor}$$

$$F_{\mu\nu} = [\underline{F}]_{\mu\nu} \qquad\qquad ;\ (\mu,\nu) = (1,2,3,4) \quad \text{(C.1)}$$

Maxwell's equations then read

$$\frac{\partial}{\partial x_\nu} F_{\mu\nu} = \frac{e}{\varepsilon_0} j_\mu \qquad ;\text{ Maxwell's equations}$$

$$\varepsilon_{\mu\nu\rho\sigma} \frac{\partial}{\partial x_\sigma} F_{\nu\rho} = 0 \qquad\qquad \mu = 1,2,3,4 \qquad \text{(C.2)}$$

The second set of equations is satisfied identically with the introduction of the vector potential

$$\frac{1}{c} F_{\mu\nu} \equiv \frac{\partial A_\nu}{\partial x_\mu} - \frac{\partial A_\mu}{\partial x_\nu} \qquad ;\ A_\mu = \left(\mathbf{A}, \frac{i}{c}\Phi\right) \qquad \text{(C.3)}$$

The fields $(\mathbf{E}, \mathbf{B})$ are related to the elements of the vector potential according to

$$\mathbf{B} = \boldsymbol{\nabla} \times \mathbf{A}$$

$$\mathbf{E} = -\boldsymbol{\nabla}\Phi - \frac{\partial \mathbf{A}}{\partial t} \qquad \text{(C.4)}$$

---

[63] We continue to work in the SI units of Vol. I; the relation to other units is presented in appendix K of Vol. I. Repeated Greek indices are again summed from 1 to 4, and repeated Latin indices from 1 to 3.

The current, studied in Probl. 8.4, satisfies the continuity equation

$$\frac{\partial j_\mu}{\partial x_\mu} = \nabla \cdot \mathbf{j} + \frac{\partial \rho}{\partial t} = 0 \qquad ; \text{ continuity equation} \quad (C.5)$$

where the four-vector current is defined by

$$j_\mu \equiv \left(\frac{1}{c}\mathbf{j}, \, i\rho\right) \tag{C.6}$$

As discussed in Probl. 8.9, Maxwell's equations are invariant under a gauge transformation

$$A_\mu \to A_\mu + \frac{\partial \Lambda}{\partial x_\mu} \qquad ; \text{ gauge transformation} \qquad (C.7)$$

C.1 Lagrangian Field Theory

We seek a lagrangian density that gives Maxwell's equations through Hamilton's principle. We take the components of the vector potential A_λ with $\lambda = 1, 2, 3, 4$ as the generalized coordinates. The second set of Maxwell's Eqs. (C.2) is then automatically satisfied. Try

$$\mathcal{L}\left(\frac{\partial A_\lambda}{\partial x_\mu}, A_\lambda\right) = -\frac{\varepsilon_0}{4} F_{\mu\nu} F_{\mu\nu} + ec\, j_\mu A_\mu \tag{C.8}$$

$$= -\frac{\varepsilon_0 c^2}{2}\left[\left(\frac{\partial A_\nu}{\partial x_\mu}\right)\left(\frac{\partial A_\nu}{\partial x_\mu}\right) - \left(\frac{\partial A_\nu}{\partial x_\mu}\right)\left(\frac{\partial A_\mu}{\partial x_\nu}\right)\right] + ec\, j_\mu A_\mu$$

This expression has the following features to recommend it:

- It leads to a Lorentz-invariant action;
- The action is also gauge-invariant;[64]
- Lagrange's equations reproduce the first set of Maxwell's equations.

We proceed to demonstrate this last point. Lagrange's equations follow from Hamilton's principle as

$$\frac{\partial}{\partial x_\mu}\frac{\partial \mathcal{L}}{\partial(\partial A_\lambda/\partial x_\mu)} = \frac{\partial \mathcal{L}}{\partial A_\lambda} \qquad ; \lambda = 1, 2, 3, 4 \qquad (C.9)$$

Compute

$$\frac{\partial \mathcal{L}}{\partial(\partial A_\lambda/\partial x_\mu)} = -\varepsilon_0 c^2\left(\frac{\partial A_\lambda}{\partial x_\mu} - \frac{\partial A_\mu}{\partial x_\lambda}\right) = \varepsilon_0 c\, F_{\lambda\mu} \tag{C.10}$$

---

[64]See Prob. C.1.

Lagrange's equations then read

$$\frac{\partial}{\partial x_\mu} F_{\lambda\mu} = \frac{e}{\varepsilon_0} j_\lambda \qquad ; \lambda = 1,2,3,4 \qquad \text{(C.11)}$$

and the result is established.

C.2   Stress Tensor

The stress tensor is given by the arguments in the text as

$$T_{\mu\nu} = \mathcal{L}\,\delta_{\mu\nu} - \frac{\partial \mathcal{L}}{\partial(\partial A_\lambda/\partial x_\mu)}\frac{\partial A_\lambda}{\partial x_\nu}$$

$$\frac{1}{\varepsilon_0}T_{\mu\nu} = \left[-\frac{1}{4}F_{\lambda\rho}F_{\lambda\rho} + \frac{ec}{\varepsilon_0}j_\lambda A_\lambda \right]\delta_{\mu\nu} - cF_{\lambda\mu}\frac{\partial A_\lambda}{\partial x_\nu} \qquad \text{(C.12)}$$

The four-momentum is obtained from the stress tensor as before

$$P_\nu = \left(\mathbf{P}, \frac{i}{c}H \right) = \frac{1}{ic}\int_\Omega d^3x\, T_{4\nu} \qquad \text{(C.13)}$$

The hamiltonian and momentum are therefore

$$H = -\int_\Omega d^3x\, T_{44}$$

$$P_k = \frac{1}{ic}\int_\Omega d^3x\, T_{4k} \qquad ; k = 1,2,3 \qquad \text{(C.14)}$$

C.3   Free Fields

Consider first free fields where $j_\mu = 0$. In this case

$$\frac{1}{\varepsilon_0}T_{44} = -\frac{1}{4}F_{\mu\nu}F_{\mu\nu} - cF_{\lambda 4}\frac{\partial A_\lambda}{\partial x_4} \qquad \text{(C.15)}$$

Now

$$c\int_\Omega d^3x\, F_{\lambda 4}\frac{\partial A_\lambda}{\partial x_4} = \frac{1}{i}\int_\Omega d^3x\, F_{k4}\frac{\partial A_k}{\partial t}$$

$$= \frac{1}{i}\int_\Omega d^3x\, F_{k4}\left(-E_k - \frac{\partial \Phi}{\partial x_k} \right)$$

$$= \frac{1}{i}\int_\Omega d^3x\left[-F_{k4}E_k + \Phi\frac{\partial F_{k4}}{\partial x_k} \right] \qquad \text{(C.16)}$$

Here a partial integration has been performed on the final term, which now vanishes by the free-field equation. With the identification of $F_{k4} = -iE_k$ from Eq. (C.1), this term reduces to

$$c \int_\Omega d^3x \, F_{\lambda 4} \frac{\partial A_\lambda}{\partial x_4} = \int_\Omega d^3x \, \mathbf{E}^2 \tag{C.17}$$

It follows from Eq. (C.1) that

$$\frac{1}{4} F_{\mu\nu} F_{\mu\nu} = \frac{1}{2} \left[(c\mathbf{B})^2 - \mathbf{E}^2 \right] \tag{C.18}$$

Hence

$$-T_{44} = \frac{\varepsilon_0}{2} \left[\mathbf{E}^2 + (c\mathbf{B})^2 \right]$$

$$H = \frac{\varepsilon_0}{2} \int_\Omega d^3x \left[\mathbf{E}^2 + (c\mathbf{B})^2 \right] \tag{C.19}$$

This is the result quoted in Vol. I.

For the momentum, one has

$$P_k = \frac{1}{ic} \int_\Omega d^3x \, T_{4k}$$

$$\frac{1}{\varepsilon_0} T_{4k} = -cF_{\lambda 4} \frac{\partial A_\lambda}{\partial x_k} \tag{C.20}$$

As above, a partial integration shows that the following integral vanishes for the free field

$$\int_\Omega d^3x \, F_{\lambda 4} \frac{\partial A_k}{\partial x_\lambda} = -\int_\Omega d^3x \, A_k \frac{\partial F_{j4}}{\partial x_j} = 0 \tag{C.21}$$

This can be used to rewrite Eq. (C.20) as

$$P_k = -\frac{\varepsilon_0}{i} \int_\Omega d^3x \left(\frac{\partial A_j}{\partial x_k} - \frac{\partial A_k}{\partial x_j} \right) F_{j4}$$

$$= -\frac{\varepsilon_0}{i} \int_\Omega d^3x \left[\varepsilon_{kjl} (\mathbf{\nabla} \times \mathbf{A})_l \right] (-iE_j) \tag{C.22}$$

Hence, in vector notation,

$$\mathbf{P} = \varepsilon_0 \int_\Omega d^3x \, \mathbf{E} \times \mathbf{B} \tag{C.23}$$

This reproduces the Poynting vector of Vol. I.[65]

---

[65] See Prob. C.2.

C.4 Quantization

One has the freedom of choosing a *gauge* for the electromagnetic potentials, and here we work in the *Coulomb gauge*. This gauge has the great advantage that there is a one-to-one correspondence of the resulting quanta with physical photons. For free fields, the Coulomb gauge is defined by

$$\mathbf{\nabla} \cdot \mathbf{A} = 0 \quad ; \Phi = 0 \quad ; \text{Coulomb gauge} \quad (C.24)$$

The electric and magnetic fields are then given by

$$\mathbf{E} = -\frac{\partial \mathbf{A}}{\partial t} \quad ; \mathbf{B} = \mathbf{\nabla} \times \mathbf{A} \quad (C.25)$$

With periodic boundary conditions, the normal modes are given by

$$q_{\mathbf{k}}(\mathbf{x}, t) = \frac{1}{\sqrt{\Omega}} e^{i(\mathbf{k} \cdot \mathbf{x} - \omega_k t)} \quad ; \mathbf{k} = \frac{2\pi}{L}(n_x, n_y, n_z)$$

$$\omega_k = |\mathbf{k}|c \quad ; n_i = 0, \pm 1, \pm 2, \cdots \quad ; i = x, y, z \quad (C.26)$$

Now introduce a set of orthogonal, transverse unit vectors $\mathbf{e}_{\mathbf{k}s}$ for each $\mathbf{k}$ as shown in Fig. C.1. They satisfy

$$\mathbf{e}_{\mathbf{k}s} \cdot \mathbf{k} = 0 \quad ; s = 1, 2$$

$$\mathbf{e}_{\mathbf{k}s} \cdot \mathbf{e}_{\mathbf{k}s'} = \delta_{ss'} \quad (C.27)$$

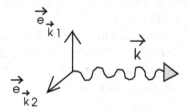

Fig. C.1   Orthogonal, transverse unit vectors $\mathbf{e}_{\mathbf{k}s}$ with $s = 1, 2$ for each $\mathbf{k}$. We shall later employ the third orthonormal unit vector $\mathbf{e}_{\mathbf{k}3} = \mathbf{k}/|\mathbf{k}|$.

The vector potential can now be expanded in normal modes as follows

$$\mathbf{A}(\mathbf{x}, t) = \sum_{\mathbf{k}} \sum_{s=1}^{2} \left(\frac{\hbar}{2\omega_k \varepsilon_0 \Omega}\right)^{1/2} \left[a_{\mathbf{k}s} \mathbf{e}_{\mathbf{k}s} e^{i(\mathbf{k} \cdot \mathbf{x} - \omega_k t)} + a_{\mathbf{k}s}^* \mathbf{e}_{\mathbf{k}s} e^{-i(\mathbf{k} \cdot \mathbf{x} - \omega_k t)}\right]$$

$$(C.28)$$

This expansion has the following features to recommend it:

- This expression is *real*, giving rise to real $(\mathbf{E}, \mathbf{B})$;
- Since only the transverse polarization vectors are used in the expansion, one has ensured that

$$\boldsymbol{\nabla} \cdot \mathbf{A} = 0 \qquad\qquad (\text{C.29})$$

- $\mathbf{A}(\mathbf{x}, t)$ satisfies the wave equation, and, since the order of partial derivatives can always be interchanged, so do the fields $(\mathbf{E}, \mathbf{B})$

$$\Box\, \mathbf{A}(\mathbf{x}, t) = 0 \qquad\qquad (\text{C.30})$$

$$\Rightarrow \qquad \Box\, \mathbf{E}(\mathbf{x}, t) = \Box\, \mathbf{B}(\mathbf{x}, t) = 0 \qquad\qquad (\text{C.31})$$

- The periodic boundary conditions are obeyed;
- There is enough freedom to match the initial conditions.

The normal-mode expansion in Eq. (C.28) can now be substituted in the expression for the hamiltonian in Eq. (C.19), making use of the definition of the fields in Eqs. (C.25). The result is (see Probl. 12.2)

$$H = \sum_{\mathbf{k}} \sum_{s=1}^{2} \hbar\omega_k \frac{1}{2} \left(a_{\mathbf{k}s}^{\star} a_{\mathbf{k}s} + a_{\mathbf{k}s} a_{\mathbf{k}s}^{\star} \right) \qquad ;\ \text{normal modes} \quad (\text{C.32})$$

The problem has again been reduced to normal modes. One again has an infinite, uncoupled set of simple harmonic oscillators, one for each value of the wavenumber $\mathbf{k}$ and polarization s.

The *quantization* of the electromagnetic field now proceeds exactly as in all the previous examples. We identify the normal-mode amplitudes $(a_{\mathbf{k}s}^{\star}, a_{\mathbf{k}s})$ with the creation and destruction operators $(a_{\mathbf{k}s}^{\dagger}, a_{\mathbf{k}s})$, and impose the usual commutation relations

$$[a_{\mathbf{k}s}, a_{\mathbf{k}'s'}^{\dagger}] = \delta_{\mathbf{k}\mathbf{k}'} \delta_{ss'}$$
$$[a_{\mathbf{k}s}, a_{\mathbf{k}'s'}] = [a_{\mathbf{k}s}^{\dagger}, a_{\mathbf{k}'s'}^{\dagger}] = 0 \qquad\qquad (\text{C.33})$$

The hamiltonian, which is just the total energy in the free E-M field, then takes the form

$$\hat{H} = \sum_{\mathbf{k}} \sum_{s=1}^{2} \hbar\omega_k \left(\hat{N}_{\mathbf{k}s} + \frac{1}{2} \right) \qquad ;\ \text{photons} \quad (\text{C.34})$$

The quanta are the familiar *photons*.

After quantization, the vector potential in Eq. (C.28) becomes a local quantum field operator that creates and destroys photons, and in the interaction picture it takes the form

$$\hat{\mathbf{A}}(\mathbf{x}, t) = \sum_{\mathbf{k}} \sum_{s=1}^{2} \left(\frac{\hbar}{2\omega_k \varepsilon_0 \Omega} \right)^{1/2} \left[a_{\mathbf{k}s} \mathbf{e}_{\mathbf{k}s} e^{i(\mathbf{k}\cdot\mathbf{x} - \omega_k t)} + a_{\mathbf{k}s}^{\dagger} \mathbf{e}_{\mathbf{k}s} e^{-i(\mathbf{k}\cdot\mathbf{x} - \omega_k t)} \right]$$

(C.35)

A parallel calculation expresses the momentum in the field as[66]

$$\hat{\mathbf{P}} = \sum_{\mathbf{k}} \sum_{s=1}^{2} \hbar \mathbf{k} \, \hat{N}_{\mathbf{k}s}$$

(C.36)

C.5 Commutation Relations

For the free field, the momentum density conjugate to A_j is obtained from Eq. (C.10) as

$$\pi_j = \frac{\partial \mathcal{L}}{\partial(\partial A_j/\partial t)} = -i\varepsilon_0 F_{j4} = \varepsilon_0 \frac{\partial A_j}{\partial t} \quad ; \text{ momentum density (C.37)}$$

The corresponding field operator in the interaction picture is then obtained from Eq. (C.35) as

$$\hat{\boldsymbol{\pi}}(\mathbf{x}, t) = \frac{1}{i} \sum_{\mathbf{k}} \sum_{s=1}^{2} \left(\frac{\hbar \omega_k \varepsilon_0}{2\Omega} \right)^{1/2} \left[a_{\mathbf{k}s} \mathbf{e}_{\mathbf{k}s} e^{i(\mathbf{k}\cdot\mathbf{x} - \omega_k t)} - a_{\mathbf{k}s}^{\dagger} \mathbf{e}_{\mathbf{k}s} e^{-i(\mathbf{k}\cdot\mathbf{x} - \omega_k t)} \right]$$

(C.38)

Consider now the equal-time commutator of $\hat{\mathbf{A}}$ and $\hat{\boldsymbol{\pi}}$. The calculation proceeds as in all of the previous cases, up to the point where the sum over polarization vectors is required. Thus

$$[\hat{A}_i(\mathbf{x}, t), \hat{\pi}_j(\mathbf{x}', t')]_{t=t'} = \frac{i\hbar}{2\Omega} \sum_{\mathbf{k}} \sum_{s=1}^{2} (\mathbf{e}_{\mathbf{k}s})_i (\mathbf{e}_{\mathbf{k}s})_j \left[e^{i\mathbf{k}\cdot(\mathbf{x}-\mathbf{x}')} + e^{-i\mathbf{k}\cdot(\mathbf{x}-\mathbf{x}')} \right]$$

(C.39)

Suppose one were summing over a complete orthonormal set of basis vectors in Fig. C.1, including a third unit vector along $\mathbf{k}$. Then one would simply

---

[66]See Prob. C.3.

obtain the unit dyadic

$$\sum_{s=1}^{3}(\mathbf{e_{ks}})_i(\mathbf{e_{ks}})_j = [\underline{\mathbf{I}}]_{ij} = \delta_{ij} \tag{C.40}$$

The unit dyadic is defined so that when dotted into a vector $\mathbf{v}$ from either side, it gives the vector $\mathbf{v}$ back again

$$\underline{\mathbf{I}} \cdot \mathbf{v} = \sum_{s=1}^{3} \mathbf{e_{ks}}(\mathbf{e_{ks}} \cdot \mathbf{v}) = \mathbf{v} \tag{C.41}$$

When only summing over the transverse unit vectors in Eq. (C.40), the result is obtained by simply taking the contribution from the third unit vector $\mathbf{e_{k3}} = \mathbf{k}/|\mathbf{k}|$ to the r.h.s. Thus

$$\sum_{s=1}^{2} \mathbf{e_{ks}}\,\mathbf{e_{ks}} = \underline{\mathbf{I}} - \mathbf{e_{k3}}\,\mathbf{e_{k3}}$$

$$\text{or;} \quad \sum_{s=1}^{2}(\mathbf{e_{ks}})_i(\mathbf{e_{ks}})_j = \delta_{ij} - \frac{k_i k_j}{\mathbf{k}^2} \tag{C.42}$$

Since this expression is even in $\mathbf{k}$, the two terms in Eq. (C.39) make identical contributions, and the commutator reduces to

$$[\hat{A}_i(\mathbf{x},t),\,\hat{\pi}_j(\mathbf{x}',t')]_{t=t'} = i\hbar\,\delta_{ij}^{\mathrm{T}}(\mathbf{x}-\mathbf{x}')$$

$$\delta_{ij}^{\mathrm{T}}(\mathbf{x}-\mathbf{x}') \equiv \frac{1}{\Omega}\sum_{\mathbf{k}} e^{i\mathbf{k}\cdot(\mathbf{x}-\mathbf{x}')}\left[\delta_{ij} - \frac{k_i k_j}{\mathbf{k}^2}\right] \tag{C.43}$$

In the Coulomb gauge, the canonical commutation relations are modified, and it is the *transverse* delta-function that appears on the r.h.s. This must be true, since $\boldsymbol{\nabla} \cdot \mathbf{A} = 0$ in the Coulomb gauge, and the canonical commutation relations must reflect this.[67]

Let us examine the properties of $\delta_{ij}^{\mathrm{T}}(\mathbf{x}-\mathbf{x}')$:

(1) It is symmetric in (ij), with $\delta_{ij}^{\mathrm{T}} = \delta_{ji}^{\mathrm{T}}$;
(2) It is divergenceless

$$\frac{\partial}{\partial x_i}\delta_{ij}^{\mathrm{T}}(\mathbf{x}-\mathbf{x}') = \frac{\partial}{\partial x_j}\delta_{ij}^{\mathrm{T}}(\mathbf{x}-\mathbf{x}') = 0 \tag{C.44}$$

---

[67]If we had been wise enough, we would have recognized the necessity for this modification of the canonical commutation relations from the outset.

(3) It projects from an arbitrary vector field $\mathbf{v}(\mathbf{x})$ that part which is divergenceless

$$\int_\Omega d^3x' \, \underline{\delta}^{\mathrm{T}}(\mathbf{x} - \mathbf{x}') \cdot \mathbf{v}(\mathbf{x}') \equiv \mathbf{v}^{\mathrm{T}}(\mathbf{x})$$

$$\boldsymbol{\nabla} \cdot \mathbf{v}^{\mathrm{T}}(\mathbf{x}) = 0 \qquad (\mathrm{C}.45)$$

To verify this last point, use a Fourier series and the completeness of the basis vectors for each $\mathbf{k}$ to write the general expansion[68]

$$\mathbf{v}(\mathbf{x}) = \frac{1}{\sqrt{\Omega}} \sum_\mathbf{k} \sum_{s=1}^{3} v_{\mathbf{k}s} \mathbf{e}_{\mathbf{k}s} \, e^{i\mathbf{k}\cdot\mathbf{x}} \qquad (\mathrm{C}.46)$$

Now compute

$$\int_\Omega d^3x' \, \underline{\delta}^{\mathrm{T}}(\mathbf{x} - \mathbf{x}') \cdot \mathbf{v}(\mathbf{x}') = \frac{1}{\sqrt{\Omega}} \sum_\mathbf{k} \sum_{s=1}^{3} v_{\mathbf{k}s} \, e^{i\mathbf{k}\cdot\mathbf{x}} \left[\underline{\mathbf{I}} - \mathbf{e}_{\mathbf{k}3} \, \mathbf{e}_{\mathbf{k}3} \right] \cdot \mathbf{e}_{\mathbf{k}s} \quad (\mathrm{C}.47)$$

Observe that

$$\left[\underline{\mathbf{I}} - \mathbf{e}_{\mathbf{k}3} \, \mathbf{e}_{\mathbf{k}3} \right] \cdot \mathbf{e}_{\mathbf{k}s} = \mathbf{e}_{\mathbf{k}s} \qquad ; \, s = 1, 2$$

$$= 0 \qquad ; \, s = 3 \qquad (\mathrm{C}.48)$$

Hence

$$\int_\Omega d^3x' \, \underline{\delta}^{\mathrm{T}}(\mathbf{x} - \mathbf{x}') \cdot \mathbf{v}(\mathbf{x}') = \frac{1}{\sqrt{\Omega}} \sum_\mathbf{k} \sum_{s=1}^{2} v_{\mathbf{k}s} \mathbf{e}_{\mathbf{k}s} \, e^{i\mathbf{k}\cdot\mathbf{x}} \equiv \mathbf{v}^{\mathrm{T}}(\mathbf{x}) \quad (\mathrm{C}.49)$$

While the sum on s in Eq. (C.46) runs from 1 to 3, because of the projector in $\underline{\delta}^{\mathrm{T}}$, the sum on s in Eq. (C.49) now only runs from 1 to 2, and this expression clearly satisfies the second of Eqs. (C.45).

The continuum limit $\Omega \to \infty$ gives

$$\delta_{ij}^{\mathrm{T}}(\mathbf{x} - \mathbf{x}') = \int \frac{d^3k}{(2\pi)^3} e^{i\mathbf{k}\cdot(\mathbf{x}-\mathbf{x}')} \left[\delta_{ij} - \frac{k_i k_j}{\mathbf{k}^2} \right] \qquad ; \, \Omega \to \infty \quad (\mathrm{C}.50)$$

C.6   Interaction With External Current

Consider the interaction of the electromagnetic field with an "external" current $j_\mu^{\mathrm{ext}}(\mathbf{x}, t)$ that is specified throughout the quantization volume and

---

[68]See Prob. C.5.

is not (yet) part of the dynamics. The lagrangian density that yields Maxwell's equations for this case has already been given in Eq. (C.8)

$$\mathcal{L}\left(\frac{\partial A_\lambda}{\partial x_\mu}, A_\lambda\right) = -\frac{\varepsilon_0}{4}F_{\mu\nu}F_{\mu\nu} + ec\,j_\mu^{\text{ext}}A_\mu$$

$$\equiv \mathcal{L}_0 + \mathcal{L}_1 \tag{C.51}$$

We still choose to work in the Coulomb gauge of Eq. (C.29); however, in contrast to the free-field case, the vector potential $A_\mu = (\mathbf{A}, i\Phi/c)$ will now have a non-zero fourth component, and that must be taken into account. To determine Φ, one can appeal to the first of Maxwell's Eqs. (C.2) whose fourth component gives

$$\frac{\partial F_{4j}}{\partial x_j} = \frac{1}{i}\frac{\partial}{\partial x_j}\frac{\partial A_j}{\partial t} - \frac{\partial}{\partial x_j}\frac{\partial}{\partial x_j}i\Phi$$

$$= \frac{ie}{\varepsilon_0}\rho^{\text{ext}}(\mathbf{x}, t) \tag{C.52}$$

Since partial derivatives can always be interchanged, the first term on the r.h.s. of the first line vanishes by Eq. (C.29), and the combination of derivatives in the second term is just the laplacian. Thus, this equation reads

$$\nabla^2\Phi(\mathbf{x}, t) = -\frac{e}{\varepsilon_0}\rho^{\text{ext}}(\mathbf{x}, t) \tag{C.53}$$

At a given instant in time, this is just the *basic equation of electrostatics*, and it can be solved using the Coulomb Green's function of freshman physics

$$\Phi(\mathbf{x}, t) = \frac{e}{4\pi\varepsilon_0}\int_\Omega \frac{d^3x'}{|\mathbf{x} - \mathbf{x}'|}\rho^{\text{ext}}(\mathbf{x}', t) \qquad ; \text{ given } t \tag{C.54}$$

Hence,

> *In the Coulomb gauge, Φ is not a dynamical field variable, but is given at any instant in time by an integral over the existing charge distribution $\rho^{\text{ext}}(\mathbf{x}, t)$.*

C.6.1 *Hamiltonian*

The stress tensor is given by Eq. (C.12)

$$T_{\mu\nu} = \mathcal{L}\,\delta_{\mu\nu} - \frac{\partial\mathcal{L}}{\partial(\partial A_\lambda/\partial x_\mu)}\frac{\partial A_\lambda}{\partial x_\nu}$$

$$\frac{1}{\varepsilon_0}T_{\mu\nu} = \left[-\frac{1}{4}F_{\lambda\rho}F_{\lambda\rho} + \frac{ec}{\varepsilon_0}j_\lambda^{\text{ext}}A_\lambda\right]\delta_{\mu\nu} - cF_{\lambda\mu}\frac{\partial A_\lambda}{\partial x_\nu} \tag{C.55}$$

The hamiltonian is then obtained from

$$\frac{1}{\varepsilon_0} T_{44} = -\frac{1}{4} F_{\mu\nu} F_{\mu\nu} + \frac{ec}{\varepsilon_0} j_\mu^{\text{ext}} A_\mu - cF_{\lambda 4} \frac{\partial A_\lambda}{\partial x_4} \qquad (C.56)$$

In previously performing an integral over this expression, the free-field equations were invoked, and those steps must now be modified. Equations (C.16)–(C.17) now read as follows[69]

$$c \int_\Omega d^3x \, F_{\lambda 4} \frac{\partial A_\lambda}{\partial x_4} = \frac{1}{i} \int_\Omega d^3x \left[-F_{k4} E_k - \Phi \frac{\partial F_{4k}}{\partial x_k} \right]$$

$$= \int_\Omega d^3x \left(\mathbf{E}^2 - \frac{e}{\varepsilon_0} \rho^{\text{ext}} \, \Phi \right) \qquad (C.57)$$

Hence Eq. (C.19) becomes

$$-T_{44} = \frac{\varepsilon_0}{2} \left[\mathbf{E}^2 + (c\mathbf{B})^2 \right] - e\rho^{\text{ext}} \, \Phi - e\mathbf{j}^{\text{ext}} \cdot \mathbf{A} + e\rho^{\text{ext}} \, \Phi$$

$$H = \frac{\varepsilon_0}{2} \int_\Omega d^3x \left[\mathbf{E}^2 + (c\mathbf{B})^2 \right] - e \int_\Omega d^3x \, \mathbf{j}^{\text{ext}} \cdot \mathbf{A} \qquad (C.58)$$

This is not yet in the form $H_0 + H_1$ since the electric field is now given by

$$\mathbf{E} = -\boldsymbol{\nabla}\Phi - \frac{\partial \mathbf{A}}{\partial t} \qquad (C.59)$$

However, in performing the integral over the quantization volume Ω, we note the following:

(1) A partial integration, and the use of Eq. (C.53), gives

$$\int_\Omega d^3x \, \boldsymbol{\nabla}\Phi \cdot \boldsymbol{\nabla}\Phi = -\int_\Omega d^3x \, \Phi\nabla^2\Phi = \frac{e}{\varepsilon_0} \int_\Omega d^3x \, \rho^{\text{ext}} \, \Phi \quad (C.60)$$

(2) Another partial integration shows that the integral of the cross term in $\mathbf{E}^2$ vanishes in the Coulomb gauge

$$\int_\Omega d^3x \, \boldsymbol{\nabla}\Phi \cdot \frac{\partial \mathbf{A}}{\partial t} = -\int_\Omega d^3x \, \Phi \frac{\partial}{\partial t} \boldsymbol{\nabla} \cdot \mathbf{A} = 0 \qquad (C.61)$$

---

[69] Note $F_{k4} = -F_{4k}$.

Hence we arrive at the hamiltonian for the electromagnetic field interacting with an external current $j_\mu^{\text{ext}}(\mathbf{x}, t)$ [recall Eq. (C.6)]

$$H = \frac{\varepsilon_0}{2} \int_\Omega d^3x \left[\left(\frac{\partial \mathbf{A}}{\partial t} \right)^2 + c^2 (\boldsymbol{\nabla} \times \mathbf{A})^2 \right] - e \int_\Omega d^3x \, \mathbf{j}^{\text{ext}} \cdot \mathbf{A}$$

$$+ \frac{e^2}{8\pi\varepsilon_0} \int_\Omega d^3x \int_\Omega d^3x' \, \frac{\rho^{\text{ext}}(\mathbf{x}, t)\rho^{\text{ext}}(\mathbf{x}', t)}{|\mathbf{x} - \mathbf{x}'|} \tag{C.62}$$

C.6.2 *Quantization*

Since there are no time derivatives in the interaction term $\mathcal{L}_1$ in Eq. (C.51), the momentum density is again given by the second of Eqs. (C.37)

$$\pi_j = \frac{\partial \mathcal{L}}{\partial(\partial A_j / \partial t)} = -i\varepsilon_0 F_{j4} \tag{C.63}$$

In the presence of the interaction, however, there is now a contribution from all four components of A_μ in Eq. (C.3), so that

$$\pi_j = -i\varepsilon_0 F_{j4} = \varepsilon_0 \left(\frac{\partial A_j}{\partial t} + \frac{\partial \Phi}{\partial x_j} \right) \tag{C.64}$$

Thus, with the interaction,

$$\boldsymbol{\pi} = \varepsilon_0 \left(\frac{\partial \mathbf{A}}{\partial t} + \boldsymbol{\nabla}\Phi \right) \tag{C.65}$$

In the Coulomb gauge, Φ is given by Eq. (C.54). Therefore, when quantized, the operator $\hat{\boldsymbol{\pi}}$ will differ from the operator $\varepsilon_0 \partial\hat{\mathbf{A}}/\partial t$ only by the addition of a c-number ("classical number"). Hence

The canonical equal-time commutation relations in the interacting theory, as written for the operators $(\hat{\mathbf{A}}, \varepsilon_0 \partial\hat{\mathbf{A}}/\partial t)$ in the Coulomb gauge, are exactly the same as in the free-field case.

Thus from Eqs. (C.37) and (C.43)

$$\left[\hat{A}_i(\mathbf{x}, t), \dot{\hat{A}}_j(\mathbf{x}', t') \right]_{t=t'} = \frac{i\hbar}{\varepsilon_0} \delta_{ij}^{\text{T}}(\mathbf{x} - \mathbf{x}') \tag{C.66}$$

where we use the notation $\dot{\hat{\mathbf{A}}}$ for the operator $\partial\hat{\mathbf{A}}/\partial t$, the time derivative of the field at a fixed position. We can then take the previous interaction-picture expansions of these quantities to hold *even in the presence of the additional interactions in* $\mathcal{L}_1$

$$\hat{\mathbf{A}}(\mathbf{x},t) = \sum_{\mathbf{k}}\sum_{s=1}^{2}\left(\frac{\hbar}{2\omega_k\varepsilon_0\Omega}\right)^{1/2}\left[a_{\mathbf{k}s}\mathbf{e}_{\mathbf{k}s}e^{i(\mathbf{k}\cdot\mathbf{x}-\omega_k t)} + a_{\mathbf{k}s}^{\dagger}\mathbf{e}_{\mathbf{k}s}e^{-i(\mathbf{k}\cdot\mathbf{x}-\omega_k t)}\right]$$

$$\frac{\partial\hat{\mathbf{A}}(\mathbf{x},t)}{\partial t} = \frac{1}{i}\sum_{\mathbf{k}}\sum_{s=1}^{2}\left(\frac{\hbar\omega_k}{2\varepsilon_0\Omega}\right)^{1/2}\left[a_{\mathbf{k}s}\mathbf{e}_{\mathbf{k}s}e^{i(\mathbf{k}\cdot\mathbf{x}-\omega_k t)} - a_{\mathbf{k}s}^{\dagger}\mathbf{e}_{\mathbf{k}s}e^{-i(\mathbf{k}\cdot\mathbf{x}-\omega_k t)}\right]$$

$$\text{(C.67)}$$

Thus the quantum field theory hamiltonian in the Coulomb gauge for the electromagnetic field in interaction with an external current is given by

$$\hat{H} = \frac{\varepsilon_0}{2}\int_{\Omega}d^3x\left[\left(\frac{\partial\hat{\mathbf{A}}}{\partial t}\right)^2 + c^2(\boldsymbol{\nabla}\times\hat{\mathbf{A}})^2\right] - e\int_{\Omega}d^3x\,\mathbf{j}^{\text{ext}}(\mathbf{x},t)\cdot\hat{\mathbf{A}}(\mathbf{x},t)$$

$$+\frac{e^2}{8\pi\varepsilon_0}\int_{\Omega}d^3x\int_{\Omega}d^3x'\,\frac{\rho^{\text{ext}}(\mathbf{x},t)\rho^{\text{ext}}(\mathbf{x}',t)}{|\mathbf{x}-\mathbf{x}'|} \qquad\text{(C.68)}$$

where the quantum fields in the interaction picture are given in Eqs. (C.67). Some comments:

- The first term in $\hat{H}$ is now just the previously calculated free-field hamiltonian

$$\hat{H}_0 = \frac{\varepsilon_0}{2}\int_{\Omega}d^3x\left[\left(\frac{\partial\hat{\mathbf{A}}}{\partial t}\right)^2 + c^2(\boldsymbol{\nabla}\times\hat{\mathbf{A}})^2\right]$$

$$= \sum_{\mathbf{k}}\sum_{s=1}^{2}\hbar\omega_k\left(a_{\mathbf{k}s}^{\dagger}a_{\mathbf{k}s} + \frac{1}{2}\right) \qquad\text{(C.69)}$$

- The eigenstates of this operator form a basis in the abstract many-particle Hilbert space

$$|n_1 n_2,\cdots,n_{\infty}\rangle = |n_1\rangle|n_2\rangle\cdots|n_{\infty}\rangle \qquad\text{(C.70)}$$

where we have simply ordered the eigenvalues $k_i = 2\pi p_i/L$ with $p_i = 0,\pm 1,\pm 2,\cdots$, and $s = (1,2)$;

- The last term in Eq. (C.68) is just a time-dependent c-number. It represents the Coulomb self-interaction of the external charge density;
- The complexity of the problem resides in the term in Eq. (C.68) that couples the external current $\mathbf{j}^{\text{ext}}(\mathbf{x},t)$ to the quantized radiation field.

In the interaction picture it is given by

$$\hat{H}_{\mathrm{I}}(t) = -e \int_\Omega d^3x \, \mathbf{j}^{\mathrm{ext}}(\mathbf{x},t) \cdot \hat{\mathbf{A}}(\mathbf{x},t) \tag{C.71}$$

where $\hat{\mathbf{A}}(\mathbf{x},t)$ is given by the first of Eqs. (C.67). This term couples the basis states in Eq. (C.70);[70]

- The dynamics of the interacting quantum field problem is governed by the Schrödinger equation

$$i\hbar \frac{\partial}{\partial t}|\Psi(t)\rangle = \hat{H}|\Psi(t)\rangle \tag{C.72}$$

- The full hamiltonian in the interaction picture is

$$\hat{H} = \hat{H}_0 - e \int_\Omega d^3x \, \mathbf{j}^{\mathrm{ext}}(\mathbf{x},t) \cdot \hat{\mathbf{A}}(\mathbf{x},t)$$
$$+ \frac{e^2}{8\pi\varepsilon_0} \int_\Omega d^3x \int_\Omega d^3x' \, \frac{\rho^{\mathrm{ext}}(\mathbf{x},t)\rho^{\mathrm{ext}}(\mathbf{x}',t)}{|\mathbf{x}-\mathbf{x}'|} \tag{C.73}$$

where $\hat{H}_0$ is given by Eq. (C.69).

- The above defines a model problem in quantum electrodynamics (QED). Since the interaction term is linear in the creation and detruction operators, this problem can actually be solved exactly with a canonical transformation for some simple configurations (see later).

---

[70]See Prob. C.4.

Irreducible Representations of SU(n)

We have had preliminary exposure to this topic: Probl. 7.9 demonstrated
the procedure for projecting the irreducible representations of SU(2) from
the direct product of fundamental representations, by symmetrizing and
antisymmetrizing in the indices; Appendix H in Vol.I, on the wave functions
for many identical particles, provided an introduction to permutations and
the symmetric group. This appendix will be a little different from the rest
of the material in this book, in that we will be content to simply quote some
general results, leaving it to the basic texts such as [Hamermesh (1989)] to
provide the details and proofs. It is sufficiently valuable, however, to have
some of these tools in one's arsenal that we here include a summary.

The special unitary group in n-dimensions SU(n) has many practical
applications in physics:

- Angular momentum is SU(2);
- Strong and weak isospin are also based on SU(2);
- The SU(3) symmetry of the three-dimensional harmonic oscillator pro-
  vides the basis for the Elliot model of deformed nuclei;
- Quantum chromodynamics (QCD) is based on SU(3) color symmetry;
- If the nuclear interactions are independent of spin and isospin, with
  $(p \uparrow, p \downarrow, n \uparrow, n \downarrow)$ all equivalent, one has Wigner's supermultiplet theory
  and SU(4);
- Unified theories of the strong and electroweak interactions have been
  proposed based on SU(5);
- A heavy quark model with equivalent $(u \uparrow, u \downarrow, d \uparrow, d \downarrow, s \uparrow, s \downarrow)$ exhibits
  SU(6) symmetry, *etc.*

There are basically three reasons to have some familiarity with the *irreducible representations* of SU(n):

(1) If the hamiltonian commutes with the generators of the group, then the states forming a basis for an irreducible representation of the group are all *degenerate*;

(2) The procedure for decomposing a direct-product representation into the direct sum of its irreducible components provides a method, in principle, for obtaining the Clebsch-Gordon coefficients for the group;

(3) The matrix elements of an irreducible tensor operator taken between states belonging to irreducible representations of the group satisfy the Wigner-Eckart theorem.

We start from the fundamental representation of SU(n)

$$e^{i\omega^a \hat{G}^a} |(n)i\rangle = r(\omega)_{i'i} |(n)i'\rangle \qquad ; (i, i') = 1, 2, \cdots, n$$

$$\underline{r}(\omega) = \exp\left\{\frac{i}{2}\omega^a \underline{\lambda}^a\right\} \qquad a = 1, 2, \cdots, n^2 - 1 \quad (D.1)$$

Here the repeated index i runs from $(1, \cdots, n)$, the repeated Latin index a runs over $(1, \cdots, n^2 - 1)$, and ω stands for a set of $n^2 - 1$ real parameters. The $n \times n$ traceless, hermitian, matrices $\underline{\lambda}^a$ provide the fundamental matrix representation of the Lie algebra of SU(n)

$$[\frac{1}{2}\underline{\lambda}^a, \frac{1}{2}\underline{\lambda}^b] = if^{abc}\frac{1}{2}\underline{\lambda}^c \qquad ; (a, b, c) = 1, \cdots, n^2 - 1 \quad (D.2)$$

The structure constants f^{abc} are real, and antisymmetric in the indices (abc). The $\hat{G}^a$ are the generators in the abstract Hilbert space of the group SU(n); they obey this same algebra.[71]

Consider a problem with two independent systems with commuting generators, as in the discussion of angular momentum in chapter 3,

$$\hat{G}^a = \hat{G}^{1a} + \hat{G}^{2a}$$

$$[\hat{G}^{1a}, \hat{G}^{2b}] = 0 \qquad (D.3)$$

Work in the direct-product space, which for the fundamental basis reads

$$|i_1 i_2\rangle = |i_1\rangle|i_2\rangle \qquad (D.4)$$

where we now suppress the dimension label (n) in the states. The simplest corresponding coordinate space wave function would then be, for example,

$$\langle \mathbf{x}_1 \mathbf{x}_2 | i_1 i_2 \rangle = \langle \mathbf{x}_1 | i_1 \rangle \langle \mathbf{x}_2 | i_2 \rangle = \psi_{i_1}(\mathbf{x}_1)\psi_{i_2}(\mathbf{x}_2) \qquad (D.5)$$

---

[71] The operators $\hat{G}^a$ can be constructed in terms of the matrices $\frac{1}{2}\underline{\lambda}^a$ as in Eq. (6.79).

Consider the transformation of this direct-product state under the action of the group

$$e^{i\omega^a \hat{G}^a} |i_1 i_2\rangle = r_{i'_1 i_1}(\omega) \, r_{i'_2 i_2}(\omega) \, |i'_1 i'_2\rangle \tag{D.6}$$

The basis for an irreducible representation consists of a set of states where the states are only transformed among themselves by the action of a group element. In Probl 7.9 it was shown how to construct those states in the present case, by symmetrizing and antisymmetrizing in the indices $(i_1 i_2)$.[72] Define a permutation operator $\hat{P}_{12}$ that interchanges the *labels on the states*

$$\hat{P}_{12} |i_1 i_2\rangle \equiv |i_2 i_1\rangle \tag{D.7}$$

Then it is readily established that the following operators project the direct-product basis onto (unnormalized) bases for the irreducible representations of the group

$$\hat{Y}_1 = 1 + \hat{P}_{12} \qquad ; \hat{Y}_2 = 1 - \hat{P}_{12} \tag{D.8}$$

We now summarize the general procedure, with any number of direct-product states, where

$$e^{i\omega^a \hat{G}^a} |i_1 i_2 \cdots i_p\rangle = r_{i'_1 i_1}(\omega) r_{i'_2 i_2}(\omega) \cdots r_{i'_p i_p}(\omega) \, |i'_1 i'_2 \cdots i'_p\rangle \tag{D.9}$$

D.1 Young Tableaux and Young Operators

With p direct-product states, arrange p boxes in rows where each row has no more boxes than the row above it, and there are no more than n boxes in any column. This forms a *Young tableau* (see Fig. D.1). Place the labels $(1, 2, \cdots, p)$ in the Young tableaux so that they increase along all rows and columns. This produces a *standard tableau*.

Denote the operator that symmetrizes along the rows by (see appendix H of Vol. I)

$$\Pi_{\text{rows}} \left(\sum \hat{P} \right) \qquad ; \text{ symmetrizes along rows} \tag{D.10}$$

Denote the operator that antisymmetrizes along columns by (again see appendix H of Vol. I)

$$\Pi_{\text{columns}} \left(\sum \delta_P \hat{P} \right) \qquad ; \text{ antisymmetrizes along columns} \tag{D.11}$$

---

[72]The opposite order of the indices on $\underline{r}$ was immaterial in Probl. 7.9.

Then with every standard tableaux, there is a *Young operator*[73]

$$\hat{Y}_T \equiv \Pi_{\text{columns}} \left(\sum \delta_P \hat{P} \right) \Pi_{\text{rows}} \left(\sum \hat{P} \right) \qquad \text{; Young operator (D.12)}$$

This Young operator projects the direct-product basis in Eq. (D.9) onto a subspace that forms a basis for an irreducible represention of SU(n).

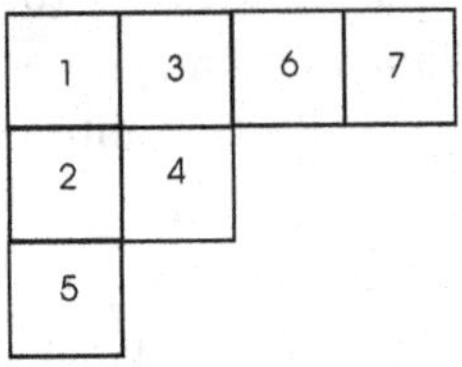

1	3	6	7
2	4		
5			

Fig. D.1   A Young tableau for SU(n) with $p = 7$ and $n \geq 3$. Also shown is a distribution of labels for one standard tableaux.

D.2   Adjoint Representation

If there are n boxes in a column in a Young tableau, then for SU(n) the corresponding indices *do not transform under a group element*. This is shown as follows. The completely antisymmetric state takes the form

$$\hat{A}|i_1 i_2 \cdots i_n\rangle = \epsilon_{i_1 i_2 \cdots i_n} |i_1 i_2 \cdots i_n\rangle \qquad (D.13)$$

where $\epsilon_{i_1 i_2 \cdots i_n}$ is the completely antisymmetric tensor in n-dimensions

$$\epsilon_{i_1 i_2 \cdots i_n} = +1 \qquad ; (i_1 i_2 \cdots i_n) \text{ an even permutation of } (1, 2, \cdots, n)$$
$$= -1 \qquad ; (i_1 i_2 \cdots i_n) \text{ an odd permutation of } (1, 2, \cdots, n)$$
$$= 0 \qquad ; \text{ otherwise} \qquad (D.14)$$

Now transform the state in Eq. (D.13)

$$e^{i\omega^a \hat{G}^a} \hat{A}|i_1 i_2 \cdots i_n\rangle = \epsilon_{i_1 i_2 \cdots i_n} r_{i'_1 i_1} r_{i'_2 i_2} \cdots r_{i'_n i_n} |i'_1 i'_2 \cdots i'_n\rangle \qquad (D.15)$$

Use (see Probl. 8.7)

$$\epsilon_{i_1 i_2 \cdots i_n} r_{i'_1 i_1} r_{i'_2 i_2} \cdots r_{i'_n i_n} = \epsilon_{i'_1 i'_2 \cdots i'_n} \det \underline{r}$$
$$= \epsilon_{i'_1 i'_2 \cdots i'_n} \qquad ; \det \underline{r} = 1 \quad (D.16)$$

---

[73]Note the order.

The last line follows since $\det \underline{r} = 1$ for SU(n). Hence

$$e^{i\omega^a \hat{G}^a} \, \hat{\mathcal{A}} |i_1 i_2 \cdots i_n\rangle = \hat{\mathcal{A}} |i_1 i_2 \cdots i_n\rangle \tag{D.17}$$

as claimed. As a result

> *Any column with n boxes can be crossed out in a Young tableau,*
> *since the corresponding indices do not transform and the corre-*
> *sponding irreducible representation is of lower dimension.*

Suppose there are $(n-1)$ boxes in a column of a Young tableau and one antisymmetrizes in $(n-1)$ indices. A new state characterized by the remaining index can then be introduced as

$$|\bar{i}_l\rangle \equiv \epsilon_{i_1 i_2 \cdots i_{n-1} i_l} \, |i_1 i_2 \cdots i_{n-1}\rangle \tag{D.18}$$

This new state transforms according to

$$e^{i\omega^a \hat{G}^a} \, |\bar{i}_l\rangle = \epsilon_{i_1 i_2 \cdots i_{n-1} i_l} \, r_{i_1' i_1} \, r_{i_2' i_2} \, \cdots \, r_{i_{n-1}' i_{n-1}} \, |i_1' \, i_2' \cdots i_{n-1}'\rangle \tag{D.19}$$

It follows from Eq. (D.16) that

$$\epsilon_{i_1 i_2 \cdots i_{n-1} i_l} \, r_{i_1' i_1} \, r_{i_2' i_2} \, \cdots \, r_{i_{n-1}' i_{n-1}} = [\underline{r}^{-1}]_{i_l i_l'} \, \epsilon_{i_1' i_2' \cdots i_{n-1}' i_l'} \tag{D.20}$$

Hence Eq. (D.19) becomes

$$\begin{aligned} e^{i\omega^a \hat{G}^a} \, |\bar{i}_l\rangle &= [\underline{r}^{-1}]_{i_l i_l'} \, |\bar{i}_l'\rangle \\ &= r_{i_l' i_l}^\star(\omega) \, |\bar{i}_l'\rangle \end{aligned} \tag{D.21}$$

The last equality follows from the unitarity of the matrix $\underline{r}$, that is $\underline{r}^{-1} = \underline{r}^\dagger$. Now the state in Eq. (D.18) corresponds to a *hole* in the fundamental representation for Dirac particles, and we have just shown that a hole state transforms according to the *complex conjugate* of the fundamental representation. For all except SU(2), the complex conjugate of the fundamental SU(n) matrix relations leads to a new, inequivalent, representation of the group. If we recall from Dirac hole theory that a hole in the Dirac sea is an *antiparticle*, then we have the two results:

- A hole in the fundamental representation for Dirac particles transforms according to the complex conjugate representation;
- Dirac antiparticles transform according to the complex conjugate representation of the internal SU(n) symmetry group.

D.3 Dimension of the Representation

As a consequence of the previous argument, one only needs to consider Young tableaux with $n-1$ rows. Label the Young tableau with the number of boxes in each row ($\lambda_1 \geq \lambda_2 \geq \cdots \geq \lambda_{n-1} \geq 0$). This is referred to as a *partition*.[74] Now for a given n, and a given partition, create Table D.1.

Table D.1 Table for finding dimension of the representation.

$[\lambda]$	l	n
λ_1	$\lambda_1 + n - 1$	$n - 1$
λ_2	$\lambda_2 + n - 2$	$n - 2$
λ_3	$\lambda_3 + n - 3$	$n - 3$
$\vdots$	$\vdots$	$\vdots$
λ_{n-1}	$\lambda_{n-1} + 1$	1
0	0	0

Define a product of decreasing differences by

$$D(x_1, x_2, \cdots, x_n) \equiv \Pi_{i<j} \left(x_i - x_j \right) \tag{D.22}$$

The dimension N of the irreducible representation of SU(n) corresponding to a Young tableau with partition $[\lambda]$ is given by

$$^{(n)}N_{[\lambda]} = \frac{D(l_1, l_2, \cdots, l_n)}{D(n-1, n-2, \cdots, 0)} \quad ; \text{ dimension of I. R.} \tag{D.23}$$

where $l_n = 0$. We shall shortly give several examples.

D.4 Outer Product

Suppose one takes the direct product of two spaces where the states in each space form a basis for an irreducible representation of the group. The direct product of the corresponding representations is again a representation of the group, but in general it is a *reducible* representation.[75] The general rules for finding the irreducible content of this direct-product representation can be found in [Hamermesh (1989)]. We give the simplest rule:

---

[74]We denote the partition generically by $[\lambda] \equiv [\lambda_1, \lambda_2, \cdots, \lambda_{n-1}, 0]$.
[75]Recall the discussion of Lie groups in chapter 6.

The direct product of the fundamental representation with that corresponding to any given Young tableau is the direct sum of the representations corresponding to Young tableaux where the block is added in all possible positions yielding an acceptable tableau (Fig. D.2).

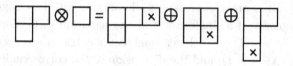

Fig. D.2 Decomposition of the direct product of an irreducible representation with the fundamental representation into a direct sum of irreducible representations.

The combination rules are both associative and distributive

$$a \otimes b \otimes c = (a \otimes b) \otimes c = a \otimes (b \otimes c) \qquad \text{; associative}$$
$$a \otimes (b \oplus c) = a \otimes b \oplus a \otimes c \qquad \text{; distributive} \qquad \text{(D.24)}$$

Now, with enough patience, one can build up all the irreducible representations of the group.

D.5 SU(n-1) Content of SU(n)

The group SU(n) always has SU(n-1) as a subgroup— just leave the last index untransformed. The rule for finding the irreducible representations of SU(n-1) contained in a given irreducible representation of SU(n) characterized by the partition $[\lambda_1, \lambda_2, \cdots, \lambda_n]$ with distinct λ_i is as follows:[76] Find all sets of acceptable $[\lambda'_1, \lambda'_2, \cdots, \lambda'_{n-1}]$ that can be interspersed in the first set, satisfying

$$\lambda_1 \geq \lambda'_1 \geq \lambda_2 \geq \lambda'_2 \geq \lambda_3 \geq \lambda'_3 \cdots \geq \lambda'_{n-1} \geq \lambda_n$$
$$\lambda_2 \geq \lambda'_1 \geq \lambda_3 \geq \lambda'_2 \geq \lambda_4 \geq \lambda'_3 \cdots \geq \lambda'_{n-1} \geq \lambda_n \qquad \text{; etc.} \qquad \text{(D.25)}$$

All these irreducible representations of SU(n-1) are then contained in the given representation of SU(n).

---

[76] Here $\lambda_n = \lambda'_{n-1} = 0$.

D.6   Some Examples

We proceed to discuss a few examples of this analysis.

D.6.1   *Angular Momentum–SU(2)*

For SU(2), since one can cross out all columns with two boxes, the Young tableaux consist of a single row and the partition $[\lambda, 0]$ merely labels the number of boxes. There is then a single set of entries above zero in Table D.1 ($l = \lambda + 1$, $n - 1 = 1$), and the dimension of the corresponding irreducible representation is given by

$$
\begin{aligned}
^{(2)}N_{[\lambda,0]} &= \lambda + 1 \\
&= 2j + 1 \qquad\qquad ; \lambda \equiv 2j
\end{aligned}
\tag{D.26}
$$

where we have defined $\lambda \equiv 2j$ in the second line.

The decomposition of the direct product of a representation with partition λ with the fundamental representation is analyzed in terms of Young tableaux as shown in Fig. D.3.

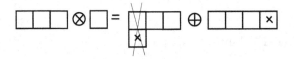

Fig. D.3 Decomposition into irreducible representations of the direct product of a representation of partition $[\lambda, 0]$ with the fundamental representation $[1, 0]$ in SU(2). The new representations have partitions $[\lambda - 1, 0]$ and $[\lambda + 1, 0]$; here $\lambda = 3$.

In terms of the *dimensions* of the representations one has

$$
[\lambda + 1] \otimes [2] = [\lambda] \oplus [\lambda + 2]
$$
$$
\text{or ;} \qquad [2j + 1] \otimes [2] = [2(j - 1/2) + 1] \oplus [2(j + 1/2) + 1] \tag{D.27}
$$

This, indeed, reproduces our previous result for the addition of angular momenta $\mathbf{j} + \mathbf{s}$ with $s = 1/2$. The states and C-G coefficients can now be explicitly constructed by applying the Young operators to the standard tableaux, of which there is here a single one for each Young tableau.

D.6.2   *Sakata Model–SU(3)*

Consider next the Sakata model where the symmetry is SU(3), and the Dirac particles belong to the fundamental representation of dimension [3]. The Young tableaux now have two rows, and there are two sets of entries above zero in Table D.1. For example, the dimensions of the irreducible representations corresponding to the Young tableaux in the bottom line of Fig. D.4 are obtained, in turn, from the following tables (the second and third tableaux have the same dimension).

Table D.2

$[\lambda]$	l	3	$[\lambda]$	l	3	$[\lambda]$	l	3
0	2	2	2	4	2	3	5	2
0	1	1	1	2	1	0	1	1
0	0	0	0	0	0	0	0	0

Thus, using the rule in Eq. (D.23), one has

$$^{(3)}N_{[0,0,0]} = \frac{2 \cdot 1 \cdot 1}{2 \cdot 1 \cdot 1} = 1 \qquad ; \text{ singlet}$$

$$^{(3)}N_{[2,1,0]} = \frac{4 \cdot 2 \cdot 2}{2 \cdot 1 \cdot 1} = 8 \qquad ; \text{ octet}$$

$$^{(3)}N_{[3,0,0]} = \frac{5 \cdot 4 \cdot 1}{2 \cdot 1 \cdot 1} = 10 \qquad ; \text{ decouplet} \qquad (D.28)$$

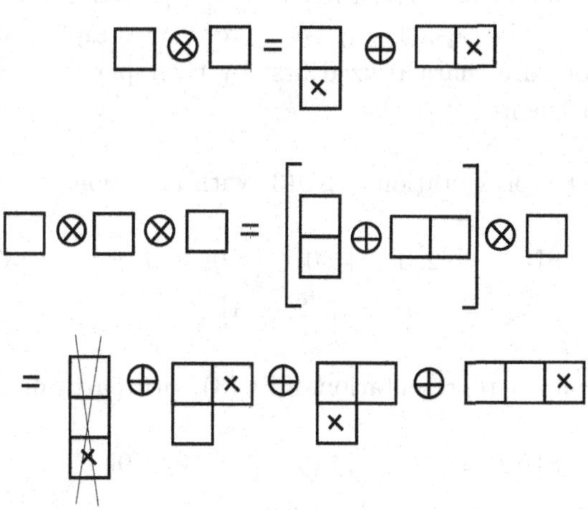

Fig. D.4   Young tableaux for $[3] \otimes [3] = [\bar{3}] \oplus [6]$ (first line) and $[3] \otimes [3] \otimes [3] = [1] \oplus [8] \oplus [8] \oplus [10]$ (last two lines) in the Sakata model.

The direct product $[3] \otimes [3]$ represents the internal SU(3) symmetry of the states of two Dirac particles in this model, and this is decomposed into irreducible representations through the Young tableaux in the first line of Fig. D.4. In terms of dimensions one has[77]

$$[3] \otimes [3] = [\bar{3}] \oplus [6] \tag{D.29}$$

Here the bar in $[\bar{3}]$ represents the *hole* (or adjoint) representation, which also has dimension 3.

The internal states available to three Dirac particles in this model are obtained from $[3] \otimes [3] \otimes [3]$, and this is decomposed into irreducible representations as indicated in the second and third lines of Fig. D.4. In terms of the dimensions of the representations, with the aid of our previous results, this gives

$$[3] \otimes [3] \otimes [3] = [1] \oplus [8] \oplus [8] \oplus [10] \tag{D.30}$$

The states and C-G coefficients can be explicitly constructed by applying the Young operators to the standard tableaux, of which there are now two for the Young tableau with partition $[2, 1, 0]$.[78]

It is clear from the first three matrices in Eqs. (6.74) that the SU(2) subgroup obtained when the third coordinate is held fixed is ordinary isospin. The isospin content of a given irreducible representation of SU(3) is then given by the rule in Eqs. (D.25). We give two examples, where the isospin representations are characterized first by their partitions $[\lambda', 0]$, and then by their dimensions $[2T + 1]$:

- The octet representation of SU(3) with partition $[2, 1, 0]$

$$\begin{aligned} SU(2) &= [2, 0] \oplus [1, 0] \oplus [1, 0] \oplus [0, 0] &&; [\lambda', 0] \\ &= [3] \oplus [2] \oplus [2] \oplus [1] &&; [2T + 1] \end{aligned} \tag{D.31}$$

- The decouplet representation of SU(3) with partition $[3, 0, 0]$

$$\begin{aligned} SU(2) &= [3, 0] \oplus [2, 0] \oplus [1, 0] \oplus [0, 0] &&; [\lambda', 0] \\ &= [4] \oplus [3] \oplus [2] \oplus [1] &&; [2T + 1] \end{aligned} \tag{D.32}$$

---

[77]The dimension of the tableau with partition $[2, 0, 0]$ is $^{(3)}N_{[2,0,0]} = 4 \cdot 3 \cdot 1/2 \cdot 1 \cdot 1 = 6$.
[78]For the C-G coefficients for SU(n), see [Chen *et al.* (1987)].

D.6.3 *Giant Resonances–SU(4)*

Recall from the discussion of the two-nucleon system in Vol. I that the spin and isospin dependence of the force between two nucleons is not very strong. To a first approximation, any difference in the force can be treated as a perturbation, at least in light nuclei. In this approximation, the states $(p \uparrow, p \downarrow, n \uparrow, n \downarrow)$ all enter on an equal footing, and the nuclear hamiltonian $\hat{H}_{\text{nucl}}$ exhibits SU(4) symmetry [Wigner (1937)]. The degenerate multiplets of $\hat{H}_{\text{nucl}}$ then come in *supermultiplets* corresponding to the irreducible representations of SU(4).

Consider light closed-shell nuclei with $N = Z$. The ground state of such nuclei will belong to the identity, or [1], representation of SU(4) in order to maximize the nuclear overlap with the attractive nuclear force.[79] Suppose these nuclei are excited through some one-body operator, for example through reactions such as $(e, e'), (\nu_\mu, \mu^-)$, or (ν_l, ν_l'). In this case, a *particle-hole* excitation will be created in the closed-shell nuclei.[80] The corresponding SU(4) states are then obtained from $[\bar{4}] \otimes [4]$, which is decomposed into irreducible representations through the Young tableaux in Fig. D.5. The dimension of the representation with partition $[2, 1, 1, 0]$ is given by

$$^{(4)}N_{[2,1,1,0]} = \frac{5 \cdot 3 \cdot 2 \cdot \cdot 3 \cdot 1 \cdot 2}{3 \cdot 2 \cdot 1 \cdot 2 \cdot 1 \cdot 1} = 15 \tag{D.33}$$

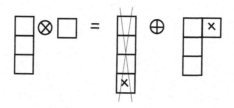

Fig. D.5 Young tableaux for $[\bar{4}] \otimes [4] = [1] \oplus [15]$ for giant resonances in light closed-shell nuclei.

In this case it is very easy to construct the state vectors since there are 16 direct-product states, and due to the unitarity of the matrix $\underline{r}$, the following

---

[79]The most antisymmetric internal state will lead to the most symmetric spatial state.

[80]See [Walecka (2004)]. Although all spatial multipoles contribute at high momentum transfer in these processes, they are dominated by the leading multipoles with $L = (0, 1)$ at low momentum transfer. A model dynamical calculation of the [15] supermultiplet is contained in this reference.

state is the one that is left invariant under the SU(4) transformation

$$|[1]\rangle = |i_n \, \bar{i}_n\rangle \qquad\qquad ; \text{ SU(4) singlet} \qquad\qquad (D.34)$$

The spin and isospin content of the [15] representation is most easily obtained by simply combining the spin and isospin of the particle and hole, exactly as was done for the Fermi-Yang model of mesons in Vol. I. The result in the form $[2S + 1 \otimes 2T + 1]$ is[81]

$$[2S + 1 \otimes 2T + 1] = [3 \otimes 3] \oplus [3 \otimes 1] \oplus [1 \otimes 3] \qquad ;$$

$$\text{spin-isospin content of [15] supermultiplet} \qquad (D.35)$$

As an example, the [15] supermultiplet of giant resonances, built upon orbital angular momentum $L = 1$, is clearly evidenced in the lightest doubly-magic nucleus $^{4}_{2}$He in [Tilley, Weller, and Hale (1992)].

---

[81]See Table 7.4 in Vol. I; formally, this is the spin and isospin SU(2)⊗SU(2) subgroup content of SU(4).

Appendix E

Lorentz Transformations in Quantum Field Theory

We again define our transformations to be *active* transformations. They take the *physical system* and give it an additional velocity $\mathbf{v}$, which is the same as giving the coordinate system a velocity $-\mathbf{v}$. Recall from Vol. I that under a homogeneous Lorentz transformation the coordinates are related by

$$x'_\mu = a_{\mu\nu}(v)x_\nu \qquad (\text{E.1})$$

The appropriate Lorentz transformation for a particle *boost* of v in the z-direction is then

$$\underline{a}(-v) = \begin{bmatrix} 1 & 0 & 0 & 0 \\ 0 & 1 & 0 & 0 \\ 0 & 0 & \dfrac{1}{\sqrt{1-v^2/c^2}} & \dfrac{-iv/c}{\sqrt{1-v^2/c^2}} \\ 0 & 0 & \dfrac{iv/c}{\sqrt{1-v^2/c^2}} & \dfrac{1}{\sqrt{1-v^2/c^2}} \end{bmatrix} \qquad (\text{E.2})$$

This can be checked by looking at a particle's wave number $k_\mu = (\mathbf{k}, ik_0)$

$$k'_\mu = a_{\mu\nu}(-v)k_\nu$$

$$k'_x = k_x \qquad\qquad ;\ k'_z = \frac{k_z + vk_0/c}{(1 - v^2/c^2)^{1/2}}$$

$$k'_y = k_y \qquad\qquad ;\ k'_0 = \frac{k_0 + vk_z/c}{(1 - v^2/c^2)^{1/2}} \qquad (\text{E.3})$$

which, indeed, corresponds to a boost.

E.1   Scalar Field

Before discussing Lorentz transformations, it is necessary to go back and make things look covariant. This is analogous to what was involved in going over from Fourier series to Fourier integrals in Vol. I.

E.1.1   *States*

In a big cubical box of volume Ω with periodic boundary conditions, we have been using the states and normalization

$$|\mathbf{k}\rangle = c_{\mathbf{k}}^{\dagger}|0\rangle \qquad ; \langle \mathbf{k}'|\mathbf{k}\rangle = \delta_{\mathbf{k}\mathbf{k}'} \qquad (E.4)$$

Consider the *completeness* statement for identical spin-zero bosons in the many-particle Hilbert space[82]

$$\sum_{n=0}^{\infty} \frac{1}{n!} \sum_{\mathbf{k}_1 \cdots \mathbf{k}_n} |\mathbf{k}_1 \cdots \mathbf{k}_n\rangle\langle\mathbf{k}_1 \cdots \mathbf{k}_n| = \hat{1} \qquad (E.5)$$

In the limit as the volume $\Omega \to \infty$, this expression becomes

$$\sum_{n=0}^{\infty} \frac{1}{n!} \left[\frac{\Omega}{(2\pi)^3}\right]^n \int d^3k_1 \cdots d^3k_n \, |\mathbf{k}_1 \cdots \mathbf{k}_n\rangle\langle\mathbf{k}_1 \cdots \mathbf{k}_n| = \hat{1} \qquad (E.6)$$

This completeness relation should be manifestly invariant and hold in any Lorentz frame. From Probl. 9.12, one can write

$$\int \frac{d^3k_1}{2k_{01}} \cdots \frac{d^3k_n}{2k_{0n}} 2k_{01} \cdots 2k_{0n} = \int d^4k_1 \cdots d^4k_n \delta(k_1^2 + m^2) \cdots \delta(k_n^2 + m^2) \times$$
$$\theta(k_{01}) \cdots \theta(k_{0n}) 2k_{01} \cdots 2k_{0n} \qquad (E.7)$$

Now define the states

$$|\tilde{\mathbf{k}}\rangle \equiv \sqrt{2k_0\Omega}\, c_{\mathbf{k}}^{\dagger}|0\rangle \equiv \tilde{c}_{\mathbf{k}}^{\dagger}|0\rangle \qquad (E.8)$$

Here $k_0 = \sqrt{\mathbf{k}^2 + m^2}$. The completeness relation then *looks invariant*

$$\sum_{n=0}^{\infty} \frac{1}{n!} \frac{1}{(2\pi)^{3n}} \int d^4k_1 \cdots d^4k_n \, \delta(k_1^2 + m^2) \cdots \delta(k_n^2 + m^2)\theta(k_{01}) \cdots \theta(k_{0n}) \times$$
$$|\tilde{\mathbf{k}}_1 \cdots \tilde{\mathbf{k}}_n\rangle\langle\tilde{\mathbf{k}}_1 \cdots \tilde{\mathbf{k}}_n| = \hat{1} \qquad (E.9)$$

---

[82]The $n = 0$ term is just $|0\rangle\langle 0|$.

What is the *normalization* of these new states? Just compute

$$\langle \tilde{\mathbf{k}} | \tilde{\mathbf{k}}' \rangle = 2k_0 \Omega \, \delta_{\mathbf{kk}'} \tag{E.10}$$

In the same limit $\Omega \to \infty$, one has

$$\Omega \delta_{\mathbf{kk}'} = \int_\Omega d^3x \, e^{i(\mathbf{k}-\mathbf{k}')\cdot\mathbf{x}} \to (2\pi)^3 \delta^{(3)}(\mathbf{k} - \mathbf{k}') \tag{E.11}$$

Hence, in the limit,

$$\langle \tilde{\mathbf{k}} | \tilde{\mathbf{k}}' \rangle = 2k_0 (2\pi)^3 \delta^{(3)}(\mathbf{k} - \mathbf{k}') \tag{E.12}$$

The normalization of these new states is *also invariant* since

$$\int \frac{d^3k}{2k_0} \langle \tilde{\mathbf{k}} | \tilde{\mathbf{k}}' \rangle = (2\pi)^3 = \int d^4k \, \delta(k^2 + m^2)\theta(k_0) \langle \tilde{\mathbf{k}} | \tilde{\mathbf{k}}' \rangle \tag{E.13}$$

The commutation relation of the original creation and destruction operators is

$$[c_{\mathbf{k}}, c_{\mathbf{k}'}^\dagger] = \delta_{\mathbf{kk}'} \tag{E.14}$$

The commutation relation of the *new* operators is then

$$[\tilde{c}_{\mathbf{k}}, \tilde{c}_{\mathbf{k}'}^\dagger] = 2k_0 \Omega \, \delta_{\mathbf{kk}'} \to 2k_0 (2\pi)^3 \delta^{(3)}(\mathbf{k} - \mathbf{k}') \tag{E.15}$$

This is again invariant, just as above.

We will work with the states $|\tilde{\mathbf{k}}\rangle$, *and define the Lorentz transformation operator* $\hat{U}(L)$ *to be a unitary operator when acting on these states. Thus it takes one complete orthonormal set of such states, relevant to the first Lorentz frame, into another such set, relevant to the new frame.*

Thus, as a physical boost on these states, the Lorentz transformation operator satisfies

$$\begin{aligned}
\hat{U}(L)\, \tilde{c}_{\mathbf{k}}^\dagger |0\rangle &= \tilde{c}_{\mathbf{k}'}^\dagger |0\rangle && ; \ k_\mu' = a_{\mu\nu}(-v)k_\nu \\
\hat{U}(L)^\dagger &= \hat{U}(L)^{-1} && ; \ \text{unitary} \\
\hat{U}(L)|0\rangle &= |0\rangle && ; \ \text{vacuum invariant}
\end{aligned} \tag{E.16}$$

The last relation states that the Lorentz transformation should leave the zero-particle state unchanged. It then follows by inserting $\hat{U}(L)^{-1}\hat{U}(L)$ in the first relation that the behavior of the new creation operators under a Lorentz transformation is

$$\hat{U}(L)\, \tilde{c}_{\mathbf{k}}^{\dagger}\, \hat{U}(L)^{-1} = \tilde{c}_{\mathbf{k}'}^{\dagger} \tag{E.17}$$

The scalar field in the interaction picture is given by

$$\hat{\phi}(x_{\mu}) = \sum_{\mathbf{k}} \left(\frac{\hbar}{2k_0 c \Omega} \right)^{1/2} \left(c_{\mathbf{k}}^{\dagger} e^{-ik\cdot x} + c_{\mathbf{k}} e^{ik\cdot x} \right) \tag{E.18}$$

$$\rightarrow \frac{\Omega}{(2\pi)^3} \int d^3k \left(\frac{\hbar}{2k_0 c \Omega} \right)^{1/2} \frac{1}{\sqrt{2k_0\Omega}} \left(\tilde{c}_{\mathbf{k}}^{\dagger} e^{-ik\cdot x} + \tilde{c}_{\mathbf{k}} e^{ik\cdot x} \right)$$

Thus, in the limit $\Omega \rightarrow \infty$, the scalar field is given by

$$\hat{\phi}(x_{\mu}) = \left(\frac{\hbar}{c} \right)^{1/2} \frac{1}{(2\pi)^3} \int d^4k\, \delta(k^2 + m^2)\theta(k_0) \left[\tilde{c}_{\mathbf{k}}^{\dagger} e^{-ik\cdot x} + \tilde{c}_{\mathbf{k}} e^{ik\cdot x} \right] \quad ;$$
$$\text{scalar field} \tag{E.19}$$

E.1.2 *Lorentz Transformation*

Consider the Lorentz transformation of the field in Eq. (E.19), and use Eq. (E.17),

$$\hat{U}(L)\hat{\phi}(x_{\mu})\hat{U}(L)^{-1} = \left(\frac{\hbar}{c} \right)^{1/2} \frac{1}{(2\pi)^3} \int d^4k\, \delta(k^2 + m^2)\theta(k_0) \times$$
$$\left[\tilde{c}_{\mathbf{k}'}^{\dagger} e^{-ik\cdot x} + \tilde{c}_{\mathbf{k}'} e^{ik\cdot x} \right] \tag{E.20}$$

Now change variables in the integral, and use[83]

$$k_{\mu}' = a_{\mu\nu}(-v)k_{\nu} \quad ; \quad d^4k = d^4k' \quad ; \quad k^2 = k'^2 \quad ; \quad \theta(k_0) = \theta(k_0') \tag{E.21}$$

Also

$$k_{\mu}x_{\mu} = k_{\mu}'x_{\mu}' \qquad ; \quad x_{\mu}' = a_{\mu\nu}(-v)x_{\nu} \tag{E.22}$$

Hence

$$\hat{U}(L)\hat{\phi}(x_{\mu})\hat{U}(L)^{-1} = \hat{\phi}(x_{\mu}') \qquad ; \text{ scalar field} \tag{E.23}$$

This is the behavior of a scalar field under Lorentz transformations. Here the unitary transformation $\hat{U}(L)$ boosts the new particle states according to Eqs. (E.16)–(E.17), and x_{μ}' is given in Eqs. (E.22).

---

[83]Remember, these are four-vectors, and scalar products of four-vectors.

E.1.3 *Generators*

The fields are now operators in the abstract many-particle Hilbert space. Let us first characterize the generators for translations and rotations, as we did at the start of chapter 3.

Suppose one looks for an active transformation that translates the particle states a distance $+a_\mu$ in four-dimensional Minkowski space. This is equivalent to translating the coordinate system by $-a_\mu$. Equation (E.23) implies that the unitary translation operator $\hat{U}(a)$ should then satisfy

$$\hat{U}(a)\hat{\phi}(x_\mu)\hat{U}(a)^{-1} = \hat{\phi}(x_\mu + a_\mu) \qquad (E.24)$$

It is clear from Eq. (E.18) that this will be the case if the effect on the creation and destruction operators is given by

$$\hat{U}(a)c_{\mathbf{k}}^\dagger\hat{U}(a)^{-1} = e^{-ik\cdot a}\, c_{\mathbf{k}}^\dagger$$
$$\hat{U}(a)^\dagger = \hat{U}(a)^{-1} \qquad ; \ \hat{U}(a)|0\rangle = |0\rangle \qquad (E.25)$$

When acting on the one-particle state, the translation operator then gives

$$\hat{U}(a)|\mathbf{k}\rangle = e^{-ik\cdot a}|\mathbf{k}\rangle \qquad (E.26)$$

This allows us to identify the generator of translations with the four-momentum operator $\hat{P}_\mu$, and write

$$\hat{U}(a) = \exp\left\{-\frac{i}{\hbar}\hat{P}_\mu a_\mu\right\} \qquad ; \ \text{translation operator} \quad (E.27)$$

An expansion of Eq. (E.24) for infinitesimal $a_\mu \equiv \epsilon_\mu \to 0$ then gives

$$\hat{\phi}(x) - \frac{i}{\hbar}\epsilon_\mu[\hat{P}_\mu, \hat{\phi}(x)] + \cdots = \hat{\phi}(x) + \epsilon_\mu\frac{\partial}{\partial x_\mu}\hat{\phi}(x) + \cdots$$

$$\text{or;} \qquad \frac{i}{\hbar}[\hat{P}_\mu, \hat{\phi}(x)] = -\frac{\partial}{\partial x_\mu}\hat{\phi}(x) \qquad (E.28)$$

where the argument x indicates the four-vector x_λ. Equation (E.28) characterizes the effect of the generator of translations on the field.

Let us try to do the same thing for spatial rotations. We will here construct the generators in a form that can easily be extended to rotations in a plane in four-dimensional Minkowski space, for such are the Lorentz transformations of interest. Return to Fig. 3.1, and in line with the above, now rotate the coordinate system by an angle $-\boldsymbol{\omega} \equiv -\omega\mathbf{n}$ about the third axis. If $(\mathbf{n}_1, \mathbf{n}_2)$ are two orthonormal unit vectors lying in the $(1, 2)$ plane, then the normal can be expressed as $\mathbf{n} = \mathbf{n}_1 \times \mathbf{n}_2$ and

$$\mathbf{n} \cdot \hat{\boldsymbol{\mathcal{J}}} = (\mathbf{n}_1 \times \mathbf{n}_2) \cdot \hat{\boldsymbol{\mathcal{J}}} = \epsilon_{ijk}(\mathbf{n}_1)_i(\mathbf{n}_2)_j \hat{\mathcal{J}}_k \qquad \text{(E.29)}$$

The last is an analytic expression for the vector triple product. This result can be rewritten as follows. Define an antisymmetric tensor characterizing the plane of rotation, and an antisymmetric tensor dual to the angular momentum vector $\hat{\boldsymbol{\mathcal{J}}}$, by

$$\begin{aligned} \alpha_{ij} &\equiv (\mathbf{n}_1)_i(\mathbf{n}_2)_j - (\mathbf{n}_2)_i(\mathbf{n}_1)_j \\ \hat{M}_{ij} &\equiv \epsilon_{ijk}\hat{\mathcal{J}}_k \end{aligned} \qquad \text{(E.30)}$$

It is then an algebraic identity that[84]

$$\frac{1}{2}\alpha_{ij}\hat{M}_{ij} = (\mathbf{n}_1 \times \mathbf{n}_2) \cdot \hat{\boldsymbol{\mathcal{J}}} = \mathbf{n} \cdot \hat{\boldsymbol{\mathcal{J}}} \qquad \text{(E.31)}$$

If $(\mathbf{n}_1, \mathbf{n}_2)$ are the basis vectors $(\mathbf{e}_1, \mathbf{e}_2)$ in Fig. 3.1, then the tensor α_{ij} is given by

$$\alpha_{ij} = \delta_{i1}\delta_{j2} - \delta_{i2}\delta_{j1} = \epsilon_{ij3} \qquad \text{(E.32)}$$

We now look for a unitary rotation operator of the form

$$\hat{U}(L) = \exp\left\{-\frac{i}{\hbar}\frac{\omega}{2}\alpha_{ij}\hat{M}_{ij}\right\} \qquad \text{; rotation operator} \qquad \text{(E.33)}$$

where $\alpha_{ij}\hat{M}_{ij}$ is hermitian. From Eq. (E.23) we require

$$\hat{U}(L)\hat{\phi}(x_i)\hat{U}(L)^{-1} = \hat{\phi}(x_i')$$
$$x_i' = a_{ij}(-\omega)x_j \qquad \text{(E.34)}$$

For a rotation about the 3-axis, one has from Eq. (3.21)

$$\underline{a}(-\omega) = \begin{pmatrix} \cos\omega & -\sin\omega & 0 \\ \sin\omega & \cos\omega & 0 \\ 0 & 0 & 1 \end{pmatrix} \qquad \text{; about 3-axis} \qquad \text{(E.35)}$$

For an infinitesimal rotation $\omega \equiv \epsilon \to 0$. It follows that for a rotation of the coordinate system by an infinitesimal angle $-\epsilon$ in the $(1,2)$ plane

$$\begin{aligned} a_{ij}(-\epsilon) &= \delta_{ij} - \epsilon\,\epsilon_{ij3} \\ &= \delta_{ij} - \epsilon\,\alpha_{ij} \end{aligned} \qquad \text{(E.36)}$$

---

[84]Note that the use of $\mathbf{n}$ and $\boldsymbol{\mathcal{J}}$ is peculiar to three dimensions.

where Eq. (E.32) has been used in obtaining the second line. In this last form, however, the relation provides a general expression for an infinitesimal rotation in the plane characterized by the two orthonormal unit vectors $(\mathbf{n}_1, \mathbf{n}_2)$

$$a_{ij}(-\epsilon) = \delta_{ij} - \epsilon\alpha_{ij}$$

$$\alpha_{ij} \equiv (\mathbf{n}_1)_i(\mathbf{n}_2)_j - (\mathbf{n}_2)_i(\mathbf{n}_1)_j \tag{E.37}$$

Reduction of Eqs. (E.34) and (E.33) to infinitesimal form, and a repetition of the translation argument above, then gives

$$-\frac{i}{\hbar}\frac{1}{2}\epsilon\alpha_{ij}[\hat{M}_{ij}, \hat{\phi}(x)] = -\epsilon\alpha_{ij}x_j\frac{\partial}{\partial x_i}\hat{\phi}(x) \tag{E.38}$$

Since α_{ij} is antisymmetric, and arbitrary, this reduces to

$$\frac{i}{\hbar}[\hat{M}_{ij}, \hat{\phi}(x)] = -\left(x_i\frac{\partial}{\partial x_j} - x_j\frac{\partial}{\partial x_i}\right)\hat{\phi}(x) \tag{E.39}$$

In summary, for the rotation by an angle $-\omega$ in the α_{ij}-plane defined in Eq. (E.37)

$$\hat{U}(L) = \exp\left\{-\frac{i}{\hbar}\frac{\omega}{2}\alpha_{ij}\hat{M}_{ij}\right\} \qquad ; \text{ rotation}$$

$$\frac{i}{\hbar}[\hat{M}_{ij}, \hat{\phi}(x)] = -\left(x_i\frac{\partial}{\partial x_j} - x_j\frac{\partial}{\partial x_i}\right)\hat{\phi}(x) \tag{E.40}$$

In this form, the results are immediately generalized to rotations by an angle $-\Omega$ in the (n_1, n_2)-plane in four-dimensional Minkowski space

$$\hat{U}(L) = \exp\left\{-\frac{i}{\hbar}\frac{\Omega}{2}\alpha_{\mu\nu}\hat{M}_{\mu\nu}\right\} \qquad ; \text{ Lorentz}$$

$$\alpha_{\mu\nu} = (n_1)_\mu(n_2)_\nu - (n_2)_\mu(n_1)_\nu \qquad \text{transformation}$$

$$\frac{i}{\hbar}[\hat{M}_{\mu\nu}, \hat{\phi}(x)] = -\left(x_\mu\frac{\partial}{\partial x_\nu} - x_\nu\frac{\partial}{\partial x_\mu}\right)\hat{\phi}(x) \tag{E.41}$$

With rotations, the angular momentum tensor $\hat{M}_{ij}$ is

$$\hat{M}_{ij} = \epsilon_{ijk}\hat{J}_k \tag{E.42}$$

$$\text{or} \qquad \hat{M}_{12} = \hat{J}_3 \qquad ; \text{ and cyclic permutations}$$

Here $\hat{\mathcal{J}}$ is the total angular momentum operator for the system, and we know that $\hat{\mathcal{J}}$ is the generator of rotations. The extension is now clear. *One identifies* $\hat{M}_{\mu\nu}$, *the angular momentum tensor in Minkowski space, as the generator of homogeneous Lorentz transformations.* The general *inhomogeneous* Lorentz transformation operator then takes the form

$$\hat{U}(L,a) = \exp\left\{-\frac{i}{\hbar}\left[\frac{\Omega}{2}\alpha_{\mu\nu}\hat{M}_{\mu\nu} + a_\mu\hat{P}_\mu\right]\right\} \quad ;$$

inhomogeneous Lorentz transformation (E.43)

where the generators $(\hat{M}_{\mu\nu}, \hat{P}_\mu)$ are the angular-momentum tensor and four-momentum, respectively.[85]

E.1.4 *Commutation Rules*

As with angular momentum, the commutation relations of the generators among themselves can be obtained by simply looking at the commutation relations of the covariant, first-quantized differential operators $(P_\mu, M_{\mu\nu})$ where

$$P_\mu \equiv \frac{\hbar}{i}\frac{\partial}{\partial x_\mu}$$

$$M_{\mu\nu} \equiv \frac{\hbar}{i}\left(x_\mu\frac{\partial}{\partial x_\nu} - x_\nu\frac{\partial}{\partial x_\mu}\right) \qquad (E.44)$$

The result is[86]

$$[\hat{P}_\mu, \hat{P}_\nu] = 0 \qquad\qquad ; \text{ inhomogeneous Lorentz group}$$

$$\frac{i}{\hbar}[\hat{M}_{\mu\nu}, \hat{P}_\lambda] = \delta_{\nu\lambda}\hat{P}_\mu - \delta_{\mu\lambda}\hat{P}_\nu$$

$$\frac{i}{\hbar}[\hat{M}_{\mu\nu}, \hat{M}_{\rho\sigma}] = \delta_{\mu\sigma}\hat{M}_{\nu\rho} + \delta_{\nu\rho}\hat{M}_{\mu\sigma} - \delta_{\mu\rho}\hat{M}_{\nu\sigma} - \delta_{\nu\sigma}\hat{M}_{\mu\rho} \qquad (E.45)$$

These are the commutation relations for the generators of the *inhomogeneous Lorentz group*. This is a Lie algebra, and all the properties of the corresponding continuous 10-parameter $(\Omega\alpha_{\mu\nu}, a_\mu)$ Lie group follow from this algebra. The Hilbert space forms a basis for an infinite-dimensional, unitary representation of this Poincaré, or inhomogeneous Lorentz group. Interested readers are referred to the classic paper of [Wigner (1939)] for further study of this most basic subject.

---

[85] In this appendix, both the scalar and Dirac fields are analyzed in the interaction picture, which is the Heisenberg picture for free fields. With interacting fields, the appropriate procedure is to pass from $\mathcal{L} \to T_{\mu\nu} \to (P_\mu, M_{\mu\nu})$ and then quantize.

[86] See Probs. E.6 and E.1.

E.2 Dirac Field

The Dirac field in a big box of volume Ω with periodic boundary conditions is given in the interaction picture by

$$\hat{\psi}(x_\mu) = \frac{1}{\sqrt{\Omega}} \sum_{\mathbf{k}\lambda} \left[a_{\mathbf{k}\lambda} u(\mathbf{k}\lambda) e^{ik\cdot x} + b_{\mathbf{k}\lambda}^\dagger v(-\mathbf{k}\lambda) e^{-ik\cdot x} \right] \qquad (E.46)$$

The first task is to convert to spinors with Lorentz-invariant norm, and from Probl. 9.11, these are obtained as

$$U(\mathbf{k}\lambda) = \sqrt{\frac{k_0}{M}}\, u(\mathbf{k}\lambda) \qquad\qquad ; \ \bar{U}U = 1$$

$$V(-\mathbf{k}\lambda) = \sqrt{\frac{k_0}{M}}\, v(-\mathbf{k}\lambda) \qquad\qquad ; \ \bar{V}V = -1 \qquad (E.47)$$

Next, in direct analogy to Eq. (E.8), we define new operators by

$$\tilde{a}_{\mathbf{k}\lambda}^\dagger \equiv \sqrt{2k_0\Omega}\, a_{\mathbf{k}\lambda}^\dagger \qquad (E.48)$$

and similarly for $\tilde{b}_{\mathbf{k}\lambda}^\dagger$. All the statements on the Lorentz invariance of the completeness, normalization, and (anti)commutation relations again apply. In the limit $\Omega \to \infty$, as above, the field in Eq. (E.46) then takes the form

$$\hat{\psi}(x_\mu) = \frac{\sqrt{2M}}{(2\pi)^3} \int d^4k\, \delta(k^2 + M^2)\theta(k_0) \times$$

$$\sum_\lambda \left[\tilde{a}_{\mathbf{k}\lambda} U(\mathbf{k}\lambda) e^{ik\cdot x} + \tilde{b}_{\mathbf{k}\lambda}^\dagger V(-\mathbf{k}\lambda) e^{-ik\cdot x} \right] \qquad (E.49)$$

E.2.1 *Lorentz Transformation*

Again, we start from rotations and generalize. First, introduce the following combination of Dirac gamma matrices

$$\sigma_{ij} \equiv \frac{1}{2i}[\gamma_i, \gamma_j] = \epsilon_{ijk}\Sigma_k \qquad (E.50)$$

Here the last relation is obtained from direct multiplication with the standard representation and the identification of the Dirac spin matrix of Vol. I [see EqI. (9.69)]. The angular momentum tensor in Dirac's theory is then augmented with a spin contribution of

$$\delta M_{ij}^{\rm spin} = \epsilon_{ijk}\frac{\hbar}{2}\Sigma_k = \frac{\hbar}{2}\sigma_{ij} \qquad (E.51)$$

The generalization to four-dimensional Minkowski space is clear

$$\delta M_{\mu\nu}^{\text{spin}} = \frac{\hbar}{2}\sigma_{\mu\nu}$$

$$\sigma_{\mu\nu} \equiv \frac{1}{2i}[\gamma_\mu, \gamma_\nu] \qquad ; \sigma_{\mu\nu}^\dagger = \sigma_{\mu\nu} \qquad (E.52)$$

Thus one now has an additional term in $M_{\mu\nu}$ in Eqs. (E.44)

$$M_{\mu\nu} = \frac{\hbar}{i}\left(x_\mu\frac{\partial}{\partial x_\nu} - x_\nu\frac{\partial}{\partial x_\mu}\right) + \frac{\hbar}{2}\sigma_{\mu\nu} \qquad ; \text{Dirac} \qquad (E.53)$$

This additional spin term provides another representation of the commutation relations of the generators of the Lorentz group in Eqs. (E.45), and hence *those equations continue to hold when spin is included.*[87]

The effect of the generators on the Dirac field is obtained from Eq. (E.28) and the extension of Eq. (E.41)

$$\frac{i}{\hbar}[\hat{P}_\mu, \hat{\psi}(x)] = -\frac{\partial}{\partial x_\mu}\hat{\psi}(x)$$

$$\frac{i}{\hbar}[\hat{M}_{\mu\nu}, \hat{\psi}(x)] = -\left(x_\mu\frac{\partial}{\partial x_\nu} - x_\nu\frac{\partial}{\partial x_\mu} + \frac{i}{2}\sigma_{\mu\nu}\right)\hat{\psi}(x) \qquad (E.54)$$

Working backwards, the corresponding finite spinor transformation law for the Dirac field is[88]

$$\hat{U}(L,a)\hat{\psi}(x_\mu)\hat{U}(L,a)^{-1} = \mathcal{S}(\Omega)\,\hat{\psi}(x_\mu')$$

$$\mathcal{S}(\Omega) \equiv \exp\left\{\frac{i}{4}\Omega\alpha_{\mu\nu}\sigma_{\mu\nu}\right\} \qquad (E.55)$$

where $\hat{U}(L,a)$ is given in Eq. (E.43) and x_μ' in Eq. (E.22). Note that $\mathcal{S}(\Omega)$ is *not* a unitary matrix (its properties are investigated in Prob. E.4). Fields form a basis for a finite, in general non-unitary, representation of the Poincaré group.

The current is a bilinear expression in $\hat{\psi}(x_\mu)$, and it now follows entirely from the commutation relations that the Dirac current transforms as a four-vector under homogeneous Lorentz transformations (see Prob. E.5)

$$\hat{j}_\mu(x) = i\hat{\bar{\psi}}(x)\gamma_\mu\hat{\psi}(x)$$

$$\hat{U}(L)\hat{j}_\mu(x)\hat{U}(L)^{-1} = a_{\mu\nu}(-v)\,\hat{j}_\nu(x') \qquad (E.56)$$

---

[87]See Prob. E.1; this is ultimately why the Dirac equation is Lorentz covariant.
[88]See Probs. E.2–E.3.

Appendix F

Green's Functions and Other Singular Functions

Our analysis of the scattering amplitude reduces it to a sum of integrals over the two-point Green's functions, or propagators. In this appendix we analyze those quantities. We work in the interaction picture, which is the same as the Heisenberg picture for the free fields

$$\hat{O}_I = e^{\frac{i}{\hbar}\hat{H}_0 t}\hat{O}e^{-\frac{i}{\hbar}\hat{H}_0 t} \qquad (F.1)$$

The fields of interest are

$$\hat{\phi}(x) = \sum_{\mathbf{k}} \left(\frac{\hbar}{2\omega_k\Omega}\right)^{1/2}\left(c_{\mathbf{k}}e^{ik\cdot x} + c_{\mathbf{k}}^\dagger e^{-ik\cdot x}\right) \qquad ; \text{ real scalar}$$

$$\hat{\psi}(x) = \frac{1}{\sqrt{\Omega}}\sum_{\mathbf{k}}\sum_{\lambda}\left[a_{\mathbf{k}\lambda}u(\mathbf{k}\lambda)e^{ik\cdot x} + b_{\mathbf{k}\lambda}^\dagger v(-\mathbf{k}\lambda)e^{-ik\cdot x}\right] \qquad ; \text{ Dirac}$$

$$\hat{\mathbf{A}}(x) = \sum_{\mathbf{k}}\sum_{s=1}^{2}\left(\frac{\hbar}{2\omega_k\varepsilon_0\Omega}\right)^{1/2}\left[a_{\mathbf{k}s}\mathbf{e}_{\mathbf{k}s}e^{ik\cdot x} + a_{\mathbf{k}s}^\dagger\mathbf{e}_{\mathbf{k}s}e^{-ik\cdot x}\right] \qquad ;$$

$$\text{transverse electromagnetic} \qquad (F.2)$$

Here $x = (\mathbf{x}, ict)$ denotes the space-time point, and $k\cdot x = \mathbf{k}\cdot\mathbf{x} - \omega_k t$. The (anti)commutation relations are those of the creation and destruction operators, where the Dirac and electromagnetic operators are here distinguished through their subscripts.

F.1   Commutator at Unequal Times

The appropriate canonical equal-time (anti)commutation relations for these fields have been given in the text. Here we evaluate the (anti)commutation relations at *unequal* times. The following properties of the Dirac delta function of a single variable will be employed (see Vol. I)

$$\delta(y) = \delta(-y)$$

$$\delta(ay) = \frac{1}{|a|}\delta(y) \qquad ; \; a \text{ real}$$

$$\delta(y^2 - a^2) = \frac{1}{2|a|}\left[\delta(y - |a|) + \delta(y + |a|)\right]$$

$$\varepsilon(y)\delta(y^2 - a^2) = \frac{1}{2|a|}\left[\delta(y - |a|) - \delta(y + |a|)\right] \qquad \text{(F.3)}$$

where

$$\varepsilon(y) = +1 \qquad ; \; \text{if } y > 0$$

$$= -1 \qquad ; \; \text{if } y < 0 \qquad \text{(F.4)}$$

Scalar Field. Consider the commutator of the real scalar field at two distinct space-time points $x^{(1)}$ and $x^{(2)}$. It is no loss of generality to take $x^{(2)}$ as the origin and then define $x^{(1)} \equiv x$.[89] The commutator then follows directly from the field expansions

$$[\hat{\phi}(x), \hat{\phi}(0)] = \frac{1}{\Omega}\sum_{\mathbf{k}}\frac{\hbar}{2\omega_k}\left[e^{ik\cdot x} - e^{-ik\cdot x}\right]$$

$$= \frac{\hbar}{\Omega}\sum_{\mathbf{k}}e^{i\mathbf{k}\cdot\mathbf{x}}\frac{1}{2\omega_k}\left[e^{-i\omega_k t} - e^{i\omega_k t}\right]$$

$$= \frac{\hbar}{i}\frac{1}{\Omega}\sum_{\mathbf{k}}e^{i\mathbf{k}\cdot\mathbf{x}}\frac{1}{\omega_k}\sin\omega_k t$$

$$\equiv \frac{\hbar}{ic}\Delta(x) \qquad \text{(F.5)}$$

This defines the function $\Delta(x)$, which clearly vanishes for $t = 0$,

$$\Delta(x) = \frac{ic}{\Omega}\sum_{\mathbf{k}}\frac{1}{2\omega_k}\left[e^{ik\cdot x} - e^{-ik\cdot x}\right]$$

$$= 0 \qquad ; \; \text{if } t = 0 \qquad \text{(F.6)}$$

In the limit $\Omega \to \infty$ this becomes

$$\Delta(x) = \frac{ic}{(2\pi)^3}\int d^3k\, dk_0\, e^{ik\cdot x}\frac{1}{2\omega_k}\left[\delta\left(k_0 - \frac{\omega_k}{c}\right) - \delta\left(k_0 + \frac{\omega_k}{c}\right)\right] \qquad ;$$

$$\Omega \to \infty \qquad \text{(F.7)}$$

---

[89]Note that now $x = x^{(1)} - x^{(2)}$; see Prob. F.1.

where $x \equiv (\mathbf{x}, ict)$, $k \equiv (\mathbf{k}, ik_0)$, and $\omega_k = c\sqrt{\mathbf{k}^2 + m^2}$.

Equation (F.7) can be rewritten with the aid of Eqs. (F.3) as

$$\Delta(x) = \frac{i}{(2\pi)^3} \int d^4 k \, \varepsilon(k_0) \delta(k^2 + m^2) \, e^{ik \cdot x} \quad ; \; d^4 k = d^3 k \, dk_0 \quad ;$$

$$x = (\mathbf{x}, ict) \tag{F.8}$$

Here $k^2 = \mathbf{k}^2 - k_0^2$. Let us discuss these results:

(1) We note that the canonical equal-time commutation relation for the real scalar field is immediately recovered from Eqs. (F.5) through

$$[\hat{\pi}(x), \hat{\phi}(0)]_{t=0} = \left[\frac{\partial \hat{\phi}(x)}{\partial t}, \hat{\phi}(0) \right]_{t=0} = \frac{\hbar}{ic} \left[\frac{\partial}{\partial t} \Delta(x) \right]_{t=0}$$

$$= \frac{\hbar}{i} \delta^{(3)}(\mathbf{x}) \tag{F.9}$$

(2) Under a proper, orthochronous, Lorentz transformation[90]

$$x'_\mu = a_{\mu\nu} x_\nu \tag{F.10}$$

Now change variables in the integral in Eq. (F.8) to

$$k'_\mu = a_{\mu\nu} k_\nu \tag{F.11}$$

The point k' is just moved along the mass hyperboloid $k^2 + m^2 = 0$ in momentum space [see Probl. 9.12 and Fig. F.1(b)]. It is also true that

$$d^4 k = d^4 k' \quad ; \; k^2 = k'^2 \quad ; \; \varepsilon(k_0) = \varepsilon(k'_0) \quad ; \; e^{ik \cdot x} = e^{ik' \cdot x'} \tag{F.12}$$

The third relation follows since the sign of k_0 is preserved. In this way one shows that

$$\Delta(x) = \Delta(x') \tag{F.13}$$

The function $\Delta(x)$ is a *Lorentz scalar*. Now the Lorentz transformation in Eq. (F.10) moves one along one sheet of the hyperboloid with constant $x^2 = \mathbf{x}^2 - c^2 t^2$ in coordinate space [Fig. F.1(a)]. The function $\Delta(x)$ takes the same value everywhere on that sheet.

(3) Consider the space-like hypersurface with $x^2 = \mathbf{x}^2 - c^2 t^2 > 0$ [see Fig. F.1(a)]. We know from Eq. (F.6) that $\Delta(x) = 0$ for $t = 0$, that is,

---

[90]A proper, orthochronous, Lorentz transformation is characterized by $a_{44} > 0$; this confines one to a given sheet of the hyperboloids in Figs. F.1(a,b).

the fields commute at equal times. It follows that $\Delta(x)$ must then vanish *everywhere* on the space-like hypersurface

$$\Delta(x) = 0 \qquad ; \text{ if } x^2 = \mathbf{x}^2 - c^2 t^2 > 0 \qquad (\text{F.14})$$

Hence we arrive at the remarkable result that $\Delta(x)$ is *non-zero only inside the forward and backward light cones.*[91]

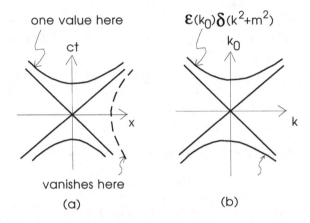

Fig. F.1 Properties of $\Delta(x)$: (a) in coordinate space, showing one component of $\mathbf{x}$, and (b) in momentum space, showing one component of $\mathbf{k}$. The asymptotes $\mathbf{x}^2 = c^2 t^2$ define the light-cone, and $k_0^2 = \mathbf{k}^2 + m^2$ defines the mass hyperboloid.

(4) We note from Eq. (F.8) that $\Delta(x; m^2)$ satisfies the Klein-Gordon equation

$$(\Box - m^2)\Delta(x; m^2) = 0 \qquad (\text{F.15})$$

where we now make the mass dependence explicit in $\Delta(x; m^2)$.

In *summary*, the commutator of the real scalar field is given for all x by

$$[\hat{\phi}(x), \hat{\phi}(0)] = \frac{\hbar}{ic}\Delta(x; m^2) \qquad ; x^2 < 0$$
$$= 0 \qquad ; x^2 > 0 \qquad (\text{F.16})$$

---

[91]The real (hermitian) scalar field is an observable. It was shown in Vol. I that if two hermitian operators fail to commute, they satisfy an uncertainty relation. This implies that a measurement of one of them can affect the other. It is only points inside the light cone that can be connected with a light signal, and hence only those points that can stand in a causal relationship with one another. The fact that the commutator of the two fields $[\hat{\phi}(x), \hat{\phi}(0)] = (\hbar/ic)\,\Delta(x)$ vanishes outside the light cone is known as *microscopic causality*.

where

$$\Delta(x; m^2) = \frac{i}{(2\pi)^3} \int d^4k \, \varepsilon(k_0) \delta(k^2 + m^2) \, e^{ik \cdot x} \tag{F.17}$$

Electromagnetic Field. The above calculation can be repeated for the transverse electromagnetic field in Eqs. (F.2). The only differences are:

- The mass of the quantum (photon) vanishes, so that $m^2 \to 0$, and we correspondingly introduce the notation

$$D(x) \equiv \Delta(x; 0) \tag{F.18}$$

- There is an additional factor of $1/\varepsilon_0$;
- The final polarization sum gives

$$\sum_{s=1}^{2} [\mathbf{e_{ks}}]_i [\mathbf{e_{ks}}]_j = \delta_{ij} - \frac{k_i k_j}{\mathbf{k}^2} \tag{F.19}$$

Thus, when the last term in Eq. (F.19) is expressed in terms of appropriate derivatives in coordinate space, and then removed from the integral, one obtains

$$[\hat{A}_i(x), \, \hat{A}_j(0)] = \frac{\hbar}{ic\varepsilon_0} \left[\delta_{ij} - \frac{1}{\nabla^2} \frac{\partial}{\partial x_i} \frac{\partial}{\partial x_j} \right] D(x)$$

$$\equiv \frac{\hbar}{ic\varepsilon_0} D_{ij}^T(x) \tag{F.20}$$

This result is obviously not covariant, which reflects the choice of the Coulomb gauge in the laboratory frame; however, we shall recover both Lorentz invariance and gauge invariance for QED once the full S-matrix is constructed.

Dirac Field. The anticommutator of the components of the Dirac field in Eqs. (F.2) at two distinct space-time points follows in a similar fashion

$$\{\hat{\psi}_\alpha(x), \, \hat{\bar{\psi}}_\beta(0)\} = \frac{1}{\Omega} \sum_{\mathbf{k}} \sum_{\lambda} \left[u_\alpha(\mathbf{k}\lambda) \bar{u}_\beta(\mathbf{k}\lambda) \, e^{ik \cdot x} + \right.$$

$$\left. v_\alpha(-\mathbf{k}\lambda) \bar{v}_\beta(-\mathbf{k}\lambda) \, e^{-ik \cdot x} \right] \tag{F.21}$$

The helicity sums over the Dirac spinors are just the positive- and negative-energy projection matrices (see Prob. F.2)

$$\sum_\lambda u_\alpha(\mathbf{k}\lambda)\bar{u}_\beta(\mathbf{k}\lambda) = \left[\frac{M - i\gamma_\mu k_\mu}{2\omega_k/c}\right]_{\alpha\beta} \qquad ; \; k_\mu = (\mathbf{k}, i\omega_k/c)$$

$$\sum_\lambda v_\alpha(-\mathbf{k}\lambda)\bar{v}_\beta(-\mathbf{k}\lambda) = \left[\frac{-M - i\gamma_\mu k_\mu}{2\omega_k/c}\right]_{\alpha\beta} \qquad (\text{F.22})$$

where $\omega_k = c\sqrt{\mathbf{k}^2 + M^2}$. Substitution of these relations into Eq. (F.21) and conversion of k_μ to an appropriate derivative with respect to x_μ, which allows one to take it outside the sum, leads to

$$\{\hat{\psi}_\alpha(x), \hat{\bar{\psi}}_\beta(0)\} = -\left[\gamma_\mu\frac{\partial}{\partial x_\mu} - M\right]_{\alpha\beta}\frac{1}{\Omega}\sum_\mathbf{k}\frac{1}{2\omega_k/c}\left(e^{ik\cdot x} - e^{-ik\cdot x}\right)$$

$$= i\left[\gamma_\mu\frac{\partial}{\partial x_\mu} - M\right]_{\alpha\beta}\Delta(x; M^2) \qquad (\text{F.23})$$

where $\Delta(x; M^2)$ has been identified from Eqs. (F.5).

Three comments:

- The equal-time canonical anticommutation relations for the Dirac field are recovered as in Eqs. (F.9)

$$\{\hat{\psi}_\alpha(x), \hat{\bar{\psi}}_\beta(0)\}_{t=0} = [\gamma_4]_{\alpha\beta}\frac{1}{c}\left[\frac{\partial}{\partial t}\Delta(x; M^2)\right]_{t=0}$$

$$= [\gamma_4]_{\alpha\beta}\,\delta^{(3)}(\mathbf{x}) \qquad (\text{F.24})$$

- Equation (F.23) again provides a Lorentz-invariant expression in the limit $\Omega \to \infty$;
- It is only bilinear combinations of the Dirac field that can be measured; the Dirac field itself is *not an observable*.[92]

F.2   Green's Functions

Let us proceed to investigate the *Green's functions* for the various four-dimensional wave equations.[93]

---

[92]The Dirac field, for example, is double-valued.

[93]We use the notation

$$d^4x = d^3x\,c\,dt = d^3x\,dx_0$$
$$d^4k = d^3k\,dk_0 = d^3x\,d\omega/c$$

Klein-Gordon Equation. The Green's function for the Klein-Gordon equation is defined by the relation

$$(\Box - m^2)G(x_\mu - x'_\mu) = -\delta^{(4)}(x_\mu - x'_\mu)$$
$$= -\delta^{(3)}(\mathbf{x} - \mathbf{x}')\delta(ct - ct') \tag{F.25}$$

This Green's function allows one to obtain a solution to the inhomogeneous Klein-Gordon equation

$$(\Box - m^2)\phi(x_\mu) = j(x_\mu) \tag{F.26}$$

If $\phi^{\text{in}}(x_\mu)$ is a solution to the homogeneous equation, then the following provides a solution to Eq. (F.26)

$$\phi(x_\mu) = \phi^{\text{in}}(x_\mu) - \int G(x_\mu - x'_\mu)j(x'_\mu)\, d^4x'$$
$$(\Box - m^2)\phi^{\text{in}}(x_\mu) = 0 \tag{F.27}$$

The Green's function in Eq. (F.25) can immediately be obtained as a four-dimensional Fourier transform

$$G(x_\mu - x'_\mu) = \frac{1}{(2\pi)^4} \int \frac{e^{ik\cdot(x-x')}}{k^2 + m^2} d^4k \tag{F.28}$$

Here $x = (\mathbf{x}, ict)$, $k = (\mathbf{k}, ik_0)$, $k^2 = \mathbf{k}^2 - k_0^2$, and $k \cdot x = \mathbf{k} \cdot \mathbf{x} - ik_0ct$. It remains to build the appropriate *boundary conditions* for the problem into the Green's function.

Dirac Equation. For the Dirac equation, one defines the Green's function by

$$\left[\gamma_\mu \frac{\partial}{\partial x_\mu} + M\right]_{\alpha\gamma} G^D_{\gamma\beta}(x_\mu - x'_\mu) = -\delta_{\alpha\beta}\, \delta^{(4)}(x_\mu - x'_\mu) \tag{F.29}$$

where the Dirac indices have been made explicit. This Green's function allows one to obtain a solution to the inhomogeneous Dirac equation

$$\left[\gamma_\mu \frac{\partial}{\partial x_\mu} + M\right]_{\alpha\gamma} \psi_\gamma(x_\mu) = j_\alpha(x_\mu) \tag{F.30}$$

If $\psi_\alpha^{\text{in}}(x_\mu)$ provides a solution to the homogeneous Dirac equation, then the following provides a solution to Eq. (F.30)

$$\psi_\alpha(x_\mu) = \psi_\alpha^{\text{in}}(x_\mu) - \int G_{\alpha\beta}^D(x_\mu - x_\mu')j_\beta(x_\mu')\,d^4x'$$

$$\left[\gamma_\mu \frac{\partial}{\partial x_\mu} + M\right]_{\alpha\gamma} \psi_\gamma^{\text{in}}(x_\mu) = 0 \qquad (F.31)$$

A solution to Eq. (F.29) can immediately be obtained in matrix form as

$$G^D(x_\mu - x_\mu') = -\frac{1}{(2\pi)^4} \int d^4k\, \frac{e^{ik\cdot(x-x')}}{i\gamma_\mu k_\mu + M} \qquad (F.32)$$

The integrand can be rationalized by multiplying both the numerator and denominator on the right by $[-i\gamma_\nu k_\nu + M]$ and then using:

- In the denominator

$$[i\gamma_\mu k_\mu + M][-i\gamma_\nu k_\nu + M] = k_\mu k_\nu \gamma_\mu \gamma_\nu + M^2$$
$$= k_\mu k_\nu \frac{1}{2}[\gamma_\mu \gamma_\nu + \gamma_\nu \gamma_\mu] + M^2$$
$$= k^2 + M^2 \qquad (F.33)$$

- In the numerator, write k_ν as a derivative with respect to x_ν, and then remove the derivative from the integral. In this way, Eq. (F.32) becomes

$$G^D(x_\mu - x_\mu') = \left[\gamma_\nu \frac{\partial}{\partial x_\nu} - M\right] \frac{1}{(2\pi)^4} \int \frac{e^{ik\cdot(x-x')}}{k^2 + M^2}\,d^4k$$

$$= \left[\gamma_\nu \frac{\partial}{\partial x_\nu} - M\right] G(x_\mu - x_\mu') \qquad (F.34)$$

where $G(x_\mu - x_\mu')$ has been identified from Eq. (F.28).[94] It again remains to build the appropriate *boundary conditions* for the problem into the Green's function, which we will now proceed to do.

---

[94]Note that both sides of this equation are now 4×4 matrices, and it is $G(x_\mu - x_\mu'; M^2)$; it is readily shown directly that this $G^D(x_\mu - x_\mu')$ satisfies Eq. (F.29) [see Prob. F.3].

F.2.1 Boundary Conditions

We wish to study

$$G(x_\mu) = \frac{1}{(2\pi)^4} \int \frac{e^{ik\cdot x}}{k^2 + m^2} d^4k \qquad (\text{F.35})$$

where x_μ is now the relative coordinate between two space-time points. In more detail

$$G(x_\mu) = \frac{1}{(2\pi)^4} \int d^3k \, e^{i\mathbf{k}\cdot\mathbf{x}} \int_{-\infty}^{\infty} dk_0 \frac{e^{-ik_0 ct}}{\mathbf{k}^2 + m^2 - k_0^2} \qquad (\text{F.36})$$

First we observe that because of the exponential damping, the k_0 integral can be closed with a large semi-circle with a radius $R \to \infty$ in the following manner (see Fig. F.2):

- Close in the upper-1/2 k_0-plane when $t < 0$;
- Close in the lower-1/2 k_0-plane when $t > 0$.

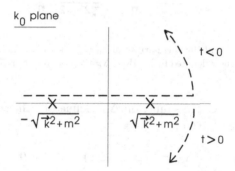

Fig. F.2   Closing the contour in the k_0 plane.

To define the Green's function, one must then specify how to go around the two singularities in the integrand at $k_0 = \pm\sqrt{\mathbf{k}^2 + m^2}$, in order to decide whether or not they are included inside of the contour. The various possibilities are illustrated in Fig. F.3. The resulting Green's functions are labeled by Δ,[95] and in the Dirac case by S.

It now follows from the analysis of complex integration in appendix B that the *retarded* Green's function Δ_R, defined by the top contour in

---

[95]The relation to the previous Δ is established below.

Fig. F.3, is given by

$$\Delta_R = 0 \qquad\qquad ; t < 0$$
$$= \Delta_+ + \Delta_- \qquad ; t > 0 \qquad\qquad (F.37)$$

where Δ_+ and Δ_- are defined by the contours in the second line — they simply encircle the singularities.

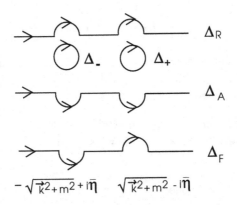

Fig. F.3   Paths around the singularities at $k_0 = \pm\sqrt{k^2 + m^2}$ used to define the various Green's functions, now denoted by Δ. Here Δ_F follows from the replacement $m \to m - i\eta$.

The *advanced* Green's function Δ_A, defined by the contour in the third line of Fig. F.3, is given by

$$\Delta_A = -(\Delta_+ + \Delta_-) \qquad ; t < 0$$
$$= 0 \qquad\qquad\qquad ; t > 0 \qquad\qquad (F.38)$$

The *Feynman* Green's function Δ_F, which plays a central role in modern physics, is defined by the contour in the last line of Fig. F.3

$$\Delta_F = -\Delta_- \qquad\qquad ; t < 0$$
$$= \Delta_+ \qquad\qquad\quad ; t > 0 \qquad\qquad (F.39)$$

Note that one immediately arrives at the Feynman Green's function by assigning an *infinitesimally small negative imaginary part to the mass*

$$m \to m - i\eta \qquad ; \text{ Feynman singularities} \qquad (F.40)$$

This moves the singularities off the real axis and into the correct position with respect to the contour integral.

The integrals for $\Delta_\pm$ can now be performed, and with careful attention to signs, one simply picks up the residue at each pole

$$
\Delta_+ = -\frac{1}{(2\pi)^4} \int d^3k \, e^{i\mathbf{k}\cdot\mathbf{x}} (-2\pi i) \frac{e^{-i\omega_k t}}{2\omega_k/c} = ic \int \frac{d^3k}{(2\pi)^3} \frac{e^{ik\cdot x}}{2\omega_k}
$$

$$
\Delta_- = -\frac{1}{(2\pi)^4} \int d^3k \, e^{i\mathbf{k}\cdot\mathbf{x}} (-2\pi i) \frac{e^{i\omega_k t}}{-2\omega_k/c} = -ic \int \frac{d^3k}{(2\pi)^3} \frac{e^{-ik\cdot x}}{2\omega_k}
$$

$$
\Delta_+ + \Delta_- = \Delta \tag{F.41}
$$

The sum of $\Delta_+ + \Delta_-$ has been identified with the previous Δ through Eq. (F.7). Since it has been shown that Δ vanishes outside the light cone, one concludes from the above that[96]

- The retarded Green's function Δ_R is non-zero only in the *forward light cone*;
- The advanced Green's function Δ_A is non-zero only in the *backward light cone*.

F.3   Time-Ordered Products

Time-ordered products occur naturally in our analysis of the perturbation series solution to the Schrödinger equation and the construction of the S-matrix. The building blocks are the vacuum expectation value of various time-ordered pairs of fields.

F.3.1   *Scalar Field*

The time-ordered product of the real scalar field at two distinct space-time points in the interaction picture is defined by

$$
T[\hat{\phi}(x_\mu), \, \hat{\phi}(x'_\mu)] \equiv \hat{\phi}(x_\mu)\hat{\phi}(x'_\mu) \qquad ; \, t > t'
$$

$$
\equiv \hat{\phi}(x'_\mu)\hat{\phi}(x_\mu) \qquad ; \, t < t' \tag{F.42}
$$

The order is important for, as we have seen, the fields do not commute at unequal times. The vacuum expectation value of these expressions is

---

[96]Note that the Feynman Green's function does *not* vanish outside the light cone; rather, it decays there over a Compton wavelength.

immediately calculated from Eqs. (F.2)[97]

$$\langle 0|T[\hat{\phi}(x_\mu), \hat{\phi}(x'_\mu)]|0\rangle = \frac{1}{\Omega}\sum_{\mathbf{k}} \frac{\hbar}{2\omega_k} e^{ik\cdot(x-x')} = \frac{\hbar}{ic}\Delta_+(x_\mu - x'_\mu) \qquad ; t > t'$$

$$= \frac{1}{\Omega}\sum_{\mathbf{k}} \frac{\hbar}{2\omega_k} e^{-ik\cdot(x-x')} = -\frac{\hbar}{ic}\Delta_-(x_\mu - x'_\mu) \quad ; t < t'$$

$$\text{(F.43)}$$

This is just the Feynman Green's function of Eq. (F.39)

$$\langle 0|T[\hat{\phi}(x_\mu), \hat{\phi}(x'_\mu)]|0\rangle = \frac{\hbar}{ic}\Delta_F(x_\mu - x'_\mu) \qquad \text{(F.44)}$$

We shall refer to this as the *Feynman propagator* for the scalar field.

F.3.2   *Electromagnetic Field*

It follows in the same manner that the vacuum expectation value of the time-ordered product of the transverse vector potential for the electromagnetic field in Eq. (F.2) is given by[98]

$$\langle 0|T[\hat{A}_i(x_\mu), \hat{A}_j(x'_\mu)]|0\rangle = \frac{\hbar}{ic\varepsilon_0}\left[\delta_{ij} - \frac{1}{\nabla^2}\frac{\partial}{\partial x_i}\frac{\partial}{\partial x_j}\right]D_F(x_\mu - x'_\mu)$$

$$\equiv \frac{\hbar}{ic\varepsilon_0}D_F^T(x_\mu - x'_\mu)_{ij} \qquad \text{(F.45)}$$

We shall show that this combines with the Coulomb interaction in the S-matrix in QED to give a covariant *photon propagator*.

F.3.3   *Dirac Field*

With a fermion field, it is more convenient to define a *P-product* that differs from the T-product only by the signature of the permutation required to get from the l.h.s. to the r.h.s. Thus for the Dirac field

$$P[\hat{\psi}_\alpha(x_\mu), \hat{\bar{\psi}}_\beta(x'_\mu)] \equiv \hat{\psi}_\alpha(x_\mu)\hat{\bar{\psi}}_\beta(x'_\mu) \qquad ; t > t'$$

$$\equiv -\hat{\bar{\psi}}_\beta(x'_\mu)\hat{\psi}_\alpha(x_\mu) \qquad ; t < t' \qquad \text{(F.46)}$$

---

[97]Use $\langle 0|c_{\mathbf{k}} c_{\mathbf{k}'}^\dagger|0\rangle = \langle 0|[c_{\mathbf{k}}, c_{\mathbf{k}'}^\dagger]|0\rangle = [c_{\mathbf{k}}, c_{\mathbf{k}'}^\dagger] = \delta_{\mathbf{kk}'}$. For finite Ω, the functions $\Delta_\pm(x)$ are defined as in Eq. (F.6).

[98]Compare Eq. (F.20), and see Prob. F.4.

The vacuum expectation value of this quantity follows immediately from Eqs. (F.2). With the aid of Eqs. (F.22) one finds for $t > t'$

$$\langle 0|P[\hat{\psi}_\alpha(x_\mu), \hat{\bar{\psi}}_\beta(x'_\mu)]|0\rangle = \frac{1}{\Omega} \sum_{\mathbf{k}} \left[\frac{M - i\gamma_\mu k_\mu}{2\omega_k/c}\right]_{\alpha\beta} e^{ik\cdot(x-x')} \qquad ; t > t'$$

$$= i\left[\gamma_\mu \frac{\partial}{\partial x_\mu} - M\right]_{\alpha\beta} \left[\frac{ic}{\Omega} \sum_{\mathbf{k}} \frac{1}{2\omega_k} e^{ik\cdot(x_\mu - x'_\mu)}\right]$$

$$= i\left[\gamma_\mu \frac{\partial}{\partial x_\mu} - M\right]_{\alpha\beta} \Delta_+(x_\mu - x'_\mu) \qquad (F.47)$$

For $t < t'$ one has

$$\langle 0|P[\hat{\psi}_\alpha(x_\mu), \hat{\bar{\psi}}_\beta(x'_\mu)]|0\rangle = -\frac{1}{\Omega} \sum_{\mathbf{k}} \left[\frac{-M - i\gamma_\mu k_\mu}{2\omega_k/c}\right]_{\alpha\beta} e^{-ik\cdot(x-x')} \qquad ; t < t'$$

$$= -i\left[\gamma_\mu \frac{\partial}{\partial x_\mu} - M\right]_{\alpha\beta} \left[\frac{-ic}{\Omega} \sum_{\mathbf{k}} \frac{1}{2\omega_k} e^{-ik\cdot(x_\mu - x'_\mu)}\right]$$

$$= -i\left[\gamma_\mu \frac{\partial}{\partial x_\mu} - M\right]_{\alpha\beta} \Delta_-(x_\mu - x'_\mu) \qquad (F.48)$$

A combination of these two results gives

$$\langle 0|P[\hat{\psi}_\alpha(x_\mu), \hat{\bar{\psi}}_\beta(x'_\mu)]|0\rangle = i\left[\gamma_\mu \frac{\partial}{\partial x_\mu} - M\right]_{\alpha\beta} \Delta_F(x_\mu - x'_\mu; M^2)$$

$$\equiv iS^F_{\alpha\beta}(x_\mu - x'_\mu) \qquad (F.49)$$

where we again make the mass dependence explicit. Here

$$\Delta_F(x - x'; M^2) = \int \frac{d^4k}{(2\pi)^4} \frac{e^{ik\cdot(x-x')}}{k^2 + M^2} \qquad ; M \to M - i\eta$$

$$S_F(x - x') = -\int \frac{d^4k}{(2\pi)^4} \left[\frac{1}{ik_\mu\gamma_\mu + M}\right] e^{ik\cdot(x-x')} \qquad (F.50)$$

We shall refer to this as the *Feynman propagator* for the Dirac field.[99]

---

[99] Note that the last expression is again a relation between 4×4 matrices.

F.3.4   *Vector Field*

The propagator for a real, massive vector meson field is calculated in Prob. F.5. The result is

$$\langle 0|T[\hat{V}_\mu(x)\,\hat{V}_\nu(0)]|0\rangle = \frac{\hbar c}{i}\Delta^F_{\mu\nu}(x;\,m^2)$$

$$\Delta^F_{\mu\nu}(x;\,m^2) = \frac{1}{(2\pi)^4}\int d^4k\,\frac{e^{ik\cdot x}}{k^2+m^2}\left(\delta_{\mu\nu}+\frac{k_\mu k_\nu}{m^2}\right)\qquad ;\ m\to m-i\eta$$

$$(\text{F.51})$$

Appendix G

Dimensional Regularization

In this appendix, a technique is developed to deal with integrals that are ill-defined in four dimensions. The trick is to start with a number of dimensions in which the integrals *are* well-defined, and where one can do mathematics. Appropriate interpretation of the theory, and analytic continuation, then allow one to ensure that everything *is* well-defined in four dimensions. This very clever technique of *dimensional regularization* is due to [t'Hooft and Veltman (1972)].[100]

The technique makes extensive use of the following properties of the gamma function $\Gamma(z)$, derived in any good book on mathematical physics[101]

- The function $\Gamma(z)$ has the following integral representation

$$\Gamma(z) = \int_0^\infty e^{-t} t^{z-1}\, dt \qquad ; \operatorname{Re} z > 0 \qquad (G.1)$$

- This is an analytic function of z for $\operatorname{Re} z > 0$;
- In this region, it satisfies the functional relationship

$$z\Gamma(z) = \Gamma(z+1) \qquad (G.2)$$

- This relation can be used to analytically continue $\Gamma(z)$ to the entire complex z-plane. The function $\Gamma(z)$ is thus analytic in the complex z-plane, except at the negative integers and zero, where it has poles;
- Two useful specific values are

$$\Gamma(1) = 1 \qquad ; \Gamma(1/2) = \sqrt{\pi} \qquad (G.3)$$

---

[100]See also [Leibbrant (1975); Itzykson and Zuber(1980)].

[101]For example [Morse and Feshbach (1953)]; see also appendix D in [Fetter and Walecka (2003)].

- At the positive integers

$$\Gamma(n) = (n-1)! \qquad ; \; n = 1, 2, \cdots \qquad (G.4)$$

We shall also make use of Euler's formula[102]

$$\int_0^1 t^{x-1} (1-t)^{y-1} \, dt = \frac{\Gamma(x)\Gamma(y)}{\Gamma(x+y)} \qquad ; \; \mathrm{Re}\,x > 0, \; \mathrm{Re}\,y > 0 \qquad (G.5)$$

G.1   Dirichlet Integral

The Dirichlet integral $I_n(R)$ represents the volume of a sphere of radius R in n-dimensional euclidian space. It is given

$$I_n(R) \equiv \int \cdots \int dx_1 \cdots dx_n \qquad ; \; x_1^2 + x_2^2 + \cdots + x_n^2 \le R^2$$

$$= \frac{(\sqrt{\pi})^n}{\Gamma(1+n/2)} R^n \qquad (G.6)$$

Before deriving this result, let us check that it reproduces some known quantities

$$I_1(R) = \frac{\sqrt{\pi}}{(1/2)\Gamma(1/2)} R = 2R \qquad ; \; \text{line}$$

$$I_2(R) = \frac{\pi}{(1)\Gamma(1)} R^2 = \pi R^2 \qquad ; \; \text{circle}$$

$$I_3(R) = \frac{\pi^{3/2}}{(3/2)(1/2)\Gamma(1/2)} R^3 = \frac{4\pi}{3} R^3 \qquad ; \; \text{sphere} \qquad (G.7)$$

Let us then derive Eq. (G.6). Introduce polar-spherical coordinates in n-dimensions, and carry out the integrations over all angles. Whatever the details, by dimensional analysis the final radial integral must be of the form

$$\left[\int \cdots \int dx_1 \cdots dx_n \right]_{\text{all angles}} = nC_n r^{n-1} \, dr \qquad (G.8)$$

where C_n is a constant. Then

$$I_n(R) = \int_0^R nC_n r^{n-1} \, dr = C_n R^n \qquad (G.9)$$

---

[102]See [Morse and Feshbach (1953)] p.486; also [Fetter and Walecka (2003)] p.545.

Now there is one such multiple integral that we know how to do

$$I_n \equiv \int_{-\infty}^{\infty} dx_1 \cdots \int_{-\infty}^{\infty} dx_n \, e^{-(x_1^2 + \cdots x_n^2)} = \left(\int_{-\infty}^{\infty} dx \, e^{-x^2} \right)^n = \left(\sqrt{\pi} \right)^n \text{(G.10)}$$

Use the result in Eq. (G.8) to write this in spherical-polar coordinates, and then change variables to $r^2 = t$ to give

$$I_n = nC_n \int_0^{\infty} e^{-r^2} r^{n-1} \, dr = \frac{n}{2} C_n \int_0^{\infty} e^{-t} t^{n/2-1} \, dt$$

$$= C_n \frac{n}{2} \Gamma \left(\frac{n}{2} \right) = C_n \Gamma \left(\frac{n}{2} + 1 \right) \tag{G.11}$$

Here the gamma function has been identified from Eq. (G.1), and then Eq. (G.2) invoked. A comparison of these two expressions for I_n then gives

$$C_n = \frac{(\sqrt{\pi})^n}{\Gamma(1 + n/2)} \tag{G.12}$$

which is the desired result.

G.2 Basic Relation

The following relation in n-dimensional euclidian space, for integer values of (p, q), is then central to the analysis[103]

$$\int d^n t \, \frac{(t^2)^p}{(t^2 + a^2)^q} = \frac{1}{(a^2)^{q-p}} \frac{n(\pi a^2)^{n/2}}{\Gamma(1 + n/2)} \frac{\Gamma(q - [p + n/2]) \, \Gamma(p + n/2)}{2\Gamma(q)} \tag{G.13}$$

This relation is proven through the following series of steps:

(1) Call this integral J. Take out the dimension a to give

$$J = \frac{a^n a^{2p}}{a^{2q}} \int d^n x \, \frac{x^{2p}}{(x^2 + 1)^q} \tag{G.14}$$

where x is now dimensionless.

(2) Do all the angular integrals, and use Eqs. (G.8) and (G.12)

$$J = \frac{a^n a^{2p}}{a^{2q}} \frac{n\pi^{n/2}}{\Gamma(1 + n/2)} \int_0^{\infty} \frac{r^{n-1+2p}}{(r^2 + 1)^q} \, dr \tag{G.15}$$

---

[103] Here $t^2 = t_1^2 + \cdots + t_n^2$.

(3) Change variables in the integral to

$$r^2 = \frac{1-u}{u} \qquad ; r^2 + 1 = \frac{1}{u} \qquad ; 2r\,dr = -\frac{1}{u^2}du \qquad \text{(G.16)}$$

Then

$$\int_0^\infty \frac{r^{n-1+2p}}{(r^2+1)^q}dr = \frac{1}{2}\int_0^1 \left(\frac{1-u}{u}\right)^{(n-2+2p)/2} u^q \frac{du}{u^2}$$

$$= \frac{1}{2}\int_0^1 u^{q-(p+n/2)-1}(1-u)^{(p+n/2)-1}\,du \qquad \text{(G.17)}$$

(4) Now make use of Euler's formula in Eq. (G.5)[104]

$$\int_0^\infty \frac{r^{n-1+2p}}{(r^2+1)^q}dr = \frac{\Gamma\left(q-[p+n/2]\right)\Gamma\left(p+n/2\right)}{2\Gamma(q)} \qquad \text{(G.18)}$$

Equation (G.15) is now just Eq. (G.13), and we have the desired result.

G.3   Complex n-plane

We now observe the following:

- The function $\Gamma(z)$ is analytic except for simple poles on the real axis at $z = 0, -1, -2, -3, \cdots$;
- The r.h.s. of Eq. (G.18) is thus an *analytic function of n except for isolated poles on the real axis*;
- One can write $r^n = e^{n\ln r}$, and where the integral exists, the l.h.s. of Eq. (G.18) is *also* an analytic function of n;
- In this case, the r.h.s provides the unique analytic continuation of this integral to the entire complex n-plane;
- In any event, the r.h.s. of Eq. (G.18) provides a *definition* of the integral in the entire complex n-plane;
- In the vicinity of $n = 4$, the integral as defined in this fashion is well-behaved and analytic in n for any real integers (p, q);

  No matter how badly the integral may diverge for $n = 4$, it is well-behaved and analytic in the vicinity of $n = 4$;

- Since the n-dependence is now explicit on the r.h.s. of the basic result in Eq. (G.13), it also defines an integral that is analytic in the vicinity

---

[104]This relation holds for $\mathrm{Re}\,(q - [p + n/2]) > 0$ and $\mathrm{Re}\,(p + n/2) > 0$.

of $n = 4$, with the possibility of an isolated singularity at the point $n = 4$;

- *Physics* is now the limit $n \to 4$ in the complex n-plane!

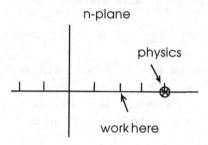

n-plane

physics

work here

Fig. G.1 Complex n-plane.

G.4 Algebra

We must be able to perform algebra within this framework. There are some requirements:

- We want to be able to do ordinary algebra — preserve current conservation, gauge invariance, Ward identities, *etc.*;
- The results should be correct at any finite, discrete n where the theory exists;
- We therefore adopt the following *algorithm*:

   Simply do all the algebra as if in discrete integer-n dimensions.

We give some examples:

$$\int d^n t \, t_\mu F(t^2) = 0 \qquad\qquad ;\text{ symmetric integration}$$

$$\int d^n t \, t_\mu t_\nu F(t^2) = I\delta_{\mu\nu} \qquad\qquad ;\text{ symmetric tensor}$$

$$\delta_{\mu\mu} = n \qquad \Rightarrow \qquad I = \frac{1}{n} \int d^n t \, t^2 F(t^2)$$

$$\int dt_1 \cdots dt_n \, G \frac{\partial}{\partial t_\mu} F = -\int dt_1 \cdots dt_n \, F \frac{\partial}{\partial t_\mu} G \qquad ;$$

$$\text{partial integration} \qquad (\text{G.19})$$

G.5   Lorentz Metric

So far, everything has been in euclidian space, but we are actually interested in Minkowski space with the Lorentz metric. How do we proceed? The square of the four-vector in Minkowski space is

$$k_\mu^2 = k_1^2 + k_2^2 + k_3^2 - k_0^2 \qquad ; \text{ Lorentz metric} \qquad (G.20)$$

If we want to change the number of dimensions, we can modify the number of *spatial* coordinates.

Assume the regular Feynman singularities in the integral in Eq. (G.13) with $a^2 \to a^2 - i\eta$, so that the integrand contains the factor

$$D = \frac{1}{(k^2 + a^2 - i\eta)^q} \qquad (G.21)$$

In this case the Feynman singularities will lie in the second and fourth quadrants in the complex k_0-plane. One generally has enough convergence to rotate the k_0 contour integral in the counter-clockwise direction from along the real axis until it runs along the imaginary axis.   From the

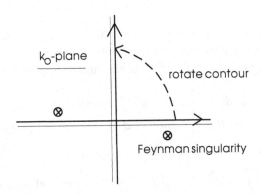

Fig. G.2   Rotation of contour in k_0-plane to go from Minkowski to euclidian space.

location of the Feynman singularities, no singularities will be encountered during this rotation (see Fig. G.2).[105] Now change variables in the k_0-

---

[105]Sufficient convergence, as well as the singularity location, can be verified in any application.

integral as follows:

$$\int_{-\infty}^{\infty} dk_0 = \int_{-i\infty}^{i\infty} dk_0 \qquad ; \text{ contour rotation}$$

$$= i \int_{-\infty}^{\infty} dt_0 \qquad ; k_0 \equiv it_0 \qquad \text{(G.22)}$$

With this change of variable, Eq. (G.20) reads

$$k_\mu^2 = k_1^2 + k_2^2 + k_3^2 + t_0^2 \qquad ; \text{ euclidian metric} \qquad \text{(G.23)}$$

and we are back to the problem in the euclidian metric! Equation (G.13) then holds for all the subsequent analysis, and in Minkowski space Eq. (G.13) therefore becomes

$$\int d^n k \, \frac{(k^2)^p}{(k^2 + a^2 - i\eta)^q} = \frac{i}{(a^2)^{q-p}} \frac{n(\pi a^2)^{n/2}}{\Gamma(1 + n/2)} \frac{\Gamma\left(q - [p + n/2]\right) \Gamma\left(p + n/2\right)}{2\Gamma(q)} \, ;$$

$$\text{Minkowski space} \qquad \text{(G.24)}$$

G.6 γ-Matrix Algebra

For the algebra of the γ matrices, we adopt the same rules as above:

- Do the γ-matrix algebra as if in discrete integer-n dimensions;
- Use $\gamma_\mu = (\gamma_1, \gamma_2, \cdots, \gamma_n)$ with

$$\{\gamma_\mu, \gamma_\nu\} = 2\delta_{\mu\nu} \qquad \text{(G.25)}$$

All the γ-matrix algebra is based on this relation;[106]
- Do the γ-matrix algebra first, and reduce the problem to k-integrals.

G.7   Examples

We give three examples of the use of dimensional regularization.

---

[106]One can use representations of different *dimension*. With these, $\mathrm{Tr}\,\underline{1} = \tau$; however, since this will always appear as an overall factor in the calculations, one can just set $\tau = 4$ at the outset.

G.7.1 *Convergent Momentum Integrals*

If the momentum integrals in Eq. (G.24) are convergent for $n = 4$, then the use of Eq. (G.4) gives the following integral in Minkowski space

$$\int d^4k \frac{(k^2)^{p-2}}{(k^2 + a^2 - i\eta)^q} = i\pi^2 (a^2)^{p-q} \frac{(q-p-1)!(p-1)!}{(q-1)!} \quad (G.26)$$

This is a very useful result.

G.7.2 *Vacuum Tadpoles*

Suppose we had not employed the normal-ordered form of the current $:\hat{\bar{\psi}}(x)\gamma_\mu\hat{\psi}(x):$ in the analysis of the scattering operator in QED, but had instead just used $\hat{\bar{\psi}}(x)\gamma_\mu\hat{\psi}(x)$. There would then be additional contractions $\hat{\bar{\psi}}^{\cdot}\gamma_\mu\hat{\psi}^{\cdot}$ contributing in Wick's theorem. These are the *vacuum tadpoles* (compare Fig. 12.1). In momentum space, this contraction is (see Prob. 7.11)

$$\hat{\bar{\psi}}^{\cdot}\gamma_\mu\hat{\psi}^{\cdot} = -\frac{1}{i}\int \frac{d^4k}{(2\pi)^4} \text{Tr}\left[\gamma_\mu \frac{1}{i\slashed{k} + M}\right] \quad (G.27)$$

Simple power counting shows that this is an ill-defined, badly divergent integral. Define the integral with dimensional regularization. In n-dimensions, the above rules then lead to

$$\int d^n k \, \text{Tr}\left[\gamma_\mu \frac{1}{i\slashed{k} + M}\right] = \int d^n k \, \text{Tr}\left[\frac{\gamma_\mu (M - i\slashed{k})}{k^2 + M^2}\right]$$
$$= -i\tau \int d^n k \frac{k_\mu}{k^2 + M^2} = 0 \quad (G.28)$$

where the last relation follows from symmetric integration. Thus *this contraction vanishes with dimensional regularization, and in this scheme, there is no need to normal-order the current.*

G.7.3 *Vacuum Polarization*

The self-mass of the photon arises from the vacuum-polarization tensor in Eq. (8.48) [see Fig. 8.3(b)], which can be re-written as

$$\Pi_{\mu\nu}(l) = -\frac{ie^2}{\hbar c\varepsilon_0}\int \frac{d^4k}{(2\pi)^4} \text{Tr}\left[\frac{1}{i(\slashed{k} - \slashed{l}/2) + M}\gamma_\mu \frac{1}{i(\slashed{k} + \slashed{l}/2) + M}\gamma_\nu\right] (G.29)$$

This is again a badly divergent, ill-defined integral. Define it through dimensional regularization, and work in n-dimensions. The current conservation statement follows directly[107]

$$l_\mu \Pi_{\mu\nu}(l) = \Pi_{\mu\nu}(l)l_\nu = 0 \tag{G.30}$$

This, and covariance, dictate the form of the vacuum-polarization tensor to be

$$\Pi_{\mu\nu}(l) = C(l^2)(l_\mu l_\nu - l^2 \delta_{\mu\nu}) \tag{G.31}$$

Hence, in n-dimensions,

$$\Pi_{\mu\mu}(l^2) = (1-n)\, l^2 C(l^2) \tag{G.32}$$

We observe that this quantity is only a function of l^2.

The starting photon in the interaction picture is massless with $l^2 = 0$, and so is a *real photon*. This reflects the underlying gauge invariance of the theory. There should be *no mass renormalization* for the photon. We shall demonstrate that with dimensional regularization, the quantity in Eq. (G.32) indeed vanishes for a real photon

$$\Pi_{\mu\mu}(0) = 0 \qquad ; \text{ real photon} \tag{G.33}$$

and thus a real photon remains massless in this scheme.[108]

The previous rules allow us to do the algebra in n-dimensions, and the integral in Eq. (G.29) becomes, for $l = 0$

$$
\begin{aligned}
I_{\mu\nu}^{(n)}(0) &\equiv \int d^n k \,\mathrm{Tr}\left[\frac{1}{i\rlap{/}k + M} \gamma_\mu \frac{1}{i\rlap{/}k + M} \gamma_\nu \right] \\
&= \int d^n k \,\mathrm{Tr}\left[\frac{(M - i\rlap{/}k)\gamma_\mu(M - i\rlap{/}k)\gamma_\nu}{(k^2 + M^2)^2} \right] \\
&= \tau \int \frac{d^n k}{(k^2 + M^2)^2} \left[M^2 \delta_{\mu\nu} - 2k_\mu k_\nu + k^2 \delta_{\mu\nu} \right]
\end{aligned}
\tag{G.34}
$$

Symmetric integration then leads to [recall Eqs. (G.19)]

$$I_{\mu\nu}^{(n)}(0) = \tau \delta_{\mu\nu} \int d^n k \left[\frac{1}{k^2 + M^2} - \frac{2}{n} \frac{k^2}{(k^2 + M^2)^2} \right] \tag{G.35}$$

---

[107]See Prob. G.1.

[108]The mass of the photon is directly related to $\Pi_{\mu\mu}(0)$ in Prob. G.2.

Now use the basic result in Eq. (G.24)

$$I_{\mu\nu}^{(n)}(0) = \tau\delta_{\mu\nu}\frac{i}{M^2}\frac{n(\pi M^2)^{n/2}}{\Gamma(1+n/2)}\Gamma(1-n/2)\left[\frac{\Gamma(n/2)}{2\Gamma(1)} - \frac{2}{n}\frac{\Gamma(1+n/2)}{2\Gamma(2)}\right] \quad (G.36)$$

However

$$\left[\frac{\Gamma(n/2)}{2\Gamma(1)} - \frac{2}{n}\frac{\Gamma(1+n/2)}{2\Gamma(2)}\right] = \Gamma(n/2)\left[\frac{1}{2} - \frac{2}{n}\frac{n}{2}\frac{1}{2}\right]$$

$$= 0 \qquad\qquad ; \text{ all } n \quad (G.37)$$

We emphasize that this relation holds for *all* n. Hence, in the limit $n \to 4$,

$$\Pi_{\mu\mu}(0) = -\frac{ie^2}{(2\pi)^4\hbar c\varepsilon_0}I_{\mu\mu}^{(4)}(0) = 0 \qquad ; \text{ massless photon} \quad (G.38)$$

and the photon remains massless as the interaction is turned on.

These examples demonstrate the great utility of dimensional regularization:

- All integrals are well-defined at the outset;
- One can do well-defined mathematics throughout;
- In particular, the algebra is carried out as if in n discrete dimensions;
- All general properties of the theory, such as current conservation, gauge invariance, Ward's identity, *etc.*, are maintained throughout;
- Contact is made with physics at the end of the calculation in the limit $n \to 4$.

Appendix H

Path Integrals and the Electromagnetic Field

Consider the lagrangian density for the free electromagnetic field

$$\mathcal{L}_0(A_\mu) = -\frac{1}{4}F_{\mu\nu}F_{\mu\nu}$$

$$F_{\mu\nu} = \frac{\partial A_\nu}{\partial x_\mu} - \frac{\partial A_\mu}{\partial x_\nu} \tag{H.1}$$

Here, to simplify the discussion in this appendix, we set $\varepsilon_0 = 1$ and work in H-L units. This lagrangian density is invariant under the local gauge transformation

$$A_\mu \rightarrow A_\mu + \frac{\partial \Lambda}{\partial x_\mu} \tag{H.2}$$

In computing path integrals over the field variables A_μ, one would like to include only *physically distinct configurations*. If the path integral is carried out over all $\mathcal{D}(A_\mu)$, many gauge-equivalent field configurations will be included.[109]

H.1 Faddeev-Popov Identity

The solution to this problem, due to [Faddeev and Popov (1967)], is to *factor* the path integral in the following manner

$$\int \mathcal{D}(A_\mu) = \int \mathcal{D}(\Lambda) \int \mathcal{D}(A_\mu) \, \delta_P[f(A_\mu) - \lambda(x)] \det [\varepsilon^4 \underline{M}_f] \tag{H.3}$$

Here $f(A_\mu) = \lambda(x)$ specifies a *particular gauge*, and the δ_P function, which

---

[109]The material in this appendix is taken from appendix D of [Serot and Walecka (1986)].

for discretized space-time (see Fig. 10.7) can be written as

$$\delta_P[f(A_\mu) - \lambda(x)] \equiv \delta(f_1 - \lambda_1) \cdots \delta(f_n - \lambda_n) \qquad (H.4)$$

ensures that this gauge-fixing condition is satisfied everywhere in space-time. In writing Eq. (H.3), we assume that the rest of the integrand (suppressed here) is *gauge invariant*.

The matrix $[\varepsilon^4 \underline{M}_f]$, whose determinant appears in Eq. (H.3), is defined as follows. Carry out a change in gauge by changing $\Lambda(x) \to \Lambda(x) + \delta\Lambda(x)$. The function $f(A_\mu)$ will also change, and to $O(\delta\Lambda)$ this change is given at each point in space-time by

$$\delta f_\alpha = \sum_\beta [\varepsilon^4 \underline{M}_f]_{\alpha\beta} \, \delta\Lambda_\beta \qquad (H.5)$$

In the continuum limit this relation takes the form

$$\delta f(x) = \int d^4y \, M_f(x,y) \delta\Lambda(y) \qquad ; \varepsilon \to 0, \quad \alpha \to x, \quad \beta \to y \quad (H.6)$$

We now demonstrate that Eq. (H.3) is, in fact, an *identity*. Work in discretized space-time, and first perform the integral $\int \mathcal{D}(\Lambda)$ over all gauge functions

$$\int \mathcal{D}(\Lambda)\delta_P[f(A_\mu) - \lambda(x)] = \int \cdots \int d\Lambda_1 \cdots d\Lambda_n \, \delta(f_1 - \lambda_1) \cdots \delta(f_n - \lambda_n)$$
$$(H.7)$$

A simple change of variables in the multiple integral on the r.h.s. gives

$$\int \mathcal{D}(\Lambda)\delta_P[f(A_\mu) - \lambda(x)] = \int \cdots \int df_1 \cdots df_n \, \delta(f_1 - \lambda_1) \cdots \delta(f_n - \lambda_n) \times$$
$$\frac{\partial(\Lambda_1 \cdots \Lambda_n)}{\partial(f_1 \cdots f_n)} \qquad (H.8)$$

The integrals can be evaluated immediately with the product of δ-functions, and what remains is the *jacobian* for the transformation from integration variables $(\Lambda_1 \cdots \Lambda_n)$ to $(f_1 \cdots f_n)$

$$\frac{\partial(\Lambda_1 \cdots \Lambda_n)}{\partial(f_1 \cdots f_n)} = \left[\frac{\partial(f_1 \cdots f_n)}{\partial(\Lambda_1 \cdots \Lambda_n)} \right]^{-1} = (\det [\varepsilon^4 \underline{M}_f])^{-1} \qquad (H.9)$$

Here the linear relation in Eq. (H.5) has been used to obtain the final equality. Note that this expression is independent of $\lambda(x)$. A combination of Eqs. (H.3), (H.8), and (H.9) then yields the claimed identity

$$\int \mathcal{D}(A_\mu) = \int \mathcal{D}(A_\mu) \tag{H.10}$$

This relation clearly holds for all ε, as $\varepsilon \to 0$.

The factored form of the measure in Eq. (H.3) can now be used to rewrite the path integral over the gauge fields A_μ.

H.2 Application

As an application, consider the following convenient form of $f(A_\mu)$

$$f(A_\mu) = \frac{\partial A_\mu}{\partial x_\mu} \tag{H.11}$$

where the repeated Greek index of the scalar product in Minkowski space is once again summed from 1 to 4. In this case, a combination of Eqs. (H.2) and (H.6) yields

$$M_f(x, y) = \Box_x \delta^{(4)}(x - y) \equiv K(x, y) \tag{H.12}$$

where $K(x, y)$ is a kernal of the form that was examined in detail in chapter 10. This $K(x, y)$ has two important properties:

(1) It is independent of $\Lambda(x)$;
(2) It is also independent of $\lambda(x)$.

We now show that the coefficient of $\int \mathcal{D}(\Lambda)$ on the r.h.s. of Eq. (H.3) is *independent of* Λ. To see this, make a gauge transformation so that the vector potential and function f change to

$$A_\mu \to \bar{A}_\mu = A_\mu + \frac{\partial \Lambda}{\partial x_\mu}$$
$$f(A_\mu) \to f(\bar{A}_\mu) \tag{H.13}$$

Now change dummy integration variables *at fixed* $\Lambda(x)$ in the inner integrals in Eq. (H.3). This leaves the measure $\mathcal{D}(A_\mu)$ in the integration over the field variables A_μ in those inner integrals unchanged

$$\int \mathcal{D}(A_\mu) = \int \mathcal{D}(\bar{A}_\mu) \tag{H.14}$$

From the previous discussion, $\det [\varepsilon^4 \underline{M}_f]$ is independent of $\Lambda(x)$, as is the remainder of the integrand, which is assumed to be gauge invariant. Thus

$$\int \mathcal{D}(A_\mu)\delta_P[f(\bar{A}_\mu) - \lambda(x)]\det [\varepsilon^4 \underline{M}_f]$$

$$= \int \mathcal{D}(\bar{A}_\mu)\delta_P[f(\bar{A}_\mu) - \lambda(x)]\det [\varepsilon^4 \underline{M}_f]$$

$$= \int \mathcal{D}(A_\mu)\delta_P[f(A_\mu) - \lambda(x)]\det [\varepsilon^4 \underline{M}_f] \qquad (H.15)$$

where the last line is simply a relabeling of variables. Hence this combination is gauge-invariant, and the assertion is proven.

The path integral over different gauge functions $\int \mathcal{D}(\Lambda)$ now factors out and cancels in the ratio of path integrals used to define the generating functional of QED.

The preceding arguments also imply that the coefficient of $\int \mathcal{D}(\Lambda)$ in Eq. (H.3) is *independent of* $\lambda(x)$. Thus, no matter what specific space-time dependence we give to the gauge-fixing function $f(A_\mu)$, we obtain the same result for the path integral over A_μ. Since we have already observed that the integral over the gauge functions $\int \mathcal{D}(\Lambda)$ factors and cancels in the ratio that defines the generating functional, we conclude that the generating functional must also be independent of $\lambda(x)$. This fact can be used to great advantage, as we illustrate shortly.

The proof of the assertion that the coefficient of $\int \mathcal{D}(\Lambda)$ in Eq. (H.3) is independent of $\lambda(x)$ follows immediately from Eqs. (H.15)

$$\int \mathcal{D}(A_\mu)\delta_P[f(\bar{A}_\mu) - \lambda(x)] \det [\varepsilon^4 \underline{M}_f]$$

$$= \int \mathcal{D}(A_\mu)\delta_P[f(A_\mu) - \bar{\lambda}(x)] \det [\varepsilon^4 \underline{M}_f]$$

$$= \int \mathcal{D}(A_\mu)\delta_P[f(A_\mu) - \lambda(x)] \det [\varepsilon^4 \underline{M}_f] \qquad (H.16)$$

where we have used Eq. (H.11) and defined

$$\bar{\lambda}(x) \equiv \lambda(x) - \Box\Lambda(x) \qquad (H.17)$$

in the second line. Since Eqs. (H.16) hold for arbitrary $\Lambda(x)$, they hold for arbitrary $\bar{\lambda}(x)$, which proves the assertion.

H.3 Generating Functional

These arguments allow us to construct the generating functional for QED

$$\tilde{W}_0[J] = \left[\int \mathcal{D}(A_\mu) \exp\left\{ \frac{i}{\hbar c} \int d^4x \, [\mathcal{L}_0 + J_\mu A_\mu] \right\} \right] \Big/ [\cdots]_{J=0} \quad (\text{H.18})$$

It is assumed here that the external source is conserved, with $\partial J_\mu / \partial x_\mu = 0$, so that the total action is gauge-invariant.[110] The Faddeev-Popov identity in Eq. (H.3) can now be employed to rewrite this expression, and consistent with our previous observations:

(1) The integration over all gauge functions $\int \mathcal{D}(\Lambda)$ factors and cancels in the ratio in Eq. (H.18);
(2) All constant factors in the measure, which arise when the ratio of path integrals is evaluated in discrete space-time, also cancel in the ratio;
(3) The resulting numerator of the generating functional is independent of $\lambda(x)$; hence, so is the denominator and their ratio.

The function $\lambda(x)$ can thus be chosen arbitrarily in evaluating Eq. (H.18). In fact, we can perform the following *path integral* in the numerator and denominator of the generating functional

$$\int \mathcal{D}(\lambda) \exp\left\{ -\frac{i}{\hbar c} \frac{1}{2\xi} \int d^4x \, [\lambda(x)]^2 \right\} \quad (\text{H.19})$$

where ξ is an arbitrary real number, *since this integration again factors and cancels in the ratio!* Observe, however, that the integration over $\int \mathcal{D}(\lambda)$ can now be carried out *first*, and the δ_P function, $\delta_P[f(A_\mu) - \lambda(x)]$, used to trivially evaluate the path integral. The result of these manipulations on the generating functional of QED, for the example of Eq. (H.11), is[111]

$$\tilde{W}_0[J] = \left(\int \mathcal{D}(A_\mu) \det[\underline{M}_f] \times \right. \quad (\text{H.20})$$

$$\left. \exp\left\{ \frac{i}{\hbar c} \int d^4x \left[\mathcal{L}_0 + J_\mu A_\mu - \frac{1}{2\xi} \left(\frac{\partial A_\mu}{\partial x_\mu} \right)^2 \right] \right\} \right) \Big/ (\cdots)_{J=0}$$

The remaining integral $\int \mathcal{D}(A_\mu)$ is over all A_μ, but the *effective lagrangian appearing in the action is now no longer gauge invariant*. The final term

---

[110] Just use partial integration together with the localized boundary conditions. As before, we define this ratio of path integrals by evaluating both numerator and denominator for discretized space-time with cell size ε^4, and then letting $\varepsilon \to 0$.

[111] The ε^4 in the determinant *also* cancels in the ratio for discretized space-time.

involving the parameter ξ "fixes" the gauge according to the choices in Eqs. (H.11) and (H.19).

The one remaining complexity in Eq. (H.20) is the Faddeev-Popov determinant, $\det[\underline{M}_f]$, that appears in the integrand. This may be evaluated through the clever *trick* of writing the determinant as a path integral over Grassmann variables $[\bar{g}(x), g(x)]$. The basic result in Eq. (10.121) gives, for the example in Eqs. (H.11)–(H.12) and discretized space-time,

$$\int \int \mathcal{D}(\bar{g})\mathcal{D}(g) \exp\left\{ \frac{i}{\hbar c} \int d^4x \; [\bar{g} \,\Box_x\, g] \right\} = \left(\frac{\varepsilon^8}{i\hbar c} \right)^n \det[\underline{M}_f] \quad \text{(H.21)}$$

Since common multiplicative factors again cancel in the ratio, one can write the generating functional of QED as

$$\tilde{W}_0[J] = \left(\int \int \int \mathcal{D}(\bar{g})\mathcal{D}(g)\mathcal{D}(A_\mu) \times \right. \tag{H.22}$$

$$\left. \exp\left\{ \frac{i}{\hbar c} \int d^4x \left[\mathcal{L}_0 + J_\mu A_\mu - \frac{1}{2\xi} \left(\frac{\partial A_\mu}{\partial x_\mu} \right)^2 - \frac{\partial \bar{g}}{\partial x_\mu} \frac{\partial g}{\partial x_\mu} \right] \right\} \right) \Big/ (\cdots)_{J=0}$$

Here a partial integration has been carried out in the last term, and the localized boundary conditions again employed.

H.4 Ghosts

There are additional scalar "ghost fields" $\bar{g}(x)$ and $g(x)$ present in the effective lagrangian and in the path integrals in the generating functional in Eq. (H.22). They obey Grassmann algebras and appear through the representation of the Faddeev-Popov determinant as a path integral over Grassmann variables. Since only the free-field lagrangian for the ghost fields enters here, their contribution *factors* and can be removed from the generating functional ratio in this simple example. This will no longer be true in non-abelian gauge theories, as interaction terms involving the ghost fields will also be present.

H.5 Photon Propagator

After cancellation of the ghosts, introduction of $\mathcal{L}_0$ from Eq. (H.1), and the performance of partial integrations similar to those used previously, the effective action in the generating functional of QED can be written

$$\int d^4x \left[-\frac{1}{4} F_{\mu\nu} F_{\mu\nu} - \frac{1}{2\xi} \left(\frac{\partial A_\mu}{\partial x_\mu} \right)^2 \right]$$

$$= \int d^4x \left[-\frac{1}{2} \left(\frac{\partial A_\mu}{\partial x_\nu} \right) \left(\frac{\partial A_\mu}{\partial x_\nu} \right) + \frac{1}{2} \left(1 - \frac{1}{\xi} \right) \left(\frac{\partial A_\mu}{\partial x_\mu} \right)^2 \right]$$

$$= \frac{1}{2} \int d^4x\, A_\mu \left[\Box_x \delta_{\mu\nu} - \left(1 - \frac{1}{\xi} \right) \frac{\partial}{\partial x_\mu} \frac{\partial}{\partial x_\nu} \right] A_\nu \qquad \text{(H.23)}$$

The source can now be included, and the path integrals in the generating functional evaluated using gaussian integration, just as in the text.[112] In the result, one needs the inverse of the following kernal

$$K_{\mu\nu}(x, y) = \left[\Box_x \delta_{\mu\nu} - \left(1 - \frac{1}{\xi} \right) \frac{\partial}{\partial x_\mu} \frac{\partial}{\partial x_\nu} \right] \delta^{(4)}(x - y) \qquad \text{(H.24)}$$

The inverse is now required both in space-time and in the Lorentz indices. If the inverse is defined by [see Eqs. (10.98)]

$$\int d^4z\, K_{\mu\rho}(x, z) D_{\rho\nu}(z - y) = -\delta^{(4)}(x - y)\delta_{\mu\nu} \qquad \text{(H.25)}$$

then it follows from Eqs. (H.24)–(H.25) that the inverse has the following four-dimensional Fourier transform in Minkowski space[113]

$$D_{\mu\nu}(x - y) = \int \frac{d^4k}{(2\pi)^4} \tilde{D}_{\mu\nu}(k)\, e^{ik\cdot(x-y)}$$

$$\tilde{D}_{\mu\nu}(k) = \frac{1}{k^2 - i\eta} \left[\delta_{\mu\nu} - (1 - \xi) \frac{k_\mu k_\nu}{k^2} \right] \qquad \text{(H.26)}$$

This is the *photon propagator*. Different choices of ξ lead to the photon propagator in different gauges:

- $\xi = 1$ gives the *Feynman gauge*, and the Feynman propagator used in our discussion of QED;
- $\xi = 0$ gives the *Landau gauge*, which has the useful property that $k_\mu \tilde{D}_{\mu\nu}(k) = \tilde{D}_{\mu\nu}(k)k_\nu = 0$.

Evidently, as emphasized repeatedly in the text, one is free to include any amount of $k_\mu k_\nu / k^2$ in the photon propagator without altering the physics.

---

[112] The evaluation of this generating functional follows as in chapter 10 (see Prob. H.1).
[113] See Prob. H.3.

Appendix I

Metric Conversion

A comparison of the metric used in [Bjorken and Drell (1964); Bjorken and Drell (1965)] with that used in the present text is shown in Table I.1, taken from [Walecka (2004)]. The position of the indices, up or down, is important when the metric $g_{\mu\nu}$ is employed (see [Walecka (2007)]).

Table I.1   Metric comparison table (here $\hbar = c = 1$).

Bjorken and Drell		Present text
$g_{\mu\nu} = \begin{pmatrix} 1 & 0 & 0 & 0 \\ 0 & -1 & 0 & 0 \\ 0 & 0 & -1 & 0 \\ 0 & 0 & 0 & -1 \end{pmatrix}$	$\leftrightarrow^a$	$\delta_{\mu\nu}$
$a^\mu = (a^0, \vec{a})$	$\leftrightarrow$	$a_\mu = (a_1, a_2, a_3, a_4) = (\mathbf{a}, ia_0)$
$a_\mu b^\mu = g_{\mu\nu} a^\mu b^\nu = a^0 b^0 - \vec{a} \cdot \vec{b}$	$\leftrightarrow$	$a_\mu b_\mu = \vec{a} \cdot \vec{b} - a_0 b_0$
$x^\mu = (t, \vec{x})$	$\leftrightarrow$	$x_\mu = (\vec{x}, it)$
$x_\mu = g_{\mu\nu} x^\nu = (t, -\vec{x})$	$\leftrightarrow$	$x^\mu \equiv x_\mu$
$\partial_\mu = \partial/\partial x^\mu = (\partial/\partial t, \vec{\nabla})$	$\leftrightarrow$	$\partial/\partial x_\mu = (\vec{\nabla}, \partial/i\partial t)$
$\gamma^\mu = (\beta, \beta\vec{\alpha})$	$\leftrightarrow$	$\gamma_\mu = (i\vec{\alpha}\beta, \beta)$
$\gamma^\mu \gamma^\nu + \gamma^\nu \gamma^\mu = 2g^{\mu\nu}$	$\leftrightarrow$	$\gamma_\mu \gamma_\nu + \gamma_\nu \gamma_\mu = 2\delta_{\mu\nu}$
$\gamma^{\mu\dagger} = \gamma^0 \gamma^\mu \gamma^0$	$\leftrightarrow$	$\gamma_\mu^\dagger = \gamma_\mu$
$(i\gamma^\mu \partial_\mu - M)\psi = 0$	$\leftrightarrow$	$(\gamma_\mu \partial/\partial x_\mu + M)\psi = 0$
$(k_\mu \gamma^\mu - M)u(k) = 0$	$\leftrightarrow$	$(i\gamma_\mu k_\mu + M)u(k) = 0$
$\gamma^5 = i\gamma^0 \gamma^1 \gamma^2 \gamma^3 = \gamma_5$	$\leftrightarrow$	$\gamma_5 = \gamma_1 \gamma_2 \gamma_3 \gamma_4$
$\sigma^{\mu\nu} = \frac{i}{2}[\gamma^\mu, \gamma^\nu]$	$\leftrightarrow$	$\sigma_{\mu\nu} = \frac{1}{2i}[\gamma_\mu, \gamma_\nu]$

^{a} Note $g_{\mu\nu} = g^{\mu\nu}$.

It follows that the conversion of expressions presented in the metric of Bjorken and Drell to results in the metric used in this text is obtained by the substitutions shown in Table I.2.

Table I.2   Metric conversion table.

Bjorken and Drell		Present metric[b]
$a_\mu b^\mu$	$\rightarrow$	$-a_\mu b_\mu$
$g_{\mu\nu}$	$\rightarrow$	$-\delta_{\mu\nu}$
γ^μ	$\rightarrow^c$	$i\gamma_\mu$
∂^μ	$\rightarrow$	$-\partial/\partial x_\mu$
γ_5	$\rightarrow$	$-\gamma_5$
$\partial_\mu J^\mu$	$\rightarrow$	$\partial J_\mu/\partial x_\mu$
$\sigma^{\mu\nu}$	$\rightarrow$	$\sigma_{\mu\nu}$
$\varepsilon_{\mu\nu\rho\sigma}$	$\rightarrow$	$i\varepsilon_{\mu\nu\rho\sigma}$

[b] Some examples:

$$a_\mu b^\mu = a^\mu g_{\mu\nu} b^\nu \xrightarrow{\text{conv}} a_\mu(-\delta_{\mu\nu})b_\nu = -a_\mu b_\mu$$

$$(i\gamma^\mu\partial_\mu - M) = (i\gamma^\mu g_{\mu\nu}\partial^\nu - M) \xrightarrow{\text{conv}} -(\gamma_\mu\partial/x_\mu + M)$$

$$\partial_\mu\phi\partial^\mu\phi - m_s^2\phi^2 \xrightarrow{\text{conv}} -[(\partial\phi/\partial x_\mu)^2 + m_s^2\phi^2]$$

$$F^{\mu\nu} = \partial^\mu V^\nu - \partial^\nu V^\mu \xrightarrow{\text{conv}} -(\partial V_\nu/\partial x_\mu - \partial V_\mu/\partial x_\nu) = -F_{\mu\nu}$$

$$\partial_\nu F^{\nu\mu} = \partial^\lambda g_{\lambda\nu}F^{\nu\mu} \xrightarrow{\text{conv}} -\partial F_{\nu\mu}/\partial x_\nu = +\partial F_{\mu\nu}/\partial x_\nu$$

$$\gamma_\mu p^\mu - M = \gamma^\nu g_{\nu\mu}p^\mu - M \xrightarrow{\text{conv}} -(i\gamma_\mu p_\mu + M)$$

$$4(a\cdot b) = \text{tr }a_\mu\gamma^\mu b_\nu\gamma^\nu \xrightarrow{\text{conv}} -\text{tr }a_\mu\gamma_\mu b_\nu\gamma_\nu = -4(a\cdot b)$$

$$4i\varepsilon_{\mu\nu\rho\sigma} = \text{tr }\gamma_\mu\gamma_\nu\gamma_\rho\gamma_\sigma\gamma_5 \xrightarrow{\text{conv}} (-i^4)\text{tr }\gamma_\mu\gamma_\nu\gamma_\rho\gamma_\sigma\gamma_5 = -4\varepsilon_{\mu\nu\rho\sigma}$$

[c] The lowering and raising of the Lorentz index on the overall vertex Γ^μ itself is controlled by the $g_{\mu\nu}$ in the propagator; thus $\Gamma_1^\mu g_{\mu\nu}\Gamma_2^\nu \equiv \Gamma_{1\mu}g^{\mu\nu}\Gamma_{2\nu}$.

The metric employed in special relativity, as was the situation with units (see appendix K of Vol. I), is largely a matter of choice. The present use of an imaginary fourth component for four-vectors in Minkowski space has the advantage that the Dirac gamma matrices are hermitian and obey a simple algebra. It has the disadvantage that one has to be especially careful when taking the complex conjugate of matrix elements. The advantage of employing an explicit metric $g_{\mu\nu}$ is that it makes the transition to general relativity more evident. It has the disadvantage that one has to carry around that extra baggage and pay careful attention to whether Lorentz indices are up or down. One simply has to pick a metric convention and set of units where calculations can be performed confidently and reliably. It is important, however, to be familiar with other sets of units and metric conventions (particularly if one has collaborators who are set in their ways!).

Bibliography

Abers, E. S., and Lee, B. W., (1973). *Phys. Rep.* **9**, 1

Banks, T., (2008). *Modern Quantum Field Theory: A Concise Introduction*, Cambridge U. Press, New York, NY

Bardeen, J., Cooper, L. N., and Schrieffer, J. R., (1957). *Phys. Rev.* **106**, 162; *Phys. Rev.* **108**, 1175

Bethe, H. A., and Goldstone, J., (1957). *Proc. Roy. Soc. (London)* **A238**, 551

Bjorken, J. D., and Drell, S. D., (1964). *Relativistic Quantum Mechanics*, McGraw-Hill, New York, NY

Bjorken, J. D., and Drell, S. D., (1965). *Relativistic Quantum Fields*, McGraw-Hill, New York, NY

Blatt, J. M., and Weisskopf, V. F., (1952). *Theoretical Nuclear Physics*, John Wiley and Sons, New York, NY

Bloch, F., and Nordsieck, A., (1937). *Phys. Rev.* **73**, 54

Bogoliubov, N. N., (1947). *J. Phys. (USSR)* **11**, 23

Bogoliubov, N. N., (1958). *Sov. Phys. JETP* **7**, 41

Bohr, A., Mottelson, B. R., and Pines, D, (1958). *Phys. Rev.* **110**, 936

Brueckner, K. A., and Sawada, K., (1957). *Phys. Rev.* **106**, 1117

Chen, J.-Q., Wang, P.-N., Lü, Z.-M., and Wu, X.-B., (1987) *Tables of Clebsch-Gordan coefficients of SU(n) groups*, World Scientific, Singapore

Cheng, T.-P., and Li, L.-F., (1984). *Gauge Theory of Elementary Particle Physics*, Clarendon Press, Oxford, UK

Cooper, L. N., (1956). *Phys. Rev.* **104**, 1189

Cvitanovik, P., and Kinoshita, T., (1974). *Phys. Rev.* **D10**, 4007

Dirac, P. A. M., (1947). *The Principles of Quantum Mechanics, 3rd ed.*, Oxford University Press, New York, NY

Donoghue, J. F., Golowich, E., and Holstein, B., (1993). *Dynamics of the Standard Model*, Cambridge University Press, New York, NY

Dyson, F. J., (1949). *Phys. Rev.* **75**, 486; *Phys. Rev.* **75**, 1736

Dyson, F. J., (1952). *Phys. Rev.* **85**, 631

Edmonds, A. R., (1974). *Angular Momentum in Quantum Mechanics*, 3rd printing, Princeton University Press, Princeton, NJ

Faddeev, L. D., and Popov, V. N., (1967). *Phys. Lett.* **25B**, 29

Fetter, A. L., and Walecka, J. D., (2003). *Theoretical Mechanics of Particles and Continua*, McGraw-Hill, New York, NY (1980); reissued by Dover Publications, Mineola, NY

Fetter, A. L., and Walecka, J. D., (2003a). *Quantum Theory of Many-Particle Systems*, McGraw-Hill, New York, NY (1971); reissued by Dover Publications, Mineola, NY

Feynman, R. P., (1949). *Phys. Rev.* **76**, 749; *Phys. Rev.* **76**, 769

Feynman, R. P., (1963). *Acta Phys. Polon.* **24**, 697

Feynman, R. P., and Hibbs, A. R., (1965). *Quantum Mechanics and Path Integrals*, McGraw-Hill, New York, NY

Gabrielse, G., (2009). *Measurements of the Electon Magnetic Moment*, to appear in *Lepton Dipole Moments: The Search for Physics Beyond the Standard Model*, eds. B. L. Roberts and W. J. Marciano, World Scientific, Singapore

Gell-Mann, M., and Goldberger, M. L., (1953). *Phys. Rev.* **91**, 398

Gell-Mann, M., and Levy, M., (1960). *Nuovo Cimento* **16**, 705

Gell-Mann, M., and Low, F. E., (1954). *Phys. Rev.* **95**, 1300

Gell-Mann, M., and Ne'eman, Y., (1963). *The Eightfold Way*, W. A. Benjamin, Reading, MA

Georgi, H., (1999). *Lie Algebras in Particle Physics: from Isospin to Unified Theories, 2nd ed.*, Westview Press, Boulder, CO

Goldberger, M. L., and Watson, K. M., (2004). *Collision Theory*, Dover Publications, Mineola, NY

Gottfried, K., (1966). *Quantum Mechanics, Vol. I*, W. A. Benjamin, New York, NY

Hamermesh, M., (1989). *Group Theory and Its Applications to Physical Problems*, Dover Publications, Mineola, NY

Itzykson, C., and Zuber, J.-B., (1980). *Quantum Field Theory*, McGraw-Hill, New York, NY

Jacob, M., and Wick, G, C., (1959) *Ann. Phys.* **7**, 404

Lee, T. D., and Yang, C. N, (1957). *Phys. Rev.* **105**, 1119

Leibbrant, G., (1975). *Rev. Mod. Phys.* **47**, 849

Lippmann, B., and Schwinger, J., (1950). *Phys. Rev.* **79**, 469

Merzbacher, M., (1998). *Quantum Mechanics, 3rd ed.*, John Wiley and Sons, New York, NY

Morse, P. M., and Feshbach, H., (1953). *Methods of Theoretical Physics, Vols. I-II*, McGraw-Hill, New York, NY

National Nuclear Data Center, (2009). *Nuclear Data*, http://www.nndc.bnl.gov/

Particle Data Group, (2009). *Particle Data Tables 2009*, http://pdg.lbl.gov/2009/tables/contents_tables.html

Pauli, W., and Weisskopf, V. F., (1934). *Helv. Phys. Acta* **7**, 709

Rotenberg, M., Bivens, R., Metropolis, N., and Wooten, J. K. Jr., (1959) *The 3-j and 6-j Symbols*, The Technology Press, M. I. T., Cambridge, MA

Schiff, L. I., (1968). *Quantum Mechanics, 3rd ed.*, McGraw-Hill, New York, NY

Schweber, S., (1961). *Relativistic Quantum Field Theory*, Row-Peterson, Chicago, IL

Schwinger, J., (1957). *Annals of Physics* **2**, 407

Schwinger, J., (1958). *Selected Papers on Quantum Electrodynamics*, ed. J. Schwinger, Dover Publications, Mineola, NY

Serot, B. D., and Walecka, J. D., (1986). *The Relativistic Nuclear Many-Body Problem, Adv. Nucl. Phys.* **16**, Plenum Press, New York, NY

t'Hooft, G., and Veltman, M., (1972) *Nucl. Phys.* **B44**, 189; *Nucl. Phys.* **B62**, 444

Tilley, D. R., Weller, H. R., and Hale, G. M., (1992). *Nucl. Phys.* **A541**, 1

Titchmarsh, E. C., (1976). *Theory of Functions, 2nd ed.*, Oxford, New York, NY

Valatin, J. G., (1958) *Nuovo Cimento* **7**, 843

Van Dyck, R. S., Schwinberg, P. B., and Dehmelt, H. G., (1977). *Phys. Rev. Lett.* **38**, 310

Walecka, J. D., (2001). *Electron Scattering for Nuclear and Nucleon Structure*, Cambridge University Press, Cambridge, UK

Walecka, J. D., (2004). *Theoretical Nuclear and Subnuclear Physics, 2nd ed.*, World Scientific Publishing Company, Singapore

Walecka, J. D., (2007). *Introduction to General Relativity*, World Scientific Publishing Company, Singapore

Ward, J. C., (1951). *Proc. Phys. Soc. London* **A64**, 54

Wentzel, G., (1949). *Quantum Theory of Fields*, Interscience, New York, NY

Wick, G. C., (1950). *Phys. Rev.* **80**, 268

Wikipedia, (2009). http://en.wikipedia.org/wiki/(topic)

Wigner, E. P., (1937). *Phys. Rev.* **51**, 106

Wigner, E. P., (1939). *On unitary representations of the inhomogeneous Lorentz group, Annals Math.* **40**, 149

Wu, T.T., (1959). *Phys. Rev.* **115**, 1390

Yang, C. N., and Mills, R. L., (1954). *Phys. Rev.* **96**, 191

Index